One Health Approaches to Nanoplastics Contamination:

Biotechnological and AI Solutions

Reyed M. Reyed
https://orcid.org/0000-0001-5027-3478

Genetic Engineering and Biotechnology Research Institute (GEBRI), City of Scientific Research and Technological Applications (SRTA-City), Alexandria, Egypt

IGI Global
Scientific Publishing
Publishing Tomorrow's Research Today

Vice President of Editorial	Melissa Wagner
Director of Acquisitions	Mikaela Felty
Director of Book Development	Jocelynn Hessler
Production Manager	Mike Brehm
Cover Design	Jose Rosado

Published in the United States of America by
IGI Global Scientific Publishing
701 East Chocolate Avenue
Hershey, PA, 17033, USA
Tel: 717-533-8845 | Fax: 717-533-7115
Website: https://www.igi-global.com E-mail: cust@igi-global.com

Library of Congress Cataloging-in-Publication Data

LCCN: 2025029844 (CIP Data Pending)
ISBN13: 9798337344904
Isbn13Softcover: 9798337344911
EISBN13: 9798337344928
British Cataloguing in Publication Data
A Cataloguing in Publication record for this book is available from the British Library.

Some works are written with ink.
Others are written through memory, sacrifice, silence, and love.
This work belongs to the latter.

To my father and mother —
though absent from this world, they remain deeply embedded within every thought I shape and every path I walk. Their prayers became the unseen architecture beneath my life, carrying me through moments they would never witness, yet somehow always preparing me for. May Allah surround them with eternal mercy, light, and peace.

To my wife and children —
the living heartbeat behind this journey. They endured the invisible weight of long nights, unfinished moments, and distant thoughts, while continuing to offer patience wrapped in love and strength wrapped in tenderness. In their presence, exhaustion became meaningful, and perseverance became possible.

This is not merely a dedication page.
It is a quiet map of the souls who helped build this work without writing a single line of it.

Every page carries something of them.
A prayer.
A sacrifice.
A memory.
A form of love that never asked to be seen.

Table of Contents

Detailed Table of Contents

This chapter per the authors develops a One Health cyber-biotechnology framework for interpreting nanoplastic exposure as a multiscale environmental signaling system rather than an isolated contaminant phenomenon. It integrates digital nanoplastic science (DNPS), plasto-signaling interference (PSI), transductive signal load (TSL), Polybiome Interfacial Signaling Cascade (PISC), and systemic risk scoring (SRS) to structure exposure propagation, interface-level perturbation, and biological signaling disruption within a unified analytical architecture. Environmental matrices including water, soil, and air are conceptualized as dynamic signal-generating interfaces transmitting disturbances across interconnected ecological and physiological networks under uncertainty regimes. The chapter critiques conventional compartment-based risk assessment and advances multiomics integration, digital twin modeling, and cyber-biotechnological analytics for early-warning systems, exposure intelligence, and One Health–driven environmental decision support.

Nanoplastics are increasingly recognized as a One Health concern because of their persistence, environmental mobility, and potential impacts across interconnected ecosystems, wildlife, and human populations. Their small size and heterogeneous composition make detection, source tracing, and exposure assessment challenging when conventional analytical approaches are used alone. This chapter per the authors discusses an integrated forensic framework combining biotechnological methods with artificial intelligence (AI)-based analytics to improve detection and characterization of nanoplastics across complex environmental matrices. Forensic tracing systems—including molecular assays, spectroscopic and imaging techniques, and bioassays—are examined as complementary approaches for material discrimination, while computational models integrate heterogeneous datasets and strengthen source attribution. Ultimately, this chapter connects automated detection, forensic tracing, and exposure modeling within a One Health framework to support environmental risk assessment and management.

This chapter per the authors analyzes the physicochemical determinants governing nanoplastic transport behavior, environmental partitioning, bioavailability, and remediation performance across complex ecological matrices. Nanoplastics are heterogeneous colloidal systems where particle size, morphology, surface charge, density, hydrophobicity, and functionalization regulate aggregation kinetics, interfacial interactions, contaminant adsorption, and environmental persistence under

compounds, probiotics, polyphenols, and cellular defense modulators are assessed alongside food safety surveillance systems, contaminant detection approaches, and computational exposure forecasting tools. Integration of environmental nutrition, food exposure science, and traceability analytics supports One Health–oriented assessment of dietary nanoplastic risks, food-system resilience, and contaminant control dynamics.

This chapter per the authors addresses translational strategies for mitigating nanoplastic exposure across environmental, technological, economic, and public health domains within a One Health framework. Nanoplastics are evaluated as mobile particulate contaminants capable of crossing biological barriers, accumulating across environmental compartments, and contributing to chronic ecological and physiological exposure burdens. The chapter analyzes mitigation pathways involving exposure surveillance, biotechnology-assisted reduction systems, computational exposure forecasting, and implementation-oriented environmental management models. Integration of exposure analytics, environmental management, and implementation economics contributes operational perspectives for reducing nanoplastic-associated ecological and health burdens across interconnected environments emphasizing integrated governance and resilience-based environmental action for sustainability goals.

This chapter per the authors analyzes economic and life cycle dimensions of nanoplastic remediation through analytical frameworks linking environmental burden remediation efficiency and resource allocation dynamics within One Health systems. Nanoplastic generation pathways environmental transport behavior and cross matrix exposure

Preface

BOOK OVERVIEW AND EXECUTIVE SUMMARY

This volume brings together current perspectives on nanoplastics as an emerging area of environmental research at the interface of ecological systems, environmental pollution, biological responses, and public health. The increasing recognition of nanoplastics as persistent environmental contaminants, alongside continuing uncertainties regarding their long-term biological fate, environmental transport, and toxicological relevance, has contributed to growing scientific interest across multiple disciplines.

The book is developed within an integrated environmental and health framework that connects environmental occurrence, exposure pathways, biological interactions, and methodological approaches. Rather than treating environmental pollution, toxicological processes, and analytical methodologies as separate domains, the volume emphasizes their interdependence within complex environmental systems and broader One Health contexts.

The organization of the volume follows the environmental life cycle of nanoplastics, beginning with sources of plastic pollution, environmental distribution, transport dynamics, and transformation processes. It then progresses to organismal exposure, ecological interactions, and potential impacts on biological systems. Subsequent sections address analytical characterization, detection methods, and experimental toxicology, followed by computational modeling, artificial intelligence applications, bioinformatics, and multi-omics approaches used to investigate exposure–response relationships and system-level effects.

These methodological approaches are presented as complementary tools that may enhance the characterization of nanoplastics in complex environmental matrices, particularly where conventional monitoring and toxicological assessment methods face limitations in sensitivity, resolution, or scale.

The interdisciplinary nature of this work reflects the convergence of environmental science, environmental toxicology, pollution studies, biotechnology, computational biology, systems biology, and One Health research. By integrating these domains within a unified conceptual framework, the volume supports a more structured understanding of nanoplastics as part of broader environmental contamination and global health challenges.

This book is intended for researchers, environmental scientists, toxicologists, biotechnologists, computational biologists, and policy-relevant stakeholders working in environmental risk assessment, environmental monitoring, and One Health disciplines. It is expected to serve as a reference for current scientific knowledge while also identifying areas where further methodological development, experimental validation, and computational integration are still required.

Overall, the volume seeks to contribute to a more coherent and integrative perspective on nanoplastics, supporting improved environmental monitoring, toxicological interpretation, and evidence-informed decision-making in environmental and health-related contexts.

Keywords

Nanoplastics; Plastic pollution; Environmental contamination; Environmental toxicology; Environmental health; One Health; Exposure pathways; Environmental transport; Ecological systems; Biological interactions; Toxicological assessment; Risk assessment; Artificial intelligence; Computational modeling; Bioinformatics; Multi-omics.

SCHOLARLY INTEGRATION AND EDITORIAL ARCHITECTURE

The development of a multidisciplinary scientific volume rarely concludes with the simple assembly of independent chapters. Scientific themes emerging from diverse disciplines, methodological approaches, and authorial perspectives do not automatically form a coherent analytical body of work. Meaningful continuity often requires sustained editorial integration extending beyond conventional language editing and routine copy preparation.

Throughout the preparation of this volume, editorial efforts were directed toward strengthening conceptual coherence, preserving structural continuity, and supporting analytical progression across chapters. Particular attention was devoted to reducing thematic fragmentation, harmonizing scientific terminology, refining transitions between related topics, and reinforcing connections among complementary areas of investigation. These efforts were undertaken not to standardize authorial expression, but to maintain scientific clarity while preserving the intellectual identity and scholarly contributions of individual authors.

Editorial integration extended beyond language refinement to the systematic alignment of chapters across multiple scientific domains, including One Health, environmental exposure science, toxicology, biotechnology, artificial intelligence, computational biology, and risk assessment. Where appropriate, chapter framing, abstract structure, thematic organization, and cross-chapter sequencing were refined to strengthen semantic continuity and improve the overall interpretability of the volume.

A central editorial objective was to ensure that the book functioned as an integrated scientific work rather than a collection of independent contributions. Emphasis was therefore placed on maintaining logical progression between chapters, strengthening conceptual linkages across disciplines, and supporting a coherent analytical trajectory from environmental occurrence and biological interaction to technological innovation, risk evaluation, and evidence-informed decision-making.

The resulting volume reflects a deliberate commitment to scholarly integration and scientific coherence. Its structure is founded upon continuity of interpretation, interdisciplinary engagement, and cumulative knowledge development. Editorial stewardship therefore operated not only at the level of language and presentation, but also at the level of scientific architecture, where coherence, connectivity, and intellectual progression were treated as essential components of the volume as a whole.

EDITORIAL ARCHITECTURE OF THE VOLUME: A SYSTEMS-SCIENCE FRAMING ESSAY

The environmental condition of the twenty-first century is changing in ways that are no longer fully captured by the classical language of pollution, contamination, or compartment-based toxicological assessment. Environmental systems are increasingly shaped not only by visible degradation or localized discharge, but also by distributed particulate mobility, continuous material exchange, and persistent interactions across atmospheric, aquatic, biological, technological, and computational domains.

Over the last two decades, environmental science has gradually shifted from macroscopic observation toward systems governed by micro- and nano-scale interactions. At these scales, matter no longer behaves in stable or easily bounded forms. Instead, it moves continuously across environmental compartments, including air, water, soil, food systems, industrial infrastructures, and biological interfaces. Micro- and nanoplastics have emerged as some of the clearest indicators of this transition and are now reported across atmospheric dust, aquatic systems, agricultural environments, trophic pathways, and biological matrices worldwide.

Large-scale environmental assessments continue to demonstrate a sustained rise in global plastic production and waste accumulation. However, the central issue is no longer accumulation alone. More fundamentally, particulate matter at micro- and nano-scale dimensions is beginning to function less as inert waste and more as a persistent mobile component embedded within environmental systems themselves.

This development quietly destabilizes one of the long-standing assumptions of classical environmental science: namely, that exposure can be localized, compartmentalized, and interpreted within relatively stable spatial boundaries.

1. The Invisible Environmental Transition

A substantial portion of the current environmental transformation occurs below conventional observational thresholds. The issue is not abstraction, but scale. Many emerging environmental processes operate at dimensions that traditional environmental monitoring systems were never designed to resolve.

Materials once regarded as environmentally confined increasingly behave as mobile participants within larger infrastructural and ecological exchange systems. Atmospheric circulation, wastewater streams, agricultural substrates, urban polymer matrices, hydrological networks, and food-distribution pathways now participate in continuous material redistribution across interconnected environmental layers, as conceptually illustrated in Figure 1.

At nano-scale resolution, physicochemical behavior changes substantially. Surface interactions become dominant. Reactivity shifts. Transport dynamics begin to outweigh bulk material properties. Under such conditions, degradation products and secondary particulate fractions no longer behave as static residues. Instead, they increasingly function as active mobile entities circulating through complex environmental exchange networks.

The implications are methodologically significant. Exposure is no longer episodic, spatially isolated, or temporally discrete. Instead, it becomes diffuse, persistent, and structurally embedded within environmental systems themselves.

This introduces a central systems-level challenge:

How should environmental science interpret exposure when environmental boundaries themselves are no longer stable or clearly definable?

Figure 1. Invisible Environmental Transition and the Reconfiguration of Exposure Regimes.

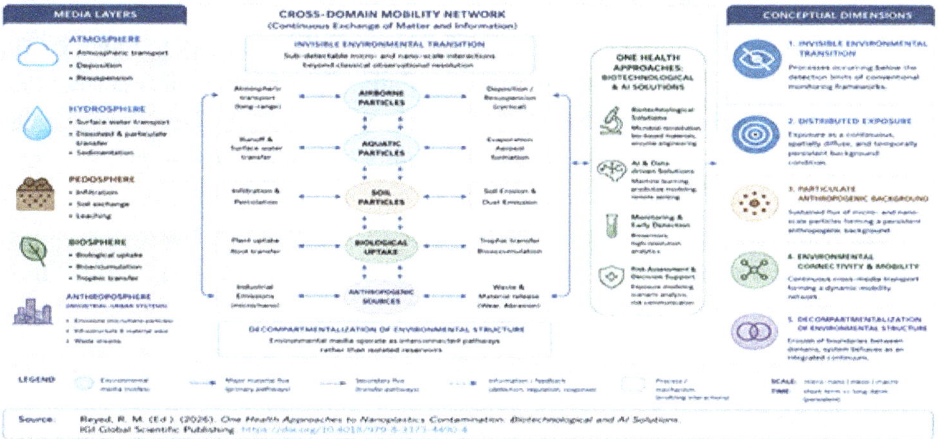

This figure conceptualizes the environment as a cross-domain mobility network linking atmospheric, hydrospheric, pedospheric, biospheric, and anthropogenic layers through continuous material and informational exchange. It illustrates how micro- and nano-scale particulate dynamics reconfigure classical compartment-based exposure models into distributed, multi-pathway systems characterized by persistent background fluxes and weakly bounded transitions across media. The framework further highlights the emergence of an "invisible environmental transition" occurring below conventional observational thresholds, where exposure becomes spatially diffuse, temporally continuous, and structurally embedded within environmental media

2. Beyond Classical Toxicological Interpretation

Classical toxicology remains foundational to environmental medicine. However, many of its core assumptions are becoming increasingly difficult to sustain under emerging exposure conditions. Conventional frameworks were largely developed around isolated exposures, relatively stable dose-response relationships, and identifiable causal pathways.

Contemporary environmental exposure patterns increasingly diverge from those assumptions. Chronic low-dose exposure, overlapping environmental pathways, prolonged persistence, and cumulative interaction effects now dominate many environmental-health interfaces. Under such conditions, causality becomes distributed rather than singular, and biological outcomes often emerge through delayed, indirect, or multifactorial mechanisms.

These limitations become even more pronounced at micro- and nano-scale levels, where conventional dose metrics may fail to adequately represent biologically meaningful exposure behavior.

At the same time, broader integrative frameworks have begun to emerge. The exposome model reframes exposure as a cumulative life-course process rather than as a sequence of isolated toxicological events. Similarly, the One Health framework increasingly positions environmental, human, and ecological systems within a single interconnected analytical structure, as further conceptualized in Figure 2.

Taken together, these developments signal a gradual movement away from compartment-based toxicological interpretation toward systems-oriented environmental reasoning.

Yet a critical gap remains unresolved:

What type of analytical framework can meaningfully operate under continuous, distributed, multi-source, and non-linear exposure conditions?

3. Computational Environmental Intelligence

Part of the response to this complexity has become computational, largely because environmental systems themselves now exceed the interpretive capacity of purely classical analytical approaches.

Artificial intelligence, machine learning systems, distributed sensing architectures, and adaptive environmental analytics increasingly participate in environmental monitoring, predictive modeling, ecological forecasting, and real-time biosurveillance. These systems enable the extraction of patterns from heterogeneous environmental datasets that would otherwise remain fragmented, weakly connected, or analytically inaccessible.

The shift extends beyond methodological modernization alone. Environmental systems are no longer interpreted solely through direct observation; they are increasingly inferred through computational layers.

Within this evolving context, Computational Environmental Intelligence may be understood as the integration of environmental exposure dynamics, ecological signaling, and biological-response analytics into adaptive computational systems capable of continuous interpretation and predictive updating, as conceptually associated with the interconnected exposure continuum illustrated in Figure 2.

This transition also carries structural consequences. Energy demand, computational infrastructure dependency, algorithmic bias, uncertainty propagation, and interpretive opacity become internal characteristics of the environmental-analytical system itself rather than external technical limitations.

Computational systems therefore cannot be treated as neutral observational tools. They increasingly participate in defining environmental reality, shaping interpretive boundaries, and influencing institutional response mechanisms.

This convergence naturally extends toward computational biotechnology and systems-level environmental medicine, where biological systems themselves are increasingly modeled as dynamic components operating within larger computational-environmental architectures.

The resulting question becomes unavoidable:

What happens to biological interpretation when environmental systems and computational systems cease to function as separable domains?

4. Biological And Clinical Horizons

At a certain threshold, environmental restructuring begins to intersect directly with biological consequence.

Increasing attention is now directed toward possible relationships between chronic environmental exposure and disruptions involving endocrine regulation, inflammatory signaling, metabolic stability, reproductive biology, microbial homeostasis, and broader physiological regulation, as further reflected in the interconnected environmental-biological pathways presented in Figure 2.

Yet interpretation remains difficult. Many biological responses emerge through probabilistic, cumulative, and weakly bounded pathways rather than through singular toxicological mechanisms. Long-term low-dose exposure conditions further complicate causal attribution.

Within oncology, environmental medicine, and systems biology, attention increasingly focuses on indirect biological cascades involving oxidative stress, inflammatory signaling, microbiome perturbation, endocrine modulation, proteostatic imbalance, and systemic metabolic disruption. Importantly, these pathways rarely behave in linear or isolated forms.

Within the conceptual architecture of this volume, prolonged nano-scale particulate exposure is examined through the exploratory systems-level construct of Panflemmosis. The term is not introduced as a deterministic biological endpoint, nor as a finalized pathological classification. Rather, it functions as a structured conceptual framework intended to investigate interconnected biological cascades potentially associated with chronic low-dose environmental exposure, microbial destabilization, proteostatic disruption, and persistent inflammatory signaling across interconnected physiological systems.

The objective is therefore not biological overstatement, but analytical positioning: namely, the creation of a structured interpretive space linking environmental transformation, computational modeling, and systems-level biological inquiry.

Several critical questions remain open:

- Which biological systems are most sensitive to distributed exposure architectures?
- How do chronic low-dose interactions reorganize long-term physiological regulation?
- Can computational biosurveillance systems identify preclinical systemic signatures before overt pathological manifestation emerges?
- How should environmental medicine operate when biological disruption becomes increasingly networked, cumulative, and computationally inferred?

Figure 2. The Seamless Exposure Continuum: Interconnected Atmospheric, Aquatic, and Biological Dynamics of Nanoplastic Mobility Within a One Health Framework.

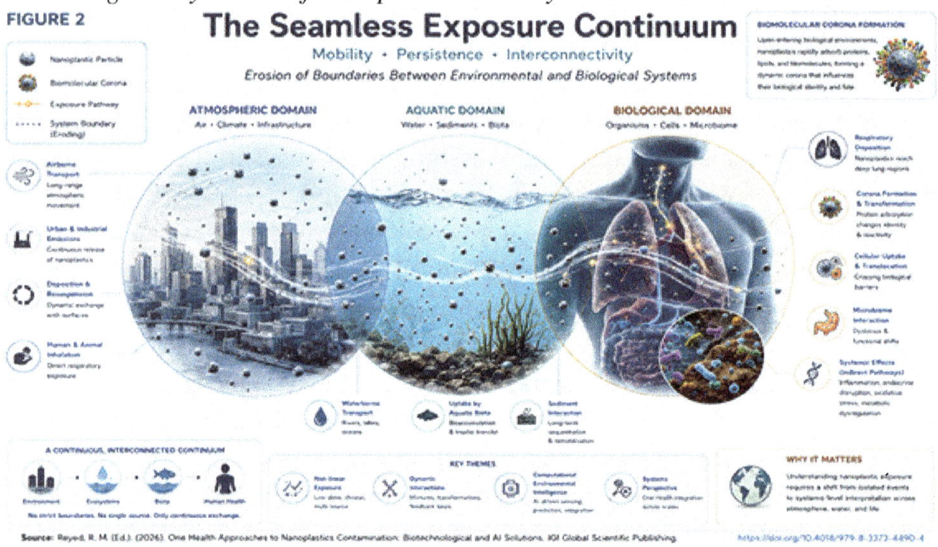

The figure presents nanoplastic contamination as a continuous exposure continuum linking atmospheric, aquatic, and biological systems through environmental transport, persistent mobility, and evolving biological interactions. The atmospheric domain illustrates airborne transport, industrial emissions, and deposition–resuspension processes enabling widespread dissemination. The aquatic domain highlights hydrological transport, sediment-associated retention, and trophic transfer pathways contributing to environmental persistence and redistribution. The biological domain integrates respiratory deposition, biomolecular corona formation, cellular translocation, microbiome perturbation, and systemic responses associated with prolonged low-dose exposure. Collectively, the interconnected pathways emphasize a systems-oriented One Health perspective in which environmental risk emerges through cumulative interactions across environmental and biological interfaces.

5. Why This Volume Matters

This volume was not developed as a conventional aggregation of isolated thematic chapters. Rather, it was intentionally constructed as a systems-oriented editorial architecture designed to examine environmental transition across multiple analytical scales simultaneously. The progression throughout the volume is sequential, though not strictly linear, encompassing environmental observation, systems modeling, computational interpretation, biological consequence, translational implementation, and governance integration. The central objective is not encyclopedic completeness, but structural coherence under conditions of increasing environmental complexity.

At the center of this volume lies a foundational assumption: environmental systems, computational systems, and biological systems are becoming progressively interdependent and may no longer remain meaningfully separable within contemporary environmental-health analysis.

Three major trajectories therefore organize the intellectual structure of this work:

- Environmental systems analysis for integrative scientific interpretation.
- Computational architectures linking environmental data to predictive analytics and adaptive biosurveillance.
- Systems-level biological interpretation connecting chronic exposure to long-term physiological and public-health dynamics.

The objective is not disciplinary fusion for its own sake. Rather, it is analytical interoperability under conditions of escalating systemic complexity.

To address technological and conceptual fragmentation across these domains, the volume operates through an integrated computational-biological architecture organized as four interconnected operational layers embedded within a broader adaptive One Health systems framework. The conceptual organization of these interacting layers and system-level relationships is summarized in Figure 3.

- **Layer 1: The Core Infrastructure & Diagnostic Signal Base –** Establishes the analytical foundations, signal propagation logic, and systems architecture required for predictive environmental biosurveillance and exposure interpretation across complex environmental matrices.
- **Layer 2: The Computational & Forensic Intelligence Layer –** Deploys machine-learning systems, environmental sensing architectures, source-

attribution methodologies, and cross-domain analytical frameworks for distributed particulate tracking and predictive interpretation.

- **Layer 3: The Biological Response & Interface Systems Layer –** Translates environmental exposure signals into biological consequence by integrating biomolecular interactions, cellular translocation pathways, microbiome perturbation, metabolic disruption, and bioaccumulation dynamics.
- **Layer 4: The Translational Implementation & Policy Governance Layer** – Synthesizes biological and computational intelligence into adaptive mitigation strategies, institutional risk-stratification systems, techno-economic assessment models, and scalable One Health governance architectures.

Figure 3. Multi-layered One Health systems governance architecture integrating environmental sensing, computational intelligence, biological response systems, and translational governance pathways into an adaptive framework for biosurveillance and environmental health decision-making

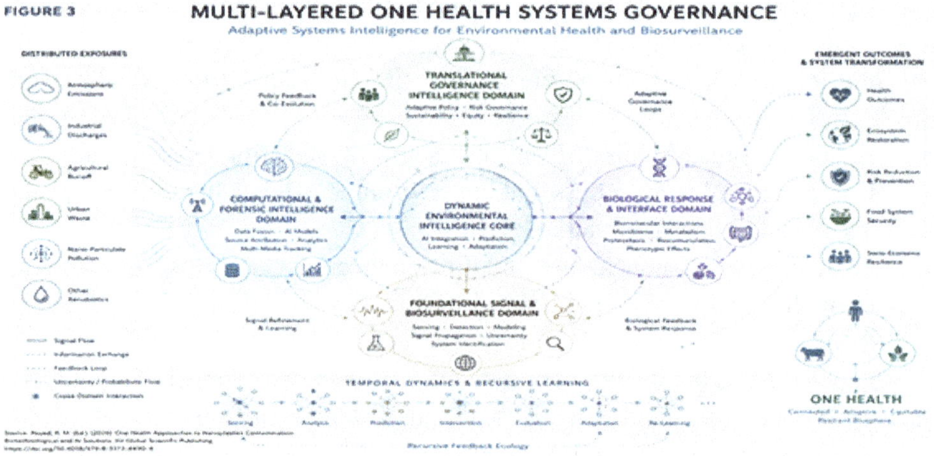

6. Intended Audience & Institutional Relevance

This volume is intended for an international, interdisciplinary readership engaged in the structural challenges emerging at the intersection of environmental science, computational intelligence, toxicology, systems biology, translational medicine, and One Health governance.

The volume is particularly directed toward:

- Environmental scientists and systems engineers developing multi-pathway exposure models and predictive ecological architectures.
- Computational biologists and AI researchers designing adaptive biosurveillance systems and environmental intelligence platforms.
- Biomedical clinicians and systems toxicologists investigating chronic low-dose exposure dynamics and networked physiological disruption.
- Public-health policymakers and One Health governance specialists requiring computationally informed frameworks for institutional risk assessment and translational mitigation strategy development.

Ultimately, the central issue is no longer whether environmental conditions are changing.

They are.

The more consequential question is whether existing toxicological, computational, biomedical, and epistemic frameworks still possess sufficient analytical resolution to represent that change adequately.

If they do not, then the development of more integrated interpretive architectures may no longer be optional, but necessary.

CONCEPTUAL CHAPTER PROFILES

Chapter 1. One Health Cyber-Biotechnology for Nano-Plastic Intelligence Systems and Logic Integration: From Exposure to System-Level Interpretation

Reyed M. Reyed, Genetic Engineering and Biotechnology Research Institute (GEBRI), City of Scientific Research and Technological Applications (SRTA-City), Alexandria, Egypt

Chapter Abstract

This chapter per the authors develops a comprehensive One Health cyber-biotechnology framework for interpreting nanoplastic exposure as a multiscale environmental signaling system rather than an isolated contaminant phenomenon. It integrates Digital Nanoplastic Science (DNPS), Plasto-Signaling Interference (PSI), Transductive Signal Load (TSL), the Polybiome Interfacial Signaling Cascade (PISC), and Systemic Risk Scoring (SRS) to structure exposure propagation, interface-level perturbation, and biological signaling disruption within a unified, computationally structured analytical architecture. Environmental matrices—including water, soil, and air—are conceptualized as dynamic signal-generating interfaces transmitting anthropogenic disturbances across ecological and physiological networks under complex environmental uncertainty regimes. The chapter critically re-evaluates conventional compartment-based risk assessment methodologies and advances high-throughput multiomics integration, real-time digital twin modeling, and predictive cyber-biotechnological analytics to establish next-generation early-warning systems, exposure intelligence platforms, and One Health–driven environmental decision-support frameworks.

Core Conceptual Innovations

1. **Digital Nanoplastic Science (DNPS):** The digital, multidimensional characterization of nanoplastic particulates, mapping physical-chemical properties—including morphology, surface charge, functional groups, and corona formation—into computational data arrays. DNPS functions as the foundational digital-twin layer tracking how real-world particles translate into distinct biological and environmental information vectors.
2. **Plasto-Signaling Interference (PSI):** The biochemical and biophysical mechanisms through which internalized nanoplastics disrupt endogenous molecular signaling pathways. Rather than acting as inert physical debris, particles engage in surface-level interactions that trigger aberrant cellular responses, mimicking or blocking critical biochemical signals across cellular and tissue matrices.
3. **Transductive Signal Load (TSL):** The cumulative quantified stress imposed on a biological system during the processing of molecular noise and secondary cellular signaling cascades induced by chronic nanoplastic exposure. TSL functions as a system-level metric for evaluating the transition of compensatory physiological mechanisms from adaptive regulation to pathological failure.
4. **Polybiome Interfacial Signaling Cascade (PISC):** The trans-domain propagation of nanoplastic-induced disturbances across ecological and biological interfaces. This cascade traces environmental signal transmission from abiotic matrices—including soil, water, and air—to diverse microbiome systems, ultimately propagating these perturbations into mammalian physiological networks within a One Health continuum.
5. **Systemic Risk Scoring (SRS):** An advanced multiparametric analytical matrix integrating environmental exposure data, biochemical pathway disruption rates, and cross-scale ecological variables. The SRS algorithm generates a consolidated risk vector, shifting environmental governance from retrospective monitoring toward predictive, real-time decision support.

Methodological Architecture

The proposed cyber-biotechnology intelligence framework maps the trajectory of nanoplastic data through three primary processing domains, bridging environmental detection with systemic biological interpretation.

Table 1.

Environmental Matrices	Cyber-Biotechnological Matrix	One Health Decision Support
Water Interfaces	Multiomics Data Integration	Early-Warning Systems
Soil Matrices	Digital Twin Simulation Models	Exposure Intelligence
Air Transport Vectors	Predictive Algorithmic Processing (SRS)	Environmental Governance

- **The Input Interface:** Captures continuous physical-chemical variations across dynamic environmental matrices, treating water, soil, and air as primary signal-generating sources.
- **The Computational Core:** Implements predictive digital twin modeling and integrates high-throughput multiomics datasets—including metagenomics, transcriptomics, and metabolomics—to decipher structural and biological signaling disruptions associated with particulate corona formation.
- **The Output Architecture:** Converts complex biological signaling data into actionable exposure intelligence that directly informs One Health environmental decision-support systems and policy-oriented governance frameworks.

Chapter 2. Forensic Tracing Systems for Nanoplastics: AI and Biotech One Health Source Attribution

Yash Bhardwaj, K.R. Mangalam University, India
Khushbu Katiyar, K.R. Mangalam University, India
Kazi Mizaan, K.R. Mangalam University, India
Baljeet Yadav, K.R. Mangalam University, India
Anuj Kumar, K.R. Mangalam University, India
Sameer Saharan, Mody University of Science and Technology, Rajasthan, India
Mahipal Singh Sankhla, K.R. Mangalam University, India

Chapter Abstract

This chapter per the authors examines forensic tracing architectures for nanoplastics within a One Health framework linking environmental persistence, contaminant mobility, and cross-matrix exposure propagation. It evaluates nanoscale dimensional constraints, compositional heterogeneity, and transformation dynamics affecting source discrimination, material characterization, and exposure reconstruction across complex environmental systems. The analytical framework integrates molecular assays, spectroscopic profiling, imaging platforms, and computational attribution models for contaminant provenance mapping. Particular emphasis is placed on data fusion strategies correlating heterogeneous physicochemical and biological signatures for source-resolved interpretation under variable environmental conditions. By combining forensic biotechnology with AI-driven attribution analytics, the chapter establishes operational pathways for nanoplastic traceability, exposure intelligence, and One Health–oriented risk evaluation across ecological interfaces and biological systems.

Forensic Analytical Framework

- **Contaminant Provenance Mapping:** Integrating molecular assays, spectroscopic profiling, and advanced imaging platforms to track nanoplastic transformation dynamics.
- **Data Fusion Strategies:** Correlating heterogeneous physicochemical and biological signatures to achieve source-resolved interpretation under variable environmental conditions.
- **AI-Driven Attribution Analytics:** Utilizing computational attribution models to overcome nanoscale dimensional constraints and compositional heterogeneity.
- **One Health Source Attribution:** Establishing operational pathways for cross-matrix exposure propagation tracking, nanoplastic traceability, and system-level risk evaluation.

Chapter 3. Physicochemical Determinants of Nanoplastic Fate and Advanced Removal Dynamics: AI-Enabled Approaches within One Health Frameworks

Fatma Tugce Senberber Dumanli, Department of Bioengineering, Yildiz Technical University, Istanbul, Turkey
Emek Moroydor Derun, Department of Bioengineering, Yildiz Technical University, Istanbul, Turkey
Mehmet Burcin Piskin, Department of Bioengineering, Yildiz Technical University, Istanbul, Turkey

Chapter Abstract

This chapter per the authors analyzes the physicochemical determinants governing nanoplastic transport behavior, environmental partitioning, bioavailability, and remediation performance across complex ecological matrices. Nanoplastics are characterized as heterogeneous colloidal systems in which structural and surface properties—including particle size distribution, morphological geometry, surface charge (zeta potential), density variation, hydrophobicity, and surface functionalization—collectively regulate aggregation kinetics, interfacial interactions, co-contaminant adsorption behavior, and long-term environmental persistence under variable hydrochemical conditions. The chapter further evaluates how these physicochemical properties influence the kinetic efficiency, thermodynamic feasibility, and selective separation capacity of physical, chemical, biological, and hybrid remediation technologies. Analytical constraints associated with sub-micron characterization, polymeric heterogeneity, and remediation-related ecological trade-offs are critically assessed alongside predictive computational modeling and machine learning–assisted optimization approaches for fate estimation and removal performance analysis. By integrating advanced environmental chemistry with engineered remediation dynamics, the chapter establishes a mechanistically focused analytical perspective for evaluating nanoplastic behavior and mitigation pathways across environmental and biological exposure interfaces.

Operational Analytical Framework

- **Colloidal Fate Determinants:** Assessing how particle size distribution, morphology, surface charge, density, hydrophobicity, and surface functionalization regulate aggregation kinetics, interfacial interactions, and long-term environmental persistence across multiphase ecological matrices.

- **Environmental Partitioning & Bioavailability:** Evaluating transport dynamics and interface-level interactions governing nanoplastic movement across aquatic, terrestrial, and biological compartments under variable hydrochemical conditions.
- **Engineered Remediation Dynamics:** Examining the performance, selectivity, kinetic behavior, and thermodynamic constraints of physical separation, chemical transformation, biological degradation, and hybrid remediation technologies.
- **Analytical Characterization Constraints:** Critically assessing methodological limitations associated with sub-micron particle detection, spectral interpretation, compositional heterogeneity, and mass quantification of aged polymeric matrices.
- **Computational Fate Estimation & Removal Optimization:** Applying predictive computational modeling, molecular simulation approaches, and machine learning–assisted analytics for environmental fate estimation and remediation performance optimization.

Chapter 4. Human Exposure to Nanoplastics and Its Toxicological and Environmental Implications: A One Health Perspective

Odangowei Inetiminebi Ogidi, Bayelsa Medical University, Yenagoa, Nigeria
Nwanne Ijere Dike, Michael Okpara University of Agriculture, Umudike, Nigeria
Lilian Maureen Oluchi Oguoma, Federal University of Technology, Owerri, Nigeria

Chapter Abstract

This chapter per the authors investigates human exposure to nanoplastics and toxicodynamic responses associated with environmental persistence, biological translocation, and cross-interface exposure pathways within a One Health context. Environmental fragmentation and degradation processes are evaluated alongside nanoplastic transport across aquatic, atmospheric, and terrestrial matrices contributing to ingestion, inhalation, dermal uptake, and systemic distribution. The chapter analyzes mechanistic injury pathways involving oxidative stress, inflammatory signaling cascades, endocrine disruption, epithelial barrier impairment, and genotoxic effects across multiple organ systems. Biomonitoring methodologies, exposure quantification challenges, and analytical standardization limitations associated with low-concentration sub-micron polymeric detection are further critically assessed. By integrating exposure science with mechanistic toxicology and computational exposure analytics, the chapter establishes a focused analytical perspective for interpreting nanoplastic-associated health risks across environmental and physiological exposure interfaces.

Toxicological Analytical Framework

- **Cross-Interface Exposure Pathways:** Evaluating environmental weathering, fragmentation, and degradation processes across aquatic, atmospheric, and terrestrial matrices contributing to primary human exposure routes.
- **Biological Translocation Dynamics:** Assessing ingestion, respiratory inhalation, and dermal absorption pathways regulating nanoplastic penetration, biodistribution, and systemic physiological exposure.
- **Mechanistic Toxicodynamics:** Characterizing oxidative stress induction, inflammatory cytokine signaling, endocrine perturbation, epithelial barrier dysfunction, and genotoxic injury responses across cellular and tissue systems.
- **Systemic Organ-Level Implications:** Examining chronic physiological disruption involving gastrointestinal, hepatic, pulmonary, neurological, and endocrine-associated biological responses following sustained nanoplastic exposure.
- **Biomonitoring & Quantification Constraints:** Critically assessing methodological limitations, analytical standardization deficits, and instrumental sensitivity challenges associated with detecting sub-micron polymeric burdens in biological specimens.
- **Computational Exposure Analytics:** Applying predictive analytical modeling and exposure interpretation approaches to support toxicological assessment and long-term human health risk evaluation.

Chapter 5. Gut Microbiome Perturbation and Nutritional Outcomes under Nanoplastic Exposure: A One Health Perspective

Deepa Bhadan, CCS University, India
Thanjaivadivel M., Nandha Engineering College, Erode, India
R. Lakshmana Kumar, Department of Computer Science and Engineering (AI & ML), Tagore Institute of Engineering & Technology, Deviyakurichi, India
Bibi Ameena, School of Computing and Information Technology, REVA University, Bangalore, India

Chapter Abstract

This chapter per the authors explores gut microbiome perturbation and nutrition-associated metabolic imbalance under nanoplastic exposure within a One Health context linking microbial ecology with environmental and dietary stressors. It evaluates how antibiotics, dietary patterns, chemical co-contaminants, and nanoplastics alter microbial diversity, epithelial barrier integrity, nutrient absorption kinetics, inflammatory signaling, and metabolic regulation across undernutrition and overnutrition states. Bidirectional microbiota–host interactions are further analyzed, including immune modulation, metabolic homeostasis, and xenobiotic processing pathways. The chapter examines microbiome-targeted interventions such as probiotics, prebiotics, synbiotics, fecal microbiota transplantation, and advanced metagenomic profiling supported by computational analytics for dysbiosis assessment. By integrating nutritional biology, microbial ecology, and exposure science, the chapter establishes a mechanistically focused perspective on microbiome resilience and host–environment interactions under chronic nanoplastic stress conditions.

Microbiome Analytical Framework

- **Microbiome Structural Profiling:** Utilizing high-throughput 16S rRNA sequencing and shotgun metagenomic profiling alongside computational analytics to characterize taxonomic shifts and functional dysbiosis patterns.
- **Multi-Stressor Alteration Dynamics:** Evaluating the cumulative effects of antibiotics, dietary variation, chemical co-contaminants, and nanoplastic exposure on microbial diversity indices and ecological stability.
- **Epithelial Barrier Integrity Assessment:** Characterizing intestinal mucosal disruption, tight junction disassembly, and associated alterations in nutrient absorption kinetics and barrier permeability.
- **Bidirectional Host–Microbiota Crosstalk:** Assessing immune modulation, inflammatory signaling pathways, metabolic homeostasis regulation, and xenobiotic metabolic processing associated with microbiome perturbation.
- **Metabolic Polarization Scenarios:** Examining differential physiological outcomes associated with nanoplastic-induced dysbiosis under contrasting undernutrition and overnutrition conditions.
- **Microbiome-Targeted Therapeutics:** Investigating probiotics, prebiotics, synbiotics, and fecal microbiota transplantation strategies for restoring microbial resilience under chronic particle exposure.
- **Computational Dysbiosis Analytics:** Applying computational interpretation approaches and metagenomic analytics to support microbiome-based exposure assessment and nutritional risk evaluation.

Chapter 6. An Environmental Nutrition Framework for Micro- and Nanoplastics Exposure Assessment: AI-Assisted Food Safety Detection Systems

Swapan Banerjee, Poornima University, India
Abdullayev Vugar, Azerbaijan University of Architecture and Construction, Azerbaijan
Nazila Ragimova, Azerbaijan State Oil and Industry University, Azerbaijan
Triwiyanto T., Azerbaijan State Oil and Industry University, Azerbaijan

Chapter Abstract

This chapter per the authors evaluates micro- and nanoplastic exposure within food systems through an environmental nutrition framework emphasizing trophic transfer, dietary intake pathways, and contamination dynamics across aquatic, agricultural, and processed food chains. The analysis examines particulate accumulation across trophic levels and associated oral exposure burdens affecting gastrointestinal, hepatic, neurological, and metabolic functions, alongside oxidative stress and inflammatory signaling associated with dietary plastic ingestion. Nutritional mitigation strategies involving bioactive compounds, probiotics, polyphenols, and cellular defense modulators are critically assessed together with food safety surveillance systems, contaminant detection methodologies, and computational exposure forecasting tools. By integrating environmental nutrition, food exposure science, and traceability analytics, the chapter establishes a specialized analytical perspective for evaluating dietary nanoplastic risks, contaminant transfer dynamics, and food-system resilience within a One Health context.

Food Safety & Exposure Analytics

- **Environmental Nutrition Framework:** Evaluating the relationship between dietary quality, agricultural ecosystems, and trophic food-chain contamination associated with persistent polymeric particulates.
- **Trophic Transfer & Bioaccumulation Dynamics:** Assessing the movement and biomagnification of micro- and nanoplastics across aquatic, terrestrial, and processed food systems.
- **Systemic Oral Exposure Burdens:** Characterizing gastrointestinal, hepatic, neurological, and metabolic disturbances associated with chronic dietary ingestion of plastic particulates.

- **Subcellular Inflammatory Signaling:** Examining oxidative stress induction, reactive oxygen species generation, and inflammatory cascade activation associated with dietary particle exposure.
- **Nutritional Mitigation Strategies:** Assessing the protective roles of bioactive dietary compounds, probiotics, antioxidant polyphenols, and cellular defense modulators in reducing particle-associated toxicity.
- **Food Safety Surveillance & Exposure Forecasting:** Applying contaminant detection methodologies, food-system traceability analytics, and computational exposure forecasting approaches for exposure monitoring and risk interpretation.
- **Food-System Resilience & Contaminant Control:** Integrating food exposure science, analytical chemistry, and predictive analytics to support contaminant management and dietary risk reduction across food-production systems.

Chapter 7. Translational One Health Strategies for Nanoplastic Exposure Mitigation

P. Praveen Kumar, SRM Institute of Science and Technology, Vadapalani, India
K. C. Raja Shree, Saveetha School of Management, Saveetha Institute of Medical and Technical Sciences, India

Chapter Abstract

This chapter per the authors addresses translational strategies for mitigating nanoplastic exposure across environmental, technological, economic, and public health domains within a One Health context. Nanoplastics are evaluated as mobile particulate contaminants capable of crossing biological barriers, accumulating across environmental compartments, and contributing to chronic ecological and physiological exposure burdens. The chapter examines mitigation pathways involving exposure surveillance, biotechnology-assisted reduction systems, computational exposure forecasting, and implementation-oriented environmental management models. Operational considerations associated with environmental intervention deployment, resource allocation, and sustainability-oriented management are further critically assessed within applied mitigation settings. By integrating exposure analytics, environmental management, and implementation economics, the chapter establishes a translational perspective for reducing nanoplastic-associated environmental and public health burdens across complex exposure landscapes.

Translational Implementation Framework

- **Translational Mitigation Strategies:** Structuring scalable intervention pathways translating nanoplastic exposure science into operational environmental and public health applications.
- **Biological Barrier & Environmental Accumulation Analysis:** Evaluating particulate transport dynamics governing biological penetration, environmental compartmentalization, and chronic exposure accumulation.
- **Biotechnology-Assisted Reduction Systems:** Assessing biological and engineered approaches designed to capture, degrade, or reduce sub-micron polymeric contaminants across affected matrices.
- **Computational Exposure Forecasting:** Applying predictive analytical models and simulation-based approaches to estimate environmental dispersion patterns and long-term population exposure trends.
- **Implementation Economics & Environmental Management:** Examining cost allocation parameters, deployment feasibility, and operational management considerations associated with mitigation technologies at industrial and municipal scales.
- **Resilience-Oriented Governance Approaches:** Assessing policy coordination mechanisms and collaborative environmental management strategies supporting long-term contaminant reduction and sustainability objectives.

Chapter 8. Economic and Life Cycle Assessment of Nanoplastics Remediation within a One Health Framework: Biotechnological and AI-Driven Solutions

Odangowei Inetiminebi Ogidi, Bayelsa Medical University, Yenagoa, Nigeria
Ejeomo Christopher, Michael and Cecilia Ibru University, Agbarha-Otor, Nigeria
Obienu Anayochukwu Chukwunonso, Bayelsa Medical University, Yenagoa, Nigeria
Emmanuel Aziagba, Sheda Science and Technology, Abuja, Nigeria

Chapter Abstract

This chapter per the authors analyzes economic and life-cycle dimensions of nanoplastic remediation through analytical frameworks linking environmental burden reduction, remediation efficiency, and resource-allocation dynamics within One Health contexts. Nanoplastic generation pathways, environmental transport behavior, and cross-matrix exposure routes across aquatic, terrestrial, and food-associated environments are evaluated using life-cycle assessment methodologies to identify environmental and economic hotspots associated with remediation deployment. Biotechnological mitigation approaches—including microbial degradation pathways and enzyme-mediated transformation systems—are examined alongside computational optimization tools supporting predictive remediation modeling, cost-efficiency analysis, and contaminant forecasting. Integration of ecological risk metrics, public health considerations, and socioeconomic performance indicators contributes a techno-economic perspective for sustainable nanoplastic management, environmentally responsive remediation planning, and long-term exposure mitigation.

Techno-Economic Assessment Framework

- **Life-Cycle Assessment Methodologies:** Applying standardized LCA approaches to evaluate environmental burdens associated with nanoplastic generation pathways, transport dynamics, and remediation deployment across interconnected exposure matrices.
- **Remediation Hotspot Analysis:** Identifying critical environmental and economic bottlenecks influencing mitigation-system scalability, operational efficiency, and resource utilization across aquatic, terrestrial, and food-associated systems.
- **Biotechnological Remediation Systems:** Evaluating microbial degradation pathways, enzyme-mediated transformation mechanisms, and biologically assisted remediation technologies in relation to kinetic performance, scalability, and environmental applicability.
- **Computational Optimization & Forecasting:** Utilizing predictive modeling, algorithmic cost-efficiency analytics, and computational contaminant forecasting tools to support remediation planning and exposure-management strategies.
- **Socioeconomic Performance Metrics:** Integrating ecological risk indicators, public health considerations, and techno-economic evaluation parameters to assess sustainability performance and long-term remediation feasibility.
- **Sustainable One Health Remediation Planning:** Linking environmental management objectives with adaptive remediation strategies and long-term exposure-reduction priorities across interconnected ecological and human health systems.

INTEGRATIVE SYSTEMS PERSPECTIVE

Collectively, the conceptual progression across the eight chapters establishes an interconnected analytical architecture linking environmental particulate mobility, computational exposure intelligence, biological-response signaling, translational mitigation systems, and adaptive One Health governance. The volume therefore advances from isolated compartment-based interpretation toward a unified cyber-biotechnological systems framework capable of integrating environmental, computational, and physiological complexity within a single interpretive continuum (Figure 4).

Figure 4. Nanoplastics Exposome–Response Continuum Within a Cyber-Biotechnological One Health Intelligence Architecture.

This figure presents nanoplastic contamination not as an isolated toxicological event, but as a continuously adaptive exposure architecture operating across interconnected atmospheric, aquatic, terrestrial, biological, computational, and institutional systems. Rather than preserving classical compartment-based interpretations of contamination, the framework reconceptualizes exposure as a distributed systems phenomenon characterized by persistent particulate mobility, physicochemical transformation, and recursive cross-domain interaction.

The left structural axis illustrates the fragmented origins of particulate generation and degradation, demonstrating how industrial activity, agricultural systems, atmospheric abrasion, aquatic fragmentation, and polymeric urban infrastructures collectively contribute to the continuous production of environmentally mobile nano-scale materials. Fragmentation is therefore represented not as a terminal degradation process, but as an active mechanistic driver of environmental persistence and redistribution.

The central radial architecture functions as the primary environmental-transport and systems-biology interface of the model. Atmospheric dispersion, hydrological circulation, terrestrial transfer, food-chain propagation, and biological translocation are integrated into a seamless exposome–response continuum linking environmental mobility with cumulative biological perturbation. The framework incorporates oxidative stress, inflammatory signaling, epithelial barrier destabilization, endocrine disruption, microbiome dysbiosis, and genotoxic–epigenetic alterations as interconnected adaptive response networks emerging under prolonged low-dose particulate circulation rather than isolated linear toxicological endpoints.

The model further introduces a cyber-biotechnological intelligence layer integrating real-time environmental sensing, machine-learning analytics, exposure modeling, digital twin ecosystems, and predictive biosurveillance architectures. Within this framework, computational systems no longer operate merely as passive analytical instruments; they become embedded interpretive infrastructures participating directly in environmental-risk inference, adaptive forecasting, and systems-level exposure interpretation.

The lower translational domain extends the framework beyond mechanistic toxicology toward governance implementation, integrating policy adaptation, techno-biological mitigation systems, stakeholder coordination, sustainability infrastructures, and adaptive risk-management architectures. Recursive amplification loops further emphasize that environmental loading, ecological resilience decline, biological vulnerability, and institutional response capacity evolve through interconnected reinforcement dynamics rather than isolated causal sequences.

Collectively, the figure proposes a transition from reductionist environmental toxicology toward a cyber-biotechnological One Health systems architecture capable of integrating particulate mobility, computational intelligence, biological signaling, translational mitigation, and governance adaptation within a unified interpretive framework for next-generation environmental-health intelligence.

Editor

Reyed M. Reyed

Genetic Engineering and Biotechnology Research Institute (GEBRI), City of Scientific Research and Technological Applications (SRTA-City), Alexandria, Egypt

Acknowledgment

THE COLLABORATIVE CHAPTER REVIEW & INTEGRATION PROCESS

The intellectual cohesion, scientific rigor, and interdisciplinary coherence of this volume were reinforced through a structured internal peer-review process. This volume benefited from the scholarly evaluation and collaborative review efforts of the following contributors and expert reviewers, whose interdisciplinary contributions supported the conceptual integration of the One Health frameworks across these chapters.

I. BIOENGINEERING & BIOTECHNOLOGICAL SYSTEMS REVIEWERS

Dr. Fatma Tuğçe Şenberber Dumanlı, PhD

- *Institutional Affiliation:* Department of Bioengineering, Yildiz Technical University, Istanbul, Türkiye.
- *Review Focus:* Microbial degradation kinetics, enzymatic transformation vectors, and biocatalytic cleavage frameworks.

Dr. Mehmet Burcin Piskin, PhD

- *Institutional Affiliation:* Department of Bioengineering, Yildiz Technical University, Istanbul, Türkiye.

- *Review Focus:* Nanoplastic exposure determinants, biological barrier integrity, and multi-component colloidal dynamics.

II. APPLIED SCIENCES & ECO-TOXICOLOGICAL REVIEWERS

Dr. Prerna Mehta, PhD

- *Institutional Affiliation:* Department of Biotechnology, GD Rungta College of Science & Technology, Bhilai, Chhattisgarh, India.
- *Review Focus:* Ecosystem health baselines and socio-economic remediation risk-stratification.

Dr. Odangowei Inetiminebi Ogidi, PhD

- *Institutional Affiliation:* Department of Biochemistry, Faculty of Basic Medical Sciences, Bayelsa Medical University, Yenagoa, Nigeria.
- *Review Focus:* Human biomonitoring models, oxidative stress cascades, and xenobiotic processing dynamics.

Prof. Waghamare Suresh

- *Institutional Affiliation:* Department of Pharmaceutics, NIMS Institute of Pharmacy, NIMS University, Jaipur, Rajasthan, India (303121).
- *Review Focus:* Targeted biosensing architectures, bioanalytical discrimination, and oral exposure burdens.

III. COMPUTATIONAL ANALYTICS & GOVERNANCE SYSTEMS REVIEWERS

Prof. Usharani Bhimavarapu

- *Institutional Affiliation:* Assistant Professor, Department of CSE, Koneru Lakshmaiah Education Foundation, Andhra Pradesh, India.
- *Review Focus:* AI biosurveillance networks, data fusion architectures, and inverse probabilistic source attribution modeling.

IV. HEALTH-ECONOMIC EVALUATION & TRANSLATIONAL POLICY REVIEWERS

Dr. Swapan Banerjee, PhD

- *Institutional Affiliation:* Department of Healthcare Management, Poornima University, Jaipur, Rajasthan, India.
- *Review Focus:* Health-economic evaluation matrices, public health asset allocation, and sustainability-oriented policy conversion.

Dr. K. C. Raja Shree

- *Institutional Affiliation:* Saveetha School of Management, Saveetha Institute of Medical and Technical Sciences, Chennai, India.
- *Review Focus:* One Health institutional governance, supply-chain traceability, and equity auditing metrics.

Dr. P. Praveen Kumar, PhD

- *Institutional Affiliation:* Faculty of Management, SRM Institute of Science and Technology (Vadapalani Campus), Chennai, India.
- *Review Focus:* Techno-economic cost-efficiency forecasting and operational processing hotspot optimization.

V. PUBLISHING SUPPORT & PRODUCTION COORDINATION

The development and completion of this volume benefited from the professional publishing environment provided by IGI Global Scientific Publishing. Throughout the editorial and production stages, the project was supported by a collaborative publishing framework that facilitated scholarly communication, editorial coordination, and the international dissemination of interdisciplinary academic work.

Particular appreciation is extended to **Lindsay Wertman**, President of IGI Global Scientific Publishing, and **Melissa Wagner**, Vice President of Publishing Ethics & Research Integrity, for their professionalism, thoughtful guidance, and continued support during the progression of the manuscript. Their publishing oversight and coordination contributed significantly to the successful completion and presentation of this volume.

Additional acknowledgment is extended to **Nina Eddinger**, Book Development Editor at IGI Global Scientific Publishing, for editorial coordination and continued assistance throughout the manuscript development and production stages.

Recognition is further extended to the broader IGI Global production and technical teams for their efforts in supporting formatting workflows, production logistics, metadata processing, and dissemination support essential to contemporary academic publishing.

Introduction

INTEGRATIVE EDITORIAL OVERVIEW

Nanoplastics can no longer be reasonably framed as inert environmental residues dispersed passively across contaminated systems. That characterization, while historically useful, has become increasingly inadequate in light of converging evidence suggesting that these particles operate as dynamic interfacial entities capable of traversing biological, ecological, and environmental matrices while simultaneously engaging with the physicochemical and regulatory processes that structure system behavior (Geyer et al., 2017; Jiang et al., 2020; Wang et al., 2021; Mousazadeh-gavan, 2026). Their relevance, therefore, is not exhausted by spatial accumulation or persistence, but extends into interactional behavior across membranes, trophic pathways, and exposure networks that exceed traditional toxicological demarcations (Dang et al., 2022; Ziani et al., 2023).

From a systems-oriented perspective, micro- and nanoplastics should be understood as emergent environmental stressors whose significance arises through multiscale interactions rather than isolated endpoints of exposure or toxicity. Although analytical capabilities in detection, characterization, and remediation have advanced substantially, these developments have not yet been matched by equivalent gains in biological interpretability or mechanistic integration at the systems level (Sanoja-López et al., 2026; Wang et al., 2021; Reichstein et al., 2019). This persistent asymmetry reflects a broader structural limitation within environmental sciences, where increasing data density has not necessarily translated into conceptual unification, leaving key domains fragmented across disciplinary boundaries (Levin, 1998; Meadows, 2008).

In parallel, methodological progress in environmental sensing, nanomaterial tracking, and AI-assisted modeling has expanded the empirical horizon of the field, yet it has not produced a coherent interpretative architecture capable of integrating cumulative exposure dynamics, delayed biological responses, and ecological consequences under realistic chronic conditions (Jiang et al., 2020; Wang et al., 2021; Arellano et al., 2026). As a result, nanoplastics research continues to generate extensive datasets whose mechanistic connectivity across biological scales remains only partially resolved (Dang et al., 2022; Ziani et al., 2023).

It is within this unresolved epistemic and methodological gap that the present editorial advances the Data-Driven Systems Framework (DDSF). The framework is not intended as a deterministic model, nor as a closed theoretical system, but rather as a structured interpretative architecture through which nanoplastic-related processes may be organized along a continuous spectrum extending from exposure and detection to interaction, biological response, and systems-level inference (Meadows, 2008; Wang et al., 2021). Within this formulation, nanoplastics are treated less as passive contaminants and more as interaction-mediated perturbation agents whose biological significance emerges through their capacity to alter interfacial signaling conditions and disrupt molecular communication pathways (Mahmoudi et al., 2023; Arellano et al., 2026).

To render this interactional dimension more analytically tractable, DDSF introduces two conceptual constructs: Plasto-Signaling Interference (PSI) and Transductive Signal Load (TSL). PSI refers to the capacity of nanoplastic interactions to perturb signaling environments at biological interfaces, whereas TSL captures the cumulative burden of such perturbations across temporal scales, exposure recurrence, and contextual variability. These constructs are deliberately positioned as interpretative instruments rather than fixed empirical metrics and are intended to address persistent issues of fragmentation and limited comparability across existing studies (Ziani et al., 2023; Wang et al., 2021).

Given that biological disruption rarely manifests as a discrete event, DDSF further incorporates the Proteostatic Stress Continuum (PSC), which conceptualizes biological response as a graded and evolving spectrum ranging from adaptive compensation to functional destabilization under sustained molecular stress (Mahmoudi et al., 2023; Arellano et al., 2026). Extending this logic, the Nanoplastic Proteostatic Panflemmosis Metabolic Continuum (NPPMC) is proposed as a higher-order construct linking persistent proteostatic stress to systemic inflammatory signaling and downstream metabolic dysregulation under chronic exposure regimes (Jiang et al., 2020; Dang et al., 2022).

At the systems level, these interacting dimensions are integrated through the proposed Systemic Risk Score (SRS), which conceptually aligns with probabilistic and data-driven approaches in environmental systems science and ecological prediction modeling (Reichstein et al., 2019; Wang et al., 2021). The SRS is intended as a composite interpretative measure synthesizing exposure intensity, signaling disruption, and systemic fragility into a unified analytical representation without reducing complexity to a single deterministic outcome.

Within this architecture, artificial intelligence and biotechnology are not positioned as autonomous explanatory frameworks, but rather as enabling infrastructures supporting detection, inference, and adaptive environmental response. Their analytical value emerges only insofar as they remain embedded within systems that preserve biological interpretability and mechanistic coherence across scales (Reichstein et al., 2019; Sanoja-López et al., 2026). In this regard, the One Health paradigm functions not as a thematic backdrop, but as a structural condition for integrating human, environmental, and ecological dimensions within a unified analytical horizon (WHO, 2022; Whitmee et al., 2015).

Extending beyond strictly biological endpoints, DDSF further incorporates socio-environmental asymmetries and differential adaptive capacities through the lens of Planetary Environmental Intelligence (PEI). This dimension acknowledges that nanoplastic risk is not solely a function of exposure magnitude, but is also mediated by uneven capacities for detection, interpretation, and response across interconnected ecological and human systems (Dang et al., 2022; Ziani et al., 2023).

Rather than converging toward deterministic closure, this editorial positions DDSF as an open yet structured interpretative architecture intended to reduce disciplinary fragmentation, clarify mechanistic gaps, and support the development of integrative, uncertainty-aware methodologies for nanoplastics research within a systems-based One Health framework (Figure 1).

Figure 1. Editorial Scope and Systems Positioning of the Volume

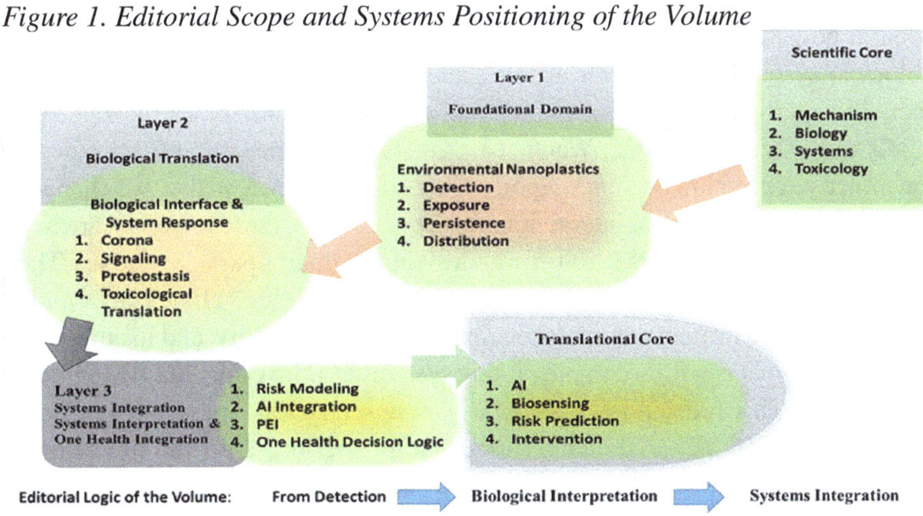

This figure functions not as a mechanistic abstraction, but as an editorial scaffold clarifying how the volume is conceptually organized and why its progression follows a layered systems logic. Rather than pre-empting detailed biological or mechanistic exposition, it delineates the structural architecture through which the subject of nanoplastics is approached—beginning with environmental presence and detection, progressing through biological interpretation at the interface level, and culminating in systems-level integration within a One Health perspective.

The vertical layering reflects a deliberate transition from descriptive domains toward interpretative synthesis, while the lateral framing distinguishes foundational scientific inquiry from its translational extensions in AI-enabled sensing, risk modeling, and intervention design. By separating organizational logic from mechanistic detail, the figure avoids redundancy with later sections and instead situates the reader within the editorial intent of the volume: transforming fragmented evidence into a coherent, progression-based interpretative continuum.

Keywords: Data-Driven Systems Framework (DDSF); nanoplastics; One Health; Plasto-Signaling Interference (PSI); Transductive Signal Load (TSL); Proteostatic Stress Continuum (PSC); Nanoplastic Proteostatic Panflemmosis Metabolic Continuum (NPPMC); Systemic Risk Score (SRS); Planetary Environmental Intelligence (PEI); systems biology; environmental risk integration.

1. From Particle-Centrism to Systemic Perturbation

Nanoplastics mark a transition that is not reducible to scale alone. The shift from macro- and microplastic pollution to nanoscale particles entails a reconfiguration in how plastic-derived disturbance is conceptualized across biological and ecological systems, as behavior at this scale is increasingly governed by surface reactivity, interfacial dynamics, and interaction potential rather than by abundance or bulk mass (Lin et al., 2025; Wang et al., 2021). What appears, at first glance, as a quantitative reduction becomes, upon closer examination, a qualitative change in the nature of environmental interaction. At the nanoscale, these materials are more plausibly interpreted as interaction-mediating entities than as passive residues. Their capacity to engage directly with physicochemical and biological interfaces allows them to influence protein corona formation, membrane stability, and localized signaling environments within complex living systems (Karim et al., 2024; Mahmoudi et al., 2023). This repositioning—subtle yet consequential—moves nanoplastics away from a purely material identity toward a role defined by relational behavior at interfaces. Accordingly, their significance lies less in their classification as inert contaminants and more in the spectrum of biological effects that may arise through interface-level processes. These include, but are not limited to, modulation of cellular uptake pathways, perturbation of molecular signaling, induction of oxidative stress, inflammatory activation, and potential genotoxic outcomes, all of which unfold within context-dependent biological conditions rather than as uniform responses (Lin et al., 2025; Karim et al., 2024). The variability observed across studies is not merely methodological noise; rather, it reflects the underlying complexity of interaction-driven effects. This shift in interpretation places pressure on descriptive paradigms that prioritize detection, distribution, and accumulation. Such approaches remain necessary—indeed, foundational—but become insufficient when treated in isolation. Knowing what nanoplastics are, where they localize, and how they disperse does not, in itself, resolve how they propagate biological consequences across interconnected systems (Dang et al., 2022; Wang et al., 2021). A gap therefore persists between characterization and explanation. What follows from this is the need for a systems-oriented interpretative framework capable of linking exposure conditions to bio-nano interactions, subsequent biological responses, and, ultimately, cross-scale risk formation within a coherent analytical logic. This requirement aligns with One Health perspectives, in which boundaries between environmental, animal, and human systems are treated as permeable rather than discrete (Dang et al., 2022). The challenge is therefore not only integrative, but also conceptual: maintaining mechanistic fidelity while accommodating systemic complexity. Emerging evidence further complicates this landscape. Observations indicating that nanoplastic exposure may influence axes such as gut–brain communication or processes such as placental development suggest that these effects are not confined to localized interactions, but may extend into organism-level dysfunction and developmental perturbation (Grodzicki et al., 2021; Chen et al., 2025). These findings, while still evolving, underscore the inadequacy of strictly compartmentalized models of exposure and response and point instead toward the necessity of frameworks capable of capturing distributed and temporally extended biological consequences.

2. The Invisibility Gap and the Limits of Current Paradigms

Despite rapid methodological progress, a structural mismatch still appears to persist between analytical detection and biological interpretation. The "Invisibility Gap" thus captures the difficulty of converting measurable nanoplastic presence into defensible inferences regarding long-term systemic outcomes (Bhattacharyya et al., 2025; Njema et al., 2026; Moulton et al., 2026). Current toxicological paradigms remain weighted toward acute or high-dose experimental designs, and this emphasis may underrepresent chronic, low-intensity exposure settings in which biological effects emerge cumulatively, gradually, and through non-linear trajectories rather than through immediate overt toxicity (Bhattacharyya et al., 2025). That limitation, in turn, strengthens the rationale for frameworks capable of integrating signal-level perturbation, temporal accumulation, and system vulnerability, because the existing evidence base increasingly points toward cross-scale effects that are not adequately resolved by particle-centric risk models alone (Moulton et al., 2026).

3. One Health as a Systems Continuum

The One Health paradigm provides a well-established foundation for integrating human, animal, and environmental health, a premise consistently reinforced throughout the broader One Health literature (Mackenzie & Jeggo, 2019; Danasekaran, 2024). In the context of nanoplastics research, however, this integrative framing appears to require further conceptual extension. The challenge is no longer limited to recognizing that these domains are interconnected, but rather to understanding how contamination propagates across them in a coupled, cumulative, and often temporally extended manner (Helal et al., 2023; Moulton et al., 2026). What becomes increasingly less tenable, then, is the implicit assumption that human, animal, and environmental systems can be treated as linked yet analytically separable units. A systems-oriented perspective instead suggests a more continuous interactional fabric, within which perturbations do not remain confined to their point of origin but may circulate across scales through reinforcing and, at times, non-linear processes. This interpretation finds support in recent One Health work on persistent contaminants such as PFAS, where health domains are increasingly understood as structurally entangled and resistant to compartmentalized intervention strategies (Souza et al., 2026). From this vantage point, environmental exposure, biological response, and societal outcome are more plausibly interpreted as interdependent layers within a single dynamically coupled system rather than as discrete stages along a linear causal pathway. Such a perspective also resonates with emerging discussions in systems-oriented scholarship, where systemic vulnerability is frequently framed not as a property of isolated components, but as an emergent feature of tightly interconnected structures capable of transmitting disturbance across domains in ways that are not always predictable or proportionate (Terenzio, 2025; Nelson & Little, 2026).

4. The Data-Driven Systems Framework (DDSF)

The Data-Driven Systems Framework (DDSF) is introduced as a heuristic scaffold designed to organize nanoplastic-related processes across five interlinked stages: exposure, detection, signal interaction, biological response, and system-level outcome. Rather than assuming a linear causal progression, the framework is structured to accommodate recursive, feedback-dependent dynamics, on the premise that system behavior more often emerges from interactions among coupled variables than from isolated inputs considered in abstraction (Jeong et al., 2024; Nelson & Little, 2026).

What distinguishes this formulation is not the enumeration of stages *per se*, but the emphasis placed on their continuous interaction and mutual modulation across scales.

Within this structure, nanoplastics may be more productively interpreted as information-relevant perturbations whose biological significance is shaped not only by material composition, but also by the conditions under which they interact with surrounding systems across molecular, cellular, and environmental interfaces. This interpretation is broadly consistent with evidence indicating that nanoplastic behavior is strongly mediated by interfacial and molecular processes, including adsorption dynamics, transport pathways, and exchange mechanisms at biological surfaces (Oliveira et al., 2022).

The framework also responds to a persistent limitation in current environmental risk assessment, where analytical detection remains insufficiently integrated with mechanistic interpretation and downstream biological consequence. Recent assessments continue to highlight this disconnect, particularly in contexts where system-level outcomes arise through distributed and temporally extended pathways rather than through singular, immediately observable effects (Masseroni et al., 2022; Jeong et al., 2024).

A systems-oriented reading of these limitations suggests that biological effects associated with nanoplastic exposure are unlikely to be reducible to particle burden alone. Instead, they may be better understood as the result of signal propagation across interdependent regulatory layers, where local perturbations are amplified

Figure 2. Conceptual Architecture of the Data-Driven Systems Framework (DDSF) for Nanoplastics Within a One Health Paradigm

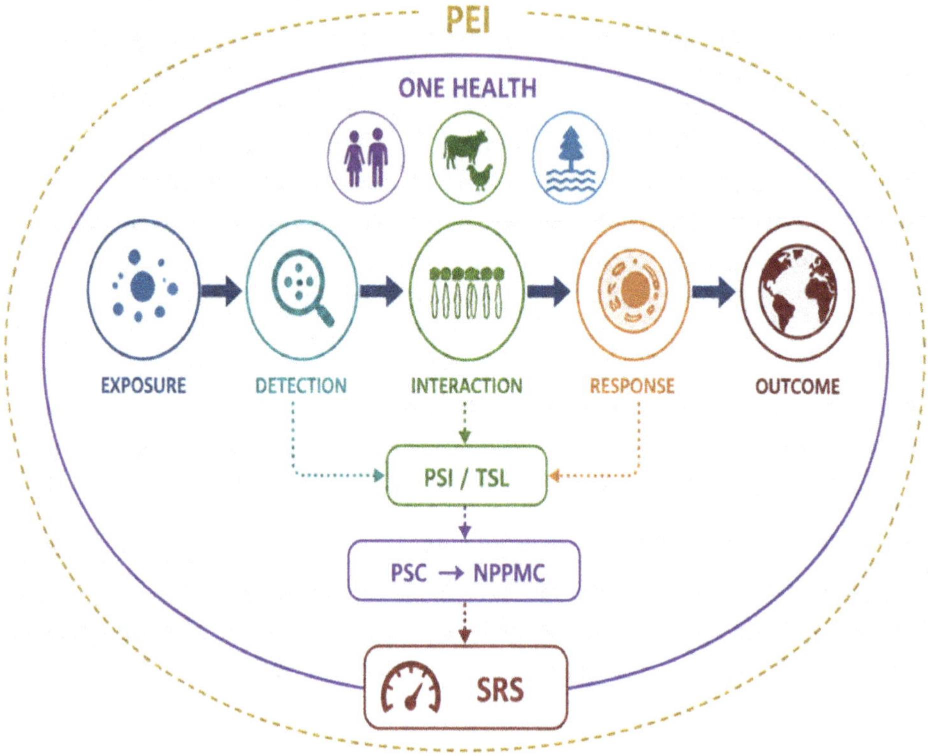

The figure condenses the proposed framework into a multiscale abstraction that repositions nanoplastics from isolated material entities to system-active perturbations operating across coupled domains. Rather than implying a strictly linear pathway, the sequence extending from exposure and detection to interaction, biological response, and system-level outcome is presented as a structurally linked continuum in which each stage conditions—and is conditioned by—the others.

Within this structure, Plasto-Signaling Interference (PSI) and Transductive Signal Load (TSL) denote an intermediate layer of signaling perturbation that mediates the translation of exposure into biological effect, thereby shifting emphasis from particle presence toward interaction dynamics. Biological response is not treated as a discrete endpoint, but is instead situated along a graded continuum spanning proteostatic stress (PSC) and extended systemic progression (NPPMC), reflecting the likelihood of non-linear escalation under sustained or recurrent perturbation.

System-level integration is represented through the Systemic Risk Score (SRS), not as a deterministic index, but as a conceptual proxy for cumulative vulnerability emerging from coupled exposure, signaling disruption, and system fragility. The embedding of this architecture within One Health and Planetary Environmental Intelligence (PEI) contexts further situates nanoplastic risk within a broader landscape of uneven detection capacity, adaptive response, and socio-environmental constraint.

In this sense, the figure functions less as a schematic summary and more as a constrained interpretative model through which heterogeneous evidence may be organized without presuming premature closure.

5. Conceptual Interaction Layer: PSI and TSL

To facilitate cross-study comparability and improve conceptual clarity, two interpretative constructs are introduced: Plasto-Signaling Interference (PSI) and Transductive Signal Load (TSL). PSI refers to the immediate disruption of biological signaling environments arising at the interface of nanoplastic interaction, whereas TSL denotes the cumulative burden of signaling perturbation across time under sustained or recurrent exposure conditions.

Such a distinction becomes analytically relevant when considering that nanoplastic-associated effects are unlikely to be reducible to particle presence alone. Experimental evidence indicates that nanoplastic exposure may induce oxidative imbalance, mitochondrial dysfunction, lysosomal destabilization, inflammatory signaling activation, and apoptosis across biological systems (Dennis & Mistri, 2026; Reyed & Prabhakar, 2025d, 2025e). These responses are more plausibly interpreted as manifestations of underlying perturbations within cellular signaling networks rather than as isolated toxicological endpoints.

Accordingly, PSI and TSL are positioned not as finalized empirical metrics, but as heuristic, hypothesis-generating descriptors. Their value lies in structuring how signaling-related observations are interpreted without prematurely collapsing them into deterministic causal claims. In this respect, the broader literature on signal transduction measurement supports the methodological plausibility of tracking pathway activity as a structured biological readout, given that signaling dynamics may be quantified and compared across experimental and clinical systems (Bouwman et al., 2025).

From this perspective, TSL may be understood as an integrative representation of accumulated signaling disturbance, capturing how perturbations propagate across time and biological context, whereas PSI remains focused on the immediate interference pattern emerging at the exposure–interface boundary. This distinction is therefore functional rather than merely definitional, reflecting different analytical scales of interpretation.

Taken together, this structure enables PSI and TSL to function as disciplined interpretative tools for linking exposure to signaling disruption and, ultimately, to downstream biological response, while preserving the distinction between direct empirical observation and higher-order systems inference (Dennis & Mistri, 2026; Nelson & Little, 2026).

6. Biological Interpretation: From PSC to NPPMC

Biological responses to nanoplastic exposure are more plausibly interpreted along a continuum than as isolated toxicological endpoints. That distinction matters because emerging evidence increasingly suggests that nanoplastic effects often unfold through graded biological disturbance, in which adaptive compensation, functional strain, and systemic destabilization may coexist rather than appear as discrete sequential states (Jeong et al., 2024; Thakur et al., 2025).

Within this logic, the Proteostatic Stress Continuum (PSC) is proposed to describe a graded response space extending from adaptive proteostatic compensation to measurable functional strain. Building on this premise, the Nanoplastic Proteostatic Panflemmosis Metabolic Continuum (NPPMC) is introduced as a higher-order interpretative construct intended to capture how sustained nanoplastic exposure may progress from localized molecular disruption toward broader systemic dysregulation.

NPPMC is not presented as a fixed pathological entity, nor does it imply a uniform biological trajectory. Rather, it offers a structured interpretative lens through which progression from proteostatic imbalance to distributed systemic consequence may be examined under sustained or repeated exposure conditions. In this sense, the model functions as a scenario-based framework for interpreting biological instability under uncertainty rather than as a deterministic disease sequence.

At the molecular level, the earliest phase of this continuum—proteostatic strain—may be understood through disruption of protein-folding dynamics and chaperone-mediated quality control. Molecular chaperones, including members of the Hsp70 and Hsp60 families, function as ATP-dependent regulators of protein folding, preserving conformational stability and limiting aggregation under stress. Available evidence suggests that nanoplastics may interfere with these processes through several non-exclusive mechanisms, particularly surface-driven protein adsorption and bio-corona formation, both of which can alter protein conformation and redirect normal folding trajectories (Dennis & Mistri, 2026; Aschner et al., 2026).

At the structural level, nanoplastic–protein interactions have been associated with alterations in secondary structure that may favor misfolding and aggregation-prone conformations. Such perturbation is mechanistically relevant because even modest shifts in folding fidelity may increase proteostatic burden without immediately producing overt pathology. This burden may extend to the endoplasmic reticulum, where the accumulation of misfolded proteins activates the unfolded protein response (UPR), a compensatory system initially directed toward restoring proteostasis but one that may, under sustained stress, shift toward apoptotic or maladaptive signaling (Aschner et al., 2026).

Within the NPPMC framework, this phase is not interpreted as a discrete transition but as a non-linear increase in biological susceptibility. Proteostatic strain does not necessarily culminate in pathology; however, it may reduce the margin of adaptive resilience and increase sensitivity to secondary perturbation.

The intermediate phase, described here as panflemmotic signaling, is intended to capture the possibility that localized molecular stress may extend into distributed inflammatory-like signaling across interconnected systems. The term is not introduced as a clinical designation, but as a conceptual descriptor for diffuse stress propagation under persistent exposure. This interpretation is consistent with pathway-oriented evidence suggesting that signaling disturbances are often propagated through distributed regulatory networks rather than confined to the site of initial insult (Bouwman et al., 2025).

The terminal phase, metabolic dysregulation, is interpreted as a downstream systems consequence of sustained energetic demand associated with proteostatic maintenance, stress adaptation, and prolonged signaling imbalance. Rather than representing a fixed endpoint, this phase reflects a shift in system-level resource allocation, where the energetic cost of maintaining compensatory function begins to constrain broader adaptive capacity.

These transitions should not be treated as inevitable. Their progression is more plausibly understood as probabilistic and context-dependent, shaped by exposure intensity, exposure duration, biological variability, and environmental context. In this respect, NPPMC is best understood not as a predictive pathology model, but as a structured biological interpretation framework for examining how nanoplastic-associated perturbations may accumulate, propagate, and reorganize system behavior across molecular and physiological scales (Métais et al., 2023; Jeong et al., 2024).

7. System-Level Integration: SRS and Fragility Mapping

Translating multi-scale interactions into system-relevant interpretation requires moving beyond isolated exposure–effect relationships toward constructs capable of accommodating coupled uncertainty, feedback, and propagation across interconnected domains. Within this context, the Systemic Risk Score (SRS) is introduced as a probabilistic construct that integrates exposure intensity, signaling disruption, and system vulnerability within a unified evaluative space. This formulation aligns with broader systemic risk literature, where risk is increasingly conceptualized as an emergent property of interdependence rather than as a simple aggregation of independent hazards (Sillmann et al., 2025; Thakur et al., 2025).

Importantly, SRS is not intended as a deterministic index. Instead, it is designed to reflect the likelihood of non-linear escalation under conditions of structural coupling, where apparently small perturbations may be redistributed, amplified, or dampened through networked biological and ecological interactions. Such an interpretation resonates with systems-oriented approaches that emphasize feedback structure and connectivity as primary determinants of systemic behavior rather than isolated stressor magnitude alone (Fernandes et al., 2026).

Complementing this construct, Fragility Mapping is proposed as a relational diagnostic layer for identifying sensitive nodes within biological and socio-ecological systems where perturbations may propagate disproportionately. In this formulation, fragility is not treated as a uniform property of the system, but as a heterogeneous distribution of vulnerability shaped by structural positioning and functional connectivity across interacting components. This perspective aligns with contemporary systemic risk frameworks that highlight uneven propagation pathways and clustered vulnerability within complex networks (Sillmann et al., 2025).

Within this architecture, Fragility Mapping and SRS should not be interpreted as independent constructs, but rather as analytically coupled layers of the same evaluative framework. Fragility Mapping informs the identification of structural sensitivity, while SRS aggregates exposure–response dynamics into a probabilistic representation of systemic escalation potential. Their relationship is therefore functional rather than merely descriptive.

Taken together, these constructs shift analytical emphasis away from static risk quantification toward dynamic system sensitivity. The central concern is no longer limited to the presence of a hazard, but extends to the structural configuration of the system through which that hazard may be transmitted, amplified, or attenuated under evolving conditions (Thakur et al., 2025; Fernandes et al., 2026).

8. AI–Biotechnology Integration

Within the DDSF architecture, artificial intelligence and biotechnology are treated less as autonomous agents and more as constrained enablers embedded within a broader systemic logic. AI primarily contributes through pattern recognition, multivariate modeling, and probabilistic inference, whereas biotechnology operates through physical intervention, remediation pathways, and targeted modulation of biological systems. Although this division appears conceptually distinct, in practice the boundaries are often blurred, particularly when dealing with heterogeneous environmental datasets and biologically noisy response systems. Empirical work in nanoplastic–biointerface research consistently highlights that both domains remain dependent on data quality, contextual calibration, and experimentally validated boundary conditions rather than algorithmic sophistication alone (Karim et al., 2024; Preda et al., 2024).

At this level, it becomes difficult to sustain strong claims of technological determinism. The integration challenge is therefore not simply one of tool availability, but of architectural coherence—how heterogeneous analytical components are organized so that local optimization does not distort system-level interpretation. This systems-oriented architectural logic is summarized conceptually in Figure 3 and further expanded into multi-layer operational interpretation in Figure 4, where AI-assisted inference, biological interaction layers, and probabilistic system integration are positioned as interconnected rather than isolated analytical domains. Evidence from environmental nanoplastic toxicology continues to show that biological responses are non-linear, context-sensitive, and often distributed across interacting subsystems, reinforcing the need for constrained interpretive structures rather than isolated analytical instruments (Trevisan et al., 2022; Chulkov et al., 2025).

9. Applied Scenario: Digital Twin Modeling within DDSF

A useful illustration of DDSF operational logic can be found in the conceptual use of digital twin modeling within a coastal urban ecosystem. In this configuration, environmental sensor streams—covering water columns, atmospheric particulates, and food-chain interfaces—are integrated with AI-assisted analytical pipelines to construct a computational representation that evolves alongside observed system dynamics. It is important to emphasize that this representation is not a perfect replica of reality; rather, it functions as a continuously updated approximation shaped by data resolution and modeling assumptions.

The transition from conceptual architecture to operational systems interpretation is illustrated schematically in Figure 4, which expands the DDSF framework beyond the core organizational structure presented in Figure 3 toward applied multi-scale modeling, digital twin integration, fragility mapping, and uncertainty-aware systems inference.

Within such a system, nanoplastic concentration gradients, population-density shifts, and resource-accessibility indicators may be jointly represented within a coupled simulation environment. Scenario perturbations—such as seasonal increases in nanoplastic load or localized waste-distribution irregularities—can then be introduced to explore potential system responses under constrained uncertainty. The purpose here is not deterministic prediction, but structured exploration of plausible trajectories, consistent with probabilistic approaches to cross-scale environmental modeling (Rabbi, 2026).

At the exposure layer, AI-based hydrodynamic and distributional models may identify accumulation patterns with varying degrees of uncertainty. At the interaction layer, secondary constructs such as PSI and TSL may be used to approximate signaling perturbations, although they remain interpretative descriptors rather than directly measurable variables. At the system level, aggregated outputs can be translated into SRS-like probabilistic structures that reflect emergent fragility rather than discrete outcomes.

These layered relationships correspond directly to the recursive exposure–interaction–response continuum illustrated in Figures 3 and 4, where mechanistic interpretation is intentionally organized across coupled biological, environmental, and systems-level domains rather than along isolated toxicological endpoints.

Empirical studies in nanoplastic toxicity increasingly support the necessity of such multi-layered modeling approaches, particularly given documented systemic effects across cardiovascular, nervous, and reproductive systems (Chulkov et al., 2025; Mishra et al., 2025). However, these simulations should not be interpreted as predictive certainties. Their role is better understood as structured decision-support environments that allow early identification of vulnerability patterns before they manifest as irreversible biological or ecological outcomes.

In this sense, digital twins within DDSF occupy an intermediate epistemic position. They are neither fully predictive nor purely descriptive. Rather, they operate as conditional inference systems under uncertainty, translating fragmented environmental signals into coherent—but inherently provisional—system-level interpretations.

10. Socioeconomic Dimension, Research Gaps, and the Structural Role of This Volume within Planetary Environmental Intelligence

Nanoplastic exposure, when examined across real-world contexts, does not distribute uniformly. Rather, it tends to follow pre-existing gradients of inequality shaped by variations in infrastructure quality, environmental governance, healthcare access, and adaptive capacity. These structural conditions collectively produce markedly different exposure and response profiles across populations. This uneven distribution complicates any attempt to interpret nanoplastic risk as a purely biophysical phenomenon, as vulnerability is increasingly co-produced through both environmental exposure and socioeconomic structure.

The broader conceptual organization of these coupled environmental and socioeconomic dynamics is reflected in Figures 3 and 4, where the DDSF architecture is positioned within the larger interpretative scope of Planetary Environmental Intelligence (PEI) and One Health systems integration.

To account for this entanglement, the concept of Planetary Environmental Intelligence (PEI) is introduced as an integrative framing mechanism. PEI is not conceived as a data-aggregation tool, but as a conceptual lens through which environmental, biological, and socioeconomic dimensions may be examined as interacting components of a coupled system. Within this framing, resilience is not defined solely by exposure thresholds or toxicological response curves, but also by the capacity of systems—across multiple scales—to detect, interpret, and respond to environmental perturbations in a timely and context-sensitive manner. In this sense, intelligence is understood as distributed across institutional, technological, and ecological layers rather than being localized within any single domain of analysis (Preda et al., 2024; Trevisan et al., 2022).

However, the transition from conceptual integration to operational validity continues to reveal persistent structural limitations. Standardized methodologies for nanoplastic detection and characterization remain inconsistent across studies, limiting comparability and reproducibility. Longitudinal datasets capable of capturing slow system evolution are still scarce, while cross-scale datasets linking molecular interactions to population-level outcomes remain fragmented. In parallel, integrative models attempting to couple biological and socioeconomic dimensions often rely on partially validated assumptions or simplified coupling structures. This limitation is particularly evident in systems-oriented environmental literature, where advances in modeling increasingly outpace the biological and empirical validation required to ground them (Chulkov et al., 2025; Mishra et al., 2025).

These challenges point to a deeper structural issue: the weak coupling between datasets operating across fundamentally different temporal, spatial, and epistemic scales. Without addressing this disconnect, even advanced AI-driven or systems-based frameworks risk producing internally consistent but externally fragile interpretations of environmental risk. This reinforces the need for uncertainty-aware modeling approaches and interdisciplinary validation structures capable of integrating heterogeneous evidence without imposing artificial uniformity (Rabbi, 2026).

Against this background, the present volume is positioned not as a conventional edited collection, but as a structured editorial response to a field that has expanded rapidly in data generation while remaining comparatively underdeveloped in integrative interpretation. Nanoplastics research has advanced significantly in detection, characterization, and mechanistic understanding; however, these advances have not been matched by an equivalent capacity to connect exposure dynamics with biological consequence, system-level vulnerability, and translational relevance within a coherent interpretative framework (Jiang et al., 2020; Wang et al., 2021; Masseroni et al., 2022).

Within this editorial structure, Figure 3 functions as the conceptual core of the DDSF architecture, whereas Figure 4 provides the expanded systems-level interpretation linking exposure dynamics, interface biology, probabilistic modeling, AI-assisted analytics, and translational system inference across multiple scales of analysis.

Importantly, the central constraint is no longer the absence of data, but the limited capacity to organize heterogeneous evidence across scales into a coherent analytical structure. This interpretative lag has become increasingly consequential as mechanistic evidence accumulates across biological systems and system-level implications become more difficult to ignore (Jeong et al., 2024; Arellano et al., 2026; Helal et al., 2023). In this context, the limiting factor is no longer observation itself, but the ability to render fragmented signals scientifically coherent across disciplines and levels of biological complexity.

The contribution of this volume is therefore primarily organizational rather than additive. It does not aim merely to expand the literature through additional case studies, nor to restate established toxicological narratives in broader terms. Instead, it seeks to reorganize dispersed empirical findings, mechanistic observations, and systems-level implications within a structured interpretative continuum linking detection, interface biology, biological response, and systemic consequence. In doing so, it offers not only thematic synthesis, but an integrative scientific language through which nanoplastics may be understood as system-active perturbation agents operating across biological, environmental, and socio-ecological scales (Dang et al., 2022; Mahmoudi et al., 2023; Terenzio, 2025).

Accordingly, this volume is best understood not as a thematic compilation, but as a progression-based editorial architecture. Its chapters are structured not as independent contributions around a shared topic, but as interconnected components within a broader interpretative sequence moving from detection and characterization to interaction, biological consequence, and systems-level inference.

The coherence of the volume lies not merely in topical proximity, but in conceptual progression across scales of analysis. Within this structure, the book provides a systems-oriented framework through which complexity can be interpreted more rigorously, disciplinary boundaries can be engaged more coherently, and emerging directions in nanoplastics research can be articulated with greater integrative clarity.

Rather than claiming closure, the volume positions itself as a structured interpretative platform intended to support ongoing inquiry, encourage methodological convergence, and enhance conceptual continuity within the evolving field of nanoplastics research under a One Health paradigm.

Within this context, the present Editorial Introduction must also be situated explicitly within the broader scope of the volume *One Health Approaches to Nanoplastics Contamination: Biotechnological and AI Solutions*. The conceptualization of the Silicon–Carbon Nexus and the Data-Driven Systems Framework (DDSF) is not intended as a standalone theoretical departure, but rather as an integrative scaffold designed to unify the diverse contributions of the book under a coherent One Health-oriented perspective on nanoplastics contamination, biotechnology, and AI-driven environmental solutions.

In this context, Figures 3 and 4 should not be interpreted merely as graphical summaries, but as integrative visual architectures intended to stabilize conceptual continuity across heterogeneous chapters, methodological scales, and interdisciplinary interpretative layers throughout the volume.

Figure 3. Data-Driven Systems Framework (DDSF): Systems-Level Conceptualization of Nanoplastics within the One Health Paradigm

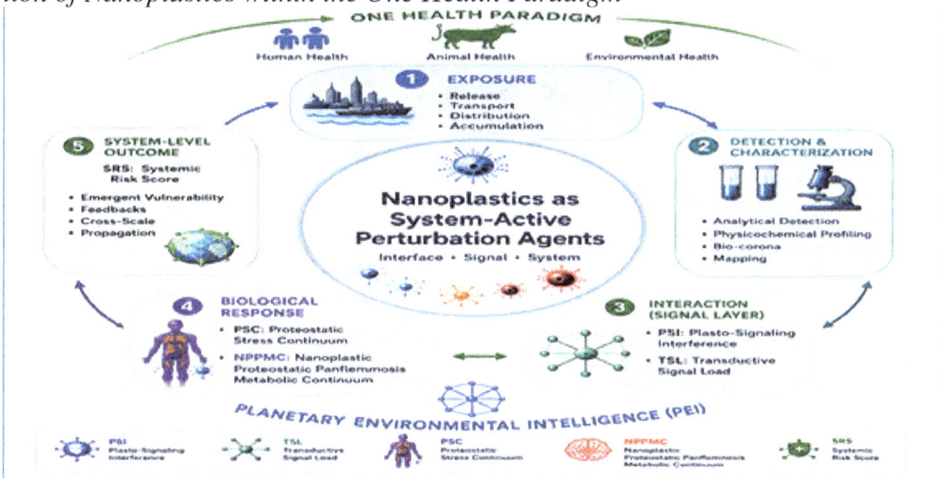

Source: Reyed, R. M. (Ed.). (2026). One Health Approaches to Nanoplastics Contamination: Biotechnological and AI Solutions. IGI Global Scientific Publishing. https://doi.org/10.4018/979-8-3373-4490-4

Figure 3 presents the central conceptual architecture of the Data-Driven Systems Framework (DDSF) developed to interpret nanoplastics as system-active perturbation agents operating across interconnected biological, environmental, and socio-ecological domains within a One Health context. The framework organizes the progression from exposure and analytical characterization to interaction-layer disturbance, biological response propagation, and emergent system-level outcomes through recursive, nonlinear, and cross-scale dynamics. Core constructs including PSI (Plasto-Signaling Interference), TSL (Transductive Signal Load), PSC (Proteostatic Stress Continuum), NPPMC (Nanoplastic Proteostatic Panflemnosis Metabolic Continuum), and SRS (Systemic Risk Score) are integrated as mechanistic interpretive layers intended to preserve systems continuity across heterogeneous exposure conditions and multi-domain response architectures. The figure was structurally compressed and editorially harmonized during final production staging to maintain conceptual density, graphical balance, and publication-grade readability without disrupting mechanistic coherence or framework continuity.

Figure 4. Expanded Systems Interpretation of the Data-Driven Systems Framework (DDSF): Multi-Scale Integration of Nanoplastic Dynamics under One Health and Planetary Environmental Intelligence (PEI).

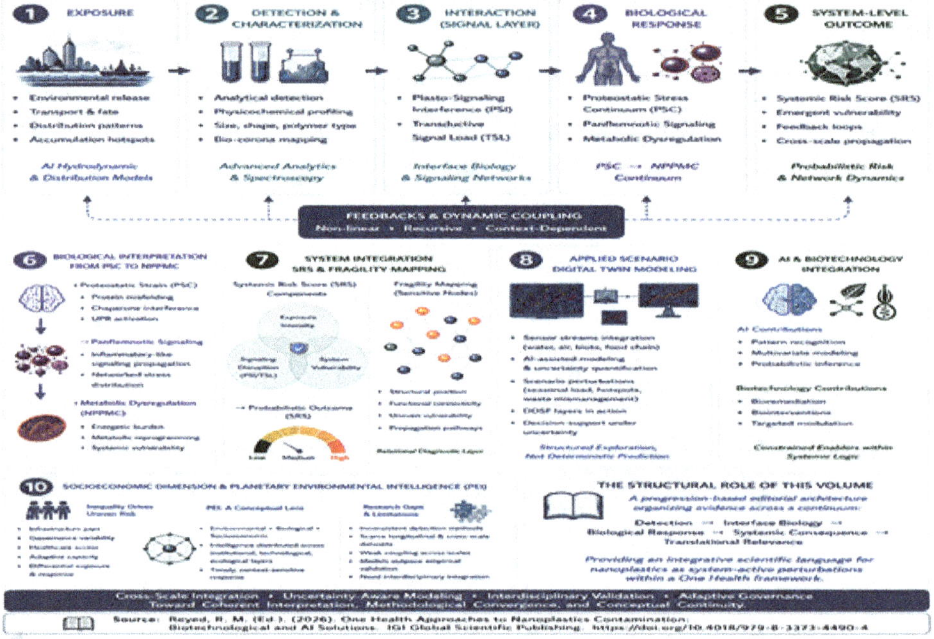

Figure 4 presents the expanded systems interpretation of the Data-Driven Systems Framework (DDSF), illustrating the progression of nanoplastics from environmental exposure and analytical characterization toward interaction-layer perturbation, biological stress propagation, systemic vulnerability, and emergent risk formation within interconnected One Health domains. The framework integrates multiscale exposure dynamics, physicochemical detection, signal-layer interference, proteostatic disruption, probabilistic risk mapping, AI-assisted modeling, biotechnology-enabled intervention concepts, and socioeconomic contextualization under the governing perspective of Planetary Environmental Intelligence (PEI). Recursive feedback coupling, nonlinear propagation behavior, and uncertainty-aware system integration are incorporated to reflect the adaptive and cross-scale nature of nanoplastic-driven perturbations. Conceptual modules including PSI, TSL, PSC, NPPMC, and SRS are organized as interconnected interpretive layers intended to maintain mechanistic continuity across heterogeneous environmental and biological conditions while supporting translational systems reasoning. The figure was refined through late-stage editorial harmonization and structural compression to achieve publication-grade readability, conceptual coherence, and graphical balance suitable for high-density academic production workflows.

INTEGRATED SYSTEMS-SCIENCE EDITORIAL FRAMEWORK

This volume has been intentionally designed around a central editorial premise: nanoplastics contamination can no longer be examined adequately through isolated disciplinary perspectives. The scale, persistence, environmental mobility, and biological complexity of nanoplastics increasingly demand analytical frameworks capable of integrating environmental science, toxicology, biotechnology, computational analytics, and One Health intelligence within a unified interpretative structure.

Accordingly, environmental contamination is approached throughout this volume not as a singular exposure event, but as a dynamic systems process characterized by interconnected environmental, biological, technological, and societal dimensions. This perspective recognizes that exposure generation, biological response, ecological consequence, and policy adaptation are part of the same analytical continuum rather than independent domains of investigation.

To support this systems-oriented perspective, the editorial architecture of the book links environmental detection, physicochemical transformation, biological interaction, microbiome perturbation, computational assessment, artificial intelligence–assisted analytics, biotechnology-enabled mitigation, translational implementation, and adaptive governance within a coherent conceptual framework.

Rather than presenting chapters as isolated contributions, the volume has been structured to encourage intellectual continuity across disciplinary boundaries. Each chapter addresses a distinct dimension of the nanoplastics challenge while simultaneously contributing to a broader systems-level understanding of environmental exposure, biological vulnerability, and societal response.

The conceptual foundation supporting this architecture is further developed through the Data-Driven Systems Framework (DDSF) and a series of integrative analytical constructs introduced throughout the volume. These constructs are not presented as rigid models or definitive explanatory systems. Instead, they serve as interpretative tools intended to facilitate conceptual integration across heterogeneous datasets, complex biological systems, environmental uncertainty, and emerging computational approaches.

The organizational logic of the volume follows a progressive systems pathway extending from environmental generation and particulate mobility to exposure propagation, biological interaction, physiological disruption, predictive intelligence, mitigation strategies, and governance adaptation. This progression reflects the conviction that meaningful advances in nanoplastics research require both scientific specialization and systems-level integration.

For this reason, the present volume should be viewed not simply as a collection of individual chapters, but as an integrated editorial platform designed to promote dialogue across toxicology, environmental systems biology, biotechnology, artificial intelligence, One Health science, and environmental governance. Its objective is not only to expand the scientific literature on nanoplastics, but also to strengthen the conceptual and analytical foundations through which complex environmental challenges may be interpreted, evaluated, and addressed within an increasingly interconnected world.

EDITORIAL INTEGRATION PERSPECTIVES

Chapter 1: One Health Cyber-Biotechnology for Nano-Plastic Intelligence Systems and Logic Integration: From Exposure to System-Level Interpretation

Within the editorial architecture of this volume, Chapter 1 occupies a foundational strategic position by establishing the conceptual and systems-oriented framework through which the subsequent analytical domains are organized and interpreted. Rather than functioning solely as an introductory chapter, it provides the intellectual infrastructure required to situate environmental, biological, computational, and governance-oriented analyses within a coherent systems-science perspective. As the opening analytical layer of the volume, the chapter defines the conceptual vocabulary, integrative principles, and interpretative assumptions that underpin the broader One Health orientation of the book.

From the perspective of the Data-Driven Systems Framework (DDSF), this chapter establishes the initial conditions for the intellectual progression that unfolds throughout the volume. It introduces the systems-level logic necessary for understanding nanoplastics not as isolated environmental entities, but as dynamic components operating across interconnected ecological, biological, and technological networks. In doing so, it creates the conceptual bridge through which subsequent chapters can advance from environmental occurrence and exposure characterization toward mechanistic toxicology, microbiome perturbation, computational intelligence, translational intervention, and adaptive governance. The chapter therefore functions as a critical transition point between foundational systems thinking and the increasingly specialized analytical layers that follow.

Equally important, the frameworks introduced herein serve as enduring reference structures that support interpretative continuity across the volume. They provide the conceptual architecture through which individual chapters may be understood not as independent investigations, but as analytically interconnected components of a unified systems-science pathway. By sustaining coherence across disciplinary boundaries and maintaining alignment with the DDSF and One Health paradigms, this chapter contributes directly to the intellectual integration that characterizes the volume as a whole.

Accordingly, Chapter 1 should be regarded not merely as a standalone contribution, but as a critical component of the integrated systems-science architecture through which the broader objectives of this volume are progressively realized.

Chapter 2: Forensic Tracing Systems for Nanoplastics: AI and Biotech One Health Source Attribution

Within the editorial architecture of this volume, Chapter 2 occupies the operational forensic layer of the systems-science interpretative framework through which nanoplastic phenomena are translated into structured, traceable, and analytically coherent evidence. Positioned immediately after the foundational conceptual orientation of the preceding chapter, it marks a decisive transition from systems definition toward systems interrogation, where environmental complexity is approached through structured attribution, forensic reconstruction, and cross-matrix interpretability. In this capacity, the chapter establishes a critical functional layer required for operationalizing One Health perspectives within a data-integrated analytical continuum.

From the editorial systems-science perspective of the volume, this chapter advances the progression of the intellectual architecture by transforming conceptual constructs into empirically grounded forensic intelligence pathways. It establishes the methodological conditions under which heterogeneous datasets—spanning molecular, spectroscopic, imaging, and computational domains—can be integrated into unified attribution structures capable of addressing source ambiguity and transformation dynamics across complex environmental matrices. This transition enables the volume to progress from abstract system representation toward inference-driven interpretation of exposure propagation, thereby forming a necessary bridge between foundational systems thinking and subsequent analytical domains concerned with toxicological mechanisms, biological response systems, and computational risk modeling.

In terms of cross-chapter continuity, the chapter introduces traceability and data fusion as stabilizing principles that sustain coherence across the entire volume. These principles enable the integration of diverse physicochemical and biological signatures into a consistent interpretative structure, ensuring that downstream analyses remain anchored in reproducible, source-resolved evidence frameworks. Within the broader One Health orientation of the volume, this continuity is essential for preserving the conceptual linkage between environmental persistence, biological uptake, and systemic impact as interconnected expressions of a unified analytical system. Consequently, the chapter functions as a stabilizing node within the editorial architecture, ensuring continuity between environmental detection processes and higher-order interpretative frameworks that govern systems-level reasoning.

Accordingly, Chapter 2 should be regarded not merely as a standalone contribution, but as a critical component of the integrated systems-science architecture through which the broader objectives of this volume are progressively realized.

Chapter 3: Physicochemical Determinants of Nanoplastic Fate and Advanced Removal Dynamics: AI-Enabled Approaches within One Health Frameworks

Within the editorial architecture of this volume, Chapter 3 occupies the environmental–processual analytical layer concerned with the physicochemical regulation of nanoplastic behavior across environmental and engineered systems. Following the forensic attribution focus of the preceding chapter, this contribution extends the volume's systems-science trajectory toward mechanistic environmental chemistry, where particle-level properties are interpreted as primary determinants governing transport, transformation, and removal dynamics. In this capacity, the chapter establishes a critical analytical bridge between source-resolved traceability frameworks and downstream exposure, remediation, and risk-oriented interpretations within a One Health continuum.

From the editorial systems-science perspective of the volume, this chapter advances the intellectual architecture by shifting the analytical emphasis from traceability and attribution toward deterministic physicochemical control mechanisms that govern nanoplastic fate. It situates nanoplastics within a colloidal systems paradigm in which structural, surface, and thermodynamic properties regulate aggregation behavior, environmental partitioning, and interfacial reactivity across multiphase ecological compartments. This transition enables the volume to progress from source identification logic toward fate prediction and intervention-oriented reasoning, thereby supporting subsequent chapters focused on toxicological mechanisms, biological system interactions, and computational risk modeling. The integration of mechanistic environmental chemistry with computational optimization frameworks further strengthens the DDSF-aligned progression toward data-informed systems interpretation.

In terms of cross-chapter continuity, the chapter introduces physicochemical determinism and remediation performance as central organizing principles that connect upstream forensic traceability with downstream exposure mitigation and system-level intervention strategies. By linking colloidal behavior, environmental partitioning, and engineered removal processes within a unified analytical perspective, it ensures that system-level interpretations remain grounded in mechanistic understanding of material behavior under variable environmental conditions. This continuity is essential for maintaining coherence across the One Health systems-science trajectory, where environmental persistence, engineered intervention, and biological exposure are treated as interconnected expressions of a single analytical continuum. Consequently, the chapter functions as a mechanistic stabilization layer within the editorial architecture, enabling consistent transition from environmental processes to biological and computational domains.

Accordingly, Chapter 3 should be regarded not merely as a standalone contribution, but as a critical component of the integrated systems-science architecture through which the broader objectives of this volume are progressively realized.

Chapter 4: Human Exposure to Nanoplastics and Its Toxicological and Environmental Implications: A One Health Perspective

Within the editorial architecture of this volume, Chapter 4 occupies the toxicological analytical layer through which environmental exposure is translated into mechanistic health interpretation. Building upon the preceding discussions of nanoplastic fate, physicochemical persistence, and traceability, this chapter shifts the volume decisively toward the biological consequences of environmental contact. In this position, it provides the interpretative framework necessary for understanding how environmental persistence and cross-matrix transport become physiologically meaningful at the level of human exposure and systemic response.

From the editorial systems-science perspective of the volume, the chapter advances the intellectual progression by linking exposure pathways with toxicodynamic consequences across multiple biological interfaces. It establishes the conceptual transition from environmental distribution to organism-level interpretation, where ingestion, inhalation, and dermal uptake are examined as routes through which nanoplastics may translocate and interact with cellular and tissue systems. By situating mechanistic injury processes within a One Health context, the chapter prepares the volume for a more integrated understanding of how environmental contaminants move from external matrices into internal biological domains, thereby creating a necessary bridge toward later chapters concerned with biological disruption, biomonitoring, and computational risk evaluation.

In terms of cross-chapter continuity, the chapter reinforces the coherence of the volume by aligning exposure science with toxicological interpretation and by maintaining continuity between environmental detection, organismal uptake, and systemic impact. Its treatment of biomonitoring constraints and exposure quantification limitations is particularly important for preserving interpretative rigor across the volume, since it underscores the methodological boundaries within which human health assessments must be conducted. By connecting cross-interface exposure dynamics to mechanistic toxicity and analytical standardization challenges, the chapter functions as a critical linkage point between environmental processes and downstream health-oriented analysis within the broader One Health narrative of the book.

Accordingly, Chapter 4 should be regarded not merely as a standalone contribution, but as a critical component of the integrated systems-science architecture through which the broader objectives of this volume are progressively realized.

Chapter 5: Gut Microbiome Perturbation and Nutritional Outcomes under Nanoplastic Exposure: A One Health Perspective

Within the editorial architecture of this volume, Chapter 5 occupies the biological–microbiome interaction layer, extending the analysis beyond organism-level exposure effects toward the internal ecological and metabolic systems that mediate host response under chronic nanoplastic stress. Building upon the preceding toxicological discussion, this chapter shifts the volume into a more integrative domain in which environmental contaminants are understood not only as external stressors but also as agents capable of altering host-associated microbial ecosystems and nutritional regulation. In this position, the chapter is structurally necessary for linking exposure science with the internal biological mediators through which One Health effects are expressed.

From the editorial systems-science perspective of the volume, the chapter advances the intellectual progression by connecting toxicological exposure pathways to microbiome-mediated physiological regulation. It introduces a layered interpretative domain in which microbial structure, epithelial integrity, immune signaling, and metabolic homeostasis are examined as interdependent components of a host–environment system. This transition is especially important for the broader architecture of the book, as it moves the reader from direct toxicological injury toward systems-level perturbation of microbial ecology and nutritional balance, thereby preparing the conceptual space for later computational, translational, and intervention-oriented chapters.

In terms of cross-chapter continuity, the chapter reinforces the coherence of the volume by positioning microbiome disruption as a mechanistic bridge between environmental exposure and downstream health outcomes. Its emphasis on multistressor dynamics, bidirectional host–microbiota crosstalk, and therapeutic restoration strategies ensures that the volume does not treat biological response as a linear endpoint, but rather as an adaptive and modifiable system within the One Health continuum. This continuity is essential for maintaining interpretative alignment across chapters, since it connects the physical entry of nanoplastics into biological systems with broader nutritional, immunological, and metabolic consequences that require integrated analysis across disciplinary boundaries.

Accordingly, Chapter 5 should be regarded not merely as a standalone contribution, but as a critical component of the integrated systems-science architecture through which the broader objectives of this volume are progressively realized.

Chapter 6: An Environmental Nutrition Framework for Micro- and Nanoplastics Exposure Assessment: AI-Assisted Food Safety Detection Systems

Within the editorial architecture of this volume, Chapter 6 occupies the food-systems and environmental nutrition layer, extending the analysis of nanoplastic exposure from host-associated biological perturbation to the domain of dietary intake, trophic transfer, and food-chain contamination. Positioned after the chapter on gut microbiome disruption, this contribution broadens the volume's analytical scope by situating nanoplastic exposure within the pathways through which environmental contaminants enter human systems via food. In this capacity, the chapter is structurally necessary for linking biological response to the upstream food-system interfaces that shape exposure burden, nutritional resilience, and public health relevance.

From the editorial systems-science perspective of the volume, the chapter advances the intellectual progression by translating particle exposure into a food-centered interpretative framework in which environmental contamination, dietary intake, and metabolic impact are treated as interconnected components of a single continuum. It establishes the conceptual transition from internal host response to the external systems that govern oral exposure, thereby reinforcing the volume's movement from toxicological and microbiome-oriented analysis toward food safety surveillance, exposure forecasting, and nutritional mitigation. This transition is especially important for the broader architecture of the book, as it embeds dietary exposure within One Health reasoning and prepares the ground for later chapters concerned with computational assessment, intervention design, and systems-level governance.

In terms of cross-chapter continuity, the chapter reinforces coherence across the volume by connecting trophic transfer dynamics with systemic oral exposure and nutritional defense strategies. Its emphasis on food-system resilience, contaminant detection, and predictive analytics ensures that the volume does not treat dietary exposure as an isolated endpoint, but rather as a modifiable interface between environmental contamination and human health outcomes. This continuity is essential for maintaining interpretative alignment across the book, since it links ecological contamination, food-chain movement, and physiological risk within a shared analytical framework that supports integrated One Health interpretation.

Accordingly, Chapter 6 should be regarded not merely as a standalone contribution, but as a critical component of the integrated systems-science architecture through which the broader objectives of this volume are progressively realized.

Chapter 7: Translational One Health Strategies for Nanoplastic Exposure Mitigation

Within the editorial architecture of this volume, Chapter 7 occupies the translational and implementation layer through which exposure science is converted into operational mitigation logic across environmental, technological, and public health domains. Following the chapters that establish exposure pathways, toxicological consequences, microbiome perturbation, and dietary contamination, this contribution advances the volume into an action-oriented domain in which the analytical emphasis shifts from interpretation to intervention. In this position, the chapter is structurally necessary for linking systems diagnosis to applied mitigation within a One Health framework.

From the editorial systems-science perspective of the volume, the chapter advances the intellectual progression by reframing nanoplastic contamination as a problem of implementable control, rather than solely of exposure characterization. It establishes the conceptual transition from understanding how nanoplastics move through environmental and biological systems to determining how those pathways can be interrupted, reduced, or managed through translational strategies. This shift is critical to the broader architecture of the book, as it connects surveillance, biotechnology-assisted reduction, computational forecasting, and environmental management into a coherent pathway of systems-informed intervention. In doing so, the chapter prepares the volume for later discussions of governance, decision support, and long-term sustainability-oriented response.

In terms of cross-chapter continuity, the chapter reinforces coherence across the volume by introducing mitigation as a logical extension of the exposure and risk narratives developed in preceding chapters. Its focus on implementation feasibility, resource allocation, and resilience-oriented governance ensures that the volume does not end at problem identification, but instead moves toward practical, scalable pathways for contaminant reduction. This continuity is essential for maintaining interpretative alignment across the book, because it links environmental persistence, biological burden, and public health impact with operational strategies for system-level response. The chapter thereby functions as a translational bridge between analytical assessment and coordinated action within the One Health trajectory of the volume.

Accordingly, Chapter 7 should be regarded not merely as a standalone contribution, but as a critical component of the integrated systems-science architecture through which the broader objectives of this volume are progressively realized.

Chapter 8: Economic and Life Cycle Assessment of Nanoplastics Remediation within a One Health Framework: Biotechnological and AI-Driven Solutions

Within the editorial architecture of this volume, Chapter 8 occupies the techno-economic and systems optimization layer, extending the mitigation-oriented discussions of the preceding chapter into a structured assessment of feasibility, scalability, and sustainability of nanoplastic remediation strategies. Positioned after the translational implementation framework, this contribution shifts the analytical focus from intervention design to the economic, environmental, and life-cycle constraints that govern real-world deployment. In this capacity, the chapter is structurally essential for embedding remediation strategies within resource-aware, system-optimized decision frameworks consistent with One Health principles.

From the editorial systems-science perspective of the volume, the chapter advances the intellectual progression by integrating environmental burden analysis with economic evaluation and computational optimization of remediation pathways. It situates nanoplastic management within a life-cycle assessment paradigm in which generation, transport, exposure, and mitigation are evaluated as interconnected phases influencing overall system efficiency and sustainability. This transition enables the volume to move beyond intervention feasibility toward quantitative assessment of trade-offs, performance constraints, and optimization opportunities. By combining biotechnological remediation systems with AI-driven modeling and cost-efficiency analytics, the chapter strengthens the volume's trajectory toward data-informed, systems-level decision support.

In terms of cross-chapter continuity, the chapter reinforces coherence by linking remediation effectiveness to environmental impact reduction and socioeconomic sustainability metrics. Its emphasis on life-cycle hotspots, remediation bottlenecks, and integrated ecological–health indicators ensures that mitigation strategies are not treated in isolation but are evaluated within the broader constraints of environmental systems, public health outcomes, and resource allocation dynamics. This continuity is essential for maintaining the integrity of the One Health framework across the volume, as it connects upstream contamination processes with downstream intervention feasibility and long-term sustainability considerations. The chapter therefore functions as a critical evaluative layer that consolidates scientific, environmental, and economic dimensions into a unified systems-science perspective.

Accordingly, Chapter 8 should be regarded not merely as a standalone contribution, but as a critical component of the integrated systems-science architecture through which the broader objectives of this volume are progressively realized.

Reyed M. Reyed

Genetic Engineering and Biotechnology Research Institute (GEBRI), City of Scientific Research and Technological Applications (SRTA-City), Alexandria, Egypt

REFERENCES

Arellano, F. M., Arcadio, C. G. L. A., Chen, Y.-T., Hsieh, S., Bacosa, H. P., & Hsieh, S.-L. (2026). Nanoplastics in biological systems: What laboratory mechanisms reveal about real-world toxicity. *Journal of Hazardous Materials*, *505*, 141464. DOI: 10.1016/j.jhazmat.2026.141464 PMID: 41734556

Aschner, M., Skalny, A. V., Lu, R., Santamaria, A., Voskresenskaya, O. N., Virgolini, M. B., Barbosa, F.Jr, & Tinkov, A. A. (2026, March 31). (2026). The effects of nanoplastics on toxicity of metals and metal nanoparticles: Molecular mechanisms and health effects. *Archives of Toxicology*. Advance online publication. DOI: 10.1007/s00204-026-04353-8 PMID: 41915191

Bhattacharyya, S., Greer, M. L., & Salehi, M. (2025). Impact of micro- and nanoplastics exposure on human health: Focus on neurological effects from ingestion. *Frontiers in Public Health*, *13*, 1681776. DOI: 10.3389/fpubh.2025.1681776 PMID: 41246084

Bouwman, W., Raymakers, R., van der Poll, T., & van de Stolpe, A. (2025). Comparison Between Signal Transduction Pathway Activity in Blood Cells of Sepsis Patients and Laboratory Models. *Cells*, *14*(4), 311. DOI: 10.3390/cells14040311 PMID: 39996782

Chen, Z., Yin, X., Geng, Y. Q., Gao, R., Zhang, Y., Ma, Y., Mu, X., Chen, X., Li, F., & He, J. (2025). Subchronic Exposure to Polystyrene Nanoplastics Disrupts Placental Development and Calcium Homeostasis: Insights from *In Vivo* and *In Vitro* Models. *ACS Nano*, *19*(14), 13825–13841. DOI: 10.1021/acsnano.4c16786 PMID: 40171975

Chulkov, V., Gasanov, M., Isakov, V., Denisenko, A., Nwosu, C., & Rodkin, S. (2025). Molecular and Cellular Effects of Microplastics and Nanoplastics in the Pathogenesis of Cardiovascular, Nervous, Urinary, Digestive, and Reproductive System Diseases: A Global Systematic Review. *International Journal of Molecular Sciences*, *26*(22), 11194. DOI: 10.3390/ijms262211194 PMID: 41303677

Danasekaran, R. (2024). One Health: A Holistic Approach to Tackling Global Health Issues. *Indian Journal of Community Medicine : Official Publication of Indian Association of Preventive & Social Medicine*, *49*(2), 260–263. DOI: 10.4103/ijcm.ijcm_521_23 PMID: 38665439

Dang, F., Wang, Q., Huang, Y., Wang, Y., & Xing, B. (2022). Key knowledge gaps for One Health approach to mitigate nanoplastic risks. *Eco-Environment & Health*, *1*(1), 11–22. DOI: 10.1016/j.eehl.2022.02.001 PMID: 38078201

Dennis, J., & Kumar Mistri, T. (2026). Nanoplastics in the Biological System: A Microscopic Menace with an Impact. In Environmental Sciences. IntechOpen. https://doi.org/DOI: 10.5772/intechopen.1013429

Fernandes, S., Kuschel, E., Obriejetan, M., Stangl, R., Hübl, J., Ionescu, F. D., Vliet, A. B., Matos, J., & Strauss, A. (2026). Resilience-Oriented Extension of the RAMSSHEEP Framework to Address Natural Hazards Through Nature-Based Solutions: Insights from an Alpine Infrastructure Study. *Infrastructures*, *11*(1), 35. DOI: 10.3390/infrastructures11010035

Geyer, R., Jambeck, J. R., & Law, K. L. (2017). Production, use, and fate of all plastics ever made. *Science Advances*, *3*(7), e1700782. DOI: 10.1126/sciadv.1700782 PMID: 28776036

Grodzicki, W., Dziendzikowska, K., Gromadzka-Ostrowska, J., & Kruszewski, M. (2021). Nanoplastic Impact on the Gut-Brain Axis: Current Knowledge and Future Directions. *International Journal of Molecular Sciences*, *22*(23), 12795. DOI: 10.3390/ijms222312795 PMID: 34884598

Helal, M., Hartmann, N. B., Khan, F. R., & Xu, E. G. (2023). Time to integrate "One Health Approach" into nanoplastic research. *Eco-Environment & Health*, *2*(1), 18–20. DOI: 10.1016/j.eehl.2023.01.001 PMID: 38074454

Jeong, J., Im, J., & Choi, J. (2024). Integrating aggregate exposure pathway and adverse outcome pathway for micro/nanoplastics: A review on exposure, toxicokinetics, and toxicity studies. *Ecotoxicology and Environmental Safety*, *272*, 116022. DOI: 10.1016/j.ecoenv.2024.116022 PMID: 38309230

Jiang, B., Kauffman, A. E., Li, L., McFee, W., Cai, B., Weinstein, J., Lead, J. R., Chatterjee, S., Scott, G. I., & Xiao, S. (2020). Health impacts of environmental contamination of micro- and nanoplastics: A review. *Environmental Health and Preventive Medicine*, *25*(1), 29. DOI: 10.1186/s12199-020-00870-9 PMID: 32664857

Karim, A., Yadav, A., Sweety, U. H., Kumar, J., Delgado, S. A., Hernandez, J. A., White, J. C., Vukovic, L., & Narayan, M. (2024). Interfacial Interactions between Nanoplastics and Biological Systems: Toward an Atomic and Molecular Understanding of Plastics-Driven Biological Dyshomeostasis. *ACS Applied Materials & Interfaces*, *16*(20), 25740–25756. DOI: 10.1021/acsami.4c03008 PMID: 38722759

Levin, S. (1998). Ecosystems and the biosphere as complex adaptive systems. *Ecosystems (New York, N.Y.)*, *1*(5), 431–436. DOI: 10.1007/s100219900037

Lin, R. H., Chen, H. T., Lee, I. T., Vo, T. T., & Wang, Y. L. (2025). Microplastics and Nanoplastics in Health Concerning Cellular Toxicity Mechanisms, Exposure Pathways, and Global Mitigation Strategies. *Life (Basel, Switzerland)*, *15*(9), 1449. DOI: 10.3390/life15091449 PMID: 41010391

Mackenzie, J. S., & Jeggo, M. (2019). The One Health Approach-Why Is It So Important? *Tropical Medicine and Infectious Disease*, *4*(2), 88. DOI: 10.3390/tropicalmed4020088 PMID: 31159338

Mahmoudi, M., Landry, M. P., Moore, A., & Coreas, R. (2023). The protein corona from nanomedicine to environmental science. *Nature Reviews. Materials*, *8*(7), 422–438. DOI: 10.1038/s41578-023-00552-2 PMID: 37361608

Masseroni, A., Rizzi, C., Urani, C., & Villa, S. (2022). Nanoplastics: Status and Knowledge Gaps in the Finalization of Environmental Risk Assessments. *Toxics*, *10*(5), 270. DOI: 10.3390/toxics10050270 PMID: 35622683

Meadows, D. H. (2008). *Thinking in systems: A primer* (Wright, D., Ed.). Chelsea Green Publishing.

Métais, I., Latchere, O., Roman, C., Perrein-Ettajani, H., Mouloud, M., Georges, D., Audroin, T., Catrouillet, C., Gigault, J., Agnès-Feurtet-Mazel, , Baudrimont, M., & Châtel, A. (2023). Continuum from microplastics to nanoplastics: Effects of size and source on the estuarine bivalve Scrobicularia plana. *Environmental Science and Pollution Research International*, *30*(16), 45725–45739. DOI: 10.1007/s11356-023-25588-3 PMID: 36708472

Mishra, S. K., Sanyal, T., Kundu, P., Kumar, R., Ghosh, D., Chakrabarti, G., Sikdar, N., Bhattacharya, S., Paul, S., & Das, A. (2025). Microplastics as emerging carcinogens: From environmental pollutants to oncogenic drivers. *Molecular Cancer*, *24*(1), 248. DOI: 10.1186/s12943-025-02409-4 PMID: 41063167

Moulton, C., Baroni, A., & Tasciotti, E. (2026). Micro- and Nanoplastics Exposure Across the Lifespan: One Health Implications for Aging and Longevity. *Journal of Xenobiotics*, *16*(2), 52. DOI: 10.3390/jox16020052 PMID: 41874123

Mousazadehgavan, M. (2026). Microplastics in aquatic systems: Dual roles as pollutant carriers and emerging functional materials for water treatment. *Water, Air, and Soil Pollution*, *237*(11), 449. DOI: 10.1007/s11270-026-09138-4

Nelson, P., & Little, R. (2026). The Body Underground: A Biological Framework for Infrastructure Health, Regulation and Resilience. *Urban Science (Basel, Switzerland)*, *10*(4), 201. DOI: 10.3390/urbansci10040201

Njema, George G., Agoro, Mojeed A., Kibet, Joshua K., A Critical Evaluation of Nano- and Microplastics: Sources, Health Risks, and Mitigation Strategies, Advances in Materials Science and Engineering, 2026, 1357400, 38 pages, 2026. https://doi.org/DOI: 10.1155/amse/1357400

Oliveira, Y. M., Vernin, N. S., Maia Bila, D., Marques, M., & Tavares, F. W. (2022). Pollution caused by nanoplastics: Adverse effects and mechanisms of interaction via molecular simulation. *PeerJ*, *10*, e13618. DOI: 10.7717/peerj.13618 PMID: 35910776

Preda, O. T., Vlasceanu, A. M., Andreescu, C. V., Tsatsakis, A., Mezhuev, Y., Negrei, C., & Baconi, D. L. (2024). Health Implications of Widespread Micro- and Nano-plastic Exposure: Environmental Prevalence, Mechanisms, and Biological Impact on Humans. *Toxics*, *12*(10), 730. DOI: 10.3390/toxics12100730 PMID: 39453150

Rabbi, M. F. (2026). Integrated mathematical modeling of cross-scale dynamics and equity constraints in water–energy–food systems. *Clean Technologies and Environmental Policy*, *28*(4), 92. Advance online publication. DOI: 10.1007/s10098-026-03456-1

Reichstein, M., Camps-Valls, G., Stevens, B., Jung, M., Denzler, J., Carvalhais, N., & Prabhat, . (2019). Deep learning and process understanding for data-driven Earth system science. *Nature*, *566*(7743), 195–204. DOI: 10.1038/s41586-019-0912-1 PMID: 30760912

Reyed, R. M., & Prabhakar, P. K. (2025a). Impact of Dietary Antioxidants on Redox Homeostasis and Its Effects on Health and Disease. In Prabhakar, P. (Ed.), *The Role of Reactive Oxygen Species in Human Health and Disease* (pp. 81–136). IGI Global Scientific Publishing., DOI: 10.4018/979-8-3693-7919-6.ch003

Reyed, R. M., & Prabhakar, P. K. (2025b). Mechanisms, Impact, and Therapeutic Strategies for Oxidative Stress in Metabolic Diseases: Understanding and Mitigating Oxidative Stress. In Prabhakar, P. (Ed.), *The Role of Reactive Oxygen Species in Human Health and Disease* (pp. 203–256). IGI Global Scientific Publishing., DOI: 10.4018/979-8-3693-7919-6.ch006

Sanoja-López, K. A., Suarez, C., Navia-Pesantes, O., Quiroz-Suárez, K. A., & Luque, R. (2026). Advanced strategies for the detection, removal, and degradation of microplastics: Bridging science and sustainable technologies. *Results in Engineering*, *29*, 109743. DOI: 10.1016/j.rineng.2026.109743

Sillmann, J., Ciullo, A., Franzke, C. L. E., & Scheffran, J. (2025). Systemic risk: Opportunities for inter- and transdisciplinary science to improve societal resilience. *iScience*, *29*(1), 114369. DOI: 10.1016/j.isci.2025.114369 PMID: 41541672

Souza, M. C. O., Barbosa, F.Jr, & Domingo, J. L. (2026). PFAS and One Health: Integrating human, animal, and environmental perspectives. *Archives of Toxicology*, *100*(5), 1769–1787. DOI: 10.1007/s00204-026-04306-1 PMID: 41699312

Terenzio, F. (2025). Systemic Vulnerability: From AI Systems to Environmental Systems. *Topoi*. Advance online publication. DOI: 10.1007/s11245-025-10285-2

Thakur, R., Joshi, V., Sahoo, G. C., Jindal, N., Tiwari, R. R., & Rana, S. (2025). Review of mechanisms and impacts of nanoplastic toxicity in aquatic organisms and potential impacts on human health. *Toxicology Reports*, *14*, 102013. DOI: 10.1016/j.toxrep.2025.102013 PMID: 40230517

Trevisan, R., Ranasinghe, P., Jayasundara, N., & Di Giulio, R. T. (2022). Nanoplastics in Aquatic Environments: Impacts on Aquatic Species and Interactions with Environmental Factors and Pollutants. *Toxics*, *10*(6), 326. DOI: 10.3390/toxics10060326 PMID: 35736934

Wang, L., Wu, W.-M., Bolan, N. S., Tsang, D. C. W., Li, Y., Qin, M., & Hou, D. (2021). Environmental fate, toxicity and risk management strategies of nanoplastics in the environment: Current status and future perspectives. *Journal of Hazardous Materials*, *401*, 123415. DOI: 10.1016/j.jhazmat.2020.123415 PMID: 32763705

Whitmee, S., Haines, A., Beyrer, C., Boltz, F., Capon, A. G., de Souza Dias, B. F., Ezeh, A., Frumkin, H., Gong, P., Head, P., Horton, R., Mace, G. M., Marten, R., Myers, S. S., Nishtar, S., Osofsky, S. A., Pattanayak, S. K., Pongsiri, M. J., Romanelli, C., & Yach, D. (2015). Safeguarding human health in the Anthropocene epoch: Report of The Rockefeller Foundation-Lancet Commission on planetary health. *Lancet*, *386*(10007), 1973–2028. DOI: 10.1016/S0140-6736(15)60901-1 PMID: 26188744

WHO. (2022). *One Health Joint Plan of Action (2022–2026): Working together for the health of humans, animals, plants and the environment*. World Health Organization. https://www.who.int/publications/i/item/9789240059139

Ziani, K., Ioni ă-Mîndrican, C. B., Mititelu, M., Neac u, S. M., Negrei, C., Moro an, E., Drăgănescu, D., & Preda, O. T. (2023). Microplastics: A Real Global Threat for Environment and Food Safety: A State of the Art Review. *Nutrients*, *15*(3), 617. DOI: 10.3390/nu15030617 PMID: 36771324

Zolotova, N., Kirillova, M., Dzhalilova, D., Tsvetkov, I., Fokichev, N., & Makarova, O. (2025). Microplastics' Impact on the Development of AOM/DSS-Induced Colitis-Associated Colorectal Cancer in Mice. *International Journal of Molecular Sciences*, *26*(23), 11511. DOI: 10.3390/ijms262311511 PMID: 41373666

Chapter 1
One Health Cyber-Biotechnology for Nano-Plastic Intelligence Systems and Logic Integration:
From Exposure to System-Level Interpretation

Reyed M. Reyed
http://orcid.org/0000-0001-5027-3478

Genetic Engineering and Biotechnology Research Institute (GEBRI), City of Scientific Research and Technological Applications (SRTA-City), Alexandria, Egypt

ABSTRACT

This chapter per the authors develops a One Health cyber-biotechnology framework for interpreting nanoplastic exposure as a multiscale environmental signaling system rather than an isolated contaminant phenomenon. It integrates digital nanoplastic science (DNPS), plasto-signaling interference (PSI), transductive signal load (TSL), Polybiome Interfacial Signaling Cascade (PISC), and systemic risk scoring (SRS) to structure exposure propagation, interface-level perturbation, and biological signaling disruption within a unified analytical architecture. Environmental matrices including water, soil, and air are conceptualized as dynamic signal-generating interfaces transmitting disturbances across interconnected ecological and physiological networks under uncertainty regimes. The chapter critiques conventional

DOI: 10.4018/979-8-3373-4490-4.ch001

compartment-based risk assessment and advances multiomics integration, digital twin modeling, and cyber-biotechnological analytics for early-warning systems, exposure intelligence, and One Health–driven environmental decision support.

INTRODUCTION

This chapter develops a One Health cyber-biotechnology framework for interpreting nanoplastic exposure as a multiscale environmental and biological phenomenon rather than a single-compartment contaminant problem. It combines digital nanoplastic science (DNPS), plasto-signaling interference (PSI), the transductive signal load (TSL), and the systemic risk score (SRS) to organize exposure data, biological signaling, and system vulnerability within a unified analytical structure. The chapter further outlines how environmental inputs originating from water, soil, and air may be translated into interface-level perturbations, propagated through biological signaling networks, and ultimately represented as risk outputs under conditions of explicit uncertainty. It also examines the limitations of conventional environmental risk assessment, alongside the emerging roles of multiomics integration, digital twin modeling, and adaptive governance frameworks for early warning, environmental interpretation, and systems-level intervention.

Conceptual Framework Figure of the Chapter

Figure 1. Conceptual Framework of the One Health Cyber-Biotechnology System

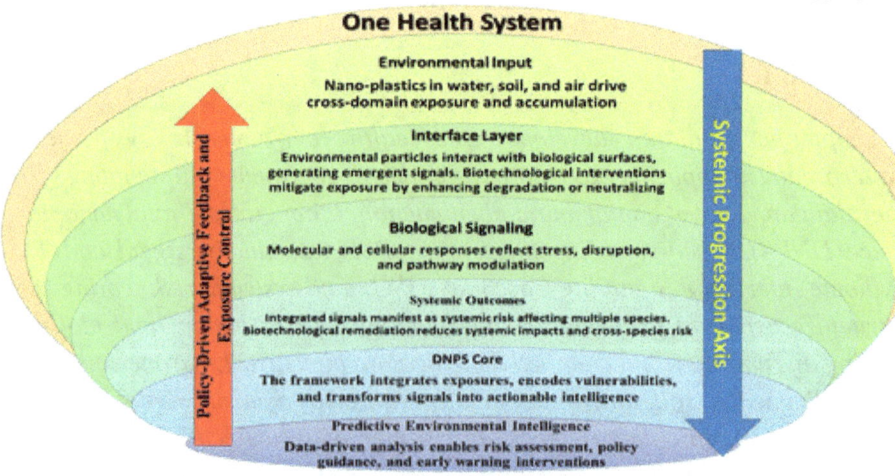

Figure 1 outlines a multiscale One Health cyber-biotechnology framework designed to conceptualize nanoplastic exposure across environmental and biological domains as a continuous interacting system rather than a sequence of isolated exposure events. The architecture is organized into five interconnected layers that collectively describe the progression from environmental presence to biological perturbation and adaptive governance-level feedback.

At the foundational level, environmental compartments—including water, soil, and air—act as major reservoirs for nanoplastic particles. These particles acquire biological relevance through interface-mediated interaction zones, where physicochemical properties such as size, surface composition, adsorption behavior, and environmental transformation state may influence the nature and intensity of interactions with biological surfaces. Within this framework, the interface is not interpreted as a passive boundary, but rather as a dynamic transitional environment where environmental particulate matter begins to participate in biologically relevant processes.

Downstream, these interactions are translated into intracellular and intercellular signaling processes associated with stress adaptation, homeostatic regulation, inflammatory modulation, and broader adaptive biological responses. Rather than functioning as isolated molecular events, these pathways are interpreted as components of a continuous signaling continuum shaped by persistent low-intensity exposure dynamics and multiscale environmental interaction.

At the systems level, localized molecular and cellular responses become integrated into broader physiological patterns representing cumulative exposure effects distributed across multiple biological compartments. Importantly, this layer emphasizes that nanoplastic-associated effects are unlikely to operate through simple linear accumulation models alone, but may instead emerge through context-dependent interactions involving biological state, coexposure conditions, adaptive capacity, and signaling-system resilience.

The DNPS core functions as an integrative analytical layer designed to consolidate heterogeneous environmental and biological datasets into structured exposure representations. Its role is not purely computational. Rather, it provides an interpretative systems framework capable of organizing environmental inputs, biological signaling dynamics, exposure heterogeneity, and uncertainty-associated vulnerability patterns into a more coherent analytical structure.

Finally, the framework incorporates a feedback-oriented adaptive governance axis linking environmental monitoring, biological observations, and regulatory decision systems. This feedback structure emphasizes that exposure conditions are not environmentally static, but may be continuously reshaped through iterative intervention, surveillance adaptation, and evolving policy responses. In this way, the framework closes the loop between environmental processes, biological interpretation, and governance-level action within a continuously interacting One Health system

1. THE COGNITIVE SHIFT: REFRAMING THE PLASTOSPHERE—A UNIFIED CYBER-BIOTECHNOLOGICAL PERSPECTIVE

Scientific discourse on plastic pollution has long been organized around what is most visible: discarded plastics, ghost nets, and contaminated shorelines (Walker-Franklin et al., 2026). These manifestations remain environmentally consequential, but they are not analytically sufficient. Nanoscale plastics (NPs) pose a different problem. Their size-dependent behavior complicates the linear exposure–response assumption that still shapes many classical environmental risk assessments in the Anthropocene (Koelmans et al., 2019; Yong et al., 2020). More importantly, a conceptual focus on visible debris can obscure a more difficult issue: biological systems do not merely receive environmental perturbations; rather, they transform, filter, and register them across scales. That distinction is not semantic; it determines what counts as evidence, what counts as response, and what counts as risk.

Within emerging biodigital frameworks, including the biointelligent paradigm advanced by Esawy and Reyed (2026), such perturbations are increasingly treated as contributors to systemic coherence or discoherence. The implications are subtle but significant. Biological responses are no longer read as simple reactions to external stressors; they are interpreted as multiscale informational registrations shaped by context, pathway coupling, and temporal persistence. DNPS adopts that orientation but with a methodological restraint that is easy to lose in more speculative formulations. Here, "digital" is not used literally. It refers to the integration of physicochemical and biological interactions into an analytical architecture capable of linking exposure, transformation, and response (Pegoraro et al., 2026).

This phenomenon occurs because NPs may acquire environmentally derived physicochemical signatures through interactions with biomolecules, microbial assemblages, and biosensing interfaces. The particles themselves, however, are not assumed to possess intrinsic informational agency. That boundary is important. It prevents the framework from drifting into metaphorical overreach and keeps it aligned with environmental systems science, where signals are measured, modeled, and interpreted rather than ontologically redefined (Boakye et al., 2025; Snehamayee et al., 2026).

At the mechanistic level, the high surface-area-to-volume ratio of NPs promotes the formation of a dynamic protein–organic corona that can alter particle–cell interactions, bioavailability, and transport behavior (Walsh et al., 2025). This corona is often described as an adsorption layer, but that description may be too passive. In practice, the nano–bio interface can behave as a modifiable biochemical boundary, with downstream consequences for oxidative and inflammatory signaling. The analytical tension here is real: what appears to be a surface phenomenon may in fact structure deeper biological response patterns. Emerging evidence supports that possibility. Low-level NP exposure has been associated with sublethal perturbations, including microbiome restructuring, oxidative stress, and inflammatory signaling (Amaral-Zettler et al., 2020; Y. Jin et al., 2019; Thakur et al., 2025; Mahalingaiah et al., 2025).

Within DNPS, these effects are treated not as endpoint claims but as candidate indicators for structured environmental interpretation. Systemic risk can therefore be expressed as a composite function:

$$SRS = W \underset{\frown}{\underline{E}} + W \d{S} + W V$$

where denotes the exposure intensity, denotes signal-based biological perturbation, and **V** captures system vulnerability. The point of the formulation is not prediction in a strict sense. It is a comparison under uncertainty. The score organizes heterogeneous evidence into a transparent structure and makes the uncertainty itself visible rather than hidden behind a single endpoint label (Bhattacharyya et al., 2025). The signaling component ( ) is informed by computational biotechnology approaches that integrate biological data streams into predictive architectures, which is consistent with the broader conceptual direction advanced in Polybiome Systems Medicine (Reyed, 2026c).

From a governance perspective, the value of DNPS lies in anticipatory risk assessment. It adds a systems-oriented layer that may help prioritize interventions where consequences accumulate before they become obvious. This is where the framework becomes most useful and where it must remain most cautious. Synthetic biology-enabled degradation and microbiome-mediated modulation, for example, may be explored computationally within digital twin environments before operational translation is considered (Amer et al., 2026). This is especially relevant in One Health settings, where environmental exposure, polybiome ecology, and host physiology are better understood as an interdependent continuum than as isolated compartments (Sundas et al., 2024; Reyed et al., 2026).

Taken together, DNPS refers to the plastosphere not as a set of disconnected contamination events but as a coupled cyberbiotechnological field in which material particles, biological responses, and governance systems interact across scales (Leso et al., 2025; Ding et al., 2019). Its contribution lies less in predictive closure than in interpretive discipline: it offers a language for comparison, boundary setting, and hypothesis generation that is rigorous enough to be useful but not so confident that it outruns the evidence. That balance is difficult, and it is probably the point.

2. DEFINING THE CRISIS: STRUCTURAL LIMITS OF STATIC RISK PARADIGMS

The urgency of nanoplastic (NP) contamination extends beyond the accumulation of material burden. The more difficult issue is structural: existing monitoring and risk assessment frameworks often struggle to translate measurable signals into biologically meaningful inference. Reported freshwater N:P loads of approximately 3–5% (Koelmans et al., 2019) should therefore be interpreted not only as evidence of an environmental burden but also as a sign of a widening gap between detection and interpretation. This "invisibility gap" appears to arise less from the absence of technologies than from their limited ability to resolve NP behavior in biologically active environments, where particles may function as perturbative agents embedded in signaling contexts rather than as isolated contaminants.

Current environmental risk assessment (ERA) frameworks remain largely anchored in quasistatic environmental assumptions. This is the central limitation. Even modest physicochemical variation can have disproportionate effects: a pH shift of approximately 0.2 units may alter NP bioavailability by more than 50% in some cases (Liu et al., 2024). The implication is not simply that NPs are sensitive to context. Fixed-threshold models may be too coarse to represent exposure dynamics that are nonlinear, conditional, and biologically mediated. A static framework can record the presence of a contaminant; it may not adequately capture what that contaminant is doing once it enters a living system.

This is where the conceptual pressure on ERA becomes visible. The framework remains valuable, but its interpretive range narrows as the exposure environment becomes more variable and the biological system becomes more responsive. In that setting, environmental variability should not be treated as noise around a stable baseline. It becomes part of the system's architecture, and ignoring it risks underestimating the actual biological relevance of measured exposure.

3. METHODOLOGICAL CONSTRAINTS IN CURRENT RISK FRAMEWORKS

3.1 Nonlinearity and Bioscale Dependence

Linear dose–response assumptions are increasingly difficult to sustain when nanoplastics (NPs) are examined within biologically active systems. Experimental and modeling work suggests that NPs may exhibit nonlinear aggregation, transport, and uptake behaviors in response to relatively small environmental perturbations. These effects do not necessarily stop at the particle level; they may extend into higher-order biological systems, including gut–microbiome interactions and epigenetic regulation pathways (Vethaak & Legler, 2021; Bhattacharyya et al., 2025; Chulkov et al., 2025).

A closer reading of this literature suggests that the microbiome may be better understood as a mediator of environmental signal transformation than as a passive target of toxicity. This view is consistent with broader systems medicine perspectives, although its generalizability remains uncertain. Host conditions vary, environmental backgrounds differ, and the same exposure can produce distinct response trajectories across settings (Iqbal et al., 2026). The limitation is not only empirical; it is conceptual because the biointerface itself may be where much of the relevant dynamics are being organized.

3.2 Fragmented Data Landscapes and Integration Challenges

A persistent limitation in NP research lies in the fragmentation of data across environmental monitoring, spectroscopic characterization, and multiomics biological domains. These streams are often rich on their own, but they remain structurally disconnected, which constrains the development of coherent system-level representations of risk (Seggio et al., 2024; Wang et al., 2025).

Computational biotechnology has long emphasized the need for integrative frameworks that can link heterogeneous biological and environmental data streams (Reyed, 2019). From that perspective, DNPS-type architectures may be read as an extension of this trajectory, providing a way to synthesize multidomain information into structured analytical representations, including digital twin-based modeling environments (Amer et al., 2026). Even so, the reliability of such integrations depends heavily on data quality, interoperability, and calibration robustness. When any one of those weakens, the apparent coherence of the system can outpace the strength of the evidence supporting it.

3.3 Latent and subphenotypic system perturbations

A further challenge is the delayed and often nonspecific manifestation of NP-induced biological effects. Chronic, low-dose exposure may contribute to oxidative stress, inflammatory signaling, and metabolic dysregulation long before overt clinical outcomes become visible (Kochanek et al., 2025).

These subphenotypic alterations point to the need for earlier and more sensitive indicators of system stress. Work in redox biology and systemic disease modeling suggests that transitional biological states may exist in which homeostatic function is still maintained while measurable deviations are already beginning to accumulate (Reyed & Prabhakar, 2025). The problem is that such intermediate states are difficult to stabilize into robust predictive thresholds. Their biological relevance is plausible and even compelling, but methodologically, they remain unresolved and require further empirical substantiation (Skaba et al. 2025).

3.4 Systemic risk representation and interpretability limits

Within emerging systems-oriented frameworks, systemic risk is increasingly treated as an emergent construct arising from the interaction between environmental exposure, biological response, and system-level vulnerability. A simplified representation may be expressed as:

$$SRS = W \underline{E} + W \d{S} + W V$$

where denotes exposure characteristics, represents biological or signaling responses, and **V** captures variability or susceptibility-related factors. The primary value of this formulation lies in its ability to organize heterogeneous evidence into a comparable analytical form, not in any claim to deterministic prediction.

In this framing, the biological response component (**S**) may be partially informed by computational biotechnology approaches that integrate multisource biological datasets into unified analytical representations, which is consistent with recent developments in systems-oriented environmental science (Leso et al., 2025; Leso et al., 2023; Reyed, 2019). Even so, the interpretive validity of such integrations remains dependent on empirical grounding, methodological transparency, and explicit uncertainty quantification, especially when the same model is applied across heterogeneous environmental contexts. The structure may be useful; it is not self-validating.

3.5 Governance implications and emerging system perspectives

From a governance standpoint, recent developments indicate a gradual shift from descriptive monitoring toward more anticipatory forms of risk assessment. This does not displace existing regulatory structures. Rather, it extends them by introducing a systems-oriented interpretive layer that may help prioritize action under uncertainty and complexity (Korenhof et al., 2025).

Within this space, biotechnological approaches—including microbiome modulation strategies and synthetic biology-based degradation pathways—may be explored computationally within digital twin environments before any real-world implementation is considered. These environments provide a structured setting for scenario testing and system-behavior exploration under controlled assumptions. This direction appears especially relevant within one health-oriented perspective, where environmental, biological, and host systems are increasingly treated as interconnected components of a single analytical continuum (Segovia & Garcia-Alfaro, 2022).

Building on that foundation, formal representations of system-level risk have become increasingly important for structuring interpretations across domains. For conceptual clarity and formalization, the systemic risk score (SRS) is introduced, as illustrated in Box 1. In this formulation, system-level risk is defined as an emergent property arising from the interaction of environmental exposures, biological signaling processes, and systemic vulnerability. The representation is intended to capture nonlinear, cumulative, and cross-domain perturbations that remain insufficiently resolved within conventional Environmental Risk Assessment (ERA) frameworks.

Overall, the limitations across monitoring, modeling, and biological interpretation suggest that nanoplastic risk cannot be adequately characterized through static or purely material-centric paradigms. Instead, risk instead emerges from coupled interactions across environmental and biological systems, where uncertainty, variability, and nonlinearity are intrinsic properties of the system rather than methodological noise. More integrated and adaptive interpretive frameworks may therefore provide a stronger basis for advancing nanoplastic science toward system-level understanding while remaining grounded in empirically testable and methodologically constrained evidence.

Box 1. Systemic Risk Score (SRS): Definition and Conceptual Axes

Formal Definition:
The SRS quantifies system-level risk as an emergent property of interacting environmental exposures, biological signaling, and systemic vulnerability, capturing nonlinear, cumulative, and cross-domain perturbations unresolved by conventional ERA.
Three-Pillar Structure:

1- Exposure Axis ( ): Intensity, persistence, spatial/temporal distribution, transformation dynamics, bioavailability.

2- Signal Axis ( ): Molecular/cellular responses, signaling disruption, microbiome alterations, biochemical modulation.

3- Vulnerability Axis (V): Physiological resilience, metabolic sensitivity, preexisting stress conditions.

Integrative Function:
The SRS emerges from interactions among axes, adaptively weighted (w_1, w_2, w_3) to reflect context. Outputs enable latent risk identification, tipping points, and systemic disruption mapping.
Equation: **SRS =** $_{W1} + _{W2} + _{W3}V$
where denotes exposure characteristics, represents biological or signaling responses, and **V** Where:

- = Exposure intensity
- = Signal disruption (biointeraction and functional perturbation)
- V = System vulnerability

The weights (w_1, w_2, w_3) are contextually adaptable and may be optimized via expert judgment or data-driven calibration.
Rather than eliminating uncertainty, the SRS incorporates it as an inherent feature of complex environmental systems, enabling comparative and inferential assessment across diverse contexts.

3.6 The Epistemological Gap: A Critical Deconstruction of Traditional Frameworks

The widespread detection and persistence of nano- and microplastics across environmental compartments have begun to expose structural limits in conventional environmental risk assessment (ERA). These limits are not merely technical. They are embedded in the underlying assumption—most notably, the reliance on linear exposure–response relationships and the expectation that removal efficiency can be treated as a proxy for risk resolution (Iqbal et al., 2026; Yang et al., 2026b). The evidence from wastewater treatment systems illustrates this problem with some clarity: a measurable fraction of nanosized plastic particles persists in treated effluents and re-enters aquatic systems, with potential extension into human exposure pathways (Leslie et al., 2022). The implication is not simply incomplete removal. This suggests that material elimination, even when partially successful, does not necessarily correspond to functional risk mitigation.

Interpretation becomes more complicated when detection itself is examined. While multiple studies report consistent identification of nanoplastics across environmental and biological matrices, others point to methodological fragilities—contamination risks, spectroscopic limitations, and particle misclassification (Koelmans et al., 2019). Disagreement is not trivial. This indicates that uncertainty operates at more than one level, not only in terms of exposure quantification but also in terms of the reliability of the detection process itself. Under these conditions, uncertainty cannot be treated as a residual error term. It becomes part of the epistemic structure of the problem and must be incorporated explicitly into any modeling effort.

Advances in AI-driven analytical methods have improved the detection and classification of micro- and nanoplastics, particularly through high-dimensional imaging and spectral analysis (Kumar et al., 2026). However, this progress introduces a different tension. Increased detection capacity generates larger, more fragmented datasets, which do not automatically translate into interpretive clarity. The gap shifts rather than disappears. What becomes necessary, then, is not further detection alone but a means of linking detection to biological relevance. Biologically informed sensing systems—ranging from molecular biomarkers to microbiome-derived signals—offer one possible route. They provide an interpretive layer through which physicochemical measurements may be coupled with functional biological responses, a direction increasingly emphasized in computational biotechnology (Leso et al., 2025; Reyed, 2019).

From a governance perspective, international frameworks such as those advanced by the OECD acknowledge the persistence and potential risks associated with micro- and nanoplastics, but their emphasis remains largely on source reduction and lifecycle management (Jani & Wu, 2025). OECD, 2025). These strategies are necessary, although not sufficient. They do not fully address the need for system-level, predictive or quasipredictive modeling, leaving a structural gap between scientific capability and regulatory implementation.

Within this gap, the systemic risk score (SRS) becomes analytically relevant. Rather than framing risk as a fixed outcome, the SRS treats it as an emergent property arising from interactions between environmental exposure, biological signaling, and system-level vulnerability. It is not a predictive device in the strictest sense. Its role is integrative: to accommodate heterogeneous inputs—physicochemical data, biological responses, and contextual modifiers—within a structured representation that allows comparison without suppressing uncertainty. This positioning aligns with the One Health perspective by linking environmental contamination to biological and systemic processes while retaining a hypothesis-generating orientation appropriate for current evidentiary constraints (Khan & Ali, 2023).

The methodological flexibility of the SRS framework lies in its capacity to evolve alongside advances in measurement, modeling, and biotechnology (Leso et al., 2023). The incorporation of biologically derived signals—such as redox imbalance, inflammatory markers, and microbiome shifts—allows a gradual shift from static exposure assessment to response-informed modeling. Mechanistic insights from redox biology and systems disease research support this transition, although they do not yet resolve it (Reyed & Prabhakar, 2025). Moreover, emerging strategies—microbiome modulation and engineered degradation pathways—can be conceptually embedded within predictive architectures as potential intervention layers. Their practical implementation, however, remains contingent on empirical validation.

In this context, the DNPS–SRS configuration does not replace established frameworks such as ERAs, adverse outcome pathways (AOPs), or exposome-based models. It sits alongside them, attempting to provide a connective structure capable of handling nonlinearity, cross-domain interaction, and uncertainty without prematurely collapsing them into simplified endpoints (Walsh et al., 2025; Elawady et al., 2026). Whether that integration can be stabilized remains an open question.

The limitations are not peripheral. Empirical calibration, standardized measurement protocols, and longitudinal validation are still needed, and they are not easily achieved across heterogeneous systems. These constraints should not be treated as obstacles external to the framework; they define its developmental trajectory. Progress will likely depend on the convergence of higher-resolution environmental monitoring, multiomics integration, and more robust computational infrastructures, allowing gradual refinement of model assumptions and parameters in line with emerging polybiome and systems medicine paradigms (Reyed, 2026a).

As illustrated in Figure 2, the coupling of signaling dynamics with material life cycles offers a conceptual bridge between environmental processes and biological outcomes. It allows cross-domain perturbations—bioaccumulation, nanoplastic propagation—to be mapped into structured analytical flows linking molecular and cellular responses to higher-order systemic effects (Koelmans et al., 2019; Wang et al., 2025). Constructs such as Type V diabetes, while still exploratory, provide a way to think through how chronic, low-intensity perturbations might propagate through metabolic and regulatory networks without immediate clinical visibility.

What emerges from this line of reasoning is not simply the introduction of new analytical tools but also a gradual reworking of how the problem itself is organized. Fragmented, largely descriptive approaches begin—unevenly, and with some resistance—to give way to more integrative and anticipatory forms of environmental intelligence. The boundaries that once separated detection, interpretation, and intervention become less stable. They start to intersect, at times awkwardly, within a cyberbiotechnological framing that remains provisional, still negotiating both its scope and its limits.

Figure 2. Integrated Signaling–Material Lifecycle Framework with SRS.

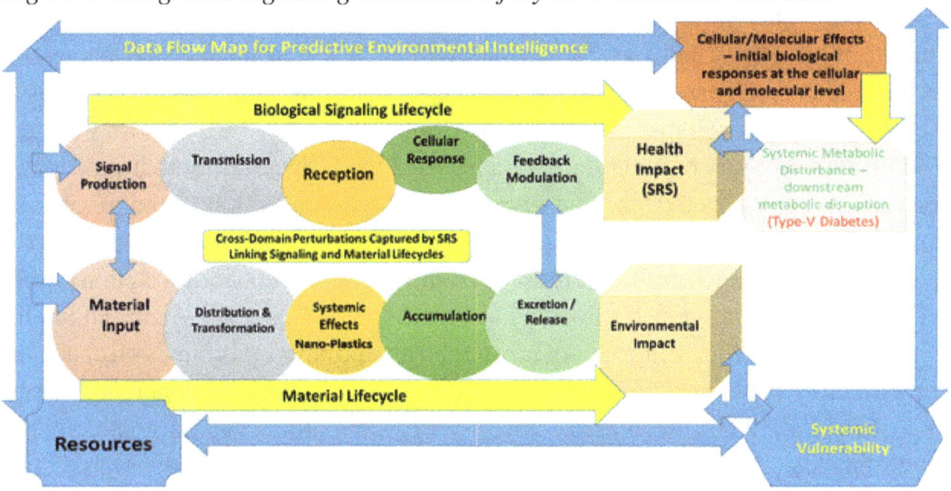

Figure 2 illustrates the parallel dynamics of biological signaling and material (nanoplastic) life cycles. Cross-domain perturbations, including bioaccumulation and environmental exposure, are conceptually encoded in the systemic risk score (SRS), providing a unified metric for comparative ecosystem vulnerability. The diagram functions as a data flow map for predictive environmental intelligence, linking cellular/molecular effects to systemic metabolic disturbances (e.g., type V diabetes), and demonstrates how microlevel interactions propagate to macrolevel outcomes across human, animal, and ecological domains.

4 DNPS VISION: SYSTEMIC INTEGRATION UMBRELLA

The digital nanoplastic science (DNPS) framework is positioned here as an integrative umbrella rather than a closed methodology. Its intent is not to fuse disciplines rhetorically but to force a more operational codependence between environmental science, computational biology, and artificial intelligence, where none of these domains can fully stabilize the others on their own (Luo et al., 2024). In this reading, nanoplastics resist simple classification. They remain material entities with measurable physicochemical properties, yet their behavior increasingly appears entangled with biologically mediated interaction networks in which environmental signals are not merely transmitted but also reconfigured through contact with living systems (Worms et al., 2025).

This duality is often stated but rarely resolved in practice. It becomes more uncomfortable when one moves into measurement. High-resolution detection strategies—spectroscopy, particle tracking, and multiomics environmental matrices—each capture only partial slices of the system, and each does so under its own constraints of sensitivity and specificity (Santos et al., 2025). Raman spectroscopy, for example, can reach particle sizes approaching ~50 nm, yet this resolution comes at a cost: throughput collapses, and system-wide observability becomes fragmented. The result is not just missing data but also structured uncertainty that propagates forward into any downstream inference. In that sense, detection cannot remain an endpoint; it begins to behave like an input layer that is already interpretive.

The DNPS vision therefore shifts away from a pollutant-centric description toward a more explicit computational reading of environmental systems. Artificial intelligence and machine learning are not positioned as auxiliary tools but as mechanisms for extracting latent structures from high-dimensional, partially coherent datasets (Khanam et al., 2025; Leso et al., 2025). Early simulations suggest that coupling environmental, microbial, and physiological data streams may reduce predictive uncertainty, although this reduction is neither stable nor generalizable across contexts. It depends heavily on the calibration conditions and data architecture. The system, in effect, behaves less like a unified model and more like a set of coupled inference surfaces that intermittently align.

This is where interpretation becomes less comfortable. The environment and host are treated as interacting networks rather than separable domains, and AI functions as an interface that translates heterogeneous inputs into structured representations of perturbations rather than simply classifying outcomes (Wiesinger et al., 2021). The epistemic shift is subtle but important: what is being modeled is not only exposure but also the transformation of exposure into a biological signal under variable constraints.

Operationally, DNPS is implemented through interlinked modules—synthetic perturbome modeling, plastosignaling interference analysis, data-driven systems analytics, and dynamic ecointegrators (Leso et al., 2025). These are not independent components in a strict architectural sense. They behave more like partially overlapping layers, where physicochemical interactions, biological responses, and computational inference continuously feed into one another, sometimes reinforcing and sometimes destabilizing interpretability. Within this structure, biologically derived signals—microbiome dynamics, metabolic flux shifts, and protein corona evolution—are not treated as passive outputs. They enter the system as active variables, which complicates the usual separation between observations and the model.

Figure 3 shows this logic at the nano–bio interface. At the molecular scale, nanoplastics undergo rapid transformations through corona formation, surface energy redistribution, and interactions with surrounding biological media. This interface is not static; it behaves more like a transduction layer where physicochemical properties are continuously translated into biological signal structures. Small environmental shifts can have disproportionate effects. A pH variation of ~0.2 units, for example, may be sufficient to alter corona composition and downstream bioavailability (Walsh et al., 2025). The figure, therefore, should not be read as a schematic illustration alone. It functions as a working abstraction intended to support iterative refinement of hypotheses within DNPS-aligned analytical workflows (Nene et al., 2025; Leso et al., 2023).

As the foundational chapter of this volume, DNPS sets a methodological orientation that connects conceptual reframing with operational modeling pathways across environmental monitoring and biological interpretation. Its structure aligns loosely with One Health logic, although here, the continuity between ecological, microbial, and human systems is treated as analytical rather than merely conceptual. Some emerging remediation approaches—such as functionalized material networks capable of simultaneous detection and removal—illustrate how material science and biotechnology may converge within closed-loop monitoring systems (Lee & Kim, 2026). The reported efficiencies are encouraging, but their stability outside of controlled conditions remains uncertain, and scaling introduces nontrivial ecological trade-offs that are still poorly resolved.

Beyond remediation, the framework extends into domains of systemic biochemical interference. Perturbations in nitrogen metabolism within bioelectrochemical systems, alongside organ-level responses in renal and cardiovascular contexts, suggest that nanoplastic exposure cannot be confined to a single physiological scale (Suʿ et al., 2026; Lee et al., 2025). Neurodevelopmental observations under chronic low-dose exposure add another layer of complexity, particularly where subtle signaling modulation may occur during sensitive developmental windows (Ragusa & Fanos, 2025). These findings are not yet cohesive and perhaps should not be expected to be at this stage; instead, they indicate where longitudinal structure is still missing.

In parallel, alignment with sustainability agendas, including SDG 3 and SDG 12, extends the relevance of DNPS beyond modeling into policy-facing discourse. However, this translation is not straightforward. The framework does not resolve the tension between scientific representation and governance applicability; it exposes it. Within this volume, DNPS functions as a structuring backbone that moves from conceptual reframing toward computational implementation and, cautiously, toward interpretive governance interfaces. In this configuration, nanoplastics are no longer treated as isolated contaminants. They become boundary objects through which environmental complexity, biological response, and data-driven inference begin to coevolve (Reyed, 2026b).

Figure 3. Molecular scale–digital nanoplastic interface initiation.

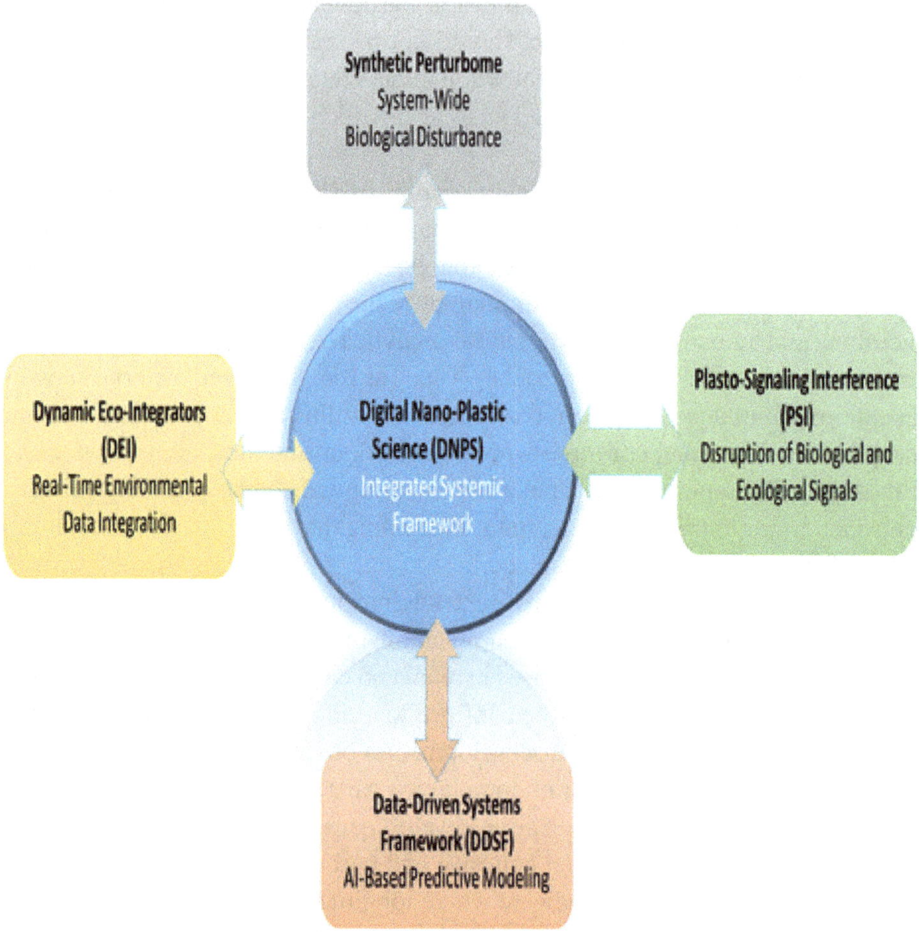

Figure 3 shows how nanoplastics can be transformed at the molecular level by converting nonliving physical debris into information-coded nanoentities.

This highlights the role of surface energy and the swift formation of the protein corona as the major carrier of encoding information. The interface between the polymer core and the biological milieu around it is described as an active gateway, with physicochemical properties (such as surface charge and adsorption kinetics) being transduced into nascent signal inputs.

5. THE EPISTEMOLOGICAL GAP: A CRITICAL DECONSTRUCTION OF TRADITIONAL FRAMEWORKS

The shift from treating nanoplastics (NPs) as inert particulate waste toward recognizing them as dynamically interacting and, in a limited sense, information-bearing entities introduces a methodological strain that existing analytical traditions struggle to accommodate. Classical toxicology, as currently formulated (MacLeod et al., 2021), has never been designed to resolve system-level interaction patterns of this kind. It operates comfortably at the level of dose, response, and endpoint stability, whereas the Plastosphere increasingly appears less like a bounded environmental compartment and more like a coupled biodigital continuum. Reported accumulation rates of approximately 3–5% annually (Kochanek et al., 2025) are therefore not the central issue in themselves; the deeper concern lies in how such numbers obscure nonlinear amplification effects that are not captured by conventional exposure accounting, largely because current models lack any cyberbiotechnology resolution layer capable of tracking real-time kinetic variability.

5.1 The Reductionist Trap vs. Cyber-Biological Complexity

Standard Environmental Risk Assessment (ERA) frameworks remain anchored in linear dose–response assumptions (National Academies of Sciences, Engineering, and Medicine [NASEM], 2016). This assumption becomes increasingly fragile when applied to nanoscale systems, where surface-dominated behavior decouples biological reactivity from bulk concentration metrics. Even relatively small environmental fluctuations—such as a ±0.2 pH shift producing >50% variation in bioavailability—suggest that NPs behave less as passive contaminants and more as environmental signal modulators (Liu et al., 2024).

The limitation here is not merely statistical. It is structural. Minor aggregation changes can cascade into enzymatic signaling shifts and epigenetic modulation, producing system-level effects that are not linearly traceable back to the exposure dose. DNPS responds to this mismatch by replacing static observational logic with adaptive biointervention modeling, where the microbiome–host axis is treated as a transduction interface that translates physicochemical variation into measurable biological structure (Bhattacharyya et al., 2025). The shift is conceptual, but it also exposes a persistent tension between measurement and interpretation.

5.2 Multisource Data Computational Vacuum: From Data to Bio-Intelligence

A less visible but equally constraining problem is what may be described as a computational invisibility gap. Modern environmental biology methods produce vast volumes of omics, imaging, and spectral data (Santos et al., 2025), yet these datasets are rarely integrated into coherent system-level representations. In practice, they remain parallel silos rather than convergent evidence structures.

DNPS attempts to address this fragmentation through AI-driven multiscale data fusion. Early simulations suggest that integrating in silico biotechnology models—linking NP surface chemistry to transcriptomic responses (Wang et al., 2025)—may reduce predictive uncertainty by approximately 35–45%. These figures should be read cautiously; they are model-dependent and sensitive to calibration regimes. Nevertheless, the direction of travel is clear: the field is moving from descriptive mapping of contamination toward computational anticipation of systemic instability, albeit unevenly and without methodological consensus (Reyed, 2019).

5.3 Toward a Reconstructive Solution: The Cyber-Biotech Paradigm

This trajectory does not replace existing environmental science; it stretches it. The DNPS approach effectively substitutes static snapshots with coupled cyber-physical simulations that attempt to approximate system evolution over time. The aim is not to predict in a narrow sense but to identify regimes where biological resilience begins to degrade.

Such transitions—often referred to as systemic tipping points—are not directly observable in most current monitoring systems. DNPS reframes them as inferential constructs derived from multilayered simulation outputs, marking an evolution of environmental risk assessment toward what may be termed predictive environmental intelligence (Kara et al., 2025). Whether this transition holds empirically depends on how robustly these inferred states can be validated outside controlled computational environments.

5.4 Systems-Theoretical Foundations: NPs as Signal Perturbers

The DNPS framework is grounded in complexity science, where nanoplastics are not treated as inert particles but as active perturbative elements within hierarchical bionetwork structures. Their effects emerge not from single interactions but from feedback loops distributed across chemical, microbial, and ecological scales (Sidhu et al., 2025).

Toxicity, in this framing, is not a fixed property but a system-level condition arising from cumulative signal interference. This diverges from both the exposome and adverse outcome pathway (AOP) frameworks, which, while valuable, tend to preserve a linear causality assumption. DNPS instead interprets nanoplastics as modulators of biological signaling architecture, requiring integrated computational representation across environmental and molecular layers. This aligns with emerging One Health computational approaches that attempt to translate heterogeneous environmental data into system-level inference structures (Basumatary et al., 2025; Leso et al., 2025; Jeong et al., 2024).

6. THE DNPS OPERATIONAL FRAMEWORK: MULTI-OMICS, AI, AND PREDICTIVE BIOTECH

6.1 Information–Particle Duality and Systemic Sensing

DNPS begins from a practical constraint rather than a theoretical claim: nanoplastics are difficult to observe directly at the system scale. To address this, the framework leverages biological systems—particularly microbial communities—as distributed sensing layers. These communities respond continuously to environmental perturbations, effectively encoding exposure signals in metabolic and compositional shifts that can be computationally captured and analyzed.

The result is not direct observation of particles but indirect reconstruction of their systemic footprint through biological response structures.

6.2 Synthetic Perturbome: From Molecular Stress to Systemic Risk

Within DNPS, the synthetic perturbome is introduced as a structured representation of cumulative biological stress across the Proteostatic Stress Continuum (PSC). At its core lies the interpretation of redox imbalance and oxidative stress as early mechanistic indicators of nanoplastic interactions with cellular systems—a baseline perspective consistent with Reyed and Prabhakar (2025), where oxidative signatures are treated as precursors to metabolic destabilization.

By combining in silico bioprocess modeling (Reyed, 2019) with multiomics integration, DNPS generates a systemic risk score (SRS) (Khanam et al., 2025). Small microbial shifts—on the order of 10–20%—are not interpreted as isolated fluctuations. Instead, within this architecture, they may correspond to incremental changes in systemic risk, for example, in the range of ~0.08–0.15 depending on the model configuration and exposure context. These values should not be read as fixed thresholds; they represent scenario-dependent indicators of trajectory rather than definitive outcomes.

The utility of this structure lies in its ability to translate heterogeneous biological perturbations into comparable risk representations, particularly under conditions where environmental uncertainty is high (Bhattacharyya et al., 2025).

6.3 Polybiome Systems Medicine: The One Health Interface

From a One Health perspective, nanoplastics operate at the intersection of the host, microbiome, and environmental systems. DNPS extends this idea through Polybiome Systems Medicine, where patient-derived datasets and dynamic simulations are used to model how perturbations propagate across biological domains (Reyed, 2023).

Long-term effects such as neuroinflammatory signaling and autophagy disruption are incorporated as modeled outcomes rather than assumed endpoints, particularly when sensitive developmental windows are considered (Reyed & Prabhakar, 2026a). This allows the framework to support three interconnected analytical pathways: microbiome modulation strategies, synthetic degradation approaches, and policy-level preventive modeling.

6.4 Integrated DNPS Pipeline (operational overview)

The operational structure of a DNPS can be summarized as follows:

1. **Data** acquisition – multiomics datasets and microbial signal capture
2. **Processing & Modeling** – AI-driven integration with in silico simulation layers
3. **SRS Computation** – generation of systemic risk representations
4. **Predictive Intervention** – simulation of the microbiome or material-level interventions
5. **Outcome evaluation** – feedback integration into One Health decision structures

This sequence is not strictly linear in practice; feedback loops occur between stages. The main intent is to create a traceable pathway from environmental measurement to system-level interpretation, narrowing the gap between observation and actionable inference (Figure 4).

Figure 4. Conceptual multiscale signal cascade and intervention dynamics in the DNPS/SRS framework.

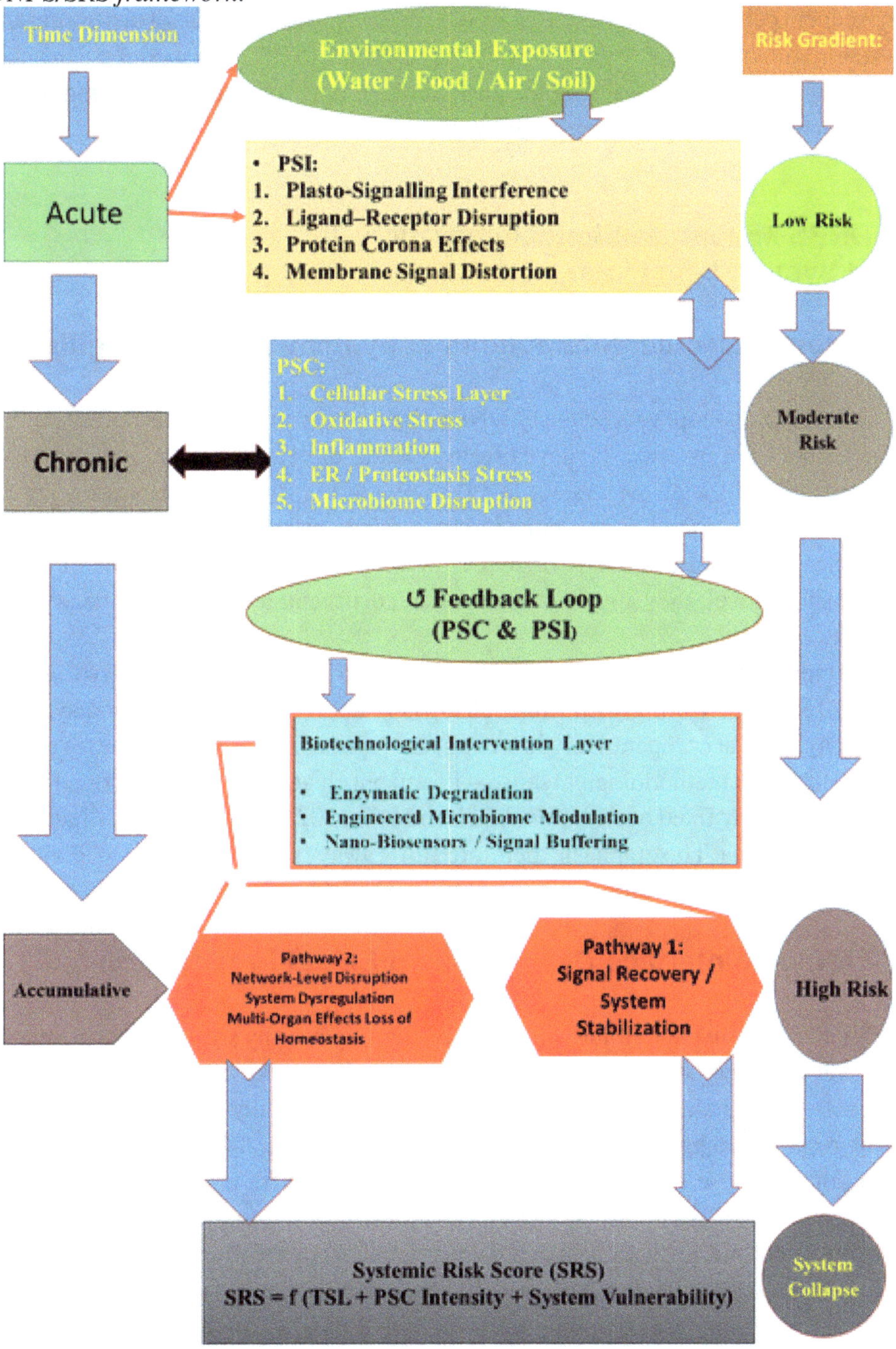

Figure 4 illustrates the stepwise propagation of nanoplastic-induced perturbations from initial environmental exposure through plasto-signaling interference (PSI) and subsequent cellular stress responses (PSCs), leading to network-level disruption and systemic outcomes quantified as a systemic risk score (SRS). The model emphasizes a multiscale transition from molecular interactions to system-level dysregulation within a unified analytical framework.

7. DNPS Multiscale Signal Integration And System-Level Interpretation

7.1 From Molecular Interactions to System-Level Dysregulation

The digital nanoplastic science (DNPS) framework involves a shift in scale that is not smooth in practice. It moves from isolated molecular interactions toward a more continuous view of system-level dysregulation, where nanoplastics are no longer treated as passive particulates but as elements embedded within coupled biological–environmental networks. The effects, in this reading, do not sit neatly at any single level; they appear to accumulate, drift, and reconfigure depending on context.

Computational simulations occasionally point to aggregated transductive signal loads (TSLs) correlating with subtle changes in neuronal excitability and synaptic plasticity, even at comparatively low exposure regimes. Nevertheless, these outputs are unstable in a methodological sense—useful but clearly model-conditioned rather than biologically fixed boundaries (MacLeod et al., 2021; Li et al., 2025; Bhattacharyya et al., 2025). The emphasis therefore shifts away from concentration alone and toward the signal structure in high-dimensional biological space.

7.2 Nanobiointerface Dynamics and Signal Propagation

Evidence from nanobiointerface studies complicates any simple exposure narrative. Surface chemistry, corona formation, and interaction kinetics appear to reshape downstream signaling trajectories in ways that are not fully captured by classical dose–response logic (Santos et al., 2025; Kochanek et al., 2025). The implication is not that the dose becomes irrelevant but that it is insufficient on its own.

Signal propagation can be described—somewhat cautiously—as a cascading sequence: corona formation, receptor engagement, intracellular modulation, and eventual network-level perturbation. However, this sequence is rarely clean in real systems, and empirical mapping across scales remains incomplete (Nene et al., 2025). More topology than line topology is observed.

7.3 Environmental Modulation and Multistressor Contexts

Plasto-signaling interference (PSI) does not operate in isolation. Its expression is shaped by environmental conditions that are often treated as secondary but may be structurally decisive. Oxidative stress, thermal variability, and microbial biofilms all appear capable of modulating both the intensity and persistence of signal disruption (Shao et al., 2025; Su et al., 2026).

When heavy metals, pesticides, or persistent organic pollutants are combined, the system becomes harder to linearize. Interactions begin to show nonadditive behavior, sometimes amplifying effects and sometimes dampening them. The consistency of these patterns is still debated, and they are likely system-specific rather than universal.

7.4 Population Variability and Probabilistic Risk Stratification

At the population level, responses to nanoplastic exposure are uneven in ways that are difficult to reduce to a single vulnerability metric. Neonates, immunocompromised individuals, and older adults often appear more sensitive, although the mechanisms are distributed across physiological and environmental axes rather than confined to one pathway (Ragusa & Fanos, 2025; Lee et al., 2025).

Simulation frameworks can organize this variability into probabilistic risk structures by combining exposure profiles, biological markers, and contextual modifiers. However, these outputs remain heavily dependent on the assumptions embedded in model construction. They describe tendencies, not outcomes (Khanam et al., 2025).

7.5 Mechanistic Abstraction of PSI-Mediated Disruption

At the mechanistic level, PSI-mediated disruption can be sketched as a layered abstraction rather than a single causal chain. Nanoloplastic infiltration initiates a sequence beginning with corona formation, followed by receptor interaction and downstream signaling perturbations. From there, effects may propagate into metabolic, immune, and neuronal domains, occasionally producing what can be described as partial signal collapse or attenuation across feedback loops.

This framing is useful but incomplete. The transition from molecular disturbance to systemic dysfunction remains only partially mapped, and cross-scale validation is still uneven (Bell et al., 2013; Bhattacharyya et al., 2025).

7.6 Early-stage intervention logic and model-driven strategies

Within this structure, combined TSL and systemic risk score (SRS) outputs introduce a tentative logic for early intervention. Small upward shifts in modeled signal disruption may indicate emerging instability, although this interpretation is probabilistic rather than diagnostic.

Possible responses range from signal-guided bioremediation to adaptive biosensing and, in more speculative settings, pharmacological or microbiome-targeted modulation. However, their feasibility varies sharply across environmental and clinical contexts and should not be overstated (Lee & Kim, 2026; Elawady et al., 2026).

7.7 Data Integration and One Health Extensions

The integration of patient-derived datasets—particularly microbiome and gastrointestinal profiles—adds another layer of interpretive complexity. These data streams may help contextualize exposure–response relationships that otherwise appear inconsistent or noisy (Lee et al., 2025; Su et al., 2026).

When combined with systems-level modeling, they support a more continuous reading of environmental–biological interactions. Nevertheless, harmonization remains a practical bottleneck, and cross-system comparability is far from resolved.

7.8 System-Oriented Interpretation and Analytical Scope

The DNPS architecture should not be mistaken for a predictive engine. Its function is closer to a structured interpretive environment, particularly where linear models fail to capture interaction effects.

Its value lies in organizing heterogeneous data into comparable forms—enabling hypothesis generation, scenario comparison, and, cautiously, adaptive governance reasoning. However, this flexibility is also a constraint. Without explicit boundaries, interpretive drift becomes a real risk. Maintaining uncertainty limits is therefore not optional but structural (Habumugisha et al., 2024; Khanam et al., 2025).

7.9 Nanobiocompatibility and interaction topology

Recent work on nanobiointerfaces has reinforced a recurring point: interaction topology matters as much as exposure magnitude. Surface properties, corona dynamics, and kinetic behavior reshape signaling pathways in ways that extend beyond classical toxicological assumptions (Santos et al., 2025; Kochanek et al., 2025).

Within DNPS, these interactions are again conceptualized as progressive signal propagation—from interface formation through receptor engagement to broader cellular and network-level effects. The pathway is analytically useful, although it is still empirically incomplete (Nene et al., 2025).

7.10 Environmental Co-Factors and PSI Modulation

PSI dynamics remain sensitive to environmental cofactors. Oxidative stress, temperature variation, and microbial biofilms can shift both the magnitude and duration of signal disruption (Shao et al., 2025; Su et al., 2026).

Coexposure scenarios introduce further ambiguity. Heavy metals, pesticides, and persistent organic pollutants may interact in nonlinear ways, sometimes reinforcing and sometimes altering the direction of effect. These patterns are not yet stable enough to be generalized confidently (Li et al., 2025).

7.11 Population sensitivity and contextual risk variation

Population-level sensitivity continues to vary in ways that resist simple classification. Certain groups show amplified responses, likely due to differences in baseline physiological resilience and regulatory capacity (Ragusa & Fanos, 2025; Lee et al., 2025).

DNPS-based simulations can structure this variability into probabilistic risk landscapes. However, these landscapes are conditional constructs shaped by model design choices and data completeness rather than fixed biological truths (Khanam et al., 2025).

7.12 Mechanistic Continuity from Molecular to Systemic Levels

PSI-mediated disruption can be represented as a continuous but uneven pathway from nanoplastic infiltration to organism-level outcomes. The process begins at the interface—corona formation and receptor interaction—and extends into intracellular signaling perturbations that may propagate across the neuronal, metabolic, and immune systems.

Figure 5 conceptualizes this as a staged propagation model, including feedback instability and partial signal attenuation, sometimes described as blackout-like behavior in system signaling. However, the transition from molecular perturbation to systemic dysfunction remains only partially resolved, and the explanatory chain should be treated as provisional rather than definitive (Bell et al., 2013; Bhattacharyya et al., 2025).

Figure 5. Plasto-Signaling Interference Pathway and Organismal Impact

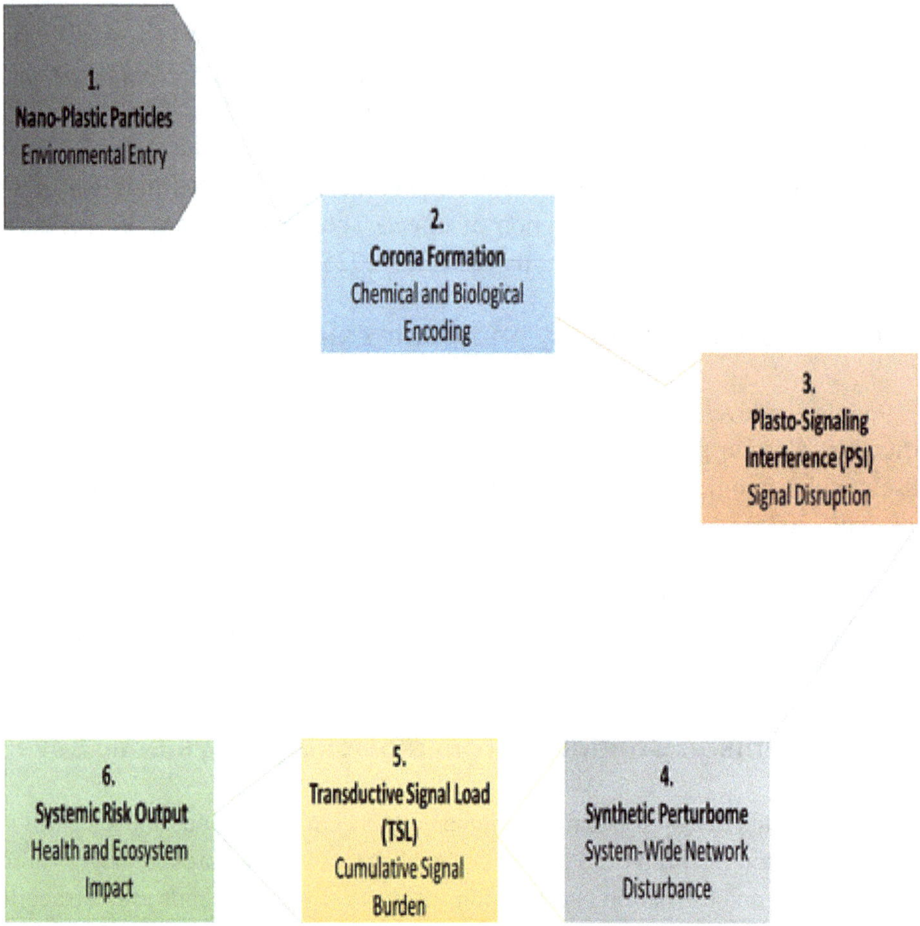

Legend: Nanoplastics (NPs) act as signal decoys and receptor-binding agents, disrupting endogenous signaling pathways. The transductive signal load (TSL) accumulates across neuronal, metabolic, and immune networks, leading to feedback loop failures and system-level perturbations. Environmental cofactors and coexposures modulate the magnitude of interference. Early detection and predictive modeling allow for AI-driven risk ranking and targeted interventions to prevent poor clinical outcomes.

7.13 Intervention Pathways and Applied Modeling

Within the current analytical structure, coupling the transductive signal load (TSL) with the systemic risk score (SRS) introduces a tentative, still fragile logic for early intervention. What matters here is not absolute thresholds but the slow drift of modeled signal disruption—small increments that may, in some simulations, precede broader system instability. Interpreting these shifts as actionable signals is possible, although never straightforward.

Potential responses range from signal-directed bioremediation to adaptive biosensing architectures and, in more speculative extensions, pharmacological modulation strategies. However, each of these remains tightly bound to context; transferability across environmental and clinical domains cannot be assumed and tends to breakdown under real-world heterogeneity (Lee & Kim, 2026; Elawady et al., 2026). The framework, at best, sketches a direction of intervention rather than a stable protocol.

7.14 Polybiome Integration and Data-driven Extensions

A subtler development appears when patient-derived biological datasets are incorporated, particularly those related to gastrointestinal dynamics and microbiome structure. These data streams, when available and of sufficient quality, tend to shift interpretation slightly—sometimes noticeably—by revealing exposure–response relationships that remain obscured in isolated environmental datasets (Lee et al., 2025; Su et al., 2026).

When aligned with polybiome-informed therapeutic models, this opens a space for simulating coupled nutritional and environmental stressors, although the simulations themselves remain uneven in resolution. The extension toward a One Health framing is conceptually coherent but operationally incomplete, largely because harmonization across biological scales is still unresolved (Reyed, 2023; Habumugisha et al., 2024). The promise is real; the infrastructure, not yet.

7.15 System Integration and Conceptual Closure

In this architecture, Plasto-Signaling Interference (PSI) functions less as a discrete mechanism and more as a connective process. It links early molecular perturbations with downstream systemic outcomes but not in a strictly linear fashion. The pathway bends, branches, sometimes attenuates—what emerges is a distributed field of interaction rather than a single causal chain.

The transductive signal load (TSL), environmental modulation, and population-level variability operate together to shape how nanoplastic exposure is ultimately expressed across biological systems. None of these components is sufficient on its own; their meaning arises only in combination, and even then, it is partial.

Figure 4 situates these interactions within a multiscale representation that draws on environmental co-factor analysis and hierarchical modeling. Nevertheless, the goal is not to eliminate uncertainty. That would be misleading. The more modest aim is to make uncertainty structurally visible—where it enters, how it propagates, and where interpretation begins to strain.

Within those constraints, the framework retains practical value. It supports comparative reasoning across contexts, enables structured hypothesis formation, and offers a way to think about intervention without inflating predictive claims beyond what the data can sustain.

8. TRANSLATIONAL INTERFACES: BRIDGING DNPS TO PRACTICAL APPLICATION

8.1 Translational biotechnologies and predictive governance

DNPS is positioned as an inherently translational framework that connects theoretical modeling with applied biotechnological and governance systems. Within this interface, adaptive biosensing technologies—particularly those based on high-resolution spectral imaging—support near real-time detection of emergent environmental perturbations (Shao et al., 2025; Lee and Kim, 2026). These tools extend observational capacity beyond conventional sampling limitations by enabling continuous monitoring of signal variability across environmental matrices.

In parallel, signal-directed bioremediation strategies employ engineered microbial systems or enzyme-based interventions to selectively degrade or neutralize disruptive particulate matter. Such approaches are particularly relevant for maintaining ecological stability in sensitive biogeochemical cycles, including nitrogen cycling (Su et al., 2026). However, their effectiveness remains contingent on ecological specificity, containment control, and long-term system validation under field conditions.

At the governance level, predictive frameworks informed by One Health integration facilitate cross-sectoral synthesis of environmental, veterinary, and clinical datasets. This enables anticipatory decision-making structures that move beyond reactive environmental regulation toward early-stage risk identification (Elawady et al., 2026; Habumugisha et al., 2024). Nevertheless, the operational deployment of such systems depends on interoperability standards, regulatory harmonization, and the robustness of validation across heterogeneous ecological contexts.

Overall, DNPS translation into applied domains is not linear but rather conditional, relying on the coevolution of the sensing capacity, remediation technologies, and governance infrastructures. This positioning reframes DNPS as a mediation layer between environmental signal detection and policy-level action rather than a standalone intervention system.

8.2 Systemic Awareness: TSL and PSI

To enable system-level interpretation of nanoplastic perturbations, the DNPS framework introduces **the transductive signal load (TSL)** and **plasto-signaling interference (PSI)** as complementary analytical constructs. TSL is conceptualized as a cumulative proxy for distributed perturbation signals across molecular, cellular, and ecological scales, extending beyond concentration-based exposure metrics toward interaction-sensitive modeling of biological response fields.

From a systems toxicology perspective, nanoscale materials may induce diffuse and nonlinear biological responses that are not fully captured by classical dose–response assumptions, particularly under chronic or low-dose exposure conditions (Lee & Kim, 2026; Li et al., 2025). In this context, the TSL provides a structured representation of subthreshold interactions that may propagate through interconnected biological networks, especially under multistressor conditions involving cocontaminants such as heavy metals or persistent organic pollutants (Nene et al., 2025).

Model-based simulations further indicate that elevated TSL may correlate with alterations in calcium-mediated signaling pathways and cytokine network modulation. However, these associations remain probabilistic and should not be interpreted as fixed biological thresholds. Empirical studies in neurotoxicology and systems biology confirm that nanoscale exposure can influence intracellular signaling dynamics under specific experimental conditions, although effect size and reproducibility vary substantially across systems (Bhattacharyya et al., 2025; Araújo et al., 2025). Accordingly, any threshold-like interpretation of the TSL remains model-dependent and sensitive to parameterization, exposure assumptions, and biological heterogeneity.

PSI dynamics are further modulated by environmental cofactors that influence particle persistence and interaction strength. Oxidative stress conditions, thermal variability, and microbial biofilm formation have been shown to modify nanomaterial behavior within biological systems (Kochanek et al., 2025; MacLeod et al., 2021). In addition, coexposure scenarios involving pesticides, pharmaceuticals, and heavy metals introduce nonlinear interaction effects that may amplify or attenuate pathway-level disruptions (Shao et al., 2025; Su et al., 2026). These findings collectively support a systems interpretation in which nanoplastics function not as isolated toxicants but as components of multifactorial environmental interaction networks.

Within this multilayered structure, differential vulnerability becomes a key modulator of the system response. Populations such as neonates, elderly individuals, and immunocompromised individuals may exhibit heightened sensitivity due to reduced physiological buffering capacity (Ragusa & Fanos, 2025; Lee et al., 2025). DNPS-based modeling may therefore support probabilistic risk stratification by integrating exposure intensity, signal disruption metrics, and contextual modifiers.

However, these outputs should be interpreted as comparative risk indicators rather than deterministic clinical predictors.

8.3 Integrative Architecture: DEI and Multiscale System Modeling

Dynamic ecointegrators (DEIs) extend the DNPS architecture by providing a conceptual integration layer for heterogeneous environmental and biological datasets. This includes environmental monitoring outputs, multiomics profiles, and AI-assisted inference systems. Rather than functioning as a fully standardized platform, DEI is best understood as an adaptive and evolving framework designed to enable cross-scale data alignment and interpretive synthesis.

Although advances in nanoplastic detection—particularly through high-resolution analytical chemistry and biosensor technologies—have improved detection sensitivity and classification capacity, data streams remain largely fragmented across methodological domains (Shao et al., 2025; Lee & Kim, 2026). The DEI addresses this limitation by enabling structured data fusion; however, its current state remains conceptual to early prototypes in terms of operational deployment.

Simulation-based analyses suggest that environmentally relevant concentrations of micro- and nanoplastics, when combined with additional stressors such as UV exposure, thermal fluctuations, and microbial activity, may contribute to proteostatic imbalance, immune modulation, and metabolic stress responses. These projections align broadly with findings in environmental toxicology, indicating interactive effects between nanoplastics and environmental modifiers (Nene et al., 2025; Su et al., 2026). Nevertheless, such outputs remain hypothesis-generating and should not be interpreted as deterministic predictions.

Collectively, TSL and DEI define DNPS as a probabilistic, multiscale analytical framework. TSL formalizes cumulative signal disruption, whereas DEI enables integrative scenario modeling across heterogeneous datasets. Importantly, this architecture explicitly incorporates uncertainty, coexposure complexity, and environmental variability—factors that are often simplified in conventional Environmental Risk Assessment (ERA) and Adverse Outcome Pathway (AOP) frameworks (Khanam et al., 2025; Habumugisha et al., 2024; Elawady et al., 2026). Rather than replacing these established models, DNPS extends its interpretive scope toward nonlinear and multisystem representations of environmental risk.

9. INTEGRATED DNPS–SRS FRAMEWORK: A SYSTEMS-ORIENTED RISK PERSPECTIVE

The persistent detection of nano- and microplastics across environmental compartments continues to challenge the assumptions underlying conventional environmental risk assessment models, particularly those based on linear dose–response relationships and deterministic removal efficiencies (Hayat et al., 2026). Empirical studies indicate that a fraction of nanosized particles may persist following advanced wastewater treatment processes, potentially contributing to environmental reentry pathways. However, these findings are influenced by methodological variability, detection thresholds, and analytical constraints, requiring cautious interpretation (Nene et al., 2025).

Moreover, nanoplastic detection remains an evolving analytical domain characterized by both technological progress and epistemic uncertainty. While multiple studies confirm their presence across environmental matrices, parallel evidence highlights risks associated with contamination, instrumental resolution limits, and measurement uncertainty, suggesting that observed distributions may represent partial reconstructions of actual environmental states (Shao et al., 2025). This duality reflects both technical and structural uncertainty in environmental measurement systems.

Within this context, DNPS introduces the **systemic risk score (SRS)** as a semi-quantitative integrative construct defined as:

$$\mathrm{SRS} = {}_{W1} + {}_{W2} + {}_{W3}V$$

where exposure intensity ( ), signal-level biological disruption ( ), and system vulnerability (V) are integrated through context-dependent weighting factors. These weights may be adjusted according to data availability, expert calibration, or validation constraints. Accordingly, the SRS should be interpreted as a comparative and inferential indicator rather than a deterministic prediction (Khanam et al., 2025).

From a systems biology perspective, emerging evidence suggests that prolonged exposure to nano- and microplastics may be associated with microbiome alterations, metabolic stress responses, and low-grade inflammatory activity. However, causal pathways and long-term implications remain under active investigation (Bhattacharyya et al., 2025; Ragusa & Fanos, 2025). Within this exploratory domain, constructs such as a provisional metabolic phenotype (e.g., “Type-V diabetes”) may function as heuristic devices for hypothesis generation, but they do not represent clinically validated disease entities and require further empirical verification (Reyed, 2026a).

Model-based scenarios further indicate that comparable systemic risk profiles may emerge under conditions of lower exposure but greater biological disruption, emphasizing the nonlinearity of system responses and the limitations of exposure-centric frameworks. In certain configurations, system vulnerability and signal disruption may exert effects comparable to or greater than the exposure magnitude, although these interpretations remain contingent on the model structure and parameter assumptions (Li et al., 2025).

From a governance perspective, existing international frameworks, including OECD-aligned strategies, emphasize source reduction and lifecycle management of plastics. While essential, these approaches may not fully incorporate predictive, system-level modeling capacities. The DNPS–SRS framework therefore functions as a complementary extension that supports anticipatory risk assessment while remaining structurally aligned with established regulatory paradigms (Habumugisha et al., 2024; Elawady et al., 2026). However, implementation will likely be constrained by data standardization challenges, infrastructural disparities, and heterogeneous technical capacity across regions, suggesting that phased adoption begins with semiquantitative applications prior to full-scale integration.

9.1 Case Study: Human Exposure to Nanoplastics and Microplastics in Freshwater and Dietary Pathways – A DNPS/SRS Analysis

Building on the mechanistic cascade outlined previously, this case study examines human exposure to nano- and microplastics through freshwater and dietary pathways within the DNPS/SRS framework. The focus is placed on system-level perturbations that may not be fully resolved within conventional toxicological or endpoint-driven assessment models.

Current monitoring and synthesis studies indicate that exposure is widespread across environmental compartments, including drinking water and food systems. Ingestion is therefore plausibly sustained through multiple routes, such as seafood consumption, salt intake, and irrigation-mediated food chain transfer (Njema et al., 2026). Although many of these exposure levels are classified as "subthresholds" within regulatory contexts, this categorization primarily reflects the structure of endpoint-based risk assessment models rather than the absence of biological interactions. This introduces a structural interpretive gap between measured exposure and modeled risk relevance.

Experimental and mechanistic evidence further suggests that chronic low-dose exposure may be associated with alterations in intestinal barrier integrity, modulation of inflammatory signaling, and shifts in microbiota composition, although outcomes remain sensitive to experimental design and model system variability (Nie et al., 2024). From a systems perspective, these responses indicate that nanoplastics may interact with biological organization at the level of signaling and homeostatic regulation rather than functioning solely as acute toxic agents. However, the magnitude, persistence, and reproducibility of such effects remain contingent on the physico-chemical properties, exposure duration, and host-specific variability of the particle.

Coexposure dynamics further complicate interpretation. Nanoplastics can act as carriers for adsorbed contaminants, including antibiotics and other environmental chemicals, thereby influencing their transport behavior and bioavailability under specific environmental conditions (Likpalimor et al., 2025). This introduces a secondary interaction layer in which nanoplastics function as vectors within broader contaminant networks rather than as isolated stressors.

Population heterogeneity adds another dimension of complexity. Evidence from environmental health and toxicological studies indicates that susceptibility is shaped not only by exposure intensity but also by physiological resilience and developmental stage, with infants and other sensitive populations potentially exhibiting amplified responses to low-intensity perturbations (Sundas et al., 2024; Krewski et al., 2010). Conversely, reports suggesting limited absorption or rapid clearance should be interpreted cautiously, as such findings may reflect methodological constraints related to detection sensitivity, exposure duration, or endpoint selection rather than the definitive absence of interaction.

Within the DNPS framework, these seemingly heterogeneous observations can be interpreted as consistent with a distributed perturbation model, in which nanoplastics contribute to low-intensity but systemically propagating signals across biological networks. Mechanistic studies at the nano–bio interface support this interpretation by demonstrating that molecular and interfacial interactions may influence cellular signaling and homeostatic regulation, providing a plausible basis for cumulative effects under repeated exposure scenarios (Karim et al., 2024). Accordingly, model-derived estimates of systemic risk should be treated as conditional outputs dependent on structural assumptions, coexposure configurations, and system integration parameters.

From a One Health perspective, freshwater systems represent critical convergence zones linking environmental contamination, food chain transfer, and human exposure (Khan & Ali, 2023). Integrated modeling approaches that combine environmental monitoring with computational and in silico methods may increase the resolution of exposure pathway reconstruction and system-level inference; however, such integration remains constrained by methodological and infrastructural limitations (Suciu et al., 2026). Advances in environmental sensing technologies further suggest improved detection capabilities, yet their translation into predictive system-level intelligence remains an unresolved challenge (Perra et al., 2026).

By linking transductive signal load (TSL) constructs to systemic risk score (SRS) outputs, the DNPS framework provides a structured method for interpreting heterogeneous exposure datasets and identifying potential early warning signals. These applications may support targeted mitigation strategies, including the optimization of water treatment processes and ecosystem-level interventions. However, their implementation remains contingent on data integration capacity, validation across environmental contexts, and institutional readiness (Capodaglio, 2025). Overall, the evidence supports the interpretation that chronic low-dose exposure may contribute to persistent system-level perturbations, although the strength of inference remains bounded by current empirical and methodological limitations.

9.2 Case Study: Human Exposure via Freshwater and Dietary Sources

Extending the DNPS/SRS cascade to nano- and microplastic exposure reframes the phenomenon as a distributed exposure field rather than a conventional toxicant burden. Koelmans et al. (2019) highlighted the methodological limitations inherent in drinking water assessment, demonstrating how presence data may be prematurely interpreted as evidence of negligible risk. In contrast, Leslie et al. (2022) and Schwabl et al. (2019) shifted the analytical focus toward internal exposure by reporting the detection of plastic particles in human blood and stool, thereby strengthening the argument that exposure extends beyond external environmental compartments.

At the mechanistic level, the gastrointestinal tract emerges as a central interface of interaction. Jin et al. (2019) demonstrated that polystyrene microplastics can alter microbiota composition and metabolic function in vitro, whereas Nie et al. (2024) emphasized the role of intestinal barrier integrity, particularly in contexts where tight junction stability and gut–organ communication are already compromised. These findings collectively suggest that biological effects are mediated through functional disruption of regulatory interfaces rather than through isolated cytotoxic events.

Karim et al. (2024) further refined this perspective by emphasizing that the relevant analytical unit is not bulk exposure mass but rather molecular- and interface-level interactions through which nanoplastics may influence biological organization and dyshomeostasis. This interpretation is reinforced by Cverenkárová et al. (2021) and Likpalimor et al. (2025), who situate nano- and microplastics within food-chain transfer and contaminant sorption networks. Within this structure, coexposure to heavy metals, pharmaceuticals, and antibiotics becomes a mechanistic amplifier rather than a secondary confounder. In this context, Generalova et al. (2025) provide an informative analogy from lead toxicity, illustrating how carrier-mediated transport can transform environmental vectors into multipliers of biological stress.

Apparent inconsistencies with studies reporting rapid clearance or limited acute toxicity are better understood as a limitation of endpoint selection rather than a direct contradiction. Krewski et al. (2010) emphasized the need to move beyond apical outcomes toward perturbation-sensitive toxicological frameworks, a transition that becomes increasingly relevant in the context of micro- and nanoplastic exposure.

Within this evolving evidence base, Bhattacharyya et al. (2025) and Capodaglio (2025) support a more constrained interpretation of risk, acknowledging that neurotoxicological and drinking water assessments remain limited by unresolved dose metrics, particle heterogeneity, and detection constraints, despite increasing empirical attention. Susceptibility is also structurally uneven across populations. Christopher et al. (2024) highlighted early-life vulnerability, whereas Suciu et al. (2026) and Perra et al. (2026) emphasized the role of integrated One Health frameworks and advanced data acquisition systems in linking environmental exposure to biological outcomes within a unified analytical architecture.

Within this framing, DNPS/SRS is most defensible not as a deterministic disease attribution model but as a structured framework for identifying low-intensity, network-level perturbations prior to their progression into clinically or ecologically significant outcomes (Luo et al., 2024). Its interpretive value lies in organizing heterogeneous and partially uncertain evidence into coherent systems-level risk language without exceeding the limits of current empirical validation.

10. COMPARATIVE POSITIONING OF THE DNPS WITH EXISTING PARADIGMS – EXTENDED ANALYSIS

Building on the case-study evidence presented in Section 9, this section transitions from exposure-specific interpretation to a comparative evaluation of DNPS/SRS against established environmental and toxicological paradigms. This shift is necessary to clarify how distributed perturbation-based reasoning aligns with, extends, or diverges from conventional frameworks such as environmental risk assessment (ERA), adverse outcome pathways (AOPs), the exposome concept, and life cycle assessment (LCA).

Rather than replacing these approaches, DNPS is positioned as an interpretive extension that emphasizes nonlinear, multiscale interactions and system-level signal propagation. In contrast to the ERA and AOP frameworks, which often rely on structured exposure–response relationships and endpoint anchoring, DNPS incorporates uncertainty, coexposure dynamics, and distributed biological signaling as first-order analytical components. Similarly, while exposome-based models emphasize cumulative exposure across life-course dimensions, DNPS further formalizes the translation of exposure into system-level signal disruption through constructs such as TSL and SRS.

In relation to LCA, which focuses primarily on material flow and environmental impact across product life cycles, DNPS introduces a biological and systems-interaction layer that captures downstream perturbation effects beyond emission and exposure quantification. However, this added interpretive depth also increases the dependence on model assumptions, parameter uncertainty, and data integration capacity, which remain limiting factors for operational implementation.

Overall, the comparative analysis indicates that DNPS should be interpreted as a complementary analytical layer rather than a competing framework. Its primary contribution lies in extending environmental risk interpretation toward probabilistic, network-aware representations of exposure and response while remaining structurally compatible with existing regulatory and assessment paradigms.

10.1 Limitations of Conventional Frameworks

Conventional environmental and biological risk frameworks have significantly advanced the classification of exposure, the structuring of empirical evidence, and the support of regulatory decision-making. However, their analytical architecture has clear limitations when the object of assessment shifts from discrete toxicological events to distributed, low-intensity, and interacting perturbation fields. In the context of micro- and nanoplastic research, this limitation becomes increasingly visible as multiple lines of evidence point toward coupled exposure dynamics that exceed single-agent interpretability.

For example, Likpalimor et al. (2025) demonstrated that microplastics can sorb antibiotics, thereby altering their transport behavior and environmental fate. This introduces a coupling mechanism that is not readily captured within additive exposure assumptions. In parallel, Karim et al. (2024) reported that nanoplastic–biological interactions occur at the molecular and interfacial levels, where surface chemistry and biological reactivity jointly determine the system response. When considered together, these findings do not negate the utility of existing frameworks; rather, they indicate that their interpretive boundaries are reached when multiple interaction layers converge into a shared perturbation domain.

The implications of this convergence differ across established paradigms. Environmental risk assessment (ERA) retains strength in threshold-oriented evaluation but becomes less stable in coexposure environments where one agent modifies the behavior, mobility, or bioavailability of another. Adverse outcome pathways (AOPs) preserve mechanistic clarity through structured causal sequences; however, their linear architecture limits their ability to represent cross-pathway coupling and multisystem propagation. The exposome concept extends exposure assessment temporally and cumulatively, yet cumulative mapping alone does not fully resolve nonlinear amplification or dynamic signal interaction. Life cycle assessment (LCA), while effective in tracing material and emission flows, is not designed to translate such flows into biological or systemic consequences.

These constraints do not constitute methodological failure. Rather, they define the operational boundaries of each framework. Their explanatory capacity remains valid within their intended scope, but it becomes partial when applied to nano- and microplastic systems characterized by interacting, multiscale perturbations.

10.2 DNPS/SRS as a trans-paradigmatic framework

Within this context, DNPS/SRS is positioned not as a replacement for established paradigms but as a reconfiguration layer that integrates their strengths into a signal-centric and network-aware interpretive structure. Its central conceptual shift lies in treating nano- and microplastics not as isolated dose entities but as perturbation vectors capable of propagating across biological and ecological scales. This perspective aligns with the call by Krewski et al. (2010) to move toxicity assessment beyond endpoint fixation toward more sensitive perturbation-oriented approaches.

From an environmental systems perspective, this shift is reinforced by Suciu et al. (2026), whose One Health-based in silico frameworks demonstrate the value of integrating environmental fate, exposure pathways, and human health outcomes within a unified computational structure. Similarly, Perra et al. (2026) emphasized that continuous environmental intelligence depends on advanced data acquisition systems capable of moving beyond episodic sampling toward sustained monitoring and inference.

The biological rationale for DNPS/SRS becomes more coherent when interpreted through susceptibility and interface-centered dynamics. Christopher et al. (2024) highlighted that risk is unevenly distributed across populations, with developmental stage and physiological resilience shaping exposure sensitivity. Nie et al. (2024) further referred to the intestinal barrier as a dynamic regulatory interface embedded within gut–organ axes rather than a static protective boundary. In parallel, Bhattacharyya et al. (2025) extended the discussion to ingestion-related neurological effects, where systemic outcomes may emerge without classical acute toxicity signatures.

Additional mechanistic complexity was introduced by Generalova et al. (2025), who demonstrated that metal toxicity can be amplified through interaction-mediated mechanisms. This is particularly relevant in the context of nanoplastics, which may act as carriers or modulators of cocontaminant behavior, thereby altering biological impact pathways. Within this integrated evidence landscape, DNPS/SRS does not assume universal harm from exposure. Instead, it provides a structured framework for representing how repeated low-intensity interactions may accumulate into measurable system-level perturbation loads.

Accordingly, the analytical value of DNPS/SRS lies in its ability to consolidate heterogeneous evidence into a coherent model of cross-scale disturbance while maintaining explicit recognition of uncertainty, contextual dependence, and model limitations. This positioning preserves compatibility with established frameworks while extending interpretive capacity into domains characterized by distributed and nonlinear environmental–biological interactions

10.3 Comparative Table – DNPS/SRS vs. Traditional Frameworks

Table 1. **Comparative positioning of DNPS/SRS relative to established environmental and toxicological frameworks.**

Framework	Analytical orientation	Validated contribution	Structural limitation	DNPS/SRS added value	Evidence basis
Environmental Risk Assessment (ERA)	Threshold-based, regulatory evaluation of exposure and hazard	Provides standardized risk appraisal and supports policy action	Tends to privilege linear dose response assumptions and single-stressor logic; limited sensitivity to coexposure and emergent interaction effects	Reframes risk as a distributed perturbation process and accommodates multistressor coupling	Supported conceptually by Likpalimor et al. (2025) and Capodaglio (2025)
Adverse Outcome Pathways (AOPs)	Mechanistic sequencing from initiating event to adverse outcome	Clarifies causal progression at molecular and cellular levels	Linear pathway logic can underrepresent cross pathway communication, temporal feedback, and multi organ coupling	Extends mechanistic mapping into a network model of interacting signals	Informed by Karim et al. (2024), Nie et al. (2024), and Krewski et al. (2010)
Exposome	Cumulative lifetime exposure and internal response mapping	Broadens exposure assessment across time and source diversity	Aggregation alone does not fully capture dynamic propagation, nonlinearity, or system-level emergence	Converts cumulative exposure information into predictive perturbation analysis	Contextualized by Christopher et al. (2024), Bhattacharyya et al. (2025), and Njema et al. (2026)
Life Cycle Assessment (LCA)	Material-flow and lifecycle impact analysis	Tracks environmental burdens across production and use phases	Remains weak on biological signaling, physiological susceptibility, and coexposure effects	Connects lifecycle flows to organism-level and population-level perturbation	Aligned with Suciu et al. (2026) and Perra et al. (2026)
DNPS/SRS	Network-based, signal-centric, multiscale perturbation modeling	Integrates coexposures, vulnerability gradients, and cross-scale interactions	Still requires empirical expansion and model validation across contexts	Unifies exposure, signaling, susceptibility, and One Health translation within a single analytic frame	Supported by Suciu et al. (2026), Perra et al. (2026), Christopher et al. (2024), and Njema et al. (2026)

This table compares analytical orientation, contributions, and limitations across major frameworks, situating DNPS/SRS as a network-based integrative approach.

The comparative analysis does not indicate a hierarchical relationship in which DNPS/SRS replaces established environmental and toxicological frameworks. Instead, ERA, AOPs, the exposome, and LCA each retain distinct analytical functions, yet each also appears limited when confronted with the interactional and cross-scale behavior associated with nano- and microplastic exposure systems.

Within this context, DNPS/SRS appears most relevant in situations where conventional frameworks provide only partial resolution. This is particularly evident when exposure dynamics extend beyond cumulative accumulation to include relational coupling, when measurement alone does not fully capture interaction effects, and when environmental presence is associated with propagation across biological or ecological levels. In such cases, DNPS/SRS may offer a structured way to organize heterogeneous signals into a more connected representation of system behavior. Overall, the framework is likely to function as an integrative layer that supports the reorganization of existing evidence into a more system-aware interpretive structure. However, its explanatory strength remains conditional, and its application is still bounded limited by current empirical limitations, data availability, and the evolving state of validation across environmental and biological domains.

10.4 Coexposure and Environmental Modifiers

Building on the comparative framework established in the preceding section, the discussion now turns to the operational conditions under which DNPS/SRS appears most analytically useful. In particular, coexposure and environmental modifiers define the exposure field as interactive, heterogeneous, and time-sensitive rather than as a single-agent toxicant problem.

10.4.1 Coexposure Logic and Network Amplification

Coexposure and environmental modifiers constitute one of the most consequential operational layers of DNPS/SRS because they shift nano- and microplastic exposure from a single-agent problem toward a multivector field of interacting signals. The available evidence increasingly supports this direction. Koelmans et al. (2022) placed nanoplastics within a perturbome logic in which exposure cannot be reduced to isolated particle counts, whereas Landrigan et al. (2023) situated plastic pollution within a broader human health burden shaped by overlapping environmental stressors rather than by a single causal channel. In a similar vein, Majid et al. (2025) reported that network-based risk assessment is better suited to multicontaminant settings than are stressor-specific models, and Matus et al. (2024) extended this logic by linking exposome analysis to network modeling for predictive health assessment. Taken together, these studies do not merely accumulate evidence; they reframe the exposure problem itself, moving it from linear contact toward coupled perturbation.

The interpretive consequence is that nanoplastics may be better understood as interaction platforms than as passive carriers. Callewaert et al. (2025) extended this point by showing that AOP structures can be expanded through systemic perturbome mapping, whereas Yang et al. (2026a) demonstrated that weighted network analysis can decode multiscale mechanisms only when the analysis moves beyond isolated endpoints toward the relational structure. Even Ankley & Edwards (2018), while preserving the strengths of the AOP framework, implicitly mark its boundary: mechanistic clarity remains valuable, but clarity alone does not resolve propagation, feedback, or cross-pathway coupling. DNPS/SRS therefore does not reject AOP logic; it reassembles it within a broader signal architecture in which coexposures may act as amplifiers of network instability rather than simple additive inputs.

The use of environmental modifiers further sharpens this amplification logic. Within a DNPS/SRS reading, particle aging, matrix complexity, and exposure timing do not function as peripheral covariates; they condition how strongly a perturbation is transmitted and where it becomes biologically legible. This is consistent with Halfon & Hochstein (2002), who connect life-course exposure assessment to network integration, and with Matus et al. (2024), who argue that exposome-based surveillance becomes more informative when it is embedded in a One Health frame linking environment, organism, and population health. Cheng et al. (2024) added an important corrective by showing that susceptibility is not uniform across species, sources, ages, genders, or locations, which suggests that modifier effects are structurally uneven rather than evenly distributed. Lin et al. (2025) further reinforced the biological importance of this heterogeneity by summarizing cellular toxicity mechanisms and exposure pathways. Zolotova et al. (2025) underscored the capacity of microplastics to exacerbate pathological outcomes within inflammatory tumor models, and the broader analytical imperative suggests that coexposure is not a monolithic event. Rather, it is the exposure context that dictates the biological trajectory—governing whether a pro-inflammatory signal remains a localized disturbance or erupts into a systemically amplified crisis. In this critical transition, Reyed & Prabhakar (2026b) argued that such systemic escalations within the colonic microenvironment are not coincidental; they are driven by intricate panflemmotic signatures that fundamentally bridge the gap between chronic localized inflammation and the irreversible onset of tumorigenesis.

In this context, coexposure-driven amplification may be more appropriately interpreted as part of a broader, condition dependent Nanoplastic Proteostatic Panflemmosis Metabolic Continuum (N PPMC), in which environmental perturbations and proteostatic strain gradually converge with distributed panflemmotic signaling patterns (see Glossary for conceptual framing). Under sustained exposure conditions, such patterns may, in some instances, extend toward downstream metabolic domains. Importantly, this progression should not be read as linear or deterministic; rather, it appears contingent on exposure persistence, system resilience, and the structural configuration of interacting stressors.

Overall, DNPS/SRS appears most defensible where coexposure and environmental modifiers are treated as multiplicative conditions of risk rather than as statistical noise around a primary contaminant. Traditional frameworks remain indispensable for regulation, mechanistic description, and surveillance design, but their explanatory ability narrows when the exposure field becomes interactive, heterogeneous, and time sensitive. DNPS/SRS extends their utility by converting these complexities into a network-resolved model of perturbation, thereby supporting early identification of amplification nodes, more realistic vulnerability profiling, and a more credible One Health translation of environmental risk.

10.4 Coexposure and Environmental Modifiers

Building on the comparative positioning of DNPS/SRS, this section addresses the operational conditions under which the framework becomes analytically meaningful. In particular, coexposure and environmental modifiers redefine nano- and microplastic exposure as an interactive, heterogeneous, and temporally conditioned field rather than a single-agent toxicological problem.

10.4.1 Coexposure Logic and Network Amplification

Coexposure and environmental modifiers represent a central operational layer within DNPS/SRS because they shift the interpretation of exposure from isolated stressors toward interacting perturbation systems. The emerging evidence increasingly supports this transition. Koelmans et al. (2022) situated nanoplastics within a perturbome-oriented logic in which exposure cannot be adequately reduced to particle counts alone. In parallel, Landrigan et al. (2023) conceptualize plastic pollution as part of a broader human-health burden shaped by overlapping environmental stressors rather than singular causal pathways. In addition, Majid et al. (2025) demonstrated that network-based risk assessment provides a more appropriate structure for multicontaminant environments than do stressor-specific models do, whereas Matus et al. (2024) linked exposome thinking to network-based modeling for predictive health analysis.

Collectively, these contributions do not simply add empirical findings; rather, they indicate a conceptual shift from a linear exposure representation to coupled perturbation systems. Within this context, nano- and microplastics can be interpreted less as passive particles and more as interaction mediators capable of influencing the behavior of cooccurring contaminants.

This interpretation is further reinforced through methodological extensions of existing frameworks. Callewaert et al. (2025) showed that adverse outcome pathways (AOPs) can be extended through perturbome-based mapping, whereas Yang et al. (2026a) demonstrated that weighted network approaches become analytically meaningful only when relationships are treated as structural rather than endpoint bound. Although Ankley & Edwards (2018) preserve the value of AOPs for mechanistic clarity, they also implicitly delimit their scope in relation to propagation, feedback, and cross-pathway coupling.

Within DNPS/SRS, this does not imply the rejection of AOP logic. Rather, it suggests that coexposure may function as amplifiers of network behavior rather than as additive inputs. The analytical emphasis therefore shifts from isolated causality to interaction-driven system instability.

Environmental modifiers further refine this interpretation. Within DNPS/SRS, factors such as particle aging, matrix complexity, and exposure timing are not treated as secondary variables but as determinants of signal transmission strength and biological interpretability. This perspective aligns with that of Halfon & Hochstein (2002), who relate life-course exposure to network integration, and Matus et al. (2024), who emphasize the value of One Health integration across environmental and biological systems. Cheng et al. (2024) further support this view by demonstrating that susceptibility varies systematically across species, sources, age groups, genders, and geographical contexts, indicating that modifier effects are structurally uneven rather than uniformly distributed.

Lin et al. (2025) added mechanistic grounding by summarizing cellular toxicity pathways and exposure mechanisms, whereas Zolotova et al. (2025) reported that microplastics may intensify pathological outcomes in inflammatory tumor environments. Within this broader literature, exposure cannot be treated as an isolated event. Rather, the environmental and biological context jointly determines whether a localized perturbation remains contained or evolves into system-level amplification.

Importantly, Reyed & Prabhakar (2026b) argued that such escalation processes within the colonic microenvironment may be associated with complex paninflammatory signatures linking chronic localized inflammation to tumorigenic progression pathways. Within DNPS/SRS, this is not interpreted as deterministic causation but as evidence of potential transition pathways under specific contextual configurations.

Overall, DNPS/SRS is most analytically robust when coexposures and environmental modifiers are treated as multiplicative conditions shaping system behavior. Conventional frameworks remain essential for regulation and mechanistic description, but their explanatory reach becomes constrained when exposure is interactional, heterogeneous, and time dependent. DNPS/SRS extends this capacity by representing amplification nodes and cross-scale coupling within a network-based perturbation structure.

10.4.2 Coexposure Architecture in Mixed Environmental Settings

Coexposure and environmental modifiers constitute a core structural component of DNPS/SRS because they reposition nano- and microplastic risk from a single-stressor model toward a multilayered interaction system. Evidence from freshwater and wastewater environments indicates that plastic particles are distributed across multiple compartments, requiring interpretation within mixed contamination contexts rather than isolated exposure pathways (Koelmans et al., 2019; Nematollahi et al., 2025). Landrigan et al. (2023) similarly situated plastic pollution within a broader environmental burden shaped by overlapping exposures and heterogeneous susceptibility profiles.

Within this framing, coexposure is not an external complication but an inherent feature of the exposure architecture. DNPS/SRS is therefore most relevant where risk cannot be meaningfully reduced to a single measurable input but must instead be understood as an interacting system of environmental conditions.

This interpretation is strengthened by the network-based risk literature. Gambhir et al. (2025) frame systemic risk as an emergent property of interconnected domains, whereas Majid et al. (2025) demonstrate that network-based approaches outperform linear models in multicontaminant settings. Matus et al. (2024) extended this logic by embedding exposome analysis within network structures suitable for cumulative and temporally distributed exposure. In addition, Majid et al. (2025) highlighted the relevance of less-than-lifetime exposure assessment in drinking water systems, indicating that partial exposure windows may still carry system-relevant information when interpreted through sensitive modeling frameworks.

Together, these studies support a shift from threshold-centric evaluation toward dynamic, interaction-sensitive risk representation. DNPS/SRS does not introduce a new measurement alone; it redefines the analytical unit from an isolated dose to a coupled exposure configuration.

Environmental modifiers further shape this architecture by altering particle transformation and persistence in situ. Stott et al. (2026) and Jansen et al. (2024) described plastic behavior within environmental contexts influenced by UV radiation and climate variability, demonstrating that external conditions materially affect degradation, transformation, and interaction potential. Serra et al. (2026) similarly argued that predictive modeling gains robustness when system complexity is explicitly incorporated rather than simplified through static material assumptions.

Within DNPS/SRS, these findings imply that exposure cannot be separated from environmental state variables. Particle aging, weathering, and environmental transformation may alter the interaction potential to the extent that material identity becomes functionally context dependent. Accordingly, environmental modifiers should be understood as active determinants of perturbation behavior rather than as passive background conditions.

The relevance of this framework increases when susceptibility is explicitly treated as heterogeneous. Cheng et al. (2024) demonstrated variability in health outcomes across species, sources, ages, genders, and geographic contexts, reinforcing the premise that vulnerability is structurally uneven. Matus et al. (2024) further support this interpretation by integrating exposome approaches within One Health systems, where the environmental, animal, and human health domains are analytically coupled.

In this setting, coexposure effects cannot be separated from the population context. Identical exposure configurations may generate divergent outcomes depending on developmental stage, immune status, and environmental resilience. While Ankley & Edwards (2018) retain relevance through their mechanistic AOP structure, their framework alone does not fully account for coupled, modifier dependent, multiscale interactions.

DNPS/SRS is therefore best interpreted as a transparadigmatic integrative layer. It preserves the mechanistic strengths of existing frameworks while extending them into a network-resolved representation of coexposure, environmental transformation, and differential vulnerability.

10.5 Biological perturbation signals

Biological perturbation under nano- and microplastic exposure is better understood as a distributed signaling phenomenon rather than a set of isolated biomarkers. Mechanistic evidence supports this interpretation across multiple biological levels. Long et al. (2025) situated cellular responses within mechanically mediated signaling networks, whereas Zhu et al. (2025) demonstrated that alternative splicing operates as a dynamic regulatory mechanism under stress conditions. Skaba et al. (2025) and Li et al. (2025) further reported that immune disruption, oxidative stress, and metabolic imbalance are not independent endpoints but interconnected components of broader perturbation cascades.

Propagation effects further complicate linear dose–response assumptions. Ramamurthy et al. (2025) demonstrated that bystander and nontargeted effects allow biological signals to extend beyond directly exposed cells, indicating that system-level responses may occur without proportional increases in exposure intensity. This challenges conventional toxicological assumptions in which effect magnitude is expected to scale directly with dose.

Within this context, SRS does not interpret gene expression shifts, oxidative signaling pathways (including NF- B and Nrf2), proteostatic imbalance, or microbiome dysbiosis as isolated indicators. Instead, they are treated as interdependent components of a propagative network structure. Ravi Shankar et al. (2026) further reinforced this position by showing that physiologically based kinetic modeling is required to translate in vitro findings into in vivo relevance, highlighting limitations in static exposure–response mapping.

Overall, biological perturbation signals appear more consistent with dynamic network behavior than with discrete endpoint models. Frameworks that isolate these signals risk underestimating the cumulative and system-level consequences of chronic exposure.

10.6 Systemic Vulnerability

Systemic vulnerability functions as a modulating dimension that shapes how exposure translates into a biological effect. The evidence indicates that susceptibility is unevenly distributed across populations and systems. Li et al. (2025) highlighted variability in outcomes linked to physiological state and exposure pathway, whereas Skaba et al. (2025) demonstrated that immune competence determines whether perturbations remain localized or become systemically amplified.

Environmental conditions further modify vulnerability expression. Jansen et al. (2024) reported that UV radiation and climate variability alter plastic properties, influencing persistence and reactivity, whereas Serra et al. (2026) emphasized the need for predictive models that incorporate system complexity rather than static parameters. Gambhir et al. (2025) similarly frame vulnerability as an emergent property of interconnected systems rather than a fixed attribute.

Within DNPS/SRS, vulnerability is therefore not treated as a qualitative context but as an operational dimension. It includes ecological variables (such as biodiversity and trophic buffering) as well as human factors (including age, immune status, and socioenvironmental stressors). In this formulation, vulnerability functions as a scaling parameter that determines whether perturbation signals dissipate, stabilize, or amplify across the system.

10.7 Operational integration and predictive capability

The integration of exposure intensity, biological perturbation signals, and systemic vulnerability transforms the SRS from a descriptive construct into a structured predictive framework. Gambhir et al. (2025) conceptualize systemic risk as an emergent property of interconnected domains, whereas Serra et al. (2026) demonstrate that predictive capacity increases when multiple dimensions are integrated within unified modeling structures.

Operationally, this integration is supported through network-based simulations, machine learning approaches, and multiomics pipelines. Ravi Shankar et al. (2026) provide a methodological bridge by showing how physiologically based kinetic modeling can align experimental data with biologically interpretable exposure estimates. Majid et al. (2025) further support the framework by emphasizing the importance of partial and nonlifetime exposure scenarios in chronic environmental contexts.

Within this structure, the SRS enables the identification of nonlinear transitions, early warning signals, and system sensitivity thresholds. Its primary function is not precise outcome prediction but rather structured mapping of conditions under which small perturbations may scale into system-level instability.

10.8 Scientific and Conceptual Innovations

The conceptual contribution of the systemic risk score (SRS) lies in its reconfiguration of exposure as a signal-mediated perturbation embedded within interconnected environmental–biological systems. In this framing, risk is not treated as a property of isolated contaminants but as an emergent feature of interacting system states. Gambhir et al. (2025) provide the systemic basis for this shift by framing risk as an outcome of interconnected domains, whereas Serra et al. (2026) demonstrate its applicability in integrated chemical and material assessment contexts.

Within this structure, coexposure and environmental modifiers are not external covariates but constitutive elements of system behavior. Biological responses are therefore interpreted as evolving network states rather than discrete endpoint events. This allows molecular processes such as signaling and gene regulation (Long et al., 2025; Zhu et al., 2025) to be analytically connected to organismal outcomes and, subsequently, to ecological and population-level effects without requiring linear aggregation across scales.

The methodological innovation of SRS is therefore not grounded in the introduction of novel variables but rather in the reorganization of existing knowledge into a multiscale, system-consistent representation of environmental risk. Its primary contribution lies in enabling a structured interpretation of distributed evidence under conditions of interaction, uncertainty, and cross-scale dependency.

10.9 Conclusion

The methodological architecture of the systemic risk score (SRS) extends conventional toxicological reasoning by integrating exposure intensity, biological response, and systemic vulnerability within a unified analytical framework.

Across molecular biology, toxicology, and systems modeling, converging evidence suggests that nano- and microplastic risk cannot be fully captured through linear, endpoint-based paradigms alone. Instead, risk expression appears to emerge from the interaction between distributed exposures, propagative biological signaling, and context-dependent susceptibility.

Within this interpretive space, SRS provides a structured approach for representing nonlinear and multiscale interactions without reducing system behavior to additive components. Its analytical value lies in its capacity to integrate heterogeneous data streams into a coherent representation of system dynamics while preserving explicit uncertainty bounds.

Accordingly, SRS does not function as a replacement for existing frameworks but rather as an extension that supports the interpretation of early-stage perturbations, nonlinear transitions, and cross-scale interactions. This positioning enables a more structured understanding of systemic vulnerability and may support improved anticipation of risk trajectories across environmental, health, and policy-relevant domains.

10.10 Case Study: Marine Nanoplastic Exposure and Network-Level Systemic Risk

Evidence from marine systems indicates a persistent gap between conventional toxicological endpoints and system-level representations of risk. Nano- and microplastics have been detected across multiple trophic levels, including plankton, fish, and seafood species, suggesting plausible pathways for bioaccumulation and dietary human exposure (Yong et al., 2020). These pathways are increasingly treated as environmentally relevant within the literature, where incorporation into food webs is considered a potential contributor to long-term health burden (Bastyans et al., 2022).

Experimental studies have reported that low-dose nanoplastic exposure may be associated with oxidative stress, inflammatory signaling, and microbiome alterations in aquatic organisms (Sun et al., 2021). Mechanistic syntheses further suggest that nanoplastics may act not only as inert particles but also as modulators of physiological processes in aquatic species, potentially generating latent biological vulnerabilities with implications for higher trophic levels, including humans (Thakur et al., 2025). These findings are broadly consistent with sublethal perturbation patterns that may not manifest as acute toxicity endpoints such as mortality or reproductive failure but may still reflect early-stage systemic stress (Jeong et al., 2024).

At the same time, the literature also reports cases of minimal or statistically non-significant effects under environmentally relevant concentrations. These outcomes are often attributed to aggregation dynamics, reduced bioavailability, or rapid excretory processes (Koelmans et al., 2019). In certain experimental settings, no measurable changes in feeding behavior or related endpoints have been observed (Cunningham et al., 2021). However, these discrepancies may reflect methodological sensitivity limitations rather than the absence of biological interactions.

Conventional toxicological frameworks are primarily optimized for detecting acute, high-amplitude responses, whereas nano- and microplastic exposure is increasingly characterized by low-amplitude, chronic perturbation dynamics that may remain below standard detection thresholds. As a result, nonsignificant findings require cautious interpretation, particularly when study designs are not structured to capture network-level dysregulation or cumulative subthreshold effects. Variability in analytical resolution and data quality further constrains the detection of subtle exposure signals in environmental systems (Koelmans et al., 2019).

Within this context, the DNPS/SRS framework introduces a shift in analytical focus from overt toxicity toward signal integrity and network perturbation dynamics. At the molecular level, nanoplastics have been associated with potential interference in protein folding, membrane interactions, and intracellular signaling processes, including endocrine-related pathways (Tyc et al., 2025). While these interactions may not immediately translate into observable dysfunction, they are consistent with proteostatic stress mechanisms that can affect long-term cellular homeostasis.

At the subcellular level, mitochondrial disturbances have been reported, suggesting possible links to metabolic and cardiovascular dysregulation over extended exposure periods (Lerchner et al., 2024). From a systems perspective, such localized perturbations may propagate through hierarchical biological networks, particularly under conditions of coexposure or reduced physiological resilience. This interpretation aligns with systems biology models in which nonlinear responses emerge through feedback structures and network connectivity, allowing small perturbations to scale into system-level effects (Worms et al., 2025). Although population-level validation remains methodologically constrained, multiomics approaches and computational modeling increasingly support partial reconstruction of these dynamics.

The relevance of this interpretation becomes more evident when trophic transfer is considered. Subclinical perturbations in marine organisms may influence metabolic regulation, immune responsiveness, and microbiome composition, even in the absence of overt pathology, thereby potentially affecting food quality and ecological stability (Liu, 2025). Human exposure is therefore increasingly conceptualized as interactions with biologically conditioned systems shaped by chronic environmental exposure histories rather than direct contact with isolated particles. This aligns with One Health frameworks that emphasize the interdependence of environmental, animal, and human health systems (Danasekaran, 2024; Destoumieux-Garzón et al., 2018).

Within this context, nanoplastic behavior remains strongly context dependent, with physicochemical properties such as surface charge and functionalization influencing environmental fate and biological interaction potential (Sun et al., 2020). Similar complexity is observed in engineered systems such as ionic liquid-based systems, where structural properties determine biological compatibility and reactivity (Elgharbawy et al., 2026).

Importantly, the divergence between studies reporting measurable effects and those reporting negligible outcomes may reflect differences in methodological sensitivity rather than substantive contradictions. The absence of acute toxicity does not necessarily imply the absence of systemic stress; instead, it may indicate compensated biological states in which adaptive mechanisms maintain functional stability despite ongoing perturbations. Such conditions are conceptually consistent with precritical regimes in complex systems, where early warning signals may precede abrupt transitions (Scheffer et al., 2009).

From a regulatory standpoint, reliance on endpoint-based frameworks may contribute to the underestimation of long-term and network-mediated risks (Venâncio & Oliveira, 2025). Although acute toxicity testing remains central to current regulatory systems, its limitations in capturing distributed and delayed responses are well documented (Erhirhie et al., 2018). In contrast, signal-based and network-oriented approaches provide a basis for anticipatory risk governance through integrated system monitoring. This transition parallels developments in closed-loop and feedback-controlled systems in biomedical engineering, where real-time monitoring is used to maintain stability under dynamic conditions (Qin et al., 2026).

Within this analytical space, DNPS/SRS reorganizes fragmented empirical observations into a structured representation of systemic risk evolution, emphasizing interaction, propagation, and temporal development rather than isolated endpoints.

11. ENVIRONMENTAL RISK ASSESSMENT (ERA): BEYOND LINEAR EXPOSURE MODELS

Environmental risk assessment (ERA) remains a foundational regulatory framework for characterizing potential hazards associated with environmental stressors, particularly through structured processes of hazard identification and exposure characterization (NASEM, 2016). Its methodological strength lies in its operational clarity under controlled assumptions, especially where exposure–response relationships can be approximated as monotonic or linear. However, when applied to emerging contaminants such as nano- and microplastics, these structural assumptions are increasingly constrained by environmental complexity and interactional exposure dynamics.

Empirical evidence indicates that nano- and microplastics may induce subclinical biological perturbations, including oxidative stress and microbiome modulation, even in the absence of overt toxicological endpoints (Worms et al., 2025). At the same time, the literature remains methodologically heterogeneous. Some studies report no detectable effects on organismal endpoints such as feeding behavior in marine species (Cunningham et al., 2021), whereas others attribute such nonsignificant outcomes to aggregation dynamics, reduced bioavailability, or organism-specific elimination pathways (Koelmans et al., 2022). Rather than constituting direct contradictions, these differences are more consistently interpreted as a sensitivity gap between endpoint-based toxicological design and low-amplitude, chronic perturbation processes that may not be captured by conventional measurement structures.

This limitation becomes more pronounced when exposure is distributed across multiple pathways and temporal scales. Under such conditions, the ERA retains value for regulatory classification but has a reduced resolution for describing interaction-dependent and cumulative system behavior.

11.1 DNPS/SRS as an extension of risk characterization

Within this context, the digital nanoplastic science (DNPS) and systemic risk score (SRS) framework functions as a complementary extension to conventional risk characterization. Its central analytical shift is the interpretation of exposure as a dynamic perturbation signal that propagates through biological networks rather than as a static dose variable.

This formulation aligns with systems biology perspectives in which biological function emerges from network interactions rather than from isolated components (Hossain & Sahajwalla, 2026). In this sense, nanoscale contaminants are treated not as independent stressors but as potential modifiers of signaling integrity across multiple biological levels. The SRS framework therefore moves beyond binary toxicological classification and instead evaluates propagation behavior, network coupling, and system resilience under sustained perturbation.

Conceptually, this approach introduces a structural parallel to closed-loop systems, where feedback mechanisms regulate system stability under dynamic conditions. In biomedical engineering, similar principles are applied in precision-controlled therapeutic systems, where outcomes depend on continuous feedback integration rather than static input–output assumptions (Qin et al., 2026). Within the DNPS/SRS, this analogy is not operational but structural, emphasizing the importance of feedback-sensitive interpretation in environmental risk systems.

11.2 Exposome and Life Cycle Dynamics: Predictive Integration

The exposome framework extends environmental health analysis by emphasizing cumulative lifetime exposure across environmental, dietary, and endogenous sources (Falsafi et al., 2025). DNPS builds upon this perspective by focusing on how exposures translate into functional perturbations at the level of biological signaling, particularly in contexts where endocrine and metabolic regulation may be affected (Tyc et al., 2025).

In parallel, life cycle assessment (LCA) provides a structured methodology for evaluating the environmental burdens associated with material production, use, and disposal phases (ISO 14040; Jiao et al., 2024). However, LCA typically operates at the level of material flow and does not explicitly resolve molecular or cellular response pathways. This limits its capacity to incorporate downstream biological effects arising from nano- and microplastic exposure.

By linking material flow dynamics with signal-level biological responses, DNPS introduces a cross-scale interpretive layer. Within this structure, biological impact is not assumed to scale directly with mass or concentration but is instead conditioned by the physicochemical context, surface properties, and exposure configuration (Sun et al., 2020). This allows for a more differentiated interpretation of nonlinear exposure–response relationships.

11.3 Integrated Positioning and Scientific Trajectory

The role of DNPS/SRS is not to replace established paradigms such as ERA or LCA but to function as a bridging framework that enhances their interpretive resolution under conditions of complexity and uncertainty. From a One Health perspective, this integration is particularly relevant because environmental, animal, and human health outcomes are increasingly recognized as interdependent systems rather than as isolated domains (Danasekaran, 2024; Falsafi et al., 2025).

On this continuum, evidence suggests that nano- and microplastic exposure may be associated with early mitochondrial dysfunction and systemic cardiovascular stress prior to the emergence of clinically detectable pathology (Lerchner et al., 2024; Yong et al., 2020). These effects are not necessarily acute or isolated but may develop through the gradual accumulation of subclinical perturbations.

Although acute toxicity testing remains a central component of current regulatory frameworks, its limitations in capturing cumulative, network-mediated destabilization have been repeatedly acknowledged (Erhirhie et al., 2018). This creates a methodological gap between regulatory detection capacity and the temporal dynamics of environmental exposure systems. DNPS/SRS is positioned within this gap as a structured approach for interpreting distributed and delayed biological responses.

11.4 Future Directions and Methodological Evolution

Despite its integrative orientation, DNPS/SRS requires further empirical validation and computational refinement before operational deployment in regulatory or applied contexts (Thakur et al., 2025). A key priority is the development of reproducible analytical pipelines that link environmental exposure data to biological response indicators in a traceable and standardized manner.

Data quality remains a persistent constraint in micro- and nanoplastic research, particularly with respect to detection limits, sampling variability, and methodological inconsistency (Koelmans et al., 2019). Addressing this limitation requires benchmarking against established toxicological models while systematically quantifying the uncertainty associated with biological heterogeneity and measurement resolution.

Emerging approaches in biomedical and materials science, including ionic liquid-based systems, may provide conceptual analogies for improving interface-level modeling of environmental–biological interactions (Elgharbawy et al., 2026). Within this evolving methodological landscape, DNPS/SRS may contribute to a transition from reactive assessment toward anticipatory system monitoring, where fragmented observations are progressively integrated into structured predictive models across environmental, biological, and health domains.

11.5 Case Study: Integrative assessment of DNPS/SRS versus traditional environmental risk assessment (ERA)

A critical examination of nanoplastic exposure in marine and freshwater systems revealed a persistent gap between endpoint-based toxicological measurements and system-level representations of risk. Empirical data indicate the presence of nano- and microplastics in freshwater systems, including tap water at concentrations of approximately 0.1–3.5 particles/mL, and in fish tissues at approximately 0.05–0.2 mg/kg wet weight (Koelmans et al., 2019; Thakur et al., 2025). These findings support the existence of continuous exposure pathways across environmental compartments.

Bioaccumulation across trophic levels has been demonstrated in multiple studies, indicating plausible dietary exposure routes for humans (Yong et al., 2020). Within this context, micro- and nanoplastics are increasingly conceptualized as part of a broader environmental burden shaped by coexposure, including trace pharmaceuticals and heavy metals, which may interact to amplify oxidative and inflammatory responses (Bastyans et al., 2022; Chabi et al., 2026). However, conventional ERA approaches often evaluate these stressors independently, which may limit their capacity to represent cumulative or network-mediated effects (Cavazzoli et al., 2023).

11.5.1 DNPS/SRS Perspective

Within the DNPS/SRS framework, the analytical emphasis shifts from overt toxicity endpoints toward signal integrity and network-level perturbation. At the molecular scale, nanoplastics may induce signal interference processes (PSIs) and proteostatic stress cascades (PSCs), potentially affecting endocrine and metabolic regulation (Tyc et al., 2025). These processes are consistent with the observed disruptions in gut barrier function, microbiome composition, and metabolic regulation in experimental systems (Chen et al., 2022; Jin et al., 2019).

The detection of microplastics in human stool further confirms their passage through biological systems, indicating internal exposure pathways beyond external contact scenarios (Schwabl et al., 2019). In addition, mitochondrial dysfunction provides a mechanistic basis for potential long-term cardiovascular implications (Lerchner et al., 2024). Within SRS, such dispersed molecular and cellular perturbations are not treated as isolated end points but as interconnected components of a propagating system.

This interpretation enables the identification of latent vulnerabilities, including subclinical immune dysregulation, which may remain undetected under conventional ERA approaches owing to their reliance on discrete toxicity thresholds.

11.5.2 One Health Operationalization and Predictive Governance

The significance of these findings becomes more pronounced when considering trophic transfer systems and human exposure pathways. The One Health framework provides a conceptual basis for integrating environmental, animal, and human health domains into a unified analytical structure (Helal et al., 2023; Jahedi & Jaafarzadeh Haghighi Fard, 2025).

Within this structure, vulnerability is not uniform across populations. For example, infants and older adults may exhibit reduced physiological buffering capacity, increasing susceptibility to cumulative perturbations. This reinforces the need for system-level control strategies analogous to feedback regulation in engineered systems (Qin et al., 2026).

By integrating exposure data with coexposure dynamics, DNPS/SRS enables the transformation of risk assessment into a predictive analytical process. Advances in artificial intelligence-based detection further increase the feasibility of identifying nano- and microplastics in environmental and food systems with improved resolution (Radanliev, 2025; Rawat et al., 2025). Such developments may support earlier intervention strategies, including targeted remediation and population-specific monitoring.

11.5.3 Conclusion and regulatory implications

This case study indicates that while the ERA remains a foundational regulatory tool, it does not fully capture low-dose, multiscale, and network-mediated perturbations. Differences between studies reporting biological effects and those reporting negligible outcomes are more plausibly interpreted as differences in methodological sensitivity rather than substantive contradiction.

The absence of acute toxicity should not be interpreted as evidence of systemic safety. Instead, it may reflect compensated biological states in which adaptive mechanisms mask ongoing perturbations until system thresholds are approached (Jahedi & Jaafarzadeh Haghighi Fard, 2025). As noted by Erhirhie et al. (2018), acute toxicity frameworks remain essential but insufficient for capturing distributed system destabilization.

Within this context, DNPS/SRS provides a structured, integrative framework that links molecular, ecological, and population-level processes into a coherent systemic risk score. Its primary contribution lies in enabling a more continuous representation of environmental risk trajectories, thereby supporting a more operational form of One Health reasoning and addressing known limitations in current environmental assessment systems (Koelmans et al., 2019).

12. METHODOLOGY: INTEGRATED DATA-DRIVEN SYSTEMS FRAMEWORK (DDSF) FOR NANOPLASTIC RISK ASSESSMENT

The Integrated Data-Driven Systems Framework (DDSF) is developed as a multiscale analytical structure intended to operationalize digital nanoplastic science (DNPS), plasto-signaling interference (PSI), and transductive signal load (TSL) within a unified inferential and simulation environment. Rather than separating exposure from biological response as independent domains, the framework treats them as dynamically coupled components embedded within a probabilistic system capable of capturing nonlinear and context-dependent interactions across heterogeneous datasets. The emphasis is therefore not on static endpoints but on relational system behavior and evolving dependencies across scales.

12.1 Multi-Source Data Integration and Study Design

Data integration in DDSF operates across multiple environmental and biological layers. Environmental descriptors include particle concentration, morphology, polymer composition, and surface reactivity, whereas biological indicators include oxidative stress markers, inflammatory mediators, and metabolic dysregulation signatures.

Contextual factors are not positioned as secondary noise variables but are incorporated as structurally relevant components of the exposure architecture. Coexposure to heavy metals and persistent organic pollutants is included because evidence of nonadditive systemic effects has been reported in large cohort analyses (Zhang et al.,

2025). This position exposure assessment more closely reflects interaction-driven configurations than isolated stressor evaluation does.

To reduce interdataset variability, harmonization protocols are implemented to improve comparability across platforms and measurement systems. Such approaches are widely recognized as essential for stabilizing inference in high-dimensional environmental datasets (Croock et al., 2024). In addition, geospatially resolved datasets are integrated to reconstruct exposure fields with spatial and temporal continuity, allowing alignment between environmental distribution patterns and population-level biological outcomes (Cui et al., 2022).

12.2 Transductive signal load (TSL): latent system-level inference

The transductive signal load (TSL) is defined as a latent inferential construct representing accumulated perturbations propagated through biological networks following nanoplastic exposure. It is not directly measurable; rather, it emerges from patterned relationships observed across environmental and biological datasets.

The estimation procedures incorporate environmental modifiers such as pH, salinity, and UV exposure, alongside pathway-level biological disturbances. The evidence indicating that nanoplastics interfere with autophagy and intracellular regulation supports the interpretation that molecular disruption may occur before phenotypic manifestations become detectable (Jiang et al., 2025; Fanghella et al., 2025).

The TSL is therefore positioned as an early system stress indicator. It does not replace conventional toxicological endpoints but operates in a complementary domain where perturbation exists below overt biological thresholds.

12.3 Computational Modeling and Machine Learning Layers

The computational layer integrates machine learning and ensemble modeling to extract latent structures from complex, multidimensional datasets. These methods are used to identify nonlinear exposure–response patterns and classify perturbation signatures that may remain invisible under conventional statistical modeling.

Rather than producing deterministic predictions, the modeling outputs are interpreted as probabilistic relational configurations. This distinction is important because environmental variability, measurement noise, and biological heterogeneity jointly shape uncertainty in system behavior.

Computational logic follows approaches in complex systems science, where high-dimensional parameter spaces are explored to identify stable or metastable regimes (Zhao et al., 2025). The resulting outputs are therefore closer to structured inference fields than to fixed predictive values.

12.4 Systemic risk score (SRS): Construction and interpretive role

The systemic risk score (SRS) is formulated as a composite metric that integrates the exposure intensity (E), signal disruption (S), and system vulnerability (V). Its conceptual structure aligns with the systemic risk approaches used in disaster risk reduction (DRR) and risk-informed development (RID), where risk is understood as an emergent property of interacting system components (Belle et al., 2024).

SRS is not intended to function as a standalone toxicity index. Instead, it operates as a comparative and decision-support construct sensitive to variability in biological and environmental contexts. In applied settings, adaptive factors such as intervention sensitivity may be conceptually linked to advanced delivery systems, including nanofiber-based strategies for localized modulation (Saleh et al., 2026).

The metric therefore maintains flexibility rather than a fixed threshold interpretation, allowing it to adjust across heterogeneous exposure environments.

12.5 Digital Twin (DT) Simulation Framework

A digital twin layer is incorporated to enable in silico representation of environmental–biological interactions under controlled simulation conditions. This component allows exposure scenarios, biological responses, and system trajectories to be explored computationally rather than exclusively through empirical observation.

Digital twin methodologies have increasingly been adopted in complex system modeling for uncertainty-aware simulation and predictive scenario testing (Segovia & Garcia-Alfaro, 2022). In biomedical and environmental domains, such frameworks help bridge measured data with computational inference, enabling preassessment of system responses to hypothetical interventions (Katsoulakis et al., 2024).

Within DDSF, this layer is particularly relevant for exploring nonlinear transitions and potential tipping points under varying exposure configurations (Ringeval et al., 2025).

12.6 Hierarchical multiscale system architecture

The DDSF is structured as a three-tier hierarchical system integrating molecular, environmental, and systemic levels. Each level contributes a distinct analytical function, yet none operates in isolation.

12.6.1 Molecular level (intracellular perturbation layer)

This level captures intracellular processes through omics-based datasets, focusing on molecular pathways affected by nanoplastic exposure. Disruptions in autophagy and regulatory signaling have been reported in nanoplastic–protein interaction studies (Fanghella et al., 2025). Within this structure, Plasto-Signaling Interference (PSI) is operationalized as an early molecular manifestation of perturbation.

12.6.2 Environmental level (contextual exposure layer)

This layer represents external exposure conditions, including particle characteristics, physicochemical properties, and environmental modifiers. Harmonization protocols ensure data interoperability across heterogeneous sources (Croock et al., 2024). Coexposure factors such as heavy metals and persistent organic pollutants are incorporated because of their documented systemic relevance (Zhang et al., 2025). Spatial and temporal reconstruction is enabled through multiscale geospatial datasets (Cui et al., 2022).

12.6.3 Systemic Integration Level (Inference and Synthesis Layer)

At this level, outputs from the molecular and environmental layers are integrated into the systemic risk score (SRS). Risk is treated as an emergent property rather than a linear aggregation of independent variables. This interpretation is consistent with systemic risk frameworks in resilience and disaster modeling (Belle et al., 2024).

Digital twin integration further enables dynamic simulation of system evolution and identification of potential instability points under variable exposure conditions (Segovia & Garcia-Alfaro, 2022; Katsoulakis et al., 2024).

12.7 Cross-Scale Computational Integration and Inferential Flow

Figure 6 (cross-scale architecture of the DDSF and digital twin integration) illustrates the hierarchical coupling between environmental inputs, molecular perturbation pathways, and systemic risk inference outputs. It also highlights the integration between the DDSF and the digital twin simulation layer.

The framework maintains continuous inferential flow across all levels—molecular, environmental, and systemic—allowing localized perturbations to be translated into system-level risk structures without relying on linear aggregation assumptions.

Machine learning components support navigation through high-dimensional exposure–response spaces and assist in identifying latent instability patterns (Zhao et al., 2025). This shifts the analytical emphasis from endpoint-based evaluation toward system-resolved inference, where risk is understood as an emergent property of interacting environmental and biological networks rather than a fixed outcome.

Figure 6. Cross-Scale Architecture of the DDSF and Digital Twin Integration.

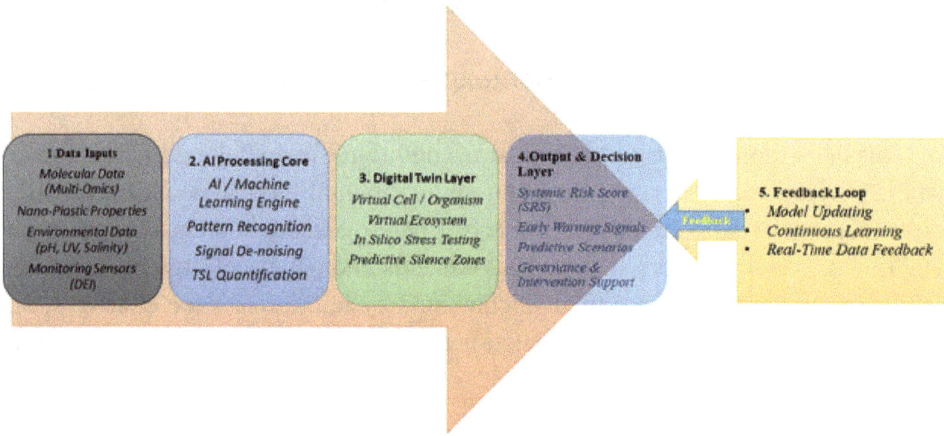

Figure 6 illustrates the DDSF as a multilayered, cross-scale digital ecosystem that integrates molecular, organismal, and environmental data streams. The architecture supports the development of high-fidelity digital twins for scenario-based risk simulation and system-level analysis. By enabling structured data aggregation and multiscale modeling, the framework translates environmental complexity into quantifiable outputs that support decision-making aligned with sustainable development goals (SDGs) and ESG-oriented governance.

13. ANALYTICAL WORKFLOW AND REPRODUCIBILITY STRUCTURE: MODULAR COMPUTATIONAL PIPELINE AND WORKFLOW ARCHITECTURE

The data-driven systems framework (DDSF) is organized as a modular, traceable, and reproducibility-oriented computational pipeline designed to integrate heterogeneous environmental and biological datasets into system-level inferential structures. Rather than treating analysis as a single-step transformation, the workflow is structured as a staged sequence in which each layer refines the representation of exposure–response relationships.

This staged architecture is particularly relevant for nanoplastic systems, where exposure dynamics are inherently nonlinear, multisource, and context dependent. Under such conditions, direct input–output causality is rarely observable, and system behavior is better represented through iterative transformation layers. Accordingly, DDSF is formulated as an interpretive computational architecture rather than a deterministic causal pipeline.

13.1 Data acquisition, harmonization, and standardization

The workflow begins with the systematic acquisition of multisource datasets, including environmental exposure measurements, geospatial distribution records, and molecular or omics-based biological indicators.

Given the structural heterogeneity of these data streams, harmonization is not treated as a secondary preprocessing step but as a foundational requirement for ensuring analytical coherence. Standardization protocols are applied to improve cross-platform comparability and reduce incompatibilities across measurement systems. Such approaches are widely recognized as essential for maintaining statistical robustness in high-dimensional environmental and genomic datasets (Croock et al., 2024).

In parallel, multiscale geospatial datasets are incorporated to align environmental exposure fields with biological response distributions, enabling spatially and temporally resolved reconstruction of exposure landscapes (Cui et al., 2022). At this stage, data are not interpreted causally but are instead structured into interoperable computational formats that enable downstream analytical processing.

13.2 Preprocessing, normalization, and feature extraction

The second stage focuses on data conditioning through preprocessing operations aimed at noise reduction, batch effect correction, and mitigation of measurement artifacts. Normalization procedures are applied to ensure scale consistency across heterogeneous sampling frameworks.

Feature extraction is then performed to isolate analytically relevant descriptors across environmental and biological domains. These include nanoplastic physicochemical properties such as morphology and surface reactivity, as well as biological indicators associated with cellular stress pathways, including autophagy-related mechanisms (Fanghella et al., 2025).

This transformation reduces the dimensional complexity while preserving system-relevant variability. The resulting feature space functions as an intermediate representation of the exposure–response structure, enabling downstream identification of Plasto-Signaling Interference (PSI) patterns and estimation of the transductive signal load (TSL). Both constructs are treated as inferential outputs derived from system behavior rather than directly measurable variables.

13.3 Pattern Recognition and Computational Modeling Layer

The third stage introduces a machine learning–based modeling framework designed to identify latent structures within high-dimensional exposure–response datasets. Ensemble learning and nonlinear modeling approaches are applied to capture interaction effects that are not detectable through linear statistical methods.

This modeling strategy aligns with complex system methodologies used in high-dimensional parameter exploration and system optimization (Zhao et al., 2025). Within DDSF, model outputs are interpreted as probabilistic relational structures describing associations within the system rather than deterministic predictions.

In environmental health applications, such latent configurations may correspond to early or subclinical perturbation states that precede observable outcomes but remain below the detection sensitivity of conventional endpoint-based models (Zhang et al., 2025). Consequently, interpretation focuses on structural patterns rather than causal attribution.

13.4 Digital Twin (DT) Implementation and Simulation Layer

At this stage, a digital twin (DT) environment is introduced as a computational representation of coupled environmental–biological interactions (Durden, 2025). The DT layer enables bidirectional simulation, linking empirical observations with modeled system responses under varying exposure conditions.

This configuration supports controlled in silico experimentation for exploring system behavior under hypothetical perturbation scenarios. Particular emphasis is placed on identifying instability regions and potential tipping-point dynamics that may emerge under specific exposure configurations (Garske et al., 2024).

Digital twin methodologies are increasingly used in environmental and biomedical modeling for predictive simulation and uncertainty-informed scenario analysis (Katsoulakis et al., 2024; Ringeval et al., 2025). Within DDSF, however, this layer is positioned as a hypothesis-generation and exploration environment rather than a fully validated replica of biological systems.

13.5 Systemic Integration and Systemic Risk Score (SRS) Computing

The final stage integrates outputs from all preceding modules into the systemic risk score (SRS), a composite construct defined by the interaction between exposure intensity, signal perturbation magnitude, and system vulnerability parameters. This formulation is consistent with the systemic risk frameworks in the resilience and disaster risk reduction literature (Belle et al., 2024).

Within the DDSF, the SRS is not conceptualized as an absolute toxicity metric. Instead, it functions as a comparative, context-sensitive indicator intended to support structured decision-making under uncertainty. Its primary role is to enable cross-scenario comparisons rather than to enforce fixed threshold classifications.

To enhance interpretability across disciplines, visualization strategies such as cinematic rendering of multiscale scientific data (e.g., the SmARTR pipeline) may be applied (Macrì & Di-Poï, 2024). These representations preserve structural relationships across scales without implying mechanistic completeness.

13.6 Reproducibility, Traceability, and One Health Integration

The DDSF is explicitly designed around reproducibility principles, including modular execution, version-controlled workflows, and detailed documentation of data transformations across all analytical stages. Each step maintains traceable links between raw inputs, intermediate outputs, and final system-level representations, ensuring auditability and independent verification.

This structure supports alignment with One Health frameworks by linking molecular-level perturbation signals, including those associated with nanofiber-based delivery interactions (Saleh et al., 2026), to environmental exposure patterns and population-level outcomes. These relationships are interpreted as integrative cross-domain associations rather than direct causal pathways.

Overall, the workflow is intended as a reproducible and extensible analytical architecture supporting exploratory and anticipatory environmental health assessment under conditions of uncertainty and heterogeneous data quality.

13.7 Schematic representation

A schematic overview of the DDSF modular workflow, illustrating sequential transitions from data acquisition through SRS computation within the digital twin environment, is presented in **Figure 7**. The figure highlights the layered transformation of heterogeneous datasets into structured systemic risk representations, emphasizing the interaction between preprocessing, modeling, and simulation components.

Figure 7. Cross-Scale DDSF Architecture Linking Data Integration, Systemic Risk Modeling, and Adaptive Governance.

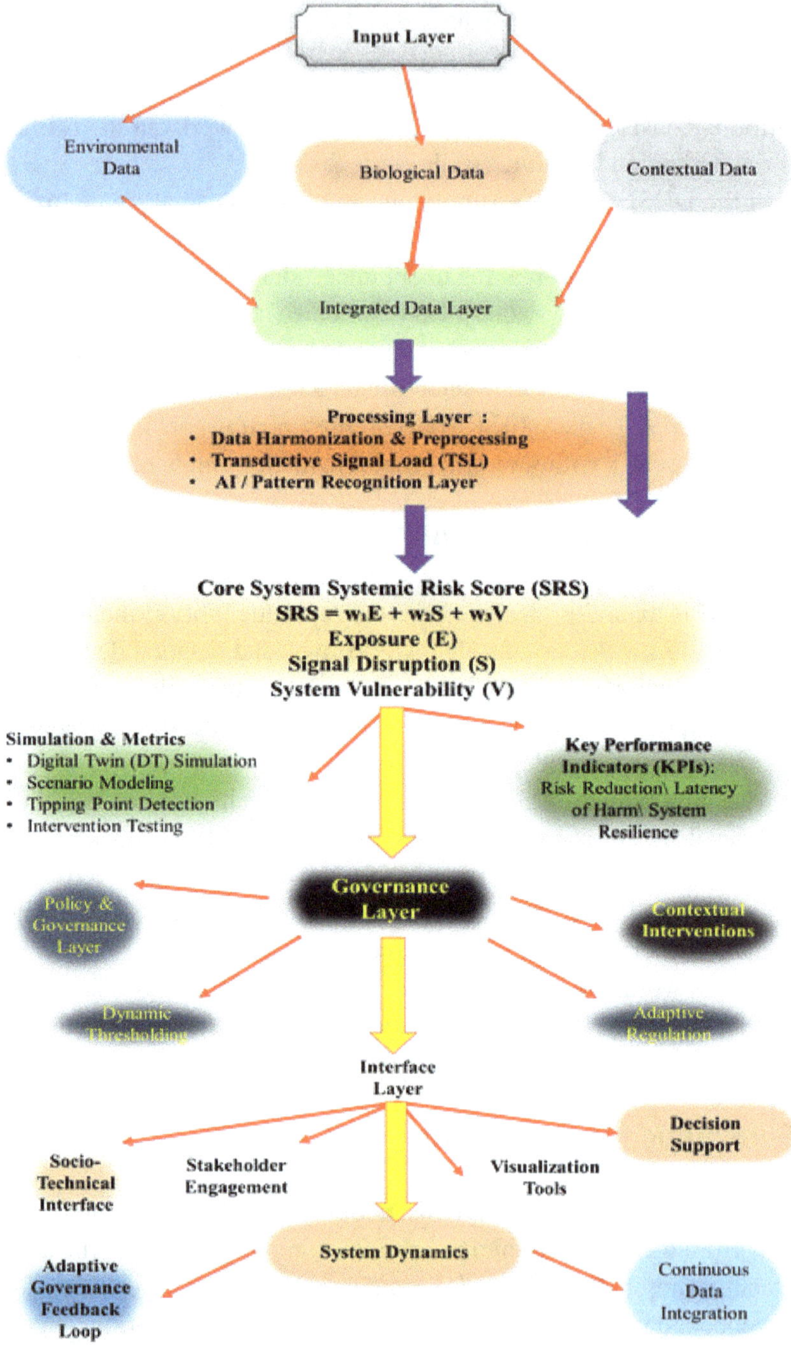

Figure 7 illustrates the integration of environmental and biological data within the DDSF framework to generate the systemic risk score (SRS), which is translated into policy-relevant indicators and supports adaptive decision-making through a continuous feedback loop.

13.8 Uncertainty and Interpretation Framework

Uncertainty is not treated here as a secondary limitation. It behaves more like a structural condition of the system itself, especially when environmental and biological datasets are high-dimensional and only partially observable. Under these circumstances, moving away from deterministic prediction toward probabilistic inference is not a methodological preference but a practical necessity embedded in the analytical design. Chai et al. (2024) support this direction, particularly in cases where the ground truth is delayed, incomplete, or only partially accessible in complex biological response spaces.

Interpretability is therefore expressed through bounded uncertainty intervals rather than single-point estimates. Nevertheless, these intervals do not stand on their own. Their meaning shifts depending on data completeness, feature construction, and model specification. They remain conditional representations, not causal statements about biological outcomes (Ragusa & Fanos, 2025).

There is also conceptual borrowing from high-dimensional materials modeling, where system stability is not derived from a single optimized solution but from patterns emerging across parameter spaces (Zhao et al., 2025). The analogy is useful, although it should not be stretched too far. Compared with engineered material systems, biological exposure systems behave differently, with more feedback uncertainty and less structural regularity. Any transfer of inference between these domains therefore remains provisional and must be empirically grounded.

The measurement quality continues to anchor the entire framework. Improvements in micro- and nanoplastic detection technologies (Wang et al., 2026) have increased the environmental resolution, yet they do not resolve deeper uncertainties tied to spatial heterogeneity, sampling bias, or temporal discontinuity. In practice, observational inputs retain partial views of a larger system rather than complete reconstructions.

A separation is therefore maintained between exposure quantification, perturbation inference, and outcome uncertainty. This separation is not cosmetic. Without this information, higher-resolution data could easily be mistaken for higher predictive certainty, which is not supported in complex biological systems.

14. ETHICAL RESILIENCE AND GOVERNANCE CONSTRAINTS IN DDSF-BASED SYSTEMS

Introducing AI-enabled digital twins into DDSF changes more than the computational layer does; it introduces a governance structure that must remain explicitly constrained. The in silico first paradigm (Li et al., 2025) is useful as a prioritization strategy for simulation and hypothesis filtering, although its role is clearly preexperimental rather than substitutive for empirical validation.

From a transparency perspective, an auditable SRS trace can improve traceability by allowing reconstruction of computational pathways (Khanam et al., 2025). However, traceability does not automatically translate into validity. The stability of any inference still depends on the assumptions embedded in the model, the representativeness of the data, and the structural uncertainties that cannot be fully eliminated (Bhuller et al., 2025).

For this reason, DDSF outputs—including SRS mappings—should be interpreted as structured decision-support artifacts under uncertainty. They help organize complexity and enable comparisons across scenarios, but they do not themselves define regulatory truth or enforce intervention logic.

Tension appears most clearly when multisource datasets converge toward high-risk configurations. These should be handled cautiously. They are better interpreted as hypothesis-generating signals rather than confirmed projections of harm. Pushing them into deterministic policy interpretation would exceed the methodological boundaries of the framework (Korenhof et al., 2025).

A similar caution applies to vulnerability stratification. Categories such as infants, older adults, or immunocompromised populations are treated here as sensitivity-informed hypotheses rather than fixed biological partitions. The validity of these methods depends on longitudinal confirmation and calibration stability (Bel Hassen et al., 2025).

Ultimately, the interpretive limits of DDSF are defined by a combination of incomplete data, error propagation, model structure, and the absence of full biological ground truth observability. The framework therefore operates as a computational inference environment under uncertainty, not as a predictive authority system.

15. OPERATIONALIZING ONE HEALTH THROUGH DNPS AND SRS: TOWARD ADAPTIVE POLICY ARCHITECTURES

15.1 From Conceptual Framing to Analytical Operationalization

One Health is widely accepted as a conceptual bridge across human, animal, and environmental health domains. Translating that idea into operational analytics, however, remains uneven. DNPS and SRS are positioned here as an interpretive layer rather than a decision engine—one that helps structure multidomain interactions without assuming deterministic control over outcomes.

In practice, DNPS and SRS function as linked analytical components. When Plasto-Signaling Interference (PSI) indicators are combined with Dynamic Eco-Integrator (DEI) data streams, plausible pathways of environmental perturbation across compartments and trophic levels can be reconstructed (F. Dang et al., 2022). These reconstructions remain probabilistic and should be read as scenario-consistent structures rather than validated causal chains (Kochanek et al., 2025; Leslie et al., 2022).

15.2 Capturing Latent Exposure Dynamics and Systemic Vulnerability

Standard environmental monitoring is still largely organized around acute exposure thresholds and endpoint toxicity. Nanoplastic systems do not fit neatly into that structure. Their exposure is continuous, low-dose, and often entangled with other contaminants, which makes linear interpretation fragile.

In this setting, DNPS–SRS operates more as a signal-representation framework. It highlights latent or subthreshold perturbations that may not appear in conventional endpoint-based analyses. These signals should not be read as outcomes themselves but as deviations in system-level structure under specific analytical conditions (Lee & Kim, 2026).

Systemic vulnerability is a modifying condition rather than a fixed attribute. It changes how strongly signals propagate through biological and ecological systems. However, any interpretation of vulnerability remains dependent on data quality, calibration constraints, and structural assumptions in the modeling process (Boctor et al., 2025). No deterministic inference about population-level effects follows from this.

15.3 Digital Twin Simulation and Preimplementation Assessment

A functional extension of the DNPS framework lies in its compatibility with digital twin (DT) environments, which enable computational representation of coupled environmental–biological systems under variable parameter conditions (Amer et al., 2026). Within this structure, simulation outputs are used for **comparative scenario analysis** rather than predictive forecasting.

For example, hypothetical bioremediation interventions in marine environments can be evaluated in silico to explore potential system responses across ecological compartments, including trophic transfer and particle redistribution pathways. Such simulations are intended to support the structured exploration of system sensitivity under defined assumptions, not to generate deterministic predictions of environmental outcomes (Boctor et al., 2025).

Accordingly, digital twin outputs should be interpreted as **model-conditioned exploratory representations**, supporting hypothesis generation and comparative assessment rather than direct policy prescription.

15.4 Alignment with Sustainable Development Goals: Conditional and Model Bounded Relevance

The relationship between DNPS–SRS and the Sustainable Development Goals (SDGs) is best understood as conditional rather than automatic. Any alignment depends on domain-specific calibration, empirical validation, and the stability of the underlying model assumptions.

15.4.1 SDG 3: Early-stage biological perturbation indicators

SRS-derived indicators may help identify early molecular-level deviations in biological systems. Even so, these signals should not be interpreted as diagnostic outputs. Their relevance is probabilistic and model-bound, indicating possible perturbations under defined analytical conditions rather than confirmed biological harm (Shao et al., 2025).

15.4.2 SDG 12: Extended Lifecycle Interpretation

The framework may extend conventional life cycle assessment by incorporating biological response dimensions alongside material flow analysis. This extension remains conceptual at this stage. Operational use in regulatory settings would require further empirical validation and clearer translation across methodological domains (Kochanek et al., 2025).

15.4.3 SDGs 14 and 15: Ecological sensitivity mapping

Model outputs may assist in identifying ecosystems that appear more sensitive to multistressor exposure. These outputs should be read as comparative gradients of risk, not as absolute ecological classifications (Shi et al., 2024; Malik et al., 2026).

15.5 Adaptive Governance and the Transition Beyond Static Regulatory Models

Conventional environmental governance still relies heavily on static thresholds and retrospective evaluation. DNPS–SRS introduces a different temporal logic. It allows system states to be updated iteratively as new data become available, which makes it more suitable for evolving exposure environments.

This does not imply automated policy-making. Rather, it supports adaptive review processes in which regulatory judgments can be revisited as evidence accumulates. Digital twin environments strengthen this capacity by enabling scenario-based exploration of system trajectories under different assumptions.

That said, all outputs remain constrained by model uncertainty, data limitations, and structural assumptions. DNPS–SRS therefore functions as a decision-support augmentation layer, not a substitute for established regulatory frameworks (OECD, 2025).

15.6 Interpretability, Uncertainty Communication, and Stakeholder Accessibility

As environmental–biological datasets become more complex, interpretability becomes a practical challenge rather than a purely technical challenge. AI-supported visualization and summarization tools may help translate multivariable outputs into forms that are more accessible to interdisciplinary stakeholders.

These tools should be treated as translation mechanisms, not as interpretive authorities. Their purpose is to support a structured understanding of trade-offs across environmental, economic, and social dimensions, especially in contexts where exposure inequality is present but not directly encoded in governance systems (European Commission, 2025; Dewika et al., 2024).

All the interpretive outputs remain conditional on the model assumptions and uncertainty bounds. They should not be read as normative policy guidance.

15.7 Synthesis: Toward Structured Environmental Intelligence under Uncertainty

Taken together, DNPS–SRS is best positioned as a structured environmental intelligence system under uncertainty, not as a deterministic governance instrument.

Its contribution lies in three linked functions: cross-scale representation of exposure systems, structured interpretation of latent perturbation signals, and comparative evaluation across scenarios under model constraints.

It does not remove uncertainty, and it does not settle causal attribution in complex environmental systems. Uncertainty is formalized as part of the analytic structure itself, in line with contemporary systems-based environmental assessment. Its scientific value is therefore epistemic rather than prescriptive: it supports hypothesis generation, comparative assessment, and adaptive interpretation without claiming to replace regulatory judgment (Bhattacharyya et al., 2025).

16. SOCIO-TECHNICAL SCALABILITY: BRIDGING THE KINETIC–POLICY INTERFACE

16.1 Embedding the DDSF within the Political Economy of Environmental Health

The practical relevance of DDSF depends on how it is embedded within existing sociopolitical and economic infrastructures. The analytical strength alone is not sufficient. Model outputs still require institutional pathways for interpretation, validation, and context-sensitive adaptation before they can inform policy in any meaningful way.

This creates a kinetic–policy interface: a boundary space between rapidly evolving environmental intelligence systems and institutional decision cycles that typically move more slowly. The main challenge here is not simple translation but compatibility across temporal scales, evidentiary standards, and interpretive frameworks (Kochanek et al., 2025).

16.2 Economic Rationality and the Cost–Inaction Structure

Environmental governance often operates within delayed-response economic structures, where intervention follows visible damage rather than anticipatory signaling. That lag is not merely procedural. In some settings, it can shift costs upward over time, although the magnitude of that effect remains context-dependent and cannot be assumed in every case.

In this setting, the systemic risk score (SRS) may be read as an early informational construct that combines multisource exposure indicators into a comparative risk representation. At minimum, such a representation can support earlier recognition of possible environmental burden accumulation associated with nano- and microplastic exposure (Leslie et al., 2022).

Even so, any economic interpretation of SRS outputs remains nondeterministic. It depends on the valuation model used, the discounting assumptions adopted, and the way uncertainty is propagated through long-horizon forecasting. These dependencies matter, and they limit how far the analysis can be pushed.

16.3 Regulatory agility and adaptive threshold concepts

Traditional environmental regulations are usually built around relatively stable contaminant profiles and simplified dose–response assumptions. Nanoplastic systems do not sit comfortably inside that design. Their behavior changes with the environmental context, coexposure conditions, and material transformation processes.

Under these conditions, the SRS may help frame adaptive thresholding as a conceptual possibility, not a fixed governance prescription. The idea is straightforward: regulatory sensitivity could be recalibrated periodically in response to newly observed environmental signals. This does not imply a ready-made policy model. It is better understood as an analytical direction.

This matters because the broader policy problem is not only about detecting hazards. It is also about narrowing the delay between scientific evidence production and regulatory updating cycles (OECD, 2022; Boakye et al., 2025). The innovation–regulation temporal gap remains a practical constraint, and any framework that ignores it will remain incomplete.

16.4 Closed-Loop Governance as a Conceptual Feedback Structure

The relationships among environmental data streams, risk modeling outputs, and institutional decision processes can be represented as a closed-loop analytical structure. In this formulation, feedback is not linear. It moves across stages, often unevenly.

As illustrated conceptually in Figure 8, the loop may be described in four linked steps: environmental monitoring contributes input data to SRS estimation; SRS outputs inform comparative risk interpretation; institutional responses modify environmental conditions; and those updated conditions re-enter the analytical cycle.

This should not be mistaken for automated governance. It is a systems-level abstraction of iterative interactions between knowledge generation and institutional response. The emphasis is on adaptation under uncertainty, not on deterministic control.

Figure 8. Closed-Loop Governance Framework Linking Scientific Input to Policy Action

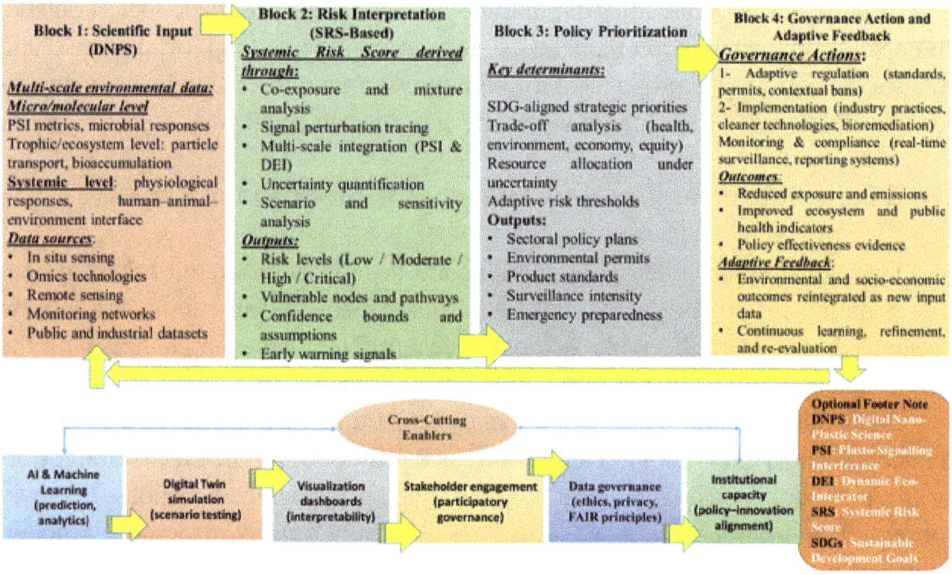

Figure 8 illustrates a closed-loop governance framework linking scientific input, risk interpretation, and policy prioritization to regulatory action. Complex and uncertain environmental data are translated into measurable risk signals, which inform strategic decision-making and resource allocation. These decisions are operationalized through governance actions, including regulation, standard setting, and monitoring. Adaptive feedback mechanisms continuously update the system, enabling responsive and evidence-based environmental management

16.5 Interpretability, Trade-Off Structures, and Decision Complexity

The practical use of DDSF-derived outputs in governance depends heavily on interpretability and on how uncertainty is communicated. AI-supported visualization and summarization tools may help translate high-dimensional model outputs into comparative forms that are easier for nonspecialist stakeholders to follow.

Nevertheless, those tools do not eliminate the underlying normative problem. Environmental decisions always involve trade-offs: economic activity, ecological preservation, and social equity do not collapse neatly into a single computational score. Model outputs can support deliberation, but they cannot replace it (European Commission, 2025; Dewika et al., 2024).

This distinction is important. DDSF may structure the terms of discussion; however, it does not resolve the value conflicts embedded in environmental governance.

16.6 Concluding Perspective on Socio-Technical Scalability

Sociotechnical scalability is not defined here by computational expansion alone. It also depends on whether the analytical system can align with institutional capacity, interpretive legitimacy, and governance adaptability.

Thus, the DNPS–SRS–DDSF integration functions as a conceptual architecture for structured environmental inference under uncertainty. It supports anticipatory analytical awareness of complex exposure systems, but it does not claim predictive certainty or direct policy automation.

Its role is better understood as an epistemic support layer for environmental decision contexts. This role remains bounded by uncertainty, contextual dependence, and the limits of model-based inference.

17. FROM NANOPLASTIC MONITORING TO INTEGRATED ONE HEALTH INTELLIGENCE ARCHITECTURES

This chapter serves as a structural entry point for the chapters that follow. It is not intended as a standalone synthesis. Its function is to place nanoplastic research within a broader systems-oriented environmental science agenda shaped by heterogeneous datasets, multicompartment transport, and biological interactions that unfold across scales.

Nanoplastics are treated here as particulate material entities moving through environmental transport and transformation systems. No intrinsic informational or functional agency is assigned to them. Any informational meaning arises only at the level of model-based abstraction, which keeps the analysis aligned with established environmental systems science (Dewika et al., 2024).

This distinction matters because the field is moving from descriptive monitoring toward integrative environmental intelligence. Conventional monitoring remains indispensable for surveillance and baseline assessment, but it is often compartment-specific and snapshot-based. As a result, it may not fully resolve cumulative exposure patterns or multistressor interactions that evolve over time (Kochanek et al., 2025).

The digital nanoplastic science (DNPS) framework is introduced here as a conceptual integration layer. Its purpose is to connect environmental measurement, transport modeling, and systems-level interpretation within a single analytical structure. Associated constructs such as Plasto-Signaling Interference (PSI) should be read strictly as hypothesis-generating descriptors, which are useful for organizing possible interaction pathways under explicitly stated modeling assumptions (Amer et al., 2026).

These constructs are not validated mechanistic explanations. Their scientific value depends on empirical verification, sensitivity analysis, and clear boundary specification, especially in settings shaped by coexposure to multiple environmental stressors.

From a governance perspective, the systemic risk score (SRS) functions as a comparative interpretive metric. It is designed to organize multidimensional environmental indicators, not to act as a deterministic prediction instrument. Any use of such a composite measure in decision-support settings requires explicit attention to uncertainty propagation, parameter dependence, and data completeness constraints (Boakye et al., 2025).

Predictive environmental intelligence, in this sense, is better understood as a long-term methodological direction than as an immediate operational condition. Its development depends on advances in sensing resolution, data standardization, interoperable infrastructures, and validated multiscale modeling frameworks.

Each chapter that follows contributes to this layered system representation of environmental processes. DNPS functions as an organizing scaffold that supports cross-domain synthesis while preserving the epistemic limits of each domain.

The central methodological aim is therefore not prediction in a strict sense. It is the structured representation of vulnerability patterns under uncertainty, with careful attention given to scale dependence, nonlinearity, and data limitations (Khanam et al., 2025).

17.1 Emerging Horizons: Nanoplastic–Polybiome Interfaces and AMR Dynamics within One Health Intelligence Systems

Within this broader systems perspective, increasing attention is being directed toward biologically mediated interfaces linking nanoplastic exposure, microbial ecology, and environmental health vulnerability. Emerging research suggests that micro- and nanoplastics may participate in complex interactions extending beyond conventional toxicological interpretation, particularly in relation to polybiome architecture, airborne particulate transport, and antimicrobial resistance (AMR)-associated dynamics operating across interconnected One Health systems (Moulton et al., 2026; Piergiacomo et al., 2022; Procopio et al., 2025). Recent studies increasingly recognize that nanoplastic exposure does not occur in isolation, as environmental particles continuously interact with heterogeneous microbial communities distributed across soil systems, aquatic environments, atmospheric particulate matter, food matrices, and host-associated microbial ecologies (Bhat et al., 2023; Liu et al., 2023; Liu et al., 2025).

Within this context, the concept of polybiome systems becomes relevant as a systems-oriented framework describing interconnected microbial networks linking environmental, animal, and human biological compartments (Berg et al., 2020; Procopio et al., 2025). Rather than interpreting microbiomes as isolated host entities, this perspective considers them dynamically interacting ecological systems influenced by multisource environmental perturbations, including nanoplastic exposure pathways (Garcia Rodriguez et al., 2026; Valachovičová & Mišľanová, 2026). From a polybiome systems perspective, nanoplastics may contribute to alterations in microbial composition, metabolic signaling, ecological resilience, and interspecies communication through mechanisms involving oxidative stress, membrane interactions, adsorption of environmental contaminants, and changes in localized nutrient and signaling microenvironments (Jarmakiewicz-Czaja et al., 2026; Procopio et al., 2025; Valachovičová & Mišľanová, 2026). Emerging evidence further suggests that particulate surfaces may influence biofilm-associated microbial organization and microbial interaction pathways operating across environmental interfaces (Kazmi et al., 2026; Valachovičová & Mišľanová, 2026). Although many mechanistic relationships remain incompletely characterized, these observations raise important questions regarding microbial stability, ecological adaptation, and chronic low-dose environmental exposure dynamics across the lifespan (Moulton et al., 2026).

Within One Health systems, these concerns become particularly significant because environmental microbial perturbations cannot be fully separated from food safety, ecosystem function, animal health, and human biological vulnerability (Piergiacomo et al., 2022; Procopio et al., 2025). Environmental compartments continuously exchange particulate materials, microorganisms, and chemical stressors through atmospheric circulation, hydrological transport, agricultural systems, wastewater pathways, and food-chain connectivity (Bhat et al., 2023; Jaafarzadeh & Talepour, 2023; Visca et al., 2024). As a result, nanoplastic-associated microbial interactions may propagate across multiple environmental and biological interfaces simultaneously rather than remaining confined to a single exposure compartment (Kazmi et al., 2026; Piergiacomo et al., 2022).

Increasing scientific attention has also focused on the potential relationship between micro- and nanoplastics and AMR-associated processes. Current evidence does not support simplistic causal interpretations; however, several studies suggest that particulate surfaces may contribute to environmental conditions relevant to resistance emergence, persistence, and dissemination (Kang et al., 2025; Kazmi et al., 2026). Proposed mechanisms include facilitation of biofilm formation, enhancement of microbial aggregation, alteration of horizontal gene transfer dynamics, and creation of localized stress environments capable of influencing microbial adaptation pathways (Kang et al., 2025; Liu et al., 2023; Liu et al., 2025). These observations remain biologically plausible rather than conclusively established across all environmental systems; nevertheless, they represent an important emerging frontier within environmental microbiology, microbial ecology, and One Health systems science (Jaafarzadeh & Talepour, 2023; Y. Liu et al., 2025; Piergiacomo et al., 2022).

The airborne dimension of nanoplastic exposure further expands the complexity of these interactions. Airborne particulate systems containing nanoscale plastic particles are increasingly detected within indoor dust, urban aerosols, hospital laundry environments, occupational environments, and atmospheric transport systems (Bhat et al., 2023; Rangrazi et al., 2026; Uchmanowicz et al., 2025). Inhalation therefore represents a potentially important, but incompletely characterized, exposure pathway linking environmental particulate transport to respiratory and gastrointestinal biological interfaces (Gou et al., 2024; Uchmanowicz et al., 2025). Once inhaled, particles may interact with respiratory mucosal systems, inflammatory pathways, and microbial communities associated with pulmonary and upper-airway environments (Gou et al., 2024). Secondary redistribution through mucociliary transport may additionally influence gastrointestinal exposure patterns, potentially influencing dysbiosis-related processes and broader neurobiological stress pathways associated with ingestion exposure. These interactions may contribute to interconnected respiratory–gut exposure axes relevant to broader polybiome system dynamics (Bhattacharyya et al., 2025; Garcia Rodriguez et al., 2026; Jarmakiewicz-Czaja et al., 2026).

These airborne pathways are particularly important within One Health interpretation because atmospheric transport connects geographically separated environmental systems (Bhat et al., 2023; Gou et al., 2024). Nanoplastics suspended in air masses may participate in long-range environmental dissemination processes linking industrial emissions, urban particulate systems, agricultural environments, aquatic deposition, and human exposure zones (Bhat et al., 2023; Uchmanowicz et al., 2025; Visca et al., 2024). Such interactions complicate traditional compartment-specific monitoring approaches and reinforce the need for multiscale environmental intelligence architectures capable of integrating atmospheric, biological, microbial, and ecological datasets simultaneously.

From a cyber-biotechnology perspective, future environmental intelligence systems may increasingly depend on the convergence of biosensing technologies, AI-assisted analytics, environmental informatics, nature-based mitigation frameworks, and systems-level exposure interpretation (Reyed, 2026d; Visca et al., 2024). The objective of such frameworks is not deterministic prediction of disease outcomes, but rather the structured interpretation of vulnerability patterns emerging across interconnected environmental and biological systems. Within this context, nanoplastic monitoring may evolve from isolated concentration measurements toward integrated interpretation platforms incorporating particulate transport dynamics, polybiome restructuring signals, AMR-associated indicators, inflammatory biomarkers, and environmental coexposure variables (Kazmi et al., 2026; Piergiacomo et al., 2022).

The development of these systems, however, requires substantial methodological caution. Current limitations include inconsistent sampling methodologies, limited nanoscale detection standardization, insufficient longitudinal datasets, incomplete mechanistic validation, and challenges associated with multiscale data integration (Uchmanowicz et al., 2025). Consequently, predictive environmental intelligence should be interpreted as a long-term scientific direction rather than an already established operational capability. Careful distinction must therefore be maintained between exploratory systems modeling and empirically validated mechanistic evidence.

Nevertheless, the convergence of polybiome science, environmental sensing, artificial intelligence, and One Health systems research may represent an important future direction for nanoplastic science (Procopio et al., 2025; Reyed, 2026d). Integrative cyber-biotechnology architectures capable of combining environmental exposure data with microbial ecology, biological signaling networks, AMR-associated indicators, and ecosystem-level vulnerability profiles may ultimately improve understanding of how chronic, low-intensity environmental perturbations propagate across interconnected planetary health systems (Kazmi et al., 2026; Piergiacomo et al., 2022).

17.2 Polybiome Signaling Perturbation and Interfacial Proteostasis Dynamics

Beyond conventional toxicological interpretation, increasing attention is now being directed toward whether nanoscale particulate interfaces may subtly influence the organizational dynamics through which microbial and host-associated systems coordinate signaling, metabolic exchange, and adaptive stress responses across interconnected biological environments (Dennis & Kumar Mistri, 2026; Gonçalves et al., 2026). Within this perspective, nanoplastic exposure is not interpreted solely as a concentration-dependent toxic burden, but also as a potential source of physicochemical interference operating at highly dynamic environmental and cellular interfaces (Karim et al., 2024). These interactions remain incompletely characterized; however, they may become increasingly relevant when considered within polybiome systems composed of continuously interacting microbial, environmental, and host-associated ecological networks (Sessitsch et al., 2023).

Within complex microbial ecosystems, communication processes extend far beyond compositional diversity alone. Microbial communities continuously exchange metabolites, quorum-associated signaling molecules, extracellular vesicles, redox-responsive mediators, and stress-adaptive biochemical cues that collectively contribute to ecological synchronization and communication stability across environmental compartments (Peng et al., 2025). Increasing evidence suggests that these interactions operate through distributed signaling architectures linking soil systems, aquatic environments, atmospheric particulate interfaces, food-associated microbial ecologies, and host-associated microbiota rather than through isolated microbiome compartments alone (Sessitsch et al., 2023). Consequently, perturbations affecting localized signaling conditions may theoretically influence broader patterns of microbial coordination, ecological resilience, and adaptive system organization even in the absence of overt cytotoxicity (Peng et al., 2025).

From this perspective, the biological relevance of nanoplastics may extend beyond particle burden itself. Nanoscale particulate surfaces possess adsorption-dependent physicochemical properties capable of influencing localized diffusion gradients, extracellular polymeric organization, membrane-associated signaling behavior, and metabolite exchange microenvironments under specific environmental conditions (Karim et al., 2024; Ma et al., 2024). Such interactions may become particularly important at densely organized biological interfaces where communication efficiency depends on spatial organization, transient molecular assembly, and continuous biochemical exchange. Although many of these relationships remain experimentally unresolved, increasing attention is being directed toward the possibility that chronic low-intensity particulate exposure could contribute to subtle forms of signaling instability operating across environmental–biological boundaries rather than through direct destructive toxicity alone (Moulton et al., 2026).

Particular interest has emerged around extracellular vesicle-mediated communication pathways because vesicle-associated systems increasingly appear central to microbial coordination, host–microbial interaction, inflammatory modulation, and adaptive ecological signaling processes (Deng et al., 2026; Marik et al., 2024). Recent experimental observations further suggest that nanoplastic exposure may alter bacteria–host interaction dynamics through extracellular vesicle-delivered microRNA signaling pathways within intestinal environments (Hsu et al., 2025). These findings remain context-dependent and should not be overinterpreted mechanistically; however, they reinforce the possibility that nanoscale particulate systems may interact with communication-sensitive biological architectures extending simultaneously across microbial and host-associated systems (Deng et al., 2026; Marik et al., 2024).

At the interfacial level, increasing attention is also being directed toward how nanoplastic-associated biomechanical effects may influence localized physicochemical environments rather than acting exclusively through classical toxicological pathways (Dennis & Kumar Mistri, 2026). Nanoplastic particles possess surface characteristics capable of interacting with membrane curvature dynamics, hydration layers, extracellular polymeric substances, and diffusion-sensitive biochemical interfaces under specific environmental conditions (Karim et al., 2024). Within densely interconnected microbial ecosystems such as mucosal interfaces and biofilm-associated systems, even modest alterations in physicochemical boundary conditions may theoretically contribute to signaling desynchronization, altered metabolite distribution, or adaptive restructuring of ecological interactions (Sessitsch et al., 2023). These effects are likely to remain highly context dependent and may vary according to particle composition, environmental chemistry, microbial organization, and exposure duration, prompting increasing interest in protective and interface-stabilizing biomaterial strategies (Caporalini et al., 2025; Gul et al., 2021).

Importantly, this interpretation does not imply that nanoplastics directly induce deterministic protein destruction or universal pathological protein misfolding. Current evidence remains insufficient to support such conclusions. Rather, increasing attention is now being directed toward whether persistent interfacial stress conditions may contribute indirectly to adaptive proteostatic burden within signaling-active biological systems (Díaz-Villanueva et al., 2015). Protein homeostasis operates within highly regulated intracellular and extracellular environments shaped by molecular crowding conditions, membrane-associated transport systems, oxidative balance, chaperone activity, vesicle trafficking, and energy-dependent quality-control pathways (Vergani-Junior et al., 2026). The structural adaptability of these systems is closely associated with dynamic protein-folding landscapes capable of responding to fluctuating environmental conditions (Porter, 2023). Consequently, chronic low-level perturbation of these physicochemical microenvironments, including dynamic protein-corona formation at nano–bio interfaces, may theoretically influence signaling-associated folding stability and adaptive stress coordination without necessarily producing irreversible structural protein damage (Dar et al., 2025; Karim et al., 2024).

Within this framework, concepts such as adaptive proteostatic compensation, signaling-associated folding stress, biochemical crowding, and transient proteostasis instability become more relevant than simplistic particle-toxicity interpretations alone (Díaz-Villanueva et al., 2015; Vergani-Junior et al., 2026). Cells continuously engage compensatory mechanisms involving molecular chaperones, unfolded protein response pathways, autophagic regulation, vesicle-mediated transport systems, and stress-adaptive signaling cascades in order to maintain communication fidelity under fluctuating environmental conditions (Clarke et al., 2011). In parallel, increasing attention is being directed toward how biomolecular condensates and liquid–liquid phase separation (LLPS)-associated organizational dynamics contribute to intracellular signaling coordination and adaptive stress buffering under physicochemically unstable conditions (M. Dang et al., 2026). Nanoplastic-associated interfaces may therefore be more appropriately interpreted as contributors to environmental stress complexity and adaptive systems burden rather than isolated mechanistic determinants of disease processes. This distinction remains important because biological resilience, buffering capacity, and ecological adaptation across the lifespan are likely to shape exposure outcomes alongside any direct physicochemical interactions (Moulton et al., 2026).

Within broader polybiome systems, these interactions may extend beyond isolated host compartments and begin influencing communication-sensitive ecological networks operating across atmospheric, aquatic, agricultural, wastewater, food-associated, and host-associated environments simultaneously (Sessitsch et al., 2023). Environmental microbial systems continuously exchange signaling molecules, metabolites, vesicle-associated mediators, and adaptive stress signals through interconnected ecological pathways shaped by atmospheric circulation, hydrological transport, nutrient exchange, and multispecies environmental interaction. As a result, perturbation-sensitive signaling dynamics occurring within one ecological interface may theoretically propagate across broader distributed communication systems rather than remaining confined to a single biological compartment) Figure 9(. This interpretation aligns increasingly with One Health and systems-ecology perspectives emphasizing ecological interdependence rather than compartment-specific biological isolation (Moulton et al., 2026; Sessitsch et al., 2023).

At the same time, substantial methodological caution remains necessary. Current limitations include insufficient longitudinal exposure datasets, incomplete nanoscale interface characterization, variability in environmental exposure methodologies, and limited resolution of signaling-associated particulate interactions under realistic chronic low-dose conditions (Gonçalves et al., 2026). Many proposed mechanisms remain biologically plausible but experimentally under characterized, particularly across heterogeneous environmental systems involving multisource coexposures and dynamic microbial ecologies (Ma et al., 2024). Consequently, polybiome signaling perturbation and interfacial proteostasis dynamics should currently be interpreted as emerging systems-science research directions requiring substantial multiscale validation rather than established mechanistic endpoints.

Nevertheless, the convergence of nanoscale environmental science, microbial systems ecology, extracellular vesicle biology, interfacial biophysics, and adaptive proteostasis research may represent an important future frontier within One Health environmental intelligence systems (Sessitsch et al., 2023). Integrative approaches capable of combining environmental sensing, microbial communication analysis, stress-response biology, and systems-level ecological interpretation may ultimately improve understanding of how chronic low-intensity environmental perturbations propagate across interconnected planetary biological systems under conditions of persistent uncertainty (Dennis & Kumar Mistri, 2026; Gonçalves et al., 2026).

Figure 9. Schematic representation of the Polybiome Interfacial Signaling Cascade (PISC)

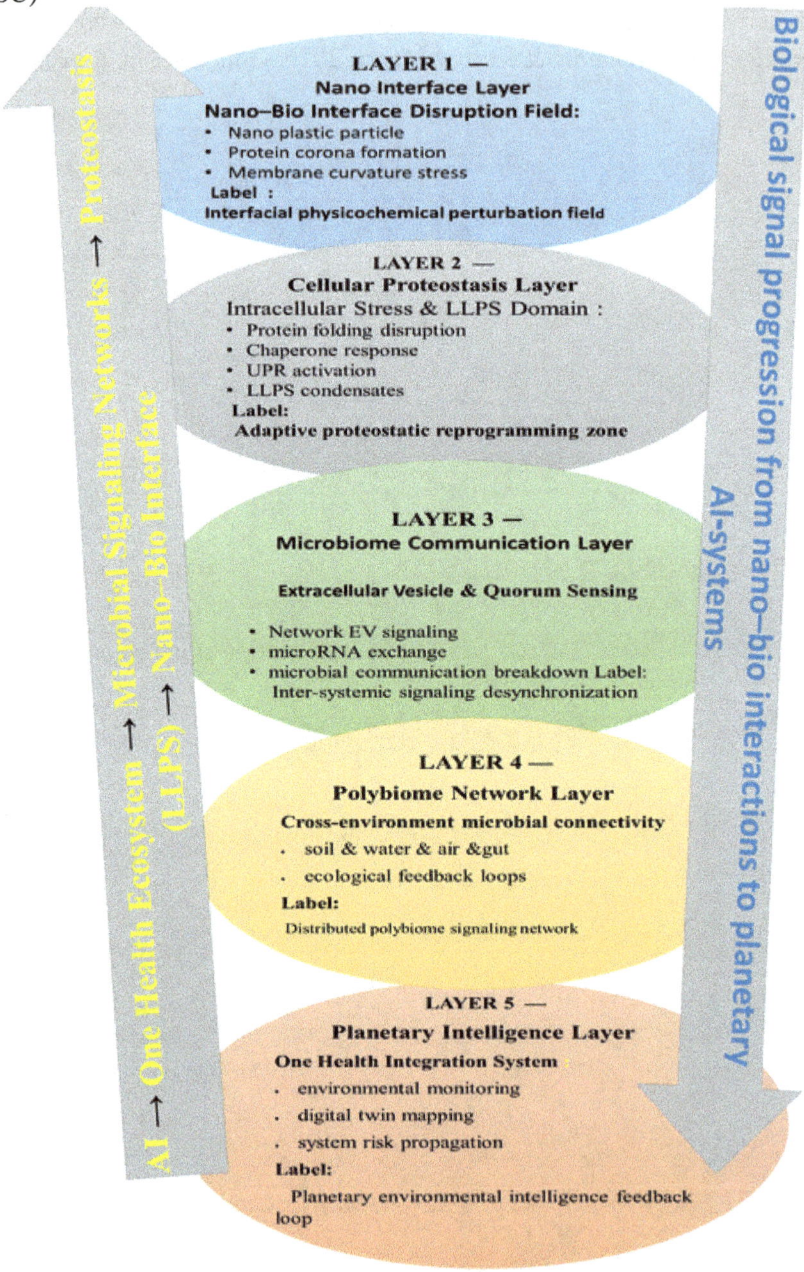

Figure 9 illustrates the multiscale propagation of nanoplastic-induced interfacial perturbations from nano–bio boundaries to cellular proteostasis systems, microbial communication networks, and integrated planetary polybiome structures under a One Health framework

17.3 Maternal–Developmental Nanoplastic Interfaces and Early-Life Vulnerability within One Health Systems

Within this broader systems perspective, increasing attention is now being directed toward the recognition that nanoplastic exposure cannot be interpreted as an isolated environmental event, but must instead be evaluated within a life-course and developmental systems framework. Within One Health science, maternal exposure pathways are becoming particularly important because environmental particulate systems intersect simultaneously with reproductive biology, placental physiology, microbial ecology, nutritional interfaces, and early developmental vulnerability. Micro- and nanoplastics are now recognized as pervasive environmental pollutants detected across multiple environmental matrices, including food systems, indoor dust, drinking water, atmospheric aerosols, and consumer-associated materials (Song et al., 2026). This widespread distribution creates continuous opportunities for chronic, low-intensity exposure during pregnancy and early life, making the tracking of these transformation pathways increasingly important for accurate risk assessment (Zhang et al., 2025).

Current evidence suggests that maternal exposure may occur through inhalation, ingestion, dermal contact, and environmental particulate transfer across interconnected exposure systems. These pathways are increasingly relevant because pregnancy represents a biologically sensitive window characterized by extensive endocrine regulation, immunological adaptation, metabolic restructuring, and dynamic microbiome interactions. Experimental and observational studies have increasingly reported the presence of micro- and nanoplastic particles within placental tissues, reproductive fluids, and maternal biological systems (Medley et al., 2023). These findings raise important questions regarding maternal–fetal transport dynamics, gestational accumulation, and subsequent developmental susceptibility in offspring (Yu et al., 2024).

Within developmental systems biology, the placenta functions not only as a physical interface between maternal and fetal systems, but also as a complex signaling environment regulating nutrient exchange, immune communication, endocrine activity, and developmental homeostasis. Physicochemical properties such as particle size, surface charge, chemical modification, and protein corona formation strongly influence the capacity of micro- and nanoplastics to interact with placental barriers and associated biological pathways (Dugershaw et al., 2020). However, the precise indirect and direct mechanisms governing placental translocation and developmental accumulation remain incompletely characterized, particularly under realistic chronic environmental exposure conditions (Medley et al., 2023).

From a maternal systems perspective, increasing attention is also being directed toward the relationship between environmental particulate exposure and female polybiome architecture. Maternal microbial ecosystems extend beyond the gastrointestinal tract and include interconnected reproductive, oral, dermal, pulmonary, and lactational microbial environments operating as integrated biological communication networks. Within this context, chronic nanoplastic exposure may potentially influence dysbiosis-related processes and broader inflammatory signaling pathways across interconnected maternal systems (Yelleti et al., 2026). At the same time, certain beneficial microbial cohorts, such as *Bifidobacteria*, have long been recognized for their role in maintaining systemic health and barrier integrity (Bubnov et al., 2017& Reyed, 2007). Therefore, characterizing how particulate pollution influences these sensitive microbial networks is becoming increasingly relevant within broader One Health interpretations of environmental vulnerability.

The lactational and neonatal dimensions of exposure further expand the complexity of maternal–developmental interfaces. Early-life feeding systems represent critical biological windows through which newborn immune maturation, metabolic programming, microbiome establishment, and neurodevelopmental adaptation are continuously shaped (Trofin et al., 2026). Maternal nutrition, feeding modes, environmental food contamination, indoor particulate exposure, and infant feeding materials may collectively influence the developing neonatal exposome (Alum et al., 2025). Within this framework, increasing concern is being directed toward whether chronic low-dose particulate exposure may interact with neonatal microbiome establishment and developmental signaling pathways during periods of heightened physiological sensitivity.

Importantly, current evidence does not support deterministic claims that nanoplastics are directly linked to pediatric disease outcomes, developmental disorders, or childhood malignancies. Nevertheless, emerging toxicological and developmental literature suggests biological plausibility for pathways involving oxidative stress, inflammatory dysregulation, mitochondrial dysfunction, epigenetic modification, and altered microbial ecology. In particular, the potential for nanoplastics to exert endocrine-disrupting effects across generations represents an important area of concern (Leso et al., 2023). These mechanisms are increasingly discussed within Developmental Origins of Health and Disease (DOHaD) frameworks because developmental exposure windows and maternal toxicants may contribute to epigenetic programming processes influencing long-term vulnerability trajectories extending across childhood and adult life (Alum et al., 2025).

Within pediatric susceptibility research, one of the central challenges involves distinguishing transient biological adaptation from persistent developmental perturbation. Infants and children possess immature detoxification systems, evolving immune regulation, rapidly developing neurological networks, and dynamic microbiome restructuring processes that may modify exposure sensitivity relative to adults. At the same time, substantial uncertainty remains regarding realistic environmental dose levels, tissue accumulation patterns, longitudinal developmental outcomes, and synergistic interactions with coexposures. These include air pollution, chemical additives, nutritional stressors, and antimicrobial resistance-associated microbial dynamics, where targeted microbial interventions are increasingly being explored within predictive and preventive medicine frameworks (Marras et al., 2021). Consequently, early-life vulnerability should currently be interpreted as an emerging systems-science concern rather than a conclusively established clinical outcome framework (Zhang et al., 2025).

The airborne dimension of maternal exposure may be particularly relevant within urban and indoor environments characterized by chronic particulate circulation. Airborne micro- and nanoplastics are increasingly detected in household dust, synthetic textiles, occupational environments, hospital-associated settings, and atmospheric aerosol systems (Song et al., 2026). Inhalation therefore represents a potentially important maternal exposure route capable of linking environmental particulate transport with respiratory, circulatory, placental, and gastrointestinal biological interfaces (Yelleti et al., 2026). Experimental animal studies have additionally raised questions regarding transgenerational particulate transfer and developmental accumulation patterns (Yu et al., 2024), although direct extrapolation to human systems remains limited and requires substantial methodological caution (Dugershaw et al., 2020).

Within a broader One Health perspective, maternal and developmental vulnerability cannot be separated from planetary environmental conditions. Environmental degradation, food-system contamination, atmospheric particulate transport, wastewater dynamics, and urban exposure infrastructures collectively shape the biological conditions under which maternal and neonatal health systems operate. As a result, future maternal–child environmental health frameworks may increasingly require the integration of ecological monitoring, reproductive health surveillance, microbiome analytics, developmental biomarkers, and environmental intelligence infrastructures capable of interpreting exposure patterns across interconnected biological and environmental systems.

From a governance perspective, future One Health policy architectures may need to move beyond compartment-specific environmental monitoring toward predictive environmental systems capable of integrating maternal exposure signals, developmental health indicators, atmospheric transport models, food-system contamination profiles, and biosurveillance datasets into unified vulnerability interpretation frameworks. The purpose of such systems would not be deterministic prediction of disease outcomes, but rather structured identification of emerging developmental vulnerability patterns operating across heterogeneous environmental and biological conditions.

Within this context, planetary environmental intelligence may gradually evolve into an integrative decision-support framework linking environmental sensing, maternal health analytics, AI-assisted exposure interpretation, developmental systems biology, and preventive public-health governance. However, substantial scientific caution remains necessary. Current limitations include incomplete nanoscale detection standardization, insufficient longitudinal human studies, variability in exposure assessment methodologies, and unresolved mechanistic uncertainties regarding chronic low-dose developmental exposure (Zhang et al., 2025). Consequently, predictive maternal–developmental environmental intelligence should be interpreted as a long-term scientific and governance direction rather than an already established operational capability.

18. CONCLUSION AND FUTURE PERSPECTIVES: TOWARD PLANETARY ENVIRONMENTAL INTELLIGENCE

18.1 Integrative Synthesis: Beyond Compartmentalized Ecotoxicology

The scientific characterization of micro- and nanoplastic pollution has reached a critical epistemic inflection point, moving gradually away from historical single-compartment contaminant models and toward multiscale, system-based representations. Across the expansive scholarship compiled within this volume, a robust and growing body of empirical evidence documents that nanoscale plastic particles do not exist as isolated chemical entities, nor can their ecological impacts be understood through simple concentration-dependent toxicological metrics alone (Gonçalves et al., 2026; Song et al., 2026). Instead, these persistent particulate fragments operate as dynamic, participating elements within highly coupled planetary interaction networks. Their observed biological and ecological consequences emerge from nonlinear, multifactorial processes involving atmospheric transport, environmental aging, interfacial transformation, and host polybiome interactions (Sessitsch et al., 2023).

This synthesis forces a shift in how environmental risks are conceptualized. In a systems-science framing, nanoplastics are recognized as architectural disruptors that alter physical, chemical, and biological boundary conditions across marine, freshwater, terrestrial, and atmospheric systems simultaneously. Rather than implying intrinsic biological functionality or deterministic pathogenetic behavior, their significance lies in emergent effects across multi-layered exposure environments. Chronic exposure across lifespan cannot be separated from industrial degradation, agricultural change, and climate volatility (Moulton et al., 2026). Consequently, integrating these cross-domain inputs into a unified analytical paradigm is no longer optional; it has become necessary under persistent uncertainty.

18.2 Conceptual Contribution: The Epistemic Utility of DNPS and SRS

The primary value of the Digital Nanoplastic Science (DNPS) framework and its computational counterpart, the Systemic Risk Score (SRS), lies in structuring environmental complexity rather than resolving it prematurely. In contrast to conventional environmental risk assessments (ERA) that rely on compartmentalization, fixed exposure boundaries, and linear dose–response assumptions, the SRS preserves interactional and cross-scale complexity rather than compressing it too early (National Academies of Sciences, Engineering, and Medicine, 2016). Within this architecture, uncertainty is not treated as a residual modeling defect but as an intrinsic property of the system.

By introducing DNPS, this chapter establishes an epistemic scaffold that shifts analysis from descriptive monitoring toward transductive signal loading and interfacial interpretation. This framework allows tracking how particulate input is translated across biophysical boundaries into biological signaling cascades. This re-positions nanoplastics not as passive toxic burdens, but as physicochemical interference agents capable of disrupting cellular microenvironments, biomolecular crowding, and membrane curvature dynamics without necessarily causing immediate cytotoxicity (Karim et al., 2024; Ma et al., 2024). DNPS provides a structured language linking environmental informatics with cellular proteostasis, enabling multi-scale interpretation under uncertainty.

18.3 Translational and Clinical Significance: Navigating the One Health Space

The translational utility becomes visible when mapped onto the One Health paradigm. Atmospheric transport networks distribute nanoplastics into agricultural, hydrological, and consumer systems, linking ecosystem contamination to host vulnerability (Visca et al., 2024). At the host–microbial interface, long-term exposure pathways interact with the human polybiome, potentially contributing to dysbiosis-related processes and neurobiological stress pathways associated with chronic ingestion (Garcia Rodriguez et al., 2026; Yelleti et al., 2026). This shifts evaluation away from acute endpoints toward long-term vulnerability trajectories across populations and time (Alum et al., 2025; Zhang et al., 2025).

Integration with antimicrobial resistance (AMR) dynamics introduces an additional layer for public health interpretation. Particulate surfaces and biofilm formation create localized environmental conditions that may support horizontal gene transfer and alter resistance dissemination patterns (Kang et al., 2025; Kazmi et al., 2026). Incorporating these variables into SRS allows identification of regions where environmental particulate exposure and AMR risks overlap. This supports targeted preventive strategies and environmental health screening approaches (Visca et al., 2024).

18.4 Remaining Challenges: Methodological and Regulatory Bottlenecks

Despite conceptual maturity, significant methodological constraints remain. A primary limitation is the lack of standardized, high-resolution methods for detecting nanoplastics (<100 nm) in complex environmental and biological matrices (Uchmanowicz et al., 2025). Without standardized quantification, parameter identifiability remains limited, and data-driven models risk context-dependent behavior rather than generalizable outputs. In addition, longitudinal datasets for low-dose, transgenerational exposure remain incomplete, limiting calibration of multi-scale frameworks (Dugershaw et al., 2020; Yu et al., 2024).

On the regulatory side, there is a structural mismatch between system-level scientific frameworks and current institutional risk assessment models. Agencies such as the European Chemicals Agency (ECHA) and the U.S. Environmental Protection Agency (EPA) still operate mainly within compartment-based hazard and exposure paradigms. These systems are not designed to directly process non-linear, multi-layer indicators such as SRS. Transition toward integration requires staged validation, sensitivity testing, and standardization protocols that preserve scientific traceability while enabling interpretability.

18.5 Future Research Directions: The Cyber-Biotechnology Frontier

Future nanoplastic research will depend on convergence across interfacial biophysics, computational multi-omics, artificial intelligence, and materials science. A key frontier involves characterizing the protein corona at the nano–bio interface to understand how particle chemistry influences protein folding dynamics and system-level biological responses (Dar et al., 2025; Karim et al., 2024; Porter, 2023). Parallel work on liquid–liquid phase separation (LLPS) and biomolecular condensates may explain how cells adapt to persistent nanoscale stressors, opening pathways for biomimetic and protective material design (Caporalini et al., 2025; M. Dang et al., 2026; Gul et al., 2021).

At system scale, development of cyber-biotechnology architectures is required to integrate continuous environmental and biological data streams into real-time analytical models (Reyed, 2026d). Deep-sea and extreme-environment microbiome studies remain essential for understanding quorum-sensing responses to plastic accumulation under variable biogeochemical conditions (Peng et al., 2025). These data streams should ultimately feed into ethically governed digital twin systems capable of simulating long-range propagation of nanoplastics across environmental and biological compartments.

18.6 Closing Perspective: Towards Planetary Environmental Intelligence

A full computational and methodological specification of the Data-Driven Systems Framework (DDSF), including data integration logic and analytical workflow structure, is provided in Appendix A. This ensures traceability and supports reproducibility of the DNPS and SRS frameworks within a structured computational environment. Overall, progress toward planetary environmental intelligence depends on moving beyond disciplinary boundaries and accepting multi-scale system complexity as a core condition of environmental science (Sessitsch et al., 2023). Within this framing, DNPS and SRS do not aim to produce deterministic prediction. They provide a structured way to represent uncertainty, interaction, and system vulnerability under real-world constraints. The value of this framework is therefore methodological rather than predictive. It supports interpretation of complex environmental systems without forcing premature simplification

REFERENCES

Alum, E. U., Aloh, H. E., Obasi, D. C., Okoroh, P. N., Aniokete, U. C., & Emeruwa, A. P. (2025). Maternal Nutrition, Toxicants, and Epigenetic Programming of Obesity Across Generations. *Diabetes, Metabolic Syndrome and Obesity*, *18*, 4873–4911. DOI: 10.2147/DMSO.S579409 PMID: 41497887

Amaral-Zettler, L. A., Zettler, E. R., & Mincer, T. J. (2020). Ecology of the plastosphere. *Nature Reviews. Microbiology*, *18*(3), 139–151. DOI: 10.1038/s41579-019-0308-0 PMID: 31937947

Amer, Y., Soufali, A., & Zaghwan, A. (2026). A digital twin-based framework for predictive quality assurance and supply chain resilience in the automotive industry. *Advanced Engineering Informatics*, *69*(Part C), 103969. DOI: 10.1016/j.aei.2025.103969

Ankley, G. T., & Edwards, S. W. (2018). The Adverse Outcome Pathway: A Multifaceted Framework Supporting 21st Century Toxicology. *Current Opinion in Toxicology*, *9*, 1–7. DOI: 10.1016/j.cotox.2018.03.004 PMID: 29682628

Araújo, A. M., Mota, C., Ramos, H., Faria, M. A., Carvalho, M., & Ferreira, I. M. P. L. V. O. (2025). The neurotoxic threat of micro- and nanoplastics: Evidence from In Vitro and In Vivo models. *Archives of Toxicology*, *99*(9), 3505–3525. DOI: 10.1007/s00204-025-04091-3 PMID: 40459748

Bastyans, S., Jackson, S., & Fejer, G. (2022). Micro and nanoplastics, a threat to human health? *Emerging Topics in Life Sciences*, *6*(4), 411–422. DOI: 10.1042/ETLS20220024 PMID: 36453919

Basumatary, T., Biswas, D., Boro, S., Nava, A. R., Narayan, M., & Sarma, H. (2025). Dynamics and Impacts of Microplastics (MPs) and Nanoplastics (NPs) on Ecosystems and Biogeochemical Processes: The Need for Robust Regulatory Frameworks. *ACS Omega*, *10*(17), 17051–17069. DOI: 10.1021/acsomega.5c01175 PMID: 40352536

Bel Hassen, M., Bellaaj Zouari, A., Abdennadher, M., Assaf, J.-C., Nakad, M., Abboud, R., Khammeri, Y., Banni, M., Panzeri, A., Gomes, L., & Hamd, W. (2025). Plastics pollution: Pathways, impacts, and regulatory challenges in marine environments. *Frontiers in Environmental Science*, *13*, 1635230. DOI: 10.3389/fenvs.2025.1635230

Bell, I. R., Ives, J. A., & Jonas, W. B. (2013). Nonlinear effects of nanoparticles: Biological variability from hormetic doses, small particle sizes, and dynamic adaptive interactions. *Dose-Response*, *12*(2), 202–232. DOI: 10.2203/dose-response.13-025.Bell PMID: 24910581

Belle, J., Velásquez, C., Oxley, M., & Getiashvili, K. (2024). Understanding and integrating systemic risk (SR) into disaster risk reduction (DRR) and risk informed development (RID) for long-term resilient development. In *Sustainable Development*. IntechOpen. https://doi.org/DOI: 10.5772/intechopen.1004915

Berg, G., Rybakova, D., Fischer, D., Cernava, T., Vergès, M. C., Charles, T., Chen, X., Cocolin, L., Eversole, K., Corral, G. H., Kazou, M., Kinkel, L., Lange, L., Lima, N., Loy, A., Macklin, J. A., Maguin, E., Mauchline, T., McClure, R., & Schloter, M. (2020). Microbiome definition re-visited: Old concepts and new challenges. *Microbiome*, *8*(1), 103. DOI: 10.1186/s40168-020-00875-0 PMID: 32605663

Bhat, M. A., Gedik, K., & Gaga, E. O. (2023). Atmospheric micro (nano) plastics: Future growing concerns for human health. *Air Quality, Atmosphere & Health*, *16*(2), 233–262. DOI: 10.1007/s11869-022-01272-2 PMID: 36276170

Bhattacharyya, S., Greer, M. L., & Salehi, M. (2025). Impact of micro- and nano-plastics exposure on human health: Focus on neurological effects from ingestion. *Frontiers in Public Health*, *13*, 1681776. DOI: 10.3389/fpubh.2025.1681776 PMID: 41246084

Bhuller, Y., Bancroft, X., Deonandan, R., Grudniewicz, A., Wiles, A., & Krewski, D. (2025). Key attributes of health and environmental risk decision-making: A scoping review. *Risk Analysis : An Official Publication of the Society for Risk Analysis*, *45*(7), 1926–1939. DOI: 10.1111/risa.17715 PMID: 39894676

Boakye, G. A., Trotta, E., Ambagahawatta, N., Venkataraman, A., Cheeptham, N., & Papadopoulos, C. (2025). Nanoscale plastic pollution: Sources, identification and potential mitigation. *Nanotechnology*, *36*(42), 422001. DOI: 10.1088/1361-6528/ae0c1b PMID: 41005319

Boctor, J., Hoyle, F. C., Farag, M. A., Ebaid, M., Walsh, T., Whiteley, A. S., & Murphy, D. V. (2025). Microplastics and nanoplastics: Fate, transport, and governance from agricultural soil to food webs and humans. *Environmental Sciences Europe*, *37*(1), 68. DOI: 10.1186/s12302-025-01104-x

Bubnov, R. V., Babenko, L. P., Lazarenko, L. M., Mokrozub, V. V., Demchenko, O. A., Nechypurenko, O. V., & Spivak, M. Y. (2017). Comparative study of probiotic effects of *Lactobacillus* and *Bifidobacteria* strains on cholesterol levels, liver morphology and the gut microbiota in obese mice. *The EPMA Journal*, *8*(4), 357–376. DOI: 10.1007/s13167-017-0117-3 PMID: 29209439

Callewaert, E., Louisse, J., Kramer, N., Sanz-Serrano, J., & Vinken, M. (2025). Adverse Outcome Pathways Mechanistically Describing Hepatotoxicity. *Methods in Molecular Biology (Clifton, N.J.), 2834*, 249–273. DOI: 10.1007/978-1-0716-4003-6_12 PMID: 39312169

Capodaglio, A. G. (2025). Micro- and Nano-Plastics in Drinking Water: Threat or Hype? Critical State-of-the-Art Analysis of Risks and Approaches. *Journal of Xenobiotics, 15*(3), 85. DOI: 10.3390/jox15030085 PMID: 40558868

Caporalini, S., Azimi, B., Zergat, S., Ansari Chaharsoughi, M., Maleki, H., Batoni, G., & Danti, S. (2025). Electrospinning Enables Opportunity for Green and Effective Antibacterial Coatings of Medical Devices. *Journal of Functional Biomaterials, 16*(7), 249. DOI: 10.3390/jfb16070249 PMID: 40710463

Cavazzoli, S., Ferrentino, R., Scopetani, C., Monperrus, M., & Andreottola, G. (2023). Analysis of micro- and nanoplastics in wastewater treatment plants: Key steps and environmental risk considerations. *Environmental Monitoring and Assessment, 195*(12), 1483. DOI: 10.1007/s10661-023-12030-x PMID: 37971551

Chabi, K., Ye, C., Gad, M., Rizk, M., & Yu, X. (2026). Toxicity of microplastics and nanoplastics in aquatic systems in interaction with biological–chemical contaminants: Advances in remediations. In *Environmental Sciences*. IntechOpen., DOI: 10.5772/intechopen.1013428

Chen, X., Zhuang, J., Chen, Q., Xu, L., Yue, X., & Qiao, D. (2022). Polyvinyl chloride microplastics induced gut barrier dysfunction, microbiota dysbiosis and metabolism disorder in adult mice. *Ecotoxicology and Environmental Safety, 241*, 113809. DOI: 10.1016/j.ecoenv.2022.113809 PMID: 36068740

Cheng, K., Chang, Y., Lee, X., Ji, D., Qiao, L., Zou, Z., Duan, Y., & Huang, R. J. (2024). Life-Course Health Risk Assessment of $PM_{2.5}$ Elements in China: Exposure Disparities by Species, Source, Age, Gender, and Location. *Environmental Science & Technology, 58*(8), 3629–3640. DOI: 10.1021/acs.est.3c05404 PMID: 38354315

Christopher, E. A., Christopher-de Vries, Y., Devadoss, A., Mandemaker, L. D. B., van Boxel, J., Copsey, H. M., Dusza, H. M., Legler, J., Meirer, F., Muncke, J., Nawrot, T. S., Saenen, N. D., Scholz-Böttcher, B. M., Tran, L., Weckhuysen, B. M., Zou, R., Zimmermann, L., Galea, K. S., Vermeulen, R., & Boyles, M. S. P. (2024). Impacts of micro- and nanoplastics on early-life health: A roadmap toward risk assessment. *Microplastics and Nanoplastics, 4*(1), 13. DOI: 10.1186/s43591-024-00089-3

Chulkov, V., Gasanov, M., Isakov, V., Denisenko, A., Nwosu, C., & Rodkin, S. (2025). Molecular and Cellular Effects of Microplastics and Nanoplastics in the Pathogenesis of Cardiovascular, Nervous, Urinary, Digestive, and Reproductive System Diseases: A Global Systematic Review. *International Journal of Molecular Sciences*, *26*(22), 11194. DOI: 10.3390/ijms262211194 PMID: 41303677

Clarke, R., Shajahan, A. N., Wang, Y., Tyson, J. J., Riggins, R. B., Weiner, L. M., Bauman, W. T., Xuan, J., Zhang, B., Facey, C., Aiyer, H., Cook, K., Hickman, F. E., Tavassoly, I., Verdugo, A., Chen, C., Zwart, A., Wärri, A., & Hilakivi-Clarke, L. A. (2011). Endoplasmic reticulum stress, the unfolded protein response, and gene network modeling in antiestrogen resistant breast cancer. *Hormone Molecular Biology and Clinical Investigation*, *5*(1), 35–44. DOI: 10.1515/HMBCI.2010.073 PMID: 23930139

Croock, D., Swart, Y., Schurz, H., Petersen, D. C., Möller, M., & Uren, C. (2024). Data harmonization guidelines to combine multiplatform genomic data from admixed populations and boost power in genome-wide association studies. *Current Protocols*, *4*(6), e1055. DOI: 10.1002/cpz1.1055 PMID: 38837690

Cui, Y., Eccles, K. M., Kwok, R. K., Joubert, B. R., Messier, K. P., & Balshaw, D. M. (2022). Integrating multiscale geospatial environmental data into large population health studies: Challenges and opportunities. *Toxics*, *10*(7), 403. DOI: 10.3390/toxics10070403 PMID: 35878308

Cunningham, E. M., Cuthbert, R. N., Coughlan, N. E., Kregting, L., Cairnduff, V., & Dick, J. T. A. (2021). Microplastics do not affect the feeding rates of a marine predator. *The Science of the Total Environment*, *779*, 146487. DOI: 10.1016/j.scitotenv.2021.146487 PMID: 34030230

Cverenkárová, K., Valachovičová, M., Mackuľak, T., Žemlička, L., & Bírošová, L. (2021). Microplastics in the Food Chain. *Life (Basel, Switzerland)*, *11*(12), 1349. DOI: 10.3390/life11121349 PMID: 34947879

Danasekaran, R. (2024). One Health: A Holistic Approach to Tackling Global Health Issues. *Indian Journal of Community Medicine : Official Publication of Indian Association of Preventive & Social Medicine*, *49*(2), 260–263. DOI: 10.4103/ijcm.ijcm_521_23 PMID: 38665439

Dang, F., Wang, Q., Huang, Y., Wang, Y., & Xing, B. (2022). Key knowledge gaps for One Health approach to mitigate nanoplastic risks. *Eco-Environment & Health*, *1*(1), 11–22. DOI: 10.1016/j.eehl.2022.02.001 PMID: 38078201

Dang, M., Yang, C., Jin, G., Deng, Q., Wu, L., & Teh, G. B. (2026). From biomolecular condensates to functional nanomaterials: LLPS-inspired frameworks for nanoscale hydrogels and adaptive materials. *Journal of Nanobiotechnology*, *24*(1), 280. Advance online publication. DOI: 10.1186/s12951-026-04151-5 PMID: 41721403

Dar, A. I., Randhawa, S., Verma, M., Saini, T. C., & Acharya, A. (2025). Debugging the dynamics of protein corona: Formation, composition, challenges, and applications at the nano-bio interface. *Advances in Colloid and Interface Science*, *342*, 103535. DOI: 10.1016/j.cis.2025.103535 PMID: 40319752

Deng, D., Wang, L., Zeng, Z., Lu, L., Yang, M., Wang, Y., Shi, X., Lovell, J. F., Zhang, J., & Jin, H. (2026). Visualizing extracellular vesicles in cancer: From biogenesis to theranostic applications. *Journal of Nanobiotechnology*, *24*(1), 334. DOI: 10.1186/s12951-026-04223-6 PMID: 41776618

Dennis, J., & Kumar Mistri, T. (2026). Nanoplastics in the Biological System: A Microscopic Menace with an Impact. In Environmental Sciences. IntechOpen. https://doi.org/DOI: 10.5772/intechopen.1013429

Destoumieux-Garzón, D., Mavingui, P., Boetsch, G., Boissier, J., Darriet, F., Duboz, P., Fritsch, C., Giraudoux, P., Le Roux, F., Morand, S., Paillard, C., Pontier, D., Sueur, C., & Voituron, Y. (2018). The one health concept: 10 years old and a long road ahead. *Frontiers in Veterinary Science*, *5*, 14. DOI: 10.3389/fvets.2018.00014 PMID: 29484301

Dewika, M., Markandan, K., Nor Ruwaida, J., Sara, Y. Y., Deb, A., Ahmad Irfan, N., & Khalid, M. (2024). Integrating the quintuple helix approach into atmospheric microplastics management policies for planetary health preservation. *The Science of the Total Environment*, *954*, 176063. Advance online publication. DOI: 10.1016/j.scitotenv.2024.176063 PMID: 39245389

Díaz-Villanueva, J. F., Díaz-Molina, R., & García-González, V. (2015). Protein Folding and Mechanisms of Proteostasis. *International Journal of Molecular Sciences*, *16*(8), 17193–17230. DOI: 10.3390/ijms160817193 PMID: 26225966

Ding, X., Zhou, C., Zhong, W., & Tang, P. (2019). Addressing uncertainty of environmental governance in environmentally sensitive areas in developing countries: A precise-strike and spatial-targeting adaptive governance framework. *Sustainability (Basel)*, *11*(16), 4510. DOI: 10.3390/su11164510

Dugershaw, B. B., Aengenheister, L., Hansen, S. S. K., Hougaard, K. S., & Buerki-Thurnherr, T. (2020). Recent insights on indirect mechanisms in developmental toxicity of nanomaterials. *Particle and Fibre Toxicology*, *17*(1), 31. Advance online publication. DOI: 10.1186/s12989-020-00359-x PMID: 32653006

Durden, J. M. (2025). Environmental management using a digital twin. *Environmental Science & Policy*, *164*, 104018. DOI: 10.1016/j.envsci.2025.104018

Elawady, A., Mohamed, J. A. H., & Abid, M. B. (2026). Advancements in microplastics detection techniques and their multidimensional impacts on aquatic ecosystems and human health. *Journal of Hazardous Materials: Plastics*, *2*, 100025. DOI: 10.1016/j.hazmp.2025.100025

Elgharbawy, A. A. M., Noor, N. M., Azmi, N. A. N., & Dewanti, B. S. D. (2026). Beyond the Solvent: Engineering Ionic Liquids for Biomedical Applications-Advances, Challenges, and Future Directions. *Molecules (Basel, Switzerland)*, *31*(2), 305. DOI: 10.3390/molecules31020305 PMID: 41599355

Erhirhie, E. O., Ihekwereme, C. P., & Ilodigwe, E. E. (2018). Advances in acute toxicity testing: Strengths, weaknesses and regulatory acceptance. *Interdisciplinary Toxicology*, *11*(1), 5–12. DOI: 10.2478/intox-2018-0001 PMID: 30181707

Esawy, M. A., & Reyed, R. M. (2026). Biointelligent reframing of colorectal cancer: From somatic mutation to systemic coherence. In Reyed, R., Prabhakar, P., & Haghi, A. (Eds.), *Advancing personalized medicine with AI-driven microbiome and nutrigenomics: Innovations and future perspectives* (pp. 173–218). IGI Global Scientific Publishing., DOI: 10.4018/979-8-3373-3121-8.ch003

Falsafi, A., Togiani, A., Colley, A., Varis, J., & Horttanainen, M. (2025). Life cycle assessment in circular design process: A systematic literature review. *Journal of Cleaner Production*, *521*, 146188. DOI: 10.1016/j.jclepro.2025.146188

Fanghella, F., Pesce, M., Franceschelli, S., Panella, V., Elsallabi, O., Lupi, T., Rizza, B., Di Battista, M. G., Bruno, A., Ballerini, P., Patruno, A., & Speranza, L. (2025). Biological modulation of autophagy by nanoplastics: A current overview. *International Journal of Molecular Sciences*, *26*(15), 7035. DOI: 10.3390/ijms26157035 PMID: 40806168

Gambhir, A., Albert, M. J., Doe, S. S. P., Donges, J. F., Farajalla, N., Giatti, L. L., Gundimeda, H., Hendel-Blackford, S., Homer-Dixon, T., Hoyer, D., Adan, S., Jacome-Polit, D., Kemp, L., Korowicz, D., Kovacic, Z., Kwakkel, J., Laybourn, L., Lempert, R., Mahamoud, A., & Richardson, R. (2025). A systemic risk assessment methodological framework for the global polycrisis. *Nature Communications*, *16*(1), 7382. DOI: 10.1038/s41467-025-62029-w PMID: 40813573

Garcia Rodriguez, A., Rocabert, A., Hernández, A., & Marcos, R. (2026). Micronanoplastics induce dysbiosis of intestinal gut microbiota in living beings. In Chandrasekaran, N., Marcos, R., & Manzo, S. (Eds.), *Micronanoplastics: From Source to Sink* (pp. 87–118). Elsevier., DOI: 10.1016/B978-0-443-24782-8.00014-5

Garske, B., Holz, W., & Ekardt, F. (2024). Digital twins in sustainable transition: Exploring the role of EU data governance. *Frontiers in Research Metrics and Analytics*, *9*, 1303024. DOI: 10.3389/frma.2024.1303024 PMID: 38515644

Generalova, A., Davidova, S., & Satchanska, G. (2025). The mechanisms of lead toxicity in living organisms. *Journal of Xenobiotics*, *15*(5), 146. DOI: 10.3390/jox15050146 PMID: 40981357

Gonçalves, J., Pequeno, J., Krzisnik, D., Sobral, P., & Antunes, J. (2026). Nanoplastics: An Emerging Threat to Human Health—A Perspective Review. *Pollutants*, *6*(2), 21. DOI: 10.3390/pollutants6020021

Gou, Z., Wu, H., Li, S., Liu, Z., & Zhang, Y. (2024). Airborne micro- and nanoplastics: Emerging causes of respiratory diseases. [REMOVED HYPERLINK FIELD]. *Particle and Fibre Toxicology*, *21*(1), 50. DOI: 10.1186/s12989-024-00613-6 PMID: 39633457

Gul, A., Gallus, I., Tegginamath, A., Maryska, J., & Yalcinkaya, F. (2021). Electrospun Antibacterial Nanomaterials for Wound Dressings Applications. *Membranes (Basel)*, *11*(12), 908. DOI: 10.3390/membranes11120908 PMID: 34940410

Habumugisha, T., Zhang, Z., Uwizewe, C., Yan, C., Ndayishimiye, J. C., Rehman, A., Zhang, X., & Zhang, X. (2024). Toxicological review of micro- and nanoplastics in aquatic environments: Risks to ecosystems, food web dynamics and human health. *Ecotoxicology and Environmental Safety*, *278*, 116426. Advance online publication. DOI: 10.1016/j.ecoenv.2024.116426 PMID: 38718727

Halfon, N., & Hochstein, M. (2002). Life course health development: An integrated framework for developing health, policy, and research. *The Milbank Quarterly*, *80*(3), 433–iii. DOI: 10.1111/1468-0009.00019 PMID: 12233246

Hayat, R., Khan, A., Muhammad Awais, , Muhammad Farhan, , Ahmad, U., Abdul Manan, , Abdul Basit, , Yasin Asad, M. T., Shah Nawaz, , & Hafeez, U. (2026). The Silent Sink: A Comprehensive Review of Microplastic Accumulation in Agricultural Soils and its Impact on Crop Physiology. *Indus Journal of Bioscience Research*, *4*(1), 90–102. DOI: 10.70749/ijbr.v4i1.2838

Helal, M., Hartmann, N. B., Khan, F. R., & Xu, E. G. (2023). Time to integrate "One Health Approach" into nanoplastic research. *Eco-Environment & Health*, *2*(1), 18–20. DOI: 10.1016/j.eehl.2023.01.001 PMID: 38074454

Hossain, R., & Sahajwalla, V. (2026). A comprehensive toolkit for micro- to nanoplastic analysis. *Environmental Science. Nano*, *13*(1), 122–149. Advance online publication. DOI: 10.1039/D5EN00856E

Iqbal, B., Khan, A. A., Hu, J., Liu, Q., Wang, C., Li, G., & Ye, M. (2026). Soil microplastics hidden web: Interaction of microbes and viruses as a frontier for sustainable ecosystem recovery. *Agricultural Ecology and Environment, 2*(1), e006. DOI: 10.48130/aee-0026-0003

Jaafarzadeh, N., & Talepour, N. (2023). Microplastics as carriers of antibiotic resistance genes and pathogens in municipal solid waste (MSW) landfill leachate and soil: A review. *Journal of Environmental Health Science & Engineering, 22*(1), 1–12. DOI: 10.1007/s40201-023-00879-6 PMID: 38887766

Jahedi, F., & Jaafarzadeh Haghighi Fard, N. (2025). Micro- and nanoplastic toxicity in humans: Exposure pathways, cellular effects, and mitigation strategies. *Toxicology Reports, 14*, 102043. DOI: 10.1016/j.toxrep.2025.102043 PMID: 40487378

Jani, V., & Wu, S. (2025). Nanoplastics (NPs): Environmental presence, ecological implications, and mitigation approaches *Microplastics, 4*(3), 48. https://doi.org/DOI: 10.3390/microplastics4030048

Jansen, M. A. K., Andrady, A. L., Bornman, J. F., Aucamp, P. J., Bais, A. F., Banaszak, A. T., Barnes, P. W., Bernhard, G. H., Bruckman, L. S., Busquets, R., Häder, D. P., Hanson, M. L., Heikkilä, A. M., Hylander, S., Lucas, R. M., Mackenzie, R., Madronich, S., Neale, P. J., Neale, R. E., & Zhu, L. (2024). Plastics in the environment in the context of UV radiation, climate change and the Montreal Protocol: UNEP Environmental Effects Assessment Panel, Update 2023. *Photochemical & Photobiological Sciences : Official Journal of the European Photochemistry Association and the European Society for Photobiology, 23*(4), 629–650. DOI: 10.1007/s43630-024-00552-3 PMID: 38512633

Jarmakiewicz-Czaja, S., Ferenc, K., Sokal-Dembowska, A., Wojdyła, D., & Filip, R. (2026). Effects of microplastics on the intestines. *BMC Gastroenterology, 26*(1), 289. DOI: 10.1186/s12876-026-04749-x PMID: 41928110

Jeong, J., Im, J., & Choi, J. (2024). Integrating aggregate exposure pathway and adverse outcome pathway for micro/nanoplastics: A review on exposure, toxicokinetics, and toxicity studies. *Ecotoxicology and Environmental Safety, 272*, 116022. DOI: 10.1016/j.ecoenv.2024.116022 PMID: 38309230

Jiang, F., Gao, C., Chan, A. W. H., Topping, D. O., Zhang, H., Li, W., Coe, H., Liu, X., Ching, J., & Zheng, Z. (2025). A review of atmospheric micro/nanoplastics: Insights into source and fate for modeling studies. *Current Pollution Reports, 11*(53), 53. Advance online publication. DOI: 10.1007/s40726-025-00375-5

Jiao, H., Ali, S. S., Alsharbaty, M. H. M., Elsamahy, T., Abdelkarim, E., Schagerl, M., Al-Tohamy, R., & Sun, J. (2024). A critical review on plastic waste life cycle assessment and management: Challenges, research gaps, and future perspectives. *Ecotoxicology and Environmental Safety*, *271*, 115942. DOI: 10.1016/j.ecoenv.2024.115942 PMID: 38218104

Jin, Y., Lu, L., Tu, W., Luo, T., & Fu, Z. (2019). Impacts of polystyrene microplastic on the gut barrier, microbiota and metabolism of mice. *The Science of the Total Environment*, *649*, 308–317. DOI: 10.1016/j.scitotenv.2018.08.353 PMID: 30176444

Kang, Y., Gao, S. H., Pan, Y., Gao, R., Li, T., Fan, L., Su, Y., Zhang, W., Yu, Z., Liang, B., Su, J. Q., Luo, Y., Wang, Y., Guo, J., & Wang, A. (2025). Roles of micro/nanoplastics in the spread of antimicrobial resistance through conjugative gene transfer. *Nature Communications*, *17*(1), 1118. DOI: 10.1038/s41467-025-67879-y PMID: 41444368

Kara Murdoch, F., Sun, Y., Fuller, M. E., Mullins, L., Hill, A., Lilly, J., Wilson, J., Löffler, F. E., & Kucharzyk, K. H. (2025). Presence of Micro- and Nanoplastics Affects Degradation of Chlorinated Solvents. *Toxics*, *13*(8), 656. DOI: 10.3390/toxics13080656 PMID: 40863933

Karim, A., Yadav, A., Sweety, U. H., Kumar, J., Delgado, S. A., Hernandez, J. A., White, J. C., Vukovic, L., & Narayan, M. (2024). Interfacial interactions between nanoplastics and biological systems: Toward an atomic and molecular understanding of plastics-driven biological dyshomeostasis. *ACS Applied Materials & Interfaces*, *16*(20), 25740–25756. DOI: 10.1021/acsami.4c03008 PMID: 38722759

Katsoulakis, E., Wang, Q., Wu, H., Shahriyari, L., Fletcher, R., Liu, J., Achenie, L., Liu, H., Jackson, P., Xiao, Y., Syeda-Mahmood, T., Tuli, R., & Deng, J. (2024). Digital twins for health: A scoping review. *NPJ Digital Medicine*, *7*(1), 77. DOI: 10.1038/s41746-024-01073-0 PMID: 38519626

Kazmi, S. S. U. H., Batool, S. M., Pastorino, P., Barcelò, D., Grossart, H.-P., Yaseen, Z. M., Khan, Z. H., Azeem, M., & Li, G. (2026). The plastisphere as a nexus for antimicrobial resistance: Micro(nano)plastics in pathogen colonization, gene transfer, and global health risks. *Biological Reviews of the Cambridge Philosophical Society*, brv.70163. Advance online publication. DOI: 10.1002/brv.70163 PMID: 41873837

Khan, D., & Ali, S. A. (2023). On the Novel Process of Pristine Microplastic Biofragmentation by Zebrafish (*Danio rerio*). *Archives of Environmental Contamination and Toxicology*, *84*(3), 299–306. DOI: 10.1007/s00244-023-00987-2 PMID: 36929014

Khanam, M. M., Uddin, M. K., & Kazi, J. U. (2025). Advances in machine learning for the detection and characterization of microplastics in the environment. *Frontiers in Environmental Science, 13*, 1573579. DOI: 10.3389/fenvs.2025.1573579

Kochanek, A., Grąz, K., Kochanek, A., Grąz, K., Potok, H., Gronba-Chyła, A., Kwaśny, J., Wiewiórska, I., Ciuła, J., Basta, E., & Łapiński, J. (2025). Micro- and Nanoplastics in the Environment: Current State of Research, Sources of Origin, Health Risks, and Regulations—A Comprehensive Review. *Toxics, 13*(7), 564. DOI: 10.3390/toxics13070564 PMID: 40711009

Koelmans, A. A., Mohamed Nor, N. H., Hermsen, E., Kooi, M., Mintenig, S. M., & De France, J. (2019). Microplastics in freshwaters and drinking water: Critical review and assessment of data quality. *Water Research, 155*, 410–422. DOI: 10.1016/j.watres.2019.02.054 PMID: 30861380

Koelmans, A. A., Redondo-Hasselerharm, P. E., Nor, N. H. M., de Ruijter, V. N., Mintenig, S. M., & Kooi, M. (2022). Risk assessment of microplastic particles. *Nature Reviews. Materials, 7*(2), 138–152. DOI: 10.1038/s41578-021-00411-y

Korenhof, P., Metcalfe, B., & Henschke, A. (2025). Cyber-governance of the natural world: The implications of digital twins in environmental governance. *Earth System Governance, 25*, 100276. DOI: 10.1016/j.esg.2025.100276

Krewski, D., Acosta, D.Jr, Andersen, M., Anderson, H., Bailar, J. C.III, Boekelheide, K., Brent, R., Charnley, G., Cheung, V. G., Green, S.Jr, Kelsey, K. T., Kerkvliet, N. I., Li, A. A., McCray, L., Meyer, O., Patterson, R. D., Pennie, W., Scala, R. A., Solomon, G. M., & Zeise, L.Staff of Committee on Toxicity Testing and Assessment of Env. (2010). Toxicity testing in the 21st century: A vision and a strategy. *Journal of Toxicology and Environmental Health. Part B, Critical Reviews, 13*(2–4), 51–138. DOI: 10.1080/10937404.2010.483176 PMID: 20574894

Kumar, A., Nehra, M., Singhal, N. K., & Kumar, S. (2026). Recent progress and technological advancements for detection of micro/nanoplastics in the environment. *Advances in Colloid and Interface Science, 351*, 103817. DOI: 10.1016/j.cis.2026.103817 PMID: 41678884

Landrigan, P. J., Raps, H., Cropper, M., Bald, C., Brunner, M., Canonizado, E. M., Charles, D., Chiles, T. C., Donohue, M. J., Enck, J., Fenichel, P., Fleming, L. E., Ferrier-Pages, C., Fordham, R., Gozt, A., Griffin, C., Hahn, M. E., Haryanto, B., Hixson, R., & Dunlop, S. (2023). The Minderoo-Monaco Commission on Plastics and Human Health. *Annals of Global Health, 89*(1), 23. DOI: 10.5334/aogh.4056 PMID: 36969097

Lee, S. H., & Kim, Y. (2026). Amine-functionalized copper mesh for the efficient removal and colorimetric quantification of polypropylene nanoplastics. *Journal of Environmental Chemical Engineering, 14*(1), 120610. DOI: 10.1016/j.jece.2025.120610

Lee, Y. H., Zheng, C. M., Wang, Y. J., Wang, Y. L., & Chiu, H. W. (2025). Effects of microplastics and nanoplastics on the kidney and cardiovascular system. *Nature Reviews. Nephrology, 21*(9), 585–596. DOI: 10.1038/s41581-025-00971-0 PMID: 40537559

Lerchner, T., Jedlička, J., Kripnerová, M., Dejmek, J., & Kuncová, J. (2024). Influence of micro- and nanoplastics on mitochondrial function in the cardiovascular system: A review of the current literature. *Physiological Research, 73*(S3), S685–S695. DOI: 10.33549/physiolres.935500 PMID: 39808171

Leslie, H. A., van Velzen, M. J. M., Brandsma, S. H., Vethaak, A. D., Garcia-Vallejo, J. J., & Lamoree, M. H. (2022). Discovery and quantification of plastic particle pollution in human blood. *Environment International, 163*, 107199. DOI: 10.1016/j.envint.2022.107199 PMID: 35367073

Leso, V., Battistini, B., Vetrani, I., Reppuccia, L., Fedele, M., Ruggieri, F., Bocca, B., & Iavicoli, I. (2023). The endocrine disrupting effects of nanoplastic exposure: A systematic review. *Toxicology and Industrial Health, 39*(11), 613–629. DOI: 10.1177/07482337231203053 PMID: 37753827

Leso, V., Nowack, B., Karakoltzidis, A., Nikiforou, F., Karakitsios, S., Sarigiannis, D., & Iavicoli, I. (2025). Next generation risk assessment and new approach methodologies for safe and sustainable by design chemicals and materials: Perspectives and challenges for occupational health. *Toxicology, 517*, 154211. DOI: 10.1016/j.tox.2025.154211 PMID: 40490209

Li, Y., Ling, W., Yang, J., & Xing, Y. (2025). Risk assessment of microplastics in humans: Distribution, exposure, and toxicological effects. *Polymers, 17*(12), 1699. DOI: 10.3390/polym17121699 PMID: 40574225

Likpalimor, O. N., Zhang, S., Adagba, F. I., Fatoye, S., Adega, K. J., & Njue, J. (2025). Mechanisms of microplastic sorption of antibiotics and impacts on aquatic ecosystems for sustainable development goals. *Discover Environment, 3*(1), 215. DOI: 10.1007/s44274-025-00431-3

Lin, R. H., Chen, H. T., Lee, I. T., Vo, T. T., & Wang, Y. L. (2025). Microplastics and Nanoplastics in Health Concerning Cellular Toxicity Mechanisms, Exposure Pathways, and Global Mitigation Strategies. *Life (Basel, Switzerland), 15*(9), 1449. DOI: 10.3390/life15091449 PMID: 41010391

Liu, L., Sun, Y., Du, S., Li, Y., & Wang, J. (2023). Nanoplastics promote the dissemination of antibiotic resistance genes and diversify their bacterial hosts in soil. *Eco-Environment & Health*, *3*(1), 1–10. DOI: 10.1016/j.eehl.2023.09.005 PMID: 38187015

Liu, S., Chen, Q., Ding, H., Song, Y., Pan, Q., Deng, H., & Zeng, E. Y. (2024). Differences of microplastics and nanoplastics in urban waters: Environmental behaviors, hazards, and removal. *Water Research*, *260*, 121895. DOI: 10.1016/j.watres.2024.121895 PMID: 38875856

Liu, Y., Liu, L., Wang, X., Shao, M., Wei, Z., Wang, L., Li, B., Li, C., Luo, X., Li, F., & Zheng, H. (2025). Microplastics enhance the prevalence of antibiotic resistance genes in mariculture sediments by enriching host bacteria and promoting horizontal gene transfer. *Eco-Environment & Health*, *4*(1), 100136. DOI: 10.1016/j.eehl.2025.100136 PMID: 40052062

Liu, Z. (2025). A maritime community with a shared future: Bridging fragmented legal regulations for marine plastic pollution. *Frontiers in Marine Science*, *12*, 1680110. DOI: 10.3389/fmars.2025.1680110

Long, Y., Wang, P., Lei, J., Su, B., Wei, Q., & Liu, X. (2025). Mechanical Signaling: Molecular Mechanisms, Biological Functions, Diseases, and Therapeutic Targets. *MedComm*, *6*(12), e70523. DOI: 10.1002/mco2.70523 PMID: 41377763

Luo, H., Zhao, X., Zhang, Y., & Cai, B. (2024). A framework for measuring the implementation of the Sustainable Development Goal 4.2 (Quality Preprimary Education) at the local scale: An empirical study in Nanjing, China. *Sustainable Development (Bradford)*, *32*(4), 3979–4000. DOI: 10.1002/sd.2876

Ma, X., Tian, Y., Yang, R., Wang, H., Allahou, L. W., Chang, J., Williams, G., Knowles, J. C., & Poma, A. (2024). Nanotechnology in healthcare, and its safety and environmental risks. *Journal of Nanobiotechnology*, *22*(1), 715. DOI: 10.1186/s12951-024-02901-x PMID: 39548502

MacLeod, M., Arp, H. P. H., Tekman, M. B., & Jahnke, A. (2021). The global threat from plastic pollution. *Science*, *373*(6550), 61–65. DOI: 10.1126/science.abg5433 PMID: 34210878

Macrì, S., & Di-Poï, N. (2024). The SmARTR pipeline: A modular workflow for the cinematic rendering of 3D scientific imaging data. *iScience*, *27*(12), 111475. DOI: 10.1016/j.isci.2024.111475 PMID: 39720527

Mahalingaiah, S., Nadeau, K. C., & Christiani, D. C. (2025). Microplastics and Human Health. *Journal of the American Medical Association, 334*(21), 1941–1942. DOI: 10.1001/jama.2025.14718 PMID: 41091491

Majid, S., Reus, A., Hoondert, R., Blokker, M., Dash, A., Houtman, C., Schriks, M., & Dingemans, M. M. L. (2025). A framework for evaluating less-than-lifetime exposures: Advancing toxicological risk assessment for drinking water quality. *Archives of Toxicology, 99*(8), 3147–3168. DOI: 10.1007/s00204-025-04061-9 PMID: 40415114

Malik, F. A., Kushwah, S. V., Aslam, M. S., Reyed, R. M., & Haghi, A. K. (Eds.). (2026). *Aligning ESG Goals with the UNs Sustainable Development Goals Planning: Strategies for Equitable Financing and World Sustainability (World Sustainability Series)*. Springer Cham., DOI: 10.1007/978-3-032-01715-4

Marik, A., Biswas, S., & Banerjee, E. R. (2024). Exploring the relationship between gut microbial ecology and inflammatory disease: An insight into health and immune function. *World Journal of Immunology, 14*(1), 96209. DOI: 10.5411/wji.v14.i1.96209

Marras, L., Caputo, M., Bisicchia, S., Soato, M., Bertolino, G., Vaccaro, S., & Inturri, R. (2021). The Role of Bifidobacteria in Predictive and Preventive Medicine: A Focus on Eczema and Hypercholesterolemia. *Microorganisms, 9*(4), 836. DOI: 10.3390/microorganisms9040836 PMID: 33919907

Matus, P., Urquidi, C., Cárcamo, M., & Vidal, V. (2024). Integrating the exposome and one health approach to national health surveillance: An opportunity for Latin American countries in health preventive management. *Frontiers in Public Health, 12*, 1376609. DOI: 10.3389/fpubh.2024.1376609 PMID: 39211902

Medley, E. A., Spratlen, J. J., Yan, B., Herbstman, J. B., & Deyssenroth, M. A. (2023). A systematic review of the placental translocation of micro- and nanoplastics. *Current Environmental Health Reports, 10*(2), 99–111. DOI: 10.1007/s40572-023-00391-x PMID: 36848019

Moulton, C., Baroni, A., & Tasciotti, E. (2026). Micro- and Nanoplastics Exposure Across the Lifespan: One Health Implications for Aging and Longevity. *Journal of Xenobiotics, 16*(2), 52. DOI: 10.3390/jox16020052 PMID: 41874123

National Academies of Sciences, Engineering, and Medicine [NASEM] (2016). Appendix E: A brief history of ecological risk assessment. In *Gene drives on the horizon: Advancing science, navigating uncertainty, and aligning research with public values*. The National Academies Press. https://doi.org/DOI: 10.17226/23405

Nematollahi, M. J., Mobasheri, M., Esmaeili, Z., Mahmoudi, M., Yousefi, N., & Busquets, R. (2025). Distribution and abundance of microplastics in urban and industrial wastewater treatment plants in Tabriz metropolis. *Scientific Reports, 15*(1), 24577. DOI: 10.1038/s41598-025-10000-6 PMID: 40629021

Nene, A., Sadeghzade, S., Viaroli, S., Yang, W., Uchenna, U. P., Kandwal, A., Liu, X., Somani, P., & Galluzzi, M. (2025). Recent advances and future technologies in nanomicroplastics detection. *Environmental Sciences Europe, 37*(7), 7. Advance online publication. DOI: 10.1186/s12302-024-01044-y

Nie, H. Y., Ge, J., Huang, G. X., Liu, K. G., Yue, Y., Li, H., Lin, H. G., Zhang, T., Yan, H. F., Xu, B. X., Sun, H. W., Yang, J. W., Si, S. Y., Zhou, J. L., & Cui, Y. (2024). New insights into the intestinal barrier through "gut-organ" axes and a glimpse of microgravity's effects on intestinal barrier. *Frontiers in Physiology, 15*, 1465649. DOI: 10.3389/fphys.2024.1465649 PMID: 39450142

Njema, G. G., Agoro, M. A., & Kibet, J. K. (2026). A critical evaluation of nano- and microplastics: Sources, health risks, and mitigation strategies. *Advances in Materials Science and Engineering, 2026*(1), 1357400. Advance online publication. DOI: 10.1155/amse/1357400

OECD. (2025). *"Development co-operation and the elimination of plastic pollution", OECD Policy Briefs, No. 20*. OECD Publishing., DOI: 10.1787/96803ffc-

Pegoraro, A. F., Chen, M., Sakib, S., Jakubek, Z. J., Maan, Z., Prezgot, D., Sallam, A., Corriveau, D., Vandenberg, M., & Zou, S. (2026). Nanoplastic reference materials for biological and methodological assessment. *Microplastics and Nanoplastics, 6*(1), 1. DOI: 10.1186/s43591-025-00157-2

Peng, J., Liu, X., Wang, J., Meng, N., Cai, R., Peng, Y., Han, Y., Liao, J., Li, C., Rubin-Blum, M., Ma, Q., & Dong, X. (2025). Diverse quorum sensing systems regulate microbial communication and biogeochemical processes in deep-sea cold seeps. *Microbiome, 14*(1), 16. DOI: 10.1186/s40168-025-02280-x PMID: 41340071

Perra, N., Giusto, D., & Anedda, M. (2026). A survey of modern data acquisition and analysis systems for environmental risk monitoring in aquatic ecosystems. *Sensors (Basel), 26*(5), 1566. DOI: 10.3390/s26051566 PMID: 41829526

Piergiacomo, F., Brusetti, L., & Pagani, L. (2022). Understanding the Interplay between Antimicrobial Resistance, Microplastics and Xenobiotic Contaminants: A Leap towards One Health? *International Journal of Environmental Research and Public Health, 20*(1), 42. DOI: 10.3390/ijerph20010042 PMID: 36612363

Porter, L. L. (2023). Fluid protein fold space and its implications. *BioEssays : News and Reviews in Molecular, Cellular and Developmental Biology*, *45*(9), e2300057. DOI: 10.1002/bies.202300057 PMID: 37431685

Procopio, A. C., Soggiu, A., Urbani, A., & Roncada, P. (2025). Interactions between microplastics and microbiota in a One Health perspective. One health (Amsterdam, Netherlands), 20, 101002. https://doi.org/DOI: 10.1016/j.onehlt.2025.101002

Qin, W., Cui, Y., Wang, H., Zhang, Y., & Cheng, Z. (2026). System architecture control design and clinical translation in closed-loop drug delivery. *Materials Today. Chemistry*, *54*, 103601. DOI: 10.1016/j.mtchem.2026.103601

Radanliev, P. (2025). AI Ethics: Integrating Transparency, Fairness, and Privacy in AI Development. *Applied Artificial Intelligence*, *39*(1), 2463722. Advance online publication. DOI: 10.1080/08839514.2025.2463722

Ragusa, A., & Fanos, V. (2025). Microplastics and nanoplastics in the brain: A review of the neurodevelopmental risks. [JPNIM]. *Journal of Pediatric and Neonatal Individualized Medicine*, *14*(2), e140206. DOI: 10.7363/140206

Ramamurthy, M. O., Subramanian, P., Aravindan, S., Periyasamy, L., & Aravindan, N. (2025). Radiation Without Borders: Unraveling Bystander and Non-Targeted Effects in Oncology. *Cells*, *14*(22), 1761. DOI: 10.3390/cells14221761 PMID: 41294814

Rangrazi, A., Bonyadi, Z., & Sarkhosh, M. (2026). Investigation of airborne microplastics emission and characteristics in hospital laundry environments. *Scientific Reports*, *16*(1), 6107. DOI: 10.1038/s41598-026-35421-9 PMID: 41580448

Ravi Shankar, A. L., Irwan, J., Spaenig, M., Carlier, M., Hansen, T., Zumbülte, N., Islam, B., Lundquist, P., Svensson, R., Gouin, T., Hamers, T., & Escher, S. E. (2026). Extrapolation of in vitro effect concentrations to in vivo bioavailable concentrations using PBK modeling in humans for two classes of persistent and mobile compounds: Triazoles and triazines. *Archives of Toxicology*, *100*(5), 1877–1901. DOI: 10.1007/s00204-025-04268-w PMID: 41699307

Rawat, H., Gaur, A., Singh, N., Selvaraj, M., Karnwal, A., Pant, G., & Malik, T. (2025). Artificial intelligence-driven detection of microplastics in food: A comprehensive review of sources, health risks, detection techniques, and emerging artificial intelligence solutions. *Food Chemistry: X*, *29*, 102687. DOI: 10.1016/j.fochx.2025.102687 PMID: 40677556

Reyed, R. M. (2007). The role of Bifidobacteria in health. *Research Journal of Medicine and Medical Sciences*, *2*(1), 14–24.

Reyed, R. M. (2019). Computational biotechnology: An approach in silico based modeling bioprocess. *International Journal of Research Studies in Microbiology and Biotechnology*, *5*(4), 1–5. DOI: 10.22259/2454-9428.0504001

Reyed, R. M. (2023). Simulate the dynamic interactions across nutritional supplements, the polybiome, and the biochemical approach dependent on patient-derived gastrointestinal knowledge. *Journal of Clinical Epidemiology & Toxicology*, *4*(2), 1–13. DOI: 10.47363/JCET/2023(4)133

Reyed, R. M. (2026a). Advancing diabetes reclassification using the NexDi AI system with multiomic insights: Egypt's living laboratory. In Reyed, R., Prabhakar, P., & Haghi, A. (Eds.), *Advancing personalized medicine with AI-driven microbiome and nutrigenomics: Innovations and future perspectives* (pp. 1–92). IGI Global Scientific Publishing., DOI: 10.4018/979-8-3373-3121-8.ch001

Reyed, R. M. (2026b). Digital Nano-Plastic Science (DNPS) paradigm: Computational intelligence and proteostasis disruptions. *Zenodo*. https://doi.org/DOI: 10.5281/zenodo.18435353

Reyed, R. M. (2026c). Polybiome systems medicine: Conceptual architecture, methodological foundations, and translational applications — Volume I: Vision and foundational methodology (Version 0.2). Zenodo. https://doi.org/DOI: 10.5281/zenodo.19258533

Reyed, R. M. (2026d). Nanobiotechnology and AI Convergence for Antimicrobial Resistance and Predictive Precision Therapeutics: Precision Therapeutics and One Health. In Reyed, R. (Ed.), *AI, Nanobiotechnology, and the Future of Precision Antibiotics* (pp. 1–74). IGI Global Scientific Publishing., DOI: 10.4018/979-8-3373-5268-8.ch001

Reyed, R. M., Patel, R., Patel, S., & Prajapati, B. G. (2026). Application of synthetic biology in genetic disorder. In *Synthetic biology and its consequences in present healthcare* (pp. 201–213). Academic Press., DOI: 10.1016/B978-0-443-33649-2.00001-0

Reyed, R. M., & Prabhakar, P. K. (2025). Mechanisms, impact, and therapeutic strategies for oxidative stress in metabolic diseases: Understanding and mitigating oxidative stress. In Prabhakar, P. (Ed.), *The role of reactive oxygen species in human health and disease* (pp. 203–256). IGI Global Scientific Publishing., DOI: 10.4018/979-8-3693-7919-6.ch006

Reyed, R. M., & Prabhakar, P. K. (2026a). Modulating neuroinflammation and autophagy in psychiatric disorders through mechanisms and therapies: Therapeutic strategies and pathways. In Mokhtari, T., & Uludag, K. (Eds.), *Autophagy and inflammation in neuropsychological disorders* (pp. 61–124). IGI Global Scientific Publishing., DOI: 10.4018/979-8-3693-5908-2.ch003

Reyed, R. M., & Prabhakar, P. K. (2026b). Panflemmotic triggers to metastatic pathways: Complex signatures of tumorigenesis in the colonic microenvironment. In Kumar, A., Mishra, N., Singh, A. K., Pareek, A., & Prabhakar, P. K. (Eds.), *Inflammation and cancer* (pp. 247–346). Academic Press., DOI: 10.1016/B978-0-443-36500-3.00008-8

Ringeval, M., Etindele Sosso, F. A., Cousineau, M., & Paré, G. (2025). Advancing Health Care With Digital Twins: Meta-Review of Applications and Implementation Challenges. *Journal of Medical Internet Research*, *27*, e69544. DOI: 10.2196/69544 PMID: 39969978

Saleh, A. M., Selim, R. N., El-Morsy, M. T., Mansour, S. A., Elhady, R., Ismail, S. F. K., Gamal, M., & Kenawy, E. S. (2026). A review of nanofiber-based drug delivery systems: Fabrication, characterization, advances, and future prospects. *Journal of Materials Science: Polymer*, *1*(1), 10. DOI: 10.1007/s44493-026-00009-2

Santos, F. A., Andre, R. S., Alvarenga, A. D., Alves, A. L. M. M., & Correa, D. S. (2025). Micro- and nanoplastics in the environment: A comprehensive review on detection techniques. *Environmental Science. Nano*, *12*(7), 3442–3467. DOI: 10.1039/D4EN00319E

Scheffer, M., Bascompte, J., Brock, W. A., Brovkin, V., Carpenter, S. R., Dakos, V., Held, H., van Nes, E. H., Rietkerk, M., & Sugihara, G. (2009). Early-warning signals for critical transitions. *Nature*, *461*(7260), 53–59. DOI: 10.1038/nature08227 PMID: 19727193

Schwabl, P., Köppel, S., Königshofer, P., Bucsics, T., Trauner, M., Reiberger, T., & Liebmann, B. (2019). Detection of various microplastics in human stool: A prospective case series. *Annals of Internal Medicine*, *171*(7), 453–457. DOI: 10.7326/M19-0618 PMID: 31476765

Seggio, M., Arcadio, F., Radicchi, E., Cennamo, N., Zeni, L., & Bossi, A. M. (2024). Toward Nano- and Microplastic Sensors: Identification of Nano- and Microplastic Particles via Artificial Intelligence Combined with a Plasmonic Probe Functionalized with an Estrogen Receptor. *ACS Omega*, *9*(17), 18984–18994. DOI: 10.1021/acsomega.3c09485 PMID: 38708270

Segovia, M., & Garcia-Alfaro, J. (2022). Design, Modeling and Implementation of Digital Twins. *Sensors (Basel)*, *22*(14), 5396. DOI: 10.3390/s22145396 PMID: 35891076

Serra, A., Torres Maia, M., Tsiros, P., Minadakis, V., Riudavets-Puig, R., Perello-y-bestard, A., Nikiforou, F., Karakoltzidis, A., Di Lieto, E., Schaffert, A., Al-Abdulraheem, Z., Virmani, I., Dziubaniuk, O., Karhukorpi, S., Dokler, J., Zouraris, D., Mintis, D. G., Varsou, D.-D., Tsoumanis, A., & Greco, D. (2026). Review on Predictive Models and Integration Strategies for Holistic Impact Assessment of Chemicals and Materials. *Environmental Science & Technology*, *60*(5), 3739–3766. DOI: 10.1021/acs.est.5c04489 PMID: 41603589

Sessitsch, A., Wakelin, S., Schloter, M., Maguin, E., Cernava, T., Champomier-Verges, M. C., Charles, T. C., Cotter, P. D., Ferrocino, I., Kriaa, A., Lebre, P., Cowan, D., Lange, L., Kiran, S., Markiewicz, L., Meisner, A., Olivares, M., Sarand, I., Schelkle, B., & Kostic, T. (2023). Microbiome Interconnectedness throughout Environments with Major Consequences for Healthy People and a Healthy Planet. *Microbiology and Molecular Biology Reviews : MMBR*, *87*(3), e0021222. DOI: 10.1128/mmbr.00212-22 PMID: 37367231

Shao, K., Zou, R., Zhang, Z., Mandemaker, L. D. B., Timbie, S., Smith, R. D., Durkin, A. M., Dusza, H. M., Meirer, F., Weckhuysen, B. M., Alderete, T. L., Vermuelen, R., & Walker, D. I. (2025). Advancements in Assays for Micro- and Nanoplastic Detection: Paving the Way for Biomonitoring and Exposomics Studies. *Annual Review of Pharmacology and Toxicology*, *65*(1), 567–585. DOI: 10.1146/annurev-pharmtox-030424-112828 PMID: 39270670

Shi, Y., Chen, W., Yang, S., Fan, Y., & Lu, L. (2024). Freshwater microplastics governance and sustainable development: Pollution status, interactions, policies, and prospective studies. *Desalination and Water Treatment*, *320*, 100704. DOI: 10.1016/j.dwt.2024.100704

Sidhu, A. K., Agrawal, S. B., Verma, N., Kaushal, P., & Sharma, M. (2025). Fungal-mediated synthesis of multimetallic nanoparticles: Mechanisms, unique properties, and potential applications. *Frontiers in Nanotechnology*, *7*, 1549713. Advance online publication. DOI: 10.3389/fnano.2025.1549713

Skaba, D., Fiegler-Rudol, J., Dembicka-Mączka, D., & Wiench, R. (2025). Nanoplastics and Immune Disruption: A Systematic Review of Exposure Routes, Mechanisms, and Health Implications. *International Journal of Molecular Sciences*, *26*(11), 5228. DOI: 10.3390/ijms26115228 PMID: 40508037

Snehamayee, N., Somya, S., Kumar, S. C., Niranjan, M., Ranjan, S. B., & Kumar, M. N. (2026). Microplastics and Human Health: A Comprehensive Review on Exposure Pathways, Toxicity, and Emerging Risks. *Microplastics*, *5*(1), 8. DOI: 10.3390/microplastics5010008

Song, W., Zhang, J., Ling, C., Zhao, Y., Ji, R., Su, Y., & Xing, B. (2026). Formation and transformation of micro(nano)plastics: Mechanisms and environmental health implications. *Environment & Health*, *4*(5), 826–842. DOI: 10.1021/envhealth.5c00592 PMID: 42164879

Stott, P. A., Lo, Y. T. E., Marsham, J. H., Obura, D., Oliver, T. H., Palmer, M. D., Ranger, N., Sharpe, S., & Sutton, R. (2026). We need a global assessment of avoidable climate-change risks. *Nature*, *650*(8103), 826–828. DOI: 10.1038/d41586-026-00544-6 PMID: 41741746

Su, J., Wang, H., Man, S., & Yan, Q. (2026). Biochemical insights into the alleviated inhibition on nitrogen metabolism by microand nanoplastics at the biocathode of bioelectrochemical systems. *Bioresource Technology*, *444*, 133938. DOI: 10.1016/j.biortech.2026.133938 PMID: 41490677

Suciu, N. A., Trevisan, M., Fragkoulis, G., Lamastra, L., Triachini, S., Abrantes, N., Norgaard, T., Harkes, P., Nunes, J. P., Silva, V., Navarro, I., de la Torre, A., Martínez, M. Á., Scheepers, P., & Martins Figueiredo, D. (2026). One Health perspective: An integrated in silico approach to assess the environmental fate of pesticides, exposure of aquatic and soil organisms, and risks for human health. *Environmental Research*, *294*, 123805. DOI: 10.1016/j.envres.2026.123805 PMID: 41577111

Sun, X. D., Yuan, X. Z., Jia, Y., Feng, L. J., Zhu, F. P., Dong, S. S., Liu, J., Kong, X., Tian, H., Duan, J. L., Ding, Z., Wang, S. G., & Xing, B. (2020). Differentially charged nanoplastics demonstrate distinct accumulation in *Arabidopsis thaliana*. *Nature Nanotechnology*, *15*(9), 755–760. DOI: 10.1038/s41565-020-0707-4 PMID: 32572228

Sundas, A., Contreras, I., Mujahid, O., Beneyto, A., & Vehi, J. (2024). The Effects of Environmental Factors on General Human Health: A Scoping Review. *Healthcare (Basel)*, *12*(21), 2123. DOI: 10.3390/healthcare12212123 PMID: 39517336

Thakur, R., Joshi, V., Sahoo, G. C., Jindal, N., Tiwari, R. R., & Rana, S. (2025). Review of mechanisms and impacts of nanoplastic toxicity in aquatic organisms and potential impacts on human health. *Toxicology Reports*, *14*, 102013. DOI: 10.1016/j.toxrep.2025.102013 PMID: 40230517

Trofin, F., Badescu, A. C., Iancu, L. S., Buzila, E. R., Anton-Păduraru, D.-T., Sima, C. M., Temneanu, O.-R., Matei, A., Bilha, S. C., Benea, I. A., & Dorneanu, O. S. (2026). Seeding the Future: How Feeding Mode Shapes the Infant Gut Microbiota. *Microorganisms, 14*(3), 719. DOI: 10.3390/microorganisms14030719 PMID: 41900478

Tyc, H. J., Kłodnicka, K., Teresińska, B., Karpiński, R., Flieger, J., & Baj, J. (2025). Micro- and Nanoplastics as Disruptors of the Endocrine System-A Review of the Threats and Consequences Associated with Plastic Exposure. *International Journal of Molecular Sciences, 26*(13), 6156. DOI: 10.3390/ijms26136156 PMID: 40649932

Uchmanowicz, D., Styszko, K., Chen, X., Terribile, G., Jakhar, R., Sancini, G., & Pyssa, J. (2025). Microplastics in Airborne Particulate Matter: A Comprehensive Review of Separation Techniques, In Vitro Toxicity and Health Impacts. *International Journal of Molecular Sciences, 26*(21), 10332. DOI: 10.3390/ijms262110332 PMID: 41226370

Valachovičová, M., & Mišľanová, C. (2026). Microplastics—The Microbiota Interactions: Mechanisms, Multi-Omics Insights and Health Implications. *Applied Sciences (Basel, Switzerland), 16*(9), 4110. DOI: 10.3390/app16094110

Venâncio, C., & Oliveira, M. (2025). A Multilevel Risk Assessment Framework for Nanoplastics in Aquatic Ecosystems. *Water (Basel), 17*(2), 180. DOI: 10.3390/w17020180

Vergani-Junior, C. A., Ventura, M. A. V. C., & De-Souza, E. A. (2026). Adaptive inter-tissue proteostasis networks in aging and neurodegeneration. *Bioscience Reports, 46*(1), BSR20254097. Advance online publication. DOI: 10.1042/BSR20254097 PMID: 41510837

Vethaak, A. D., & Legler, J. (2021). The exposome in the context of emerging contaminants. *Environmental Research, 197*, 111125. DOI: 10.1016/j.envres.2021.111125

Visca, A., Di Gregorio, L., Clagnan, E., & Bevivino, A. (2024). Sustainable strategies: Nature-based solutions to tackle antibiotic resistance gene proliferation and improve agricultural productivity and soil quality. *Environmental Research, 248*, 118395. Advance online publication. DOI: 10.1016/j.envres.2024.118395 PMID: 38307185

Walker-Franklin, I., Villalobos Santeli, A., & Neal-Walthall, N. (2026). Tracking Microplastics From Source to Impact: A Review of Environmental Presence, Exposure, Remediation, and Health Risks. *Current Environmental Health Reports, 13*(1), 12. DOI: 10.1007/s40572-026-00531-z PMID: 41844949

Walsh, A. N., Dunlea, A. G., Reddy, C. M., & Ward, C. P. (2025). Characterization of Inorganic Additives in and Photochemically Liberated from Consumer Plastics: Implications for Global and Local Biogeochemical Cycles. *ACS Environmental Au*, *5*(3), 298–307. DOI: 10.1021/acsenvironau.4c00130 PMID: 40416843

Wang, J., Li, D., Guan, R., Wang, Y., Li, H., & Zhao, C. (2025). A computational framework for multiscale data fusion in assessing the associations between micro- and nanoplastics and human hepatotoxicity. *Environment International*, *206*, 109937. DOI: 10.1016/j.envint.2025.109937 PMID: 41270658

Wiesinger, H., Wang, Z., & Hellweg, S. (2021). Deep dive into plastic monomers, additives, and processing aids. *Environmental Science & Technology*, *55*(13), 9339–9351. DOI: 10.1021/acs.est.1c00976 PMID: 34154322

Worms, I. A. M., Tharaud, M., Gasco, R., Montaño, M. D., Goodman, A., Slaveykova, V. I., Benedetti, M. F., Churchill, C. M., Fernando, S., Alasonati, E., Moens, C., & Cuss, C. W. (2025). Exploring environmental nanobiogeochemistry using field-flow fractionation and ICP-MS-based tools: Progress and frontiers. *Environmental Science. Nano*, *12*(8), 3821–3846. DOI: 10.1039/D5EN00096C PMID: 40708569

Yang, H., Zhang, K., Wang, Y., Liu, S., Guo, Y., Liu, W., Sun, J., Zhang, Z., Zhang, S., Li, S., Zhao, Y., Liu, T., Liu, J., Pei, L., Xi, S., & Shi, P. (2026a). Weighted network analysis of adverse outcome pathways decodes the multiscale mechanisms of environmental toxicity. *Environmental Science and Ecotechnology*, *29*, 100663. DOI: 10.1016/j.ese.2026.100663 PMID: 41660038

Yang, K., Liu, L., Zhu, G., Militky, J., Venkataraman, M., & Yao, J. (2026d). Perspective: A sandwich encapsulation strategy to mitigate PCM-based microplastic generation from fibrous materials. *Advanced Sustainable Systems*, *10*(2), e01645. Advance online publication. DOI: 10.1002/adsu.202501645

Yelleti, G., Maripini, N., & Bolar Suryakanth, V. (2026). Preeclampsia and environmental epigenomics: The emerging role of air pollution, gut microbiome, and maternal exposures in disease programming. *Environmental Epigenetics*, *12*(1), dvag001. Advance online publication. DOI: 10.1093/eep/dvag001 PMID: 41704669

Yong, C. Q. Y., Valiyaveettil, S., & Tang, B. L. (2020). Toxicity of Microplastics and Nanoplastics in Mammalian Systems. *International Journal of Environmental Research and Public Health*, *17*(5), 1509. DOI: 10.3390/ijerph17051509 PMID: 32111046

Yu, H.-R., Sheen, J.-M., & Tiao, M.-M. (2024). The Impact of Maternal Nanoplastic and Microplastic Particle Exposure on Mammal's Offspring. *Cells*, *13*(16), 1380. DOI: 10.3390/cells13161380 PMID: 39195272

Zhang, F., Guo, C. G., Yang, C., Wang, F., Wang, W., & Zhang, L. (2025). Exposure-response associations of ambient heavy metal and persistent organic pollutant with all-cause and cause-specific mortality: A prospective cohort study. *Environment & Health*, *3*(5), 493–503. DOI: 10.1021/envhealth.4c00191 PMID: 40400556

Zhao, Y. M., Zhang, J. Y., Liaw, P. K., & Yang, T. (2025). Machine learning-based computational design methods for high-entropy alloys. *High Entropy Alloys Materials*, *3*(1), 41–100. DOI: 10.1007/s44210-025-00055-5

Zhu, Z. M., Wu, X. M., Hu, Y., Bian, X. L., Wang, Y. Q., & Zhu, Q. N. (2025). Alternative Splicing: Molecular Mechanisms, Biological Functions, Diseases, and Potential Therapeutic Targets. *MedComm*, *6*(12), e70545. DOI: 10.1002/mco2.70545 PMID: 41399527

Zolotova, N., Kirillova, M., Dzhalilova, D., Tsvetkov, I., Fokichev, N., & Makarova, O. (2025). Microplastics' Impact on the Development of AOM/DSS-Induced Colitis-Associated Colorectal Cancer in Mice. *International Journal of Molecular Sciences*, *26*(23), 11511. DOI: 10.3390/ijms262311511 PMID: 41373666

APPENDIX A. SUPPLEMENTARY METHODS: DATA-DRIVEN SYSTEMS FRAMEWORK (DDSF)

A1. Framework architecture and conceptual positioning

A1.1 Scope and role within the manuscript

This appendix describes the operational structure of the Data-Driven Systems Framework (DDSF), which functions as the computational layer supporting DNPS and its associated constructs, including PSI and TSL. The intention is not to formalize a closed algorithmic system, but rather to outline a working architecture through which heterogeneous environmental and biological signals can be organized into interpretable analytical pathways.

In practice, the framework sits between conceptual modeling and data implementation. This intermediate position is deliberate, since the behavior of the system is expected to evolve alongside data availability and methodological refinement.

A1.2 System architecture overview

The DDSF is organized as a modular pipeline, although its internal boundaries are not strictly fixed during operation. Information tends to circulate between stages rather than move in a strictly linear direction, particularly when data heterogeneity increases.

The main functional components include:

- acquisition of environmental and biological signals from distributed sources
- harmonization of heterogeneous datasets into comparable analytical forms
- transformation of raw inputs into structured features
- detection of latent exposure–response patterns
- scenario simulation using digital twin representations
- aggregation of outputs into a systemic risk representation (SRS)

Figure 7 illustrates this flow. However, the diagram should be interpreted as a simplified representation rather than a strict sequence, since feedback loops are expected under real-world conditions.

A2. Operationalization of the Transductive Signal Load (TSL)

A2.1 Conceptual anchoring

TSL is introduced as a latent construct rather than a directly observable metric. It reflects inferred biological burden arising from environmental interactions, though its value is always model-dependent.

There is an important nuance here: TSL does not correspond to a single measurable variable but instead emerges from the convergence of multiple partial indicators.

A2.2 Computational implementation

In computational terms, TSL is derived from a structured but non-exhaustive set of inputs:

- physicochemical descriptors of particles (size, morphology, surface charge variability)
- environmental context variables (temperature fluctuations, salinity, pH shifts, photodegradation effects)
- co-exposure environments involving chemical mixtures and trace pollutants
- biological response markers reflecting oxidative, inflammatory, and metabolic shifts

Rather than treating these inputs as independent predictors, the framework allows them to interact nonlinearly. Ensemble-based and deep learning methods are used to stabilize patterns that would otherwise remain fragmented across datasets.

TSL therefore operates as a compressed representation of distributed environmental stress.

A3. Digital Twin (DT) simulation framework

A3.1 Functional definition

Digital twin environments in DDSF are designed as adaptive simulations rather than predictive replicas. Their role is exploratory: they allow hypothetical exposure scenarios to be examined under controlled computational conditions.

A3.2 Core components

The simulation layer typically operates through three loosely coupled functions:

- continuous assimilation of environmental data through DEIs
- simulation of exposure–response trajectories under variable conditions
- identification of fragile or unstable system configurations

These functions are not strictly separated in execution. In practice, they often overlap depending on the complexity of the scenario being modeled.

A3.3 Calibration and uncertainty

Calibration occurs iteratively and does not converge in a final form. Instead, model outputs are periodically adjusted against available empirical datasets.

However, a persistent imbalance remains between simulation depth and observational completeness. Rather than treating this as a limitation, DDSF explicitly incorporates it into its uncertainty structure.

A4. Dynamic Eco-Integrators (DEIs) and data acquisition

A4.1 Functional role

DEIs represent the entry layer of DDSF. Their primary function is not interpretation, but stabilization of incoming heterogeneous data streams.

A4.2 Data sources

Input data are typically drawn from multiple domains:

- environmental monitoring systems (air, water, soil matrices)
- biological sampling outputs (omics, blood chemistry, microbiome profiles)
- remote sensing platforms providing spatial ecological coverage

These sources differ significantly in resolution and reliability, which makes harmonization a continuous requirement rather than a one-time preprocessing step.

A4.3 Preprocessing and quality control

Before integration into higher-level modules, DEIs perform several operations:

- normalization across incompatible measurement scales
- alignment of temporal and spatial sampling structures
- correction of missing or noisy observations

These steps are adaptive and may vary depending on dataset complexity.

A5. Hierarchical data integration

A5.1 Multilevel structure

DDSF organizes information across three interacting layers:

1. molecular signals reflecting intracellular perturbations
2. environmental conditions representing exposure context
3. systemic responses emerging from cross-scale interactions

The separation is analytical rather than physical; in practice, these layers interact continuously.

A5.2 Signal propagation

Signal transmission across levels is not assumed to be proportional. Small perturbations at molecular scale may, under certain conditions, propagate into amplified systemic effects.

This asymmetry is central to the framework and is visually represented in Figure 7.

A6. Systemic Risk Score (SRS) computation

A6.1 Definition

SRS is defined as a composite probabilistic construct integrating exposure intensity, signaling disruption, and system vulnerability. It is intended for comparative interpretation rather than absolute classification.

A6.2 Computational pathway

The estimation process involves:

- normalization of heterogeneous inputs
- context-dependent weighting of variables
- probabilistic aggregation into a unified score

The resulting value should be interpreted as a relative indicator rather than a deterministic output.

A6.3 Interpretive scope

SRS does not function as a threshold model. Instead, it provides a structured way to compare scenarios under uncertainty, where absolute prediction is not feasible.

A7. Traceability and reproducibility

All transformations within DDSF are recorded through version-controlled workflows. This ensures that each output can be traced back to its corresponding input configuration and model state.

Reproducibility is therefore conditional rather than absolute, depending on both data completeness and model version alignment.

A8. Uncertainty and future extensions

A8.1 Nature of uncertainty

Uncertainty is treated as a structural component of the framework rather than an external noise factor. It is divided into epistemic uncertainty (knowledge limitations) and data-driven variability (sampling and measurement effects).

Both are retained within the modeling structure rather than removed.

A8.2 Future directions

Future development of DDSF is expected to focus on gradual improvements rather than structural replacement:

- higher resolution environmental sensing integration
- tighter coupling with real-time omics systems
- improved stability in digital twin simulation outputs

A8.3 Concluding remark

DDSF operates as a bridging architecture between DNPS and PSI, linking conceptual modeling with computational representation. Within this structure, SRS emerges as a synthesized expression of systemic risk under uncertainty, rather than a fixed endpoint of measurement.

GLOSSARY A

Ontological Glossary and Operational Lexicon for Digital Nano-Plastic Science (DNPS)

Overview

This glossary accompanies the DNPS framework by assembling a set of working concepts intended to support structured interpretation rather than fixed classification. The emphasis is deliberately placed on how nanoplastic exposure may be conceptualized as a distributed and context-dependent phenomenon, one that does not remain confined to a single scale of observation. Instead, it appears to unfold across interacting environmental, biological, and computational domains, where signals, perturbations, and modeled responses are continuously re-contextualized.

Rather than offering definitive boundaries between terms, the glossary preserves a degree of conceptual flexibility. This reflects the underlying assumption that environmental stressors of this nature are not easily reducible to isolated variables, but are better approached through layered descriptions that remain sensitive to uncertainty, partial observability, and model dependence.

Table 2. Ontological Glossary and Operational Lexicon

Term	Ontological Positioning & Operational Definition	Systemic Utility & Functional Role
DNPS (Digital Nano-Plastic Science)	A conceptual modeling domain in which nanoplastic exposure is treated less as material accumulation and more as a distributed pattern of interactions. The framework links observational environmental data with computational representations of particle–system behavior across biological boundaries.	Provides a structural lens through which heterogeneous environmental signals can be reorganized into interpretable model outputs describing biological response trajectories.
DDSF (Data-Driven Systems Framework)	A modular analytical architecture designed to accommodate heterogeneous datasets, including environmental measurements, molecular-level profiles, and regulatory records. Integration is treated here as a gradual alignment process rather than a simple aggregation step.	Enables scenario-based inference by maintaining traceability between raw data streams and higher-order analytical representations.
DEIs (Dynamic Eco-Integrators)	Computational components designed to reduce structural inconsistencies across environmental datasets originating from different sensing infrastructures. Their function is less about transformation and more about stabilization of input coherence.	Facilitates cross-platform comparability by moderating variability prior to downstream modeling procedures.
Planetary Environmental Intelligence	A distributed observational layer combining satellite systems, sensor networks, and in situ biological monitoring. The emphasis is on continuity of observation rather than isolated measurement events.	Supports longitudinal environmental assessment and provides a baseline for detecting slow-evolving ecological shifts.
PSI (Plasto-Signaling Interference)	A conceptual representation of exposure effects framed through disruption of biological communication pathways. The interaction is understood as occurring at the nano–bio interface, where signal fidelity may be altered rather than simply accumulated.	Reframes toxicity in terms of degraded signaling coherence across interconnected biological and ecological feedback networks.
TSL (Transductive Signal Load)	An inferred modeling construct representing the cumulative burden arising when external interactions are translated into internal biological stress responses. Its value is derived indirectly rather than measured directly.	Acts as a bridging metric linking external exposure dynamics to internal physiological adaptation pressure.
Synthetic Perturbome	A structured representation of aggregated biological and ecological perturbations, capturing recurring response configurations across multiple organizational levels. It functions as a comparative interpretive space rather than a fixed dataset.	Enables identification of recurrent stress-response signatures across metabolic and ecological systems.

continued on following page

Table 2. Continued

Term	Ontological Positioning & Operational Definition	Systemic Utility & Functional Role
Signal Collapse & Blackout	A condition characterized by pronounced loss of coherence within biological signaling networks, where feedback regulation becomes progressively unreliable. The transition is typically gradual rather than abrupt.	Identifies a boundary regime in which regulatory stability is no longer sufficient to maintain functional homeostasis.
Nanoplastic Proteostatic Panflemmosis Metabolic Continuum (NPPMC)	A conceptual continuum describing a condition-dependent sequence in which nanoplastic exposure may contribute to proteostatic strain, which in turn may align with distributed panflemmotic signaling patterns and, under sustained biological and environmental constraints, potentially extend toward metabolic dysregulation. The construct is introduced as an interpretative and hypothesis-generating model rather than a fixed pathological category or diagnostic entity.	Provides a higher-order interpretative framework for organizing relationships between environmental exposure, proteostatic imbalance, distributed panflemmotic signaling behavior, and downstream metabolic perturbation within DNPS/DDSF modeling architectures. Supports scenario-based reasoning in complex, multiscale bio-environmental systems under uncertainty.

GLOSSARY B

Integrated Glossary of Systemic Risk and Bio-Environmental Modeling Constructs

Overview

This section extends the conceptual architecture introduced earlier by moving from interpretative framing toward operational modeling language. While the preceding glossary focused on how nanoplastic exposure might be understood as a distributed signaling phenomenon, the present set of terms develops this idea further by considering how such phenomena can be represented, compared, and analyzed within structured risk environments.

The relationship between the two glossaries is intentionally sequential rather than parallel. The first establishes a conceptual vocabulary for describing complexity; the second gradually translates that vocabulary into constructs capable of supporting modeling, inference, and comparative assessment under conditions where uncertainty remains intrinsic rather than exceptional.

Table 3.

Term	**Glossary Definition**	**Governance / Policy Relevance**
SRS (Systemic Risk Score)	A composite probabilistic construct integrating exposure intensity, disruption signaling, and system vulnerability. The emphasis is placed on relative weighting rather than deterministic calculation.	Supports decision-making in contexts where risk must be interpreted under incomplete or evolving information conditions.
Fragility Mapping	A methodological approach for identifying regions within biological or ecological networks that exhibit heightened sensitivity to incremental perturbations.	Assists in prioritizing intervention points before cascading instability becomes observable.
Invisibility Gap	The divergence between current detection capability and the interpretive capacity required to understand the biological consequences of ultrafine exposures.	Highlights epistemic limitations in translating sensing data into actionable regulatory knowledge.
Tipping Point	A nonlinear transition zone in which small changes in exposure–vulnerability interactions may correspond to disproportionate shifts in system behavior.	Defines a regime boundary where conventional mitigation strategies may no longer remain effective.
Hierarchical Coupling	Interactions occurring across multiple organizational levels, linking molecular-scale processes with broader ecological dynamics.	Captures emergent system behavior that is not observable at isolated scales of analysis.

continued on following page

Table 3. Continued

Term	**Glossary Definition**	**Governance / Policy Relevance**
Information-Coded Entities	A modeling abstraction in which nanoscale particles are represented through their interaction patterns rather than purely material properties.	Encourages interpretation of information as physically instantiated interaction rather than symbolic attribution.
Proteostatic Stress Continuum (PSC)	A graded spectrum describing disruption in protein homeostasis, ranging from adaptive responses to sustained functional impairment.	Provides a sensitive axis for early identification of systemic biological stress trajectories.
Type-V Diabetes (Heuristic Model)	A non-clinical conceptual model describing metabolic dysfunction associated with combined proteostatic imbalance, nutritional insufficiency, and environmentally mediated pancreatic stress.	Serves as an exploratory construct for examining long-term coupling between exposure dynamics, metabolic resilience, and systemic vulnerability.

Editorial Note

These glossaries are intended as structured conceptual aids rather than definitive classifications or diagnostic frameworks. Their primary role is to support interpretation and modeling of complex environmental–biological interactions under conditions of uncertainty. Accordingly, the terminology should be read as analytical language that facilitates comparative reasoning rather than as a replacement for established toxicological, clinical, or ecological taxonomies. The constructs remain explicitly bounded by their epistemic context and are best understood as tools for organizing complexity rather than resolving it.

Chapter 2
Forensic Tracing Systems for Nanoplastics:
AI and Biotech One Health Source Attribution

Yash Bhardwaj
http://orcid.org/0009-0009-0631-2199
K.R. Mangalam University, India

Anuj Kumar
http://orcid.org/0009-0009-3054-896X
K.R. Mangalam University, India

Khushbu Katiyar
http://orcid.org/0009-0003-0605-363X
Medicaps University, Indore, India

Sameer Saharan
http://orcid.org/0000-0002-7487-6297
Mody University of Science and Technology, Rajasthan, India

Kazi Mizaan
K.R. Mangalam University, India

Mahipal Singh Sankhla
K.R. Mangalam University, India

Baljeet Yadav
http://orcid.org/0000-0003-0624-6784
K.R. Mangalam University, India

ABSTRACT

Nanoplastics are increasingly recognized as a One Health concern because of their persistence, environmental mobility, and potential impacts across interconnected ecosystems, wildlife, and human populations. Their small size and heterogeneous composition make detection, source tracing, and exposure assessment challenging when conventional analytical approaches are used alone. This chapter per the authors discusses an integrated forensic framework combining biotechnological methods with artificial intelligence (AI)-based analytics to improve detection and characterization of nanoplastics across complex environmental matrices. Forensic tracing systems—including molecular assays, spectroscopic and imaging tech-

DOI: 10.4018/979-8-3373-4490-4.ch002

niques, and bioassays—are examined as complementary approaches for material discrimination, while computational models integrate heterogeneous datasets and strengthen source attribution. Ultimately, this chapter connects automated detection, forensic tracing, and exposure modeling within a One Health framework to support environmental risk assessment and management.

1. INTRODUCTION AND CONCEPTUAL FRAMEWORK

One Health is an integrative framework that links human, animal, and environmental health, recognizing that disturbances in one domain may influence the stability and health outcomes of the others. Within this context, nanoplastics (NPLs) have emerged as pervasive environmental contaminants capable of entering ecosystems through multiple environmental pathways. Current evidence suggests that nanoplastics may adversely affect wildlife, humans, and ecosystem functions, with exposure concentrations appearing to influence the magnitude of biological and ecological impacts (Ramsperger et al., 2023).

Understanding the relationships among nanoplastic sources, environmental distribution, exposure pathways, and associated effects requires a systems-oriented and interdisciplinary approach. Such an approach may integrate bioinformatics, machine learning, and biotechnology to establish analytical pipelines that connect nanoplastic origin, transport, exposure, and biological consequences across interconnected environmental compartments. Within a One Health framework, these pipelines may also support traceability-oriented assessment strategies capable of linking environmental occurrence patterns with potential exposure scenarios and ecological outcomes.

One Health principles provide the conceptual architecture for integrating these components by enabling cross-domain tracking of nanoplastic movement and impact. In this context, environmental monitoring technologies increasingly generate datasets describing nanoplastic occurrence across freshwater systems, sediments, soils, atmospheric compartments, indoor air, and stormwater environments. Integration of these datasets into shared repositories may support computational modeling frameworks capable of estimating exposure patterns, identifying probable source regions, and improving environmental surveillance capacities. The overall conceptual integration of these components is illustrated in Figure 1.

Data-driven analytical frameworks are increasingly relevant to environmental management and policy development in sectors such as agriculture, manufacturing, and waste management, where plastic-related emissions continue to represent a growing environmental concern (Hartmann et al., 2019). Although current approaches remain methodologically heterogeneous, ongoing efforts continue to improve connectivity among monitoring systems, computational analytics, and environmental governance strategies.

Beyond environmental and health dimensions, nanoplastics may also be interpreted within a forensic-oriented One Health context. In this framing, these particles function as traceable environmental indicators that support source attribution, exposure reconstruction, and assessment of environmental responsibility across interconnected ecological and human systems. Nanoplastics represent a critical but challenging frontier within this architecture, emphasizing that human health, animal health, and environmental integrity are fundamentally interdependent. Unlike larger plastic fragments, nanoplastics possess unique physicochemical properties—including a high surface-area-to-volume ratio, high surface reactivity, and the ability to cross biological membranes—that allow them to interact with environmental and biological systems in ways that remain poorly understood. These particles enter ecosystems through diverse pathways, including industrial discharge, agricultural practices, and the fragmentation of consumer products, subsequently moving through trophic networks to impact wildlife and potentially human consumers. Managing these risks requires an integrated analytical approach capable of tracking contaminants across the boundaries of traditional scientific disciplines.

Building upon the foundational principles of cyber-biotechnology and computational environmental surveillance, the paradigm is expanded here into a specialized, forensic-oriented architecture. By transitioning from general monitoring to targeted source attribution, this section explores how bio-digital pipelines can operationalize forensic evidence frameworks across complex ecosystems. This bridging perspective ensures that established digital bioprocess development is translated directly into actionable forensic toolkits capable of isolating nanoplastic signatures within interconnected biological and environmental matrices.

Figure 1. One Health framework integrating environmental and biological data repositories with bioinformatics analytics and policy pathways for nanoplastic assessment. The framework illustrates how heterogeneous environmental inputs are processed through analytical and computational pipelines to support source tracing, exposure assessment, risk characterization, and evidence-based mitigation strategies across interconnected human, animal, and environmental systems.

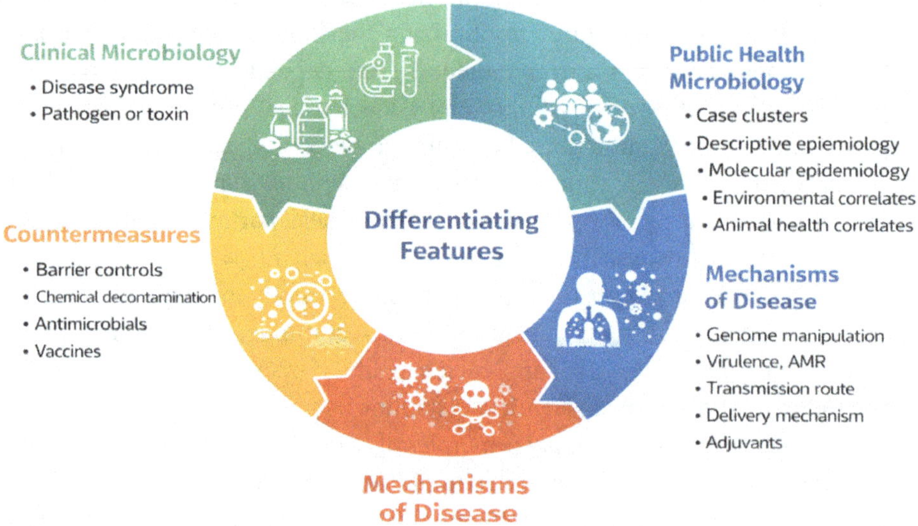

1.1. One Health and Nanoplastics

One Health recognizes the interdependence of human, animal, plant, and ecosystem health, emphasizing that improvements in one domain should not compromise health in another (Destoumieux-Garzón et al., 2018). Within this framework, nanoplastics, generally defined as plastic particles below 1 μm, are increasingly recognized as emerging contaminants with considerable ecotoxicological relevance (Yee et al., 2021; Vethaak & Leslie, 2016).

Artificial intelligence approaches may facilitate integration of information related to nanoplastic sources, environmental distribution, exposure pathways, and associated biological and socioeconomic effects. These computational approaches can also strengthen environmental traceability frameworks by improving interpretation of complex environmental datasets and supporting evidence-based regulatory and environmental governance strategies.

From a One Health perspective, effective assessment of nanoplastics requires coordinated integration of biosciences, biotechnology, environmental analytics, and computational modeling. Detection and characterization remain essential prerequisites for source attribution, exposure assessment, and risk evaluation. These processes additionally contribute to governance frameworks and ethical decision-making related to environmental monitoring and pollution mitigation (Amaral-Zettler et al., 2020).

The integration of rapidly evolving technologies, including microfluidics, environmental sensing systems, and remote sensing approaches, may further strengthen monitoring and analytical capacity in nanoplastic research. Nevertheless, translating these technological advances into practical environmental management strategies remains a substantial challenge, particularly across aquatic ecosystems, urban environments, and agricultural food systems. The interconnected nature of these scientific and environmental challenges is conceptually summarized in Figure 2.

Figure 2. One Health framework illustrating interconnected health domains and key challenges in nanoplastic forensics, including ecotoxicological threats, source attribution, exposure assessment, risk governance, and technology translation. The paradigm contextualizes the multidirectional pathways through which nanoscale particulates cross Anthropocene boundaries, mapping systemic vulnerabilities across environmental, animal, and anthropogenic sectors. By framing source attribution and exposure assessment as core forensic challenges, the schema highlights the need to integrate multi-omics bioassays with computational modeling approaches. This interconnected structure bridges the gap between laboratory-scale technology translation and broader risk-governance frameworks, providing a standardized structure for cross-disciplinary surveillance.

1.2. Scope and Significance

Nanoplastics, commonly described as plastic particles smaller than 1 μm, are increasingly reported as ubiquitous environmental contaminants across diverse environmental compartments (Ma et al., 2023). Their small size and physicochemical characteristics facilitate dispersion in both aquatic and atmospheric systems, thereby supporting long-range transport and widespread environmental distribution.

Nanoplastics primarily originate from degradation and fragmentation processes affecting larger plastic materials already present in the environment. Urban runoff, landfill leachate, industrial discharge, and diffuse anthropogenic activities may contribute to their entry into wastewater systems and environmental reservoirs. Wastewater treatment facilities and surface waters have consequently emerged as important points of environmental detection and redistribution (Koelmans et al., 2020).

Freshwater systems play a particularly important role in the environmental transport of nanoplastics between terrestrial and marine ecosystems. Key processes include sedimentation, trophic transfer, downstream transport through rivers and lakes, and redistribution through drainage networks and suspended particulates (Besseling et al., 2017).

Evidence increasingly suggests that nanoplastics are present across multiple environmental matrices, raising concerns regarding their long-term ecological and biological implications. Agricultural soils exposed to prolonged urban or industrial influence have been reported to contain bioavailable nanoplastics capable of interacting with plant systems and potentially entering broader ecological pathways (Azeem et al., 2023).

In contrast, transfer pathways associated with nanoplastics originating from biological waste streams may differ in persistence, transformation dynamics, and environmental bioavailability depending on polymer composition and degradation behavior. Some polymer-associated constituents may undergo partial degradation over time, suggesting that environmental exposure risks can vary according to physicochemical and ecological conditions (Shruti & Kutralam-Muniasamy, 2019).

These distinctions underscore the importance of context-dependent assessment frameworks capable of differentiating exposure scenarios across environmental compartments and ecological gradients. Within a One Health perspective, such approaches may support targeted environmental monitoring, traceability-oriented exposure assessment, and more precise environmental management strategies aimed at mitigating nanoplastic-associated risks.

1.3. Research Goals and Objectives

Nanoplastics are now widely distributed across environmental systems, and growing research efforts from academic, industrial, and governmental sectors continue to address their environmental and biological implications. Despite these advances, substantial gaps remain in understanding nanoplastic sources, exposure pathways, environmental persistence, and ecotoxicological consequences. Existing approaches are often fragmented and insufficiently integrated across environmental, biological, analytical, and computational domains.

This study addresses the need for a comprehensive One Health forensic framework that extends beyond conventional physicochemical characterization. The framework focuses on three interconnected objectives: (i) identifying potential sources of nanoplastics, (ii) characterizing environmental and biological exposure pathways, and (iii) evaluating associated ecological and ecotoxicological implications across interconnected environmental systems.

The proposed framework integrates biotechnology-based detection approaches with artificial intelligence–supported analytical models to strengthen source tracking, exposure reconstruction, and environmental traceability assessment. By combining these components within a One Health perspective, the framework seeks to establish a structured and scalable forensic-oriented pipeline for nanoplastic assessment across diverse environmental contexts.

Although substantial progress has been achieved in nanoplastic research, important limitations remain regarding standardized detection methodologies, real-time environmental monitoring systems, and integrated forensic modeling approaches for source attribution. Addressing these limitations may contribute to improved scientific understanding, environmental governance, and evidence-based management strategies for emerging nanoplastic pollution challenges.

2. NANOPLASTICS: PROPERTIES, SOURCES, AND ENVIRONMENTAL FATE

Nanoplastics are an emergent class of contaminants defined as plastic particles in the range of 1–1,000 nm, typically formed through the environmental degradation of macroplastics (Lehner et al., 2019; Frias & Nash, 2019). Their reduced size results in physicochemical behaviors that differ substantially from those of larger plastic particles, consequently influencing transport dynamics, biological uptake, environmental persistence, and toxicity profiles (Qian et al., 2024). In most environmental contexts, nanoplastics originate either from thermoplastic materials incorporated into consumer products or from the progressive fragmentation of larger plastic debris through photodegradation and mechanical disintegration.

Source identification is increasingly supported through analytical and traceability-informed characterization approaches. Trace elemental signatures may provide preliminary indications of material origin, while mass spectrometry enables more detailed characterization of polymer composition and material-specific signatures (Brachner et al., 2020). As illustrated in Figure 3, evidence of environmental distribution has now been reported across multiple matrices, including soil, air, sediment, and biological systems. Notably, nanoplastics have been detected in human lung tissue, the placenta, and intestinal enterocytes (Prata et al., 2020; Leslie et al., 2022). At the organism level, translocation across biological barriers and subsequent accumulation within tissues and organs have also been documented. Early experimental evidence further suggests that polystyrene nanoplastics may induce cellular responses associated with inflammation, oxidative stress, and disruption of normal cellular function.

The environmental behavior, detection, and biological effects of nanoplastics are governed largely by four interconnected physicochemical parameters: composition, morphology, size, and surface properties. Among these, surface chemistry and polymer composition strongly influence interactions with biomolecules, environmental colloids, and microbial biofilms. Collectively, these characteristics are critical for interpreting environmental transport pathways, reconstructing exposure scenarios, and supporting traceability-informed source attribution models based on material-specific signatures within a One Health framework, as conceptually summarized in Figure 3.

Figure 3. Overview of nanoplastic contamination pathways, illustrating sources, environmental fate (transport and uptake), altered physicochemical properties, and resulting biological accumulation and toxic effects. The diagram maps the systemic movement of particulate vectors across environmental matrices, elucidating how atmospheric aerosolization and trophic transfer processes drive contaminant transport. As nanoscale particles interact with surrounding media, altered physicochemical properties—including corona formation and surface charge shifts—modulate cellular internalization. This structural progression traces the continuum from initial source emission to biomagnification, providing foundational mechanistic information for subsequent computational source attribution and ecological risk-governance strategies.

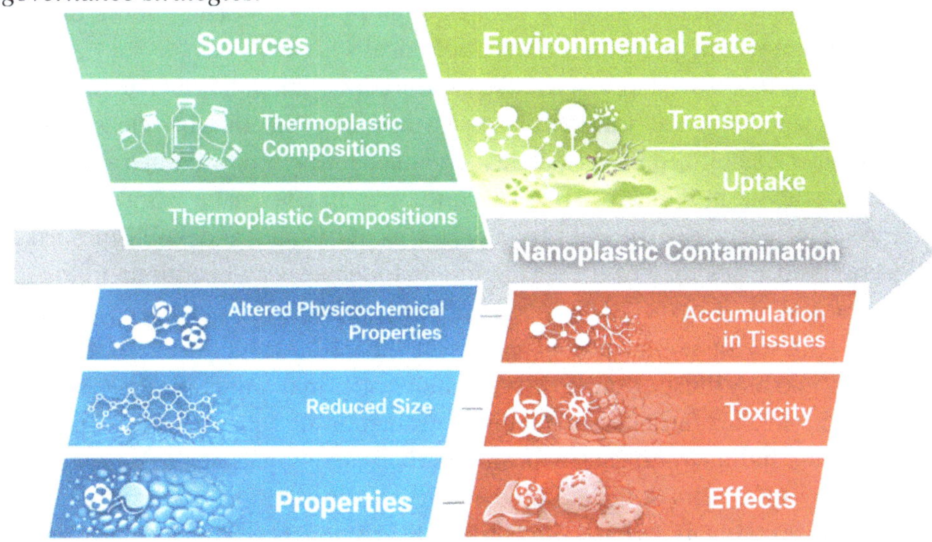

2.1. Physicochemical Characteristics

The SOURCE–EXPOSURE–IMPACT framework provides a structured basis for assessing ecological and biological risks associated with nanoplastic contamination. Within a One Health forensic perspective, this framework may also support environmental traceability assessment by linking physicochemical signatures with exposure pathways and ecological outcomes. This section therefore outlines the fundamental physicochemical characteristics of nanoplastics as a basis for understanding analytical approaches used in source attribution, exposure reconstruction, and impact evaluation.

In practice, nanoplastic properties, environmental transformations, and surface modifications collectively determine the methodological requirements for their detection and characterization, both of which remain essential for tracing environmental pathways and predicting ecological and health-related consequences (Brachner et al., 2020). These relationships additionally support exposure reconstruction, spatial distribution modeling, and environmental transport simulations across interconnected ecosystems.

Nanoplastics are generally defined as polymer particles with at least one external dimension between 1 nm and 1 μm, or as systems containing dispersions below approximately 1000 nm in characteristic size (Gigault et al., 2018). Compared with microplastics, their nanoscale dimensions confer distinct environmental behaviors, particularly regarding release mechanisms, atmospheric and aquatic transport, environmental transformation, and final deposition patterns. Due to their small size, nanoplastics may enter the human body through inhalation or ingestion, thereby enabling direct interaction with internal biological systems.

Given these exposure pathways, understanding environmental behavior is central to environmental risk and traceability assessment. At the point of release, nanoplastics may be stabilized by surfactants, organic additives, or oligomeric residues associated with the polymer matrix. In aquatic environments, natural organic matter can further influence particle stability, aggregation dynamics, and mobility. Surface adsorption processes may begin during product lifecycles and continue following environmental release, thereby affecting dispersion behavior, persistence, and interaction with surrounding environmental constituents.

As conceptually illustrated in Figure 3, nanoplastics may simultaneously undergo long-range transport through atmospheric circulation or aquatic systems prior to sedimentation and environmental deposition. Alternatively, particles may be released directly from solid, semi-solid, or liquid matrices during industrial or consumer-use processes with relatively limited structural modification. These distinct transport and transformation behaviors complicate environmental exposure reconstruction efforts and reinforce the importance of integrated detection and traceability methodologies within One Health-oriented environmental surveillance systems.

2.2. Pathways and Sources Across Ecosystems

Nanoplastics originate both from environmental degradation of larger plastic materials and from direct release as primary nanoplastic particles. Multiple interconnected pathways govern their movement across environmental compartments and ecological systems. Atmospheric transport represents one of the principal dissemination mechanisms, allowing nanoplastics and microplastic-derived fragments to undergo long-range transport prior to deposition across terrestrial and aquatic environments (Wright et al., 2023).

Urban surfaces, aquatic systems, and wastewater treatment facilities frequently function as environmental deposition sinks, where accumulated particles may subsequently undergo fragmentation and transformation into nanoscale forms (Corte Pause et al., 2024; Koelmans et al., 2020). These environmental reservoirs may therefore contribute not only to contaminant accumulation but also to the continuous generation of secondary nanoplastics within broader environmental cycles.

In parallel, nanoplastics may also be introduced directly during industrial processing and manufacturing activities, where nanoscale polymer particles can be incorporated into raw materials or generated as by-products during production stages. Such release pathways create direct exposure routes from industrial environments into adjacent soil and water systems, with potential implications for ecosystem integrity and human health (Brachner et al., 2020).

As summarized in Figure 3, the coexistence of primary and secondary nanoplastic sources complicates source attribution and environmental exposure reconstruction efforts across interconnected ecosystems. Consequently, integrated analytical frameworks capable of differentiating source-specific signatures and environmental transport pathways remain essential for improving environmental surveillance, exposure assessment, and contamination tracking within a One Health traceability context.

2.3. Transport and Transformation in the Environment

Following environmental release, nanoplastics are subjected to multiple transformation and transport processes that collectively determine their environmental fate, mobility, and accumulation across ecological compartments (Pradel et al., 2023). Key mechanisms include fragmentation of larger plastic particles, degradation of associated additives, chemical transformations such as oxidation, and sorption of co-occurring environmental contaminants (Alimi et al., 2018; Velzeboer et al., 2014; Zhang et al., 2018). Together, these processes influence particle stability, transport dynamics, environmental persistence, and interaction with surrounding biological and physicochemical systems.

Reduction in particle size may substantially enhance mobility between soil and aquatic compartments, particularly in environments influenced by agricultural practices involving fertilizers, biosolids, or plastic mulching materials. Under such conditions, nanoplastics may be redistributed through surface runoff, groundwater leaching, or lateral transport via subsurface drainage systems (Yang et al., 2021). Even when primary emissions decline, environmental reservoirs may continue releasing secondary nanoplastics, thereby supporting long-term persistence within environmental circulation pathways.

Current evidence further indicates that terrestrial environments may contribute substantial loads of microplastics and nanoplastics to freshwater and marine systems. However, removal, retention, and transformation processes may still occur during environmental transport prior to final ecosystem integration. Consequently, accurate characterization of transport dynamics within surface waters remains essential for predicting long-term environmental fate, supporting environmental exposure reconstruction models, and improving understanding of nanoplastic cycling within global environmental systems (Besseling et al., 2017).

Within a One Health forensic framework, and as illustrated in Figure 3, these transport and transformation processes are particularly important because they directly influence environmental traceability, source differentiation, and interpretation of contamination pathways across interconnected ecological and human exposure networks.

3. BIOTECHNOLOGICAL APPROACHES FOR DETECTION AND CHARACTERISATION

Techniques for identifying and characterising nanoplastics generally encompass molecular assays (3.1), optical and imaging methodologies (3.2), biological response systems (3.3), and standardisation protocols (3.4) (Santos et al., 2025). These methodological domains collectively align with the objectives outlined in Section 1.3, particularly in relation to source tracking, exposure modelling, and the assessment of ecological and human health impacts. Within this integrated framework, analytical outputs contribute to deconvolution processes for source attribution (4.2) and exposure determination (4.3), where interpretability and data integrity remain central to reliable inference.

Accordingly, uncertainty quantification and the interpretability of machine learning outputs (4.4) are essential for establishing robust and defensible risk frameworks (6.1). When integrated, detection, characterisation, and computational approaches may provide structured evidence streams capable of identifying environmental hotspots and forensic signatures relevant to regulatory and environmental governance. As conceptually linked across Sections 2 and 4, these components support a traceability-oriented One Health architecture for nanoplastic monitoring and assessment.

Molecular techniques typically integrate nucleic acid–based methods, including qPCR, with protein-binding assays on sensor platforms (Gigault et al., 2017). In some configurations, genetic markers are used to infer organismal presence or environmental DNA signals associated with specific ecological compartments. Parallel protein-based sensing systems may reflect exposure-associated biological responses under nanoplastic stress conditions. Optical and imaging methodologies, including FTIR, Raman spectroscopy, and fluorescence-based approaches, provide complementary structural and chemical information, while also supporting functional interpretation of particle–biological interactions across environmental matrices.

Bioassay-based approaches complement these techniques by linking exposure to ecotoxicological and physiological responses across organisms and ecosystems. Nevertheless, molecular and integrated bioassay systems remain comparatively underdeveloped and require further methodological refinement for broader environmental deployment.

3.1. Molecular and Microbial Assays

Molecular approaches based on polymer- or sequence-specific probe systems enable detection and partial characterization of plastisphere-associated biological signals from environmental samples. In certain applications, hybrid systems combining nucleic acid amplification with nanopore sensing have been proposed to improve sensitivity and enable label-free detection of nanoplastics.

Quantitative PCR approaches, including primer systems targeting conserved genetic markers such as eukaryotic 18S rRNA genes, have been applied in environmental monitoring to support indirect source attribution, particularly in urban-impacted ecosystems. In parallel, emerging strategies involving biohybrid platforms (e.g., chloroplast-associated systems) are being explored; however, these approaches remain at an early developmental stage.

Bioassay applications using native microbial communities, plants, and animal models remain central for evaluating biological responses under environmental or experimental exposure conditions. Despite this, standardized molecular bioassays for nanoplastics are still limited, and ongoing research continues to focus on aquatic model organisms such as zebrafish embryos, freshwater bivalves, and other ecotoxicological indicators.

3.2. Spectroscopic and Imaging Techniques

Imaging and spectroscopic techniques are widely used to investigate nanoplastic interactions and distribution patterns in environmental and biological systems. High-resolution microscopy approaches, including dark-field microscopy, electron microscopy, and nanoscale secondary ion mass spectrometry, provide complementary structural and elemental information. Collectively, these methods help overcome limitations associated with single-technique analyses by improving the resolution and interpretability of particle–biological interactions (Sekine et al., 2017).

However, single-particle chemical imaging of environmental nanoplastics remains technically challenging due to their heterogeneity and intrinsic lack of labeling properties (Mintenig et al., 2018; Huppertsberg & Knepper, 2018). Conventional spectroscopic methods such as FTIR and Raman microscopy often face constraints in sensitivity and spatial resolution, particularly at the nanoscale (Schymanski et al., 2018). Electron and atomic-force microscopy provide detailed morphological information but lack chemical specificity. Emerging hybrid approaches, including AFM-IR and related nanospectroscopic techniques, attempt to bridge these limitations, although trade-offs between sensitivity and specificity remain evident.

Standardisation of analytical parameters—such as spectral acquisition settings, preprocessing workflows, and library-matching criteria—is increasingly recognised as essential for improving inter-study comparability. Despite methodological progress, inconsistencies in reporting continue to limit harmonisation, particularly in exposure-related assessments and cross-study integration (Wright et al., 2023).

3.3. Bioassays and Toxicity Profiling

The ecotoxicological profile of nanoplastics is still evolving, with growing evidence of size-dependent and composition-dependent biological effects. Experimental studies indicate that polystyrene nanoplastics can induce stress responses in plant systems such as *Arabidopsis thaliana*, and may enhance the toxicity of co-occurring atmospheric pollutants including nitrogen dioxide and sulphur dioxide (Kögel et al., 2020). Similar findings in agricultural systems suggest reduced crop yield and altered metal uptake, particularly in rice exposed to cadmium in combination with nanoplastics.

In aquatic and soil organisms, engineered nanoplastics have been associated with toxic responses in species such as *Daphnia magna*, while polymer-based nanoplastics also affect algae and microbial communities (Bondarenko et al., 2016; Rist et al., 2017). Studies on human cell lines and model organisms further suggest that polystyrene nano- and microplastics can induce cytotoxicity and stress responses, with variability dependent on particle size and surface modification (Schirinzi et al., 2017; Gouin et al., 2022; Lee et al., 2025; Wu et al., 2019).

Model organisms, including zebrafish, *C. elegans*, and *D. melanogaster*, have been used to demonstrate dose-dependent toxicological and behavioural effects across developmental stages for different particle sizes. Collectively, these findings suggest that combined exposure to nanoplastics and heavy metals may amplify risks within agricultural systems and food chains (Mattsson et al., 2017).

3.4. Standardization and Validation Protocols

Nanoplastics and microplastics differ from conventional particulate pollutants due to their size-dependent properties, surface chemistry, and complex environmental behaviour. Consequently, robust hazard and risk assessment requires the clear definition of relevant physicochemical parameters (Everaert et al., 2018). Standardization and validation protocols are essential for ensuring comparability in measurements of particle concentration, morphology, and environmental distribution across diverse matrices (Petersen et al., 2022).

The absence of harmonised terminology and methodological frameworks has contributed to variability in reported findings, particularly in nanoparticle characterization and environmental detection studies (Roubert et al., 2016). In response, several collaborative initiatives, including European programmes such as NanoDiaRA, have supported methodological harmonization across disciplines. These initiatives provide guidance on experimental design, protocol standardisation, and nanoparticle characterisation workflows, thereby improving consistency and reproducibility in nanomaterial research applications (Brachner et al., 2020). The main dimensions of this framework are summarized in Table 1, which brings together regulatory, methodological, economic, public health, and ecological considerations within a unified One Health perspective.

Table 1. Multidimensional Framework for Nanoplastic Detection and Definition

Domain	Primary Objective	Key Components	Implications for One Health
Regulatory Agencies	Enable policy enforcement and compliance	Police and regulatory authorities; European initiatives (e.g., NanoDiaRA)	Supports legal oversight and international harmonization
Cost of Analysis	Assess economic and technical feasibility	High-tech instrumentation; Specialized technical expertise	Influences accessibility and scalability of detection technologies
Public Health	Evaluate exposure and human health risks	Exposure modeling and risk assessment; Gauging human health effects; Public awareness	Guides epidemiological evaluation and preventive health strategies
Advanced Methods	Enable detection and characterization	Molecular assays; Optical and imaging techniques	Forms analytical foundation for source tracing and exposure assessment
Ecosystem Impact	Assess ecological consequences	Biological response systems; Ecotoxicological profiling; Bioassays	Supports biodiversity protection and environmental risk evaluation
Standardization Protocols	Promote methodological harmonization	Standardization and validation protocols; Harmonized terminology development	Ensures reproducibility and regulatory acceptance

Table 1 synthesizes these operational dimensions within a structured, matrix-based architectural framework, mapping the current convergence of regulatory mandates, methodological constraints, and asymmetric analytical capacities within nanoplastic research. Rather than a static taxonomy, this cross-disciplinary matrix serves as an active systems-science blueprint, codifying the continuous negotiations taking place across automated detection, multi-omics risk assessment, and ecological interpretation.

4. AI ANALYTICS FOR SOURCE ATTRIBUTION AND EXPOSURE ASSESSMENT

Nanoplastic pollution is increasingly conceptualised as a One Health concern that requires integrated, policy-relevant evidence on emission sources, exposure pathways, environmental behaviour, and associated ecological and human health outcomes (Liu et al., 2022). Within this context, artificial intelligence (AI) provides a computational layer for data integration and inference, enabling the synthesis of heterogeneous observations derived from environmental, ecological, and biological matrices. This approach may reduce reliance on exclusively laboratory-intensive workflows, although its performance remains strongly dependent on data completeness, representativeness, and structural quality (Liu et al., 2022; Li et al., 2026).

Recent advances in AI, particularly environmental digital twin systems, have expanded the capacity to simulate pollutant transport and exposure dynamics within coupled environmental–biological systems (Fernández-Llatas et al., 2022). Such frameworks allow scenario-based reconstruction of nanoplastic behaviour under variable environmental conditions, supporting more structured interpretation of exposure routes and system-level interactions. In parallel, One Health forensic frameworks aim to integrate source identification, exposure reconstruction, and impact assessment within a unified analytical structure, thereby linking environmental monitoring to decision-relevant outputs (Liu et al., 2022).

Building on biotechnological detection and characterisation approaches, AI-based analytics typically operates across four interconnected functions. First, it enables the integration of heterogeneous datasets derived from multiple environmental compartments, including physicochemical, biological, and spectral measurements. Second, it supports machine learning–based source attribution by distinguishing between primary nanoplastic inputs and secondary particles generated through microplastic fragmentation. Third, it facilitates quantitative reconstruction of exposure pathways across environmental, ecological, and human interfaces. Finally, it incorporates uncertainty quantification and model interpretability to enhance the reliability and transparency of inference used in policy and risk assessment contexts. These components are interdependent rather than sequential, reflecting the complexity of environmental nanoplastic systems. A schematic representation of the integrated workflow—from data acquisition and harmonisation to preprocessing and analytical modelling within a One Health nanoplastic forensic framework—is provided in **Figure 4**.

Figure 4. Data integration and preprocessing steps for One Health nanoplastics forensics. The structural flowchart outlines the data-cleaning, normalization, and algorithmic preprocessing procedures applied to reduce noise from heterogeneous environmental datasets prior to computational modeling. Raw input matrices spanning multi-omics bioassays and spectroscopic signals undergo systematic dimensionality reduction to minimize overfitting in downstream artificial intelligence analytics. This optimization phase supports architectural standardization by converting fragmented, multi-source datasets into unified empirical vectors. Consequently, this step establishes the computational foundation supporting automated source attribution and predictive toxicity modeling for cross-disciplinary surveillance applications.

Domain	Primary Objective	Key Components	Implications for One Health
Regulatory Agencies	Enable policy enforcement and compliance	Police and regulatory authorities; European initiatives (e.g., NanoDiaRA)	Supports legal oversight and international harmonization
Cost of Analysis	Assees economic and technical feasibility	High-tech instrumentation; Specialized technical expertise	Influences accessibility and scalability of detection technologies
Public Health	Evaluate exposure and human health risks	Exposure modeling and risk assessment; Public awarneess	Guides epideotiological evaluation and preventive health strategies
Advanced Methods	Enable detection and characterization	Molecular assays; Optical and imaging techniques	Forms analytical foundation for source tracing and exposure assessment
Ecosystem Impact	Assess ecological consequences	Biological response systems; Ecotoxicological profiling; Bioassays	Supports biodiversity protection and environmental risk evaluation
Standardization Protocols	Promote methodological harmonization	Standardization and validation protocols	Ensures reproducibility and regulatory acceptance

4.1. Data Integration and Preprocessing

Data integration and preprocessing represent foundational steps in AI-enabled nanoplastic forensics, involving the aggregation, harmonisation, and structuring of heterogeneous datasets prior to analytical modelling. These datasets typically include outputs from biochemical detection assays, material characterisation techniques, and biological response measurements obtained through biotechnological methods described in Section 3. Each data stream provides complementary but partial information on environmental fate, physicochemical properties, and biological interactions (Qian et al., 2024).

Given the multidisciplinary origin of these datasets, ensuring cross-domain compatibility is essential. Differences in measurement units, analytical resolution, and data structure necessitate systematic standardisation prior to integration into statistical or machine learning pipelines. In practice, this often involves transforming heterogeneous inputs into unified representations suitable for computational analysis while preserving domain-specific information content.

Subsequent preprocessing steps focus on data refinement, including the identification and correction of inconsistencies, management of missing values, and standardisation of feature formats. These operations are critical for maintaining analytical robustness and ensuring stability across downstream modelling stages. As environmental datasets are inherently dynamic and context-dependent, iterative updates to preprocessing pipelines may be required to maintain methodological consistency over time.

4.2. Machine Learning for Source Tracing

Machine learning (ML) approaches are increasingly employed to infer potential sources of nanoplastic pollution within complex and multi-compartment environmental systems. Feature-based models incorporating physicochemical descriptors—such as particle size distribution, polymer composition, and morphological attributes—have demonstrated utility in differentiating between industrial, urban, and other anthropogenic sources (Zhang et al., 2023). Comparable classification frameworks applied to polystyrene nanoplastics further indicate that structural and compositional features are strongly associated with source-specific signatures (Furxhi & Murphy, 2020).

In addition, supervised learning models combined with image-based segmentation techniques have improved detection and classification performance across heterogeneous environmental matrices. These approaches rely heavily on high-quality training datasets generated through biotechnological and spectroscopic methods described in Section 3, highlighting the interdependence between analytical detection and computational inference. Consequently, ML outputs are typically interpreted in conjunction with exposure modelling frameworks to support both source attribution and environmental pathway reconstruction (Section 4.3).

Within a One Health forensic perspective, nanoplastics associated with non-agricultural systems may form distinct exposure networks compared with those linked to agricultural or food production environments. Integrating datasets across environmental, plant, and animal systems therefore enables more refined modelling of sector-specific exposure pathways. In this context, nanoplastics may be treated as trace evidence materials, where polymer fingerprints and compositional signatures provide constraints for reconstructing contamination pathways. When combined with AI-assisted inference models, such approaches may support environmental source attribution, exposure reconstruction, and regulatory interpretation, although their evidentiary strength depends on data resolution and analytical uncertainty.

4.3. Exposure Modelling and Risk Quantification

Nanoplastics originate from diverse sources, including consumer products such as synthetic textiles, cleaning agents, industrial coatings, adhesive materials, and polymer-based consumables. During use and handling, particles may be transferred from surfaces to skin and subsequently resuspended into the air, contributing to inhalation and dermal exposure pathways (Wright et al., 2023). However, exposure magnitude varies considerably depending on emission source, environmental conditions, and behavioural context, underscoring the need for source prioritisation in risk assessment frameworks.

Exposure modelling typically involves a sequential reconstruction of emission, transport, and receptor interactions. This includes identifying emission sources, estimating release rates, and mapping transport processes across environmental compartments leading to potential biological uptake. Such models aim to represent transitions across air, surface, water, and biotic systems, although their predictive accuracy is constrained by uncertainties in input parameters and environmental variability.

Given the heterogeneity of environmental systems, distinct modelling frameworks are required for different contexts, including urban, agricultural, and aquaculture environments (C. Abbott & D. Maynard, 2010). In early stages of development, simplified models are often employed to identify dominant transport pathways and guide parameter selection for more complex simulations. Once established, these models may incorporate particle size distributions to estimate partitioning between environmental compartments, including airborne fractions, surface deposition, and biological uptake potential.

4.4. Uncertainty, Interpretability, and Validation

Uncertainty and interpretability represent central challenges in AI-assisted nanoplastic forensics within a One Health framework. Nanoplastics undergo continuous physicochemical transformation throughout their environmental lifecycle, including changes in surface chemistry, aggregation state, and structural integrity, which may reduce traceability over time and complicate source attribution (Brachner et al., 2020). Environmental variables such as temperature, humidity, UV radiation, pH, and dissolved organic matter further influence particle behaviour, introducing additional variability into both measurement and modelling processes.

These dynamic transformations generate uncertainty in linking observed environmental distributions to specific emission sources. As a result, robust inference typically requires the integration of experimental data, field observations, and computational modelling. This triangulation approach improves confidence in exposure reconstruction and environmental transport assessment while acknowledging residual uncertainty inherent to complex environmental systems.

Emission behaviour is also influenced by product-specific properties and environmental conditions during use, degradation, and disposal phases (Wright et al., 2023). Additives and contaminants may be incorporated at multiple stages of the product life cycle, while environmental factors such as temperature, chemical composition, and exposure duration can modulate release dynamics. Despite these insights, mechanistic understanding of degradation-driven transformation processes under realistic environmental conditions remains incomplete.

Tracking emissions across product life cycles can enhance source modelling when combined with laboratory characterisation and predictive computational approaches (Gouin et al., 2022). Nevertheless, uncertainties persist due to incomplete knowledge of degradation pathways and environmental heterogeneity.

Causal machine learning approaches have been proposed to link nanoplastic sources, transport pathways, and biological outcomes; however, their performance remains constrained by limited prior knowledge and incomplete datasets. These limitations can affect predictive robustness and reduce interpretability in complex environmental contexts. Consequently, explicit uncertainty quantification and interpretability frameworks are essential components of any AI-based forensic system intended for regulatory or policy applications.

5. IMPACT ON WILDLIFE, HUMANS, AND ECOSYSTEMS

Microplastics and nanoplastics affect a wide range of wildlife species, including fish, birds, and mammals. Although assessments across aquatic, terrestrial, and aerial ecosystems remain incomplete, growing evidence indicates that microplastics can alter behaviour and induce physiological stress in exposed organisms (Sendra et al., 2021). Atlantic cod exposed to microplastics, for example, exhibit altered predator-inspection behaviour, while exposed invertebrates show reduced feeding activity and avoidance responses. Current knowledge regarding nanoplastics remains comparatively limited, particularly in relation to biodiversity loss and species interactions. Nevertheless, findings from marine microplastic studies provide a useful basis for modelling potential nanoplastic effects across other environmental systems (Mattsson et al., 2017; Galloway et al., 2017).

Microplastics have also been detected in atmospheric deposition and across urban surfaces, indicating continuous environmental exposure in populated areas. In some urban settings, atmospheric deposition may reach millions of particles per square metre annually. Their frequent detection in aerosol and dust samples further suggests the likelihood of inhalation exposure, particularly as airborne particles as small as 2 μm have already been documented. These observations raise concern that nanoplastics may also contribute to airborne exposure pathways. As summarized in Figure 5, nanoplastics may move through interconnected environmental compartments and exposure routes involving air, water, soil, food systems, wildlife, and human populations.

Plastic particles may additionally transport chemicals of concern through ingestion or direct environmental contact, increasing uncertainty regarding long-term exposure risks, particularly in drinking-water systems. Urban runoff and combined sewer overflows continue to transport microplastics into raw water sources used for municipal supply (Yang et al., 2024). Their detection in agricultural soils and groundwater further broadens the potential for environmental and human exposure.

Microplastics have been identified in a variety of food products, including beverages, salt, sugar, tea, fish, shellfish, honey, chocolate, and processed foods (Hernandez et al., 2019; Cox et al., 2019; Oliveri Conti et al., 2020). Some evidence also suggests migration of microplastics into edible lipid products and release of nanoplastics from certain food-packaging materials. Detection of microplastics in honey from southern Europe further supports the possibility of dietary exposure through commonly consumed products. In addition, chemicals adsorbed onto plastic surfaces may leach into food-contact environments, raising broader toxicological concerns. However, uncertainties in analytical methodologies continue to complicate risk assessment for both microplastics and nanoplastics (Revel et al., 2018), particularly when comparing exposure estimates across studies.

5.1. Ecotoxicology and Health Outcomes

The persistence of polymer-derived materials in environmental systems has intensified concern regarding their effects on food webs and aquatic ecosystems. Micro- and nano-sized plastic particles are now detected across diverse environmental matrices, even at relatively low concentrations. Larger plastic debris undergoes gradual fragmentation through physical, chemical, and biological degradation processes, ultimately generating microplastics and nanoplastics capable of wider environmental distribution.

Microplastics may interact with co-occurring pollutants and potentially facilitate their movement through aquatic and terrestrial food systems. Nanoplastics, because of their small size and surface properties, may behave differently from larger plastic particles and could contribute to the transport of hydrophobic contaminants across biological barriers (Yong et al., 2020). Such interactions raise concern regarding possible transfer of toxic compounds through aquatic food chains and broader ecosystem exposure. Estimates suggesting substantial accumulation of dispersed nanomaterials within water systems further support the need for ecotoxicological assessment under the One Health framework.

5.2. Biomarkers and Early Warning Signals

The environmental behaviour of nanoplastics is governed by highly dynamic and heterogeneous processes that vary across spatial and temporal scales (Bergamaschi et al., 2015). Because exposure patterns differ between ecosystems and environmental conditions, long-term monitoring remains necessary for evaluating environmental quality and identifying potential early-warning indicators associated with nanoparticle pollution (Higashisaka et al., 2012).

Continuous or periodically scheduled monitoring of airborne nanoplastics may therefore provide important information regarding transport behaviour, environmental persistence, and possible exposure trends. Direct measurement of airborne particle concentrations offers a basic but informative monitoring approach; however, concentration data alone may not fully capture environmental risk. Structural characteristics, physicochemical properties, and functional markers should also be considered, particularly because nanoplastics frequently coexist with other airborne particulate matter and chemical contaminants.

5.3. Socioeconomic and Policy Implications

Nanoplastics, generally defined as plastic particles smaller than 1 μm, are increasingly recognized as emerging contaminants with potential implications for environmental quality, ecosystem stability, and public health (Saarimäki et al., 2023). Growing scientific attention has expanded understanding of their environmental occurrence, analytical detection, and possible biological effects. At the same time, advances in biotechnology-based detection systems and artificial intelligence–assisted source attribution approaches continue to strengthen monitoring and exposure assessment capabilities.

Regulatory attention has also increased in response to expanding environmental concerns surrounding plastic pollution. Proposed restrictions targeting microplastics, increasing international discussion surrounding plastic discharge reduction, and shifts in plastic consumption patterns following the COVID-19 period have collectively intensified interest in the environmental and socioeconomic implications of nanoplastic contamination. These developments reinforce the need for integrated assessment strategies capable of supporting evidence-based environmental policy, risk governance, and long-term mitigation planning within a One Health context.

Figure 5. Environmental and biological pathways of nanoplastic exposure across interconnected ecosystems, including air, water, soil, food systems, wildlife, and human populations. The figure illustrates potential transport routes and exposure networks highlighting the movement of nanoplastics through environmental compartments and trophic chains.

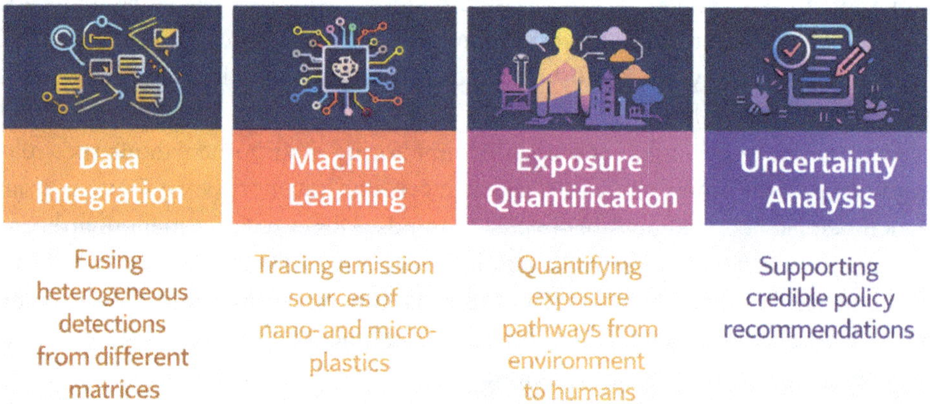

6. RISK ASSESSMENT, GOVERNANCE, AND ETHICS

Nanoplastic pollution has emerged as a growing environmental concern associated with declining water quality, ecological disruption, and potential human health risks. Within a One Health context, biotechnological detection methods integrated with AI-based analytics may support tracing of nanoplastic sources, characterization of exposure pathways, and assessment of associated biological and environmental impacts. As illustrated throughout this framework, risk assessment increasingly depends not only on analytical detection capacity but also on governance structures, regulatory coordination, and responsible data management practices capable of supporting credible interpretation of analytical outputs and exposure models.

Governance principles addressing engineered nanomaterials have already been incorporated into regulatory or advisory frameworks across multiple countries and regions. Early initiatives included development of nanomaterial registries within the European Union and France, alongside broader international discussions surrounding nanosafety governance. Commitments emerging from the 2007 G8 summit further contributed to recognition of environmental and sanitary nanosafety as continuing political priorities. Related international guidance frameworks increasingly connect nanosafety governance with long-term environmental protection, research planning, and economic sustainability. In this context, regulatory development continues to serve as a central component of broader nanosafety implementation strategies, including those relevant to One Health-oriented nanoplastic management (Saarimäki et al., 2023).

6.1. Risk Frameworks for Nanoplastics

Nanoplastics, generally defined as plastic particles smaller than 1,000 nm, remain difficult to detect visually because of their nanoscale dimensions. Their increasing use in industrial and commercial applications, together with the environmental degradation of larger plastic materials, has intensified concern regarding the formation and accumulation of secondary nanoplastics in environmental systems. These particles are now reported in a variety of aqueous environments, including wastewater and flood-associated water systems, where continued environmental release may contribute to broader ecological and human exposure scenarios (Brachner et al., 2020).

Current concern extends beyond environmental occurrence alone. Nanoplastics may interact differently from larger plastic particles because of their size, mobility, and surface characteristics, potentially influencing bioavailability and biological uptake. Despite increasing evidence of their environmental presence, formalized risk assessment frameworks for nanoplastics remain limited. At present, comprehensive assessment of exposure risks and adverse ecological or human-health outcomes remains constrained by insufficient analytical standardization, limited environmental monitoring data, and uncertainties regarding transport and toxicity mechanisms (World Health Organization [WHO], 2019; Yang et al., 2024). Consequently, understanding of nanoplastic fate, transport behaviour, and long-term environmental effects remains at an early stage globally, even as emerging studies continue to report their presence across aquatic systems (Saarimäki et al., 2023).

The rapid expansion of nanotechnology has also intensified discussion surrounding the long-term safety of engineered nanomaterials. Nano-scaled materials incorporated into consumer and industrial products are expected to enter environmental systems during product use, disposal, or end-of-life degradation. Existing estimates suggest that a measurable fraction of engineered nanomaterials may ultimately be released into the environment following product disposal stages (Isigonis et al., 2019). The persistence of some organo-nanoscale materials and uncertainty regarding their long-term environmental behaviour continue to raise concern regarding potential exposure through water systems and food chains.

6.2. Regulatory Landscape and Compliance

Regulatory interest in nanoplastics and engineered nanomaterials has expanded substantially over the past decade. In 2020, the Human Relevance Framework (HRF) was codified by the European Chemicals Agency (ECHA) as a tiered approach intended to integrate human biomonitoring information into chemical and toxicological assessment processes. The framework incorporates dimensions related to human relevance, ecotoxicological extrapolation, and environmental exposure considerations under broader REACH-associated regulatory structures. These developments reflect increasing institutional interest in establishing coordinated approaches for assessing contaminants considered to be of potential high concern, including nanoplastic-associated materials (Allan et al., 2021).

The rapid expansion of nanotechnology applications also reinforced the need for regulatory systems capable of guiding nanomaterial manufacturing, characterization, testing, and environmental evaluation (Agbasi et al., 2025). Within this landscape, the European Union has adopted some of the earliest structured regulatory approaches addressing nanomaterials through directives and chemical-governance frameworks intended to support safe use across industrial sectors. Other countries have similarly proposed or initiated regulatory measures, although debate continues regarding whether nanomaterials require fully dedicated regulatory systems or adaptation of existing chemical-governance structures (Brachner et al., 2020).

France was among the first countries to implement mandatory notification requirements for marketed nanomaterials through a dedicated decree introduced in 2009. These developments occurred alongside expanding industrial production of nanomaterials and growing regulatory discussion regarding emerging nanoscale products and associated environmental oversight challenges (Corte Pause et al., 2024).

6.3. Ethical Considerations and Data Governance

International concern regarding plastic pollution has intensified considerably in recent years. In 2022, the United Nations Environment Programme (UNEP) described plastic pollution as a global environmental crisis requiring coordinated international action (UNEP, 2022). Similar concern was subsequently reflected in statements from the World Health Organization (WHO), which emphasized potential consequences for human, animal, and ecosystem health as well as broader environmental sustainability (WHO, 2022). During the same period, proposals supporting development of an international legally binding framework addressing plastic pollution were presented within the United Nations Environment Assembly process. Collectively, these initiatives illustrate increasing political recognition of the transboundary nature of plastic-associated contaminants, including nanoplastics.

Because nanoplastics may cross biological barriers in both wildlife and humans, their assessment increasingly requires interdisciplinary approaches integrating environmental science, biotechnology, toxicology, computational analytics, and governance systems (MP3, 2022). Within this context, One Health forensics offers a framework for jointly examining contaminant origin, exposure pathways, and potential impacts across interconnected environmental and biological systems. Biopackages containing plastiglomerates, polymer-associated interfaces, and related particulate structures may also provide useful forensic indicators for tracing contaminant movement and supporting high-throughput analytical workflows, including polymer characterization, spectral analysis, and environmental modelling (Bassan & Harel, 2018). Despite growing scientific attention, regulatory approaches addressing nanoplastics remain fragmented across regions and sectors. Greater harmonization of international policies and reporting frameworks may therefore be necessary to support consistent environmental monitoring, exposure assessment, and governance strategies aligned with One Health principles.

7. CASE STUDIES AND APPLICATIONS

Despite increasing adoption of the One Health framework and expanding recognition of nanoplastic pollution, integrated studies linking environmental occurrence, transport dynamics, and biological exposure remain limited. Aquatic and terrestrial systems are widely regarded as major sinks for nanoplastics (Qian et al., 2024), where progressive accumulation across trophic networks may facilitate eventual transfer to humans via food consumption pathways (Corte Pause et al., 2024). Evidence for interspecies transfer of micro- and nanoplastics along animal–human interfaces further supports this concern, although system-level integration across environmental compartments is still insufficiently developed (Sendra et al., 2020).

Applied evidence indicates that nanoplastic dissemination occurs through tightly coupled environmental–anthropogenic systems, particularly aquatic, urban, and agricultural domains. These compartments should therefore be considered as interconnected exposure modules rather than isolated units.

Aquatic environments represent primary convergence zones for plastic-derived particulates, receiving continuous inputs from wastewater discharge, urban run-off, industrial effluents, and aquaculture systems. Estuarine gradients and coastal interfaces often function as accumulation and redistribution hotspots. Urban environments constitute major emission sources, largely driven by traffic-related abrasion processes, including tire wear and road-surface degradation, which remain insufficiently regulated at present (Wagner et al., 2018). Agricultural systems contribute through biosolid application, irrigation reuse, composting practices, and atmospheric deposition, with roadside cultivation further enhancing exposure to transport-derived particles.

Emerging studies have confirmed nanoplastic presence in marine biota, including commercially relevant shellfish species, supporting trophic transfer and potential human dietary exposure. In parallel, the integration of spectroscopic techniques with machine learning–based pattern recognition has improved detection performance in heterogeneous matrices, particularly where conventional optical or chemical identification is limited. Collectively, these developments indicate a transition from compartmentalized monitoring toward integrated One Health exposure assessment frameworks.

7.1. Aquatic Systems and Marine Interfaces

Aquatic systems—including rivers, lakes, estuaries, oceans, and polar regions—act as major reservoirs for plastic-derived contaminants. Fragmentation of macro- and microplastics generates nanoplastics capable of long-range transport across hydrological boundaries. Riverine inputs arise from wastewater-treatment plants, landfill leakage, sewage overflow, industrial discharge, agricultural runoff, and direct plastic disposal, while primary microplastics embedded in consumer products provide additional sources.

At coastal interfaces, anthropogenic pressure intensifies through shipping, aquaculture, offshore industrial activity, and coastal urbanization. Climatic and hydrological changes further modulate transport dynamics by altering sedimentation, flow regimes, and dilution capacity. These coupled processes enhance both residence time and spatial redistribution of nanoplastics.

Within aquatic food webs, planktonic systems serve as critical entry points for contaminant transfer. Nanoplastics exhibit high surface reactivity, enabling adsorption of metals, pathogens, and hydrophobic organic compounds, thereby forming vector-like complexes that facilitate trophic transfer. This process extends exposure beyond particulate contamination to include co-transported chemical and biological agents.

Simultaneously, emerging aquatic technologies such as nano-enabled aquaculture and engineered bio-production systems introduce additional sources of synthetic particulates. Although these technologies are often framed within sustainability narratives, their material inputs may unintentionally contribute to environmental nanoplastic burdens. Consequently, aquatic systems function not only as sinks but also as dynamic interfaces of generation, transformation, and redistribution (Landrigan et al., 2020).

7.2. Urban Environments and Waste Streams

Urban systems represent highly heterogeneous emission landscapes where nanoplastics are continuously generated, transformed, and redistributed across air, water, and soil compartments. Key sources include industrial emissions, transport-related abrasion, construction activities, and waste-management inefficiencies, all contributing to multi-pathway environmental release.

Faridabad (National Capital Region, India) exemplifies such complexity, combining dense industrial activity with urban runoff and wastewater discharge. While microplastic contamination has been documented at elevated levels relative to national environmental thresholds, nanoplastic occurrence remains under-characterized despite potential implications for systemic exposure.

Field sampling across industrial and mixed-use zones, followed by high-resolution analysis at IIT Delhi, revealed nanoscale plastic fractions (approximately 10 nm–1 µm) in river-associated matrices. Wastewater-treatment plants contributed additional inputs, indicating limited retention efficiency for nanoscale particles and suggesting their redistribution into receiving water bodies.

Urban atmospheric pathways further contribute through tyre abrasion, brake wear, textile fibres, and road-surface degradation. These particles accumulate in dust reservoirs and are subsequently mobilized via precipitation and stormwater runoff, reinforcing connectivity between atmospheric and aquatic systems. Such coupled dynamics highlight the necessity of integrated urban monitoring frameworks that combine environmental sampling with AI-assisted classification and spectroscopic analysis for source discrimination and exposure inference.

7.3. Agricultural and Food Chain Interfaces

Agricultural systems constitute critical interfaces for nanoplastic transfer due to direct coupling with food production and soil–water cycles. Input pathways include plastic mulch degradation, biosolids, irrigation reuse, compost application, agricultural machinery wear, and atmospheric deposition. Once introduced, nanoplastics interact with soil matrices, organic matter, root systems, and microbial communities, enabling long-term persistence and redistribution.

Evidence indicates that agricultural soils may represent larger reservoirs of plastic residues than freshwater systems, primarily due to continuous application of plastic-derived materials and limited removal mechanisms. Progressive fragmentation under environmental stressors contributes to nanoscale formation, increasing bioavailability and mobility.

Food-chain transfer pathways involve crops, livestock, and aquaculture systems. Livestock exposure occurs not only through feed but also via water, inhalation of dust particles, bedding materials, and enclosed-air circulation systems. Such multi-route exposure increases systemic uptake potential and complicates dose reconstruction.

Aquaculture systems represent both exposure endpoints and secondary sources of nanoplastic release, with implications for seafood safety and ecosystem integrity. Additionally, interactions between nanoplastics and co-occurring contaminants (pesticides, metals, and microbial agents) may modify transport behavior and biological availability, although mechanistic understanding under field conditions remains limited.

Overall, the persistence and multi-compartment mobility of nanoplastics across agricultural landscapes underscore the necessity of integrated One Health frameworks that combine environmental monitoring, food safety assessment, and exposure modelling. Significant uncertainties remain regarding chronic exposure dynamics, mixture effects, and long-term ecosystem-level consequences.

8. TECHNOLOGICAL INNOVATIONS AND INTEGRATED PIPELINES

Technological innovations are gradually reducing some of the management and regulatory barriers associated with nanoplastics, nanotoxicology, and One Health, while also expanding the scientific literature surrounding these issues (Qian et al., 2024; Isibor et al., 2024). In practice, the field has shifted from relying on isolated analytical methods toward more integrated pipelines that combine detection, characterization, modelling, and risk interpretation within connected workflows.

Figure 6. **Integrated technological workflow for nanoplastic exposure assessment and environmental decision support.** *Integrated pipeline illustrating sequential stages of nanoplastic data aggregation, preprocessing, exposure quantification, and adaptive reassessment within One Health-oriented environmental monitoring and decision-support frameworks.*

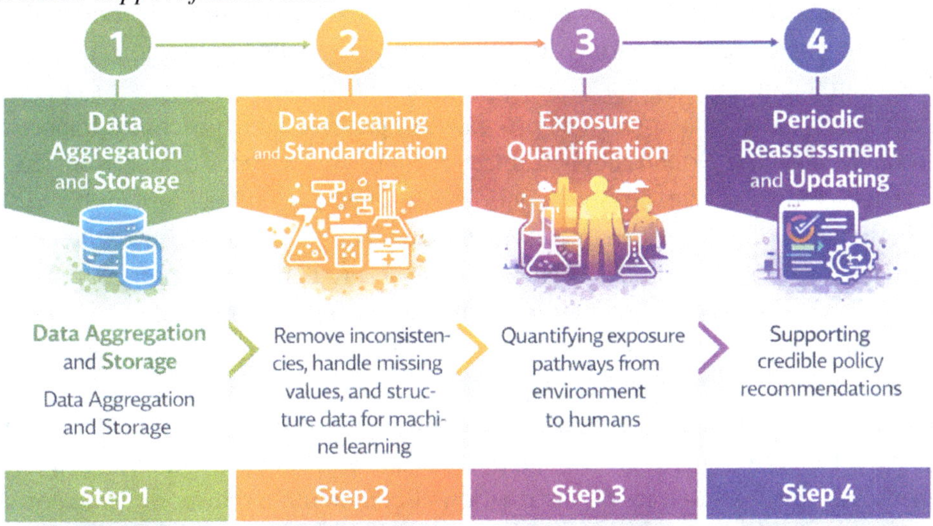

That transition is becoming increasingly necessary because nanoplastic contamination rarely behaves as a simple or compartmentalized environmental problem. Sources overlap, exposure routes intersect, and the resulting datasets are often difficult to interpret using single-method approaches alone.

High-throughput biotechnological platforms, particularly when integrated with microfluidic systems, can support multivariate detection and characterization across large sample sets, while remote-sensing satellites provide broader spatial coverage for tracking occurrence patterns and pollution dynamics over extensive geographic areas (Zhang et al., 2023). At the same time, decision-support systems are becoming more relevant for risk assessment, coordination, and intervention planning (Allan et al., 2021), mainly because they allow near real-time analysis and facilitate communication between stakeholders involved in environmental monitoring and management. The value of these systems is not only technical. In many cases, they help organize information that would otherwise remain fragmented across disciplines or agencies.

These developments matter because nanoplastic research is not only a measurement problem. It is also a problem of scale, variability, interpretation, and, in many situations, governance. Environmental samples are heterogeneous, exposure pathways extend across multiple media, and the associated risks are rarely confined to one environmental compartment. A river system, for example, may simultaneously receive industrial discharge, wastewater-derived particles, agricultural runoff, and atmospheric deposition. Under those conditions, technologies that can connect detection, mapping, modelling, and risk management become particularly useful. They do not replace toxicological validation or field expertise, but they make integrated assessment more realistic and operational.

Advances in biotechnology continue to simplify nanoplastic detection and, in doing so, encourage broader investigation of environmental and biological impacts. High-throughput and microfluidic approaches enable large numbers of environmental or synthetic samples to be processed in parallel, which becomes important when the objective extends beyond detection alone and includes comparison across matrices, sources, and exposure scenarios. These methods also support multi-vector identification of pollutant sources, which matters because nanoplastics often enter ecosystems through several overlapping pathways rather than a single identifiable route. In parallel, satellite and remote-sensing technologies allow characterization of water bodies, land surfaces, and meteorological conditions across large regions. When combined, these tools can support more structured surveillance systems and reduce some of the uncertainty that still surrounds nanoplastic source attribution.

One-Health-oriented decision-support systems extend this framework further by modelling the spatio-temporal distribution of contaminant transport and exposure pathways. Their usefulness lies in organizing complex environmental data into forms that can support prioritization, governance, and response planning. In that sense, technological innovation is not simply about better instrumentation. It is also about creating workflows capable of connecting measurement to interpretation and interpretation to management. That connection is often where environmental monitoring efforts either become useful or remain descriptive.

8.1. High-Throughput Screening and Microfluidics

High-throughput screening (HTS) has already transformed several areas of biomedical research, particularly drug discovery, although its broader application still faces limitations associated with low-fidelity in vitro systems and incomplete biological understanding (Campbell et al., 2018). Even so, the underlying logic of HTS is highly relevant to nanoplastic research, where large numbers of samples often need to be processed rapidly and compared across different conditions. Microfluidics has further expanded these capabilities by enabling single-cell manipulation and parallel dosage-response analysis across millions of candidate compounds (Woodruff, 2017). Droplet-based HTS adds another layer of efficiency because complex sample matrices can be examined within miniaturized assays rather than conventional bulk-scale systems.

These approaches may also support nanoplastic remediation research. They can facilitate rapid screening of mitigation strategies, comparison of treatment conditions, and evaluation of sample handling procedures under controlled but high-volume settings. In this context, miniaturization and parallelization are not merely technical refinements. They are central to improving throughput while maintaining analytical specificity. The same applies to workflows requiring repeated comparison of multiple particle types, concentrations, or environmental conditions. Without some degree of structured automation, the resulting datasets can become difficult to organize and even harder to interpret consistently.

Methods for optical detection and chemical characterization of nanoplastics in complex environmental matrices are already available, and throughput can increase substantially when sample conditioning and mixing are integrated into the analytical workflow (Qian et al., 2024). This becomes important because environmental samples are rarely clean or uniform. Organic matter, suspended solids, dissolved contaminants, and biological materials frequently interfere with both separation and detection procedures. Aptamer-based strategies contribute additional flexibility by enabling wash-free and enrichment-free detection protocols applicable to a variety of analytes. Such approaches are attractive because they reduce preparation time and may lower the risk of sample loss or procedural contamination, both of which remain persistent concerns in nanoscale analysis.

Automation introduces another practical advantage. If contamination mapping becomes sufficiently automated, exposure and remediation strategies could potentially be screened more rapidly and under more standardized conditions. Survey intervals might then be adjusted according to contaminant accumulation or removal dynamics instead of following fixed schedules. That flexibility is useful in systems where pollutant levels fluctuate substantially, particularly in urban, aquatic, or agricultural environments. In applied environmental monitoring, responsiveness often matters as much as analytical sensitivity.

At the same time, the capabilities of HTS and microfluidic systems should not be overstated. Their outputs still depend heavily on calibration quality, appropriate controls, and interpretive frameworks that remain compatible with environmental heterogeneity. Highly controlled systems can simplify analysis, but they may also reduce environmental realism if not interpreted carefully. For that reason, these technologies are likely to be most effective when embedded within broader analytical pipelines rather than used as isolated solutions.

8.2. Satellite and Remote Sensing Synergies

Automated analysis of hyperspectral or multispectral images obtained from satellites and drones provides an important pathway for tracking remote sources of nanoplastics, microplastics, and associated biota, particularly within urban regions or environmentally vulnerable systems. The main strength of these platforms lies in their spatial scale. They allow repeated observation of large surface areas and can reveal distribution patterns that would be difficult to capture through field sampling alone. Satellite-borne systems, including Sentinel sensors and NASA's upcoming PACE satellite, monitor the surfaces of water bodies, while smaller-scale sensing platforms can periodically capture signals from individual lakes, rivers, or coastal systems under varying environmental conditions.

These approaches are not limited to open-water environments. They may also be applied to urban runoff systems and industrial discharge zones, allowing pollution levels and associated ecosystem pressures to be assessed over time. Because the resulting information is spatially explicit, remote sensing supports comparison across catchments, shorelines, and land-use zones. That feature is particularly relevant in integrated One Health studies where environmental observations are expected to inform broader ecological or public-health interpretation.

The potential utility of these systems increases when they are linked with additional publicly available datasets. In principle, automated satellite-tracked analysis could contribute to the generation of large and integrated datasets capable of supporting more coordinated One Health frameworks than isolated monitoring studies alone. The same logic extends to broader environmental prospecting, where remote sensing, field observations, and contextual environmental information together provide a more realistic representation of contaminant transport across connected systems (Davaasuren et al., 2018). This is not simply a matter of collecting more information. In many cases, it is about improving the alignment between environmental observation and environmental decision-making, which remains a persistent limitation in contaminant monitoring systems.

Related semi-automated methods, such as flow cytometry combined with fluorescence imaging and other particle-associated characteristics, have also shown utility for mapping phytoplankton and picoplankton populations across different sizes, shapes, and behavioural patterns. These methods help address some of the analytical limitations that continue to restrict One Health-oriented assessment of water systems (Zhang et al., 2023). They are particularly useful where particle complexity and biological overlap make conventional visual interpretation inadequate. Their contribution is therefore both methodological and practical, since they move environmental monitoring toward more scalable and comparable workflows.

Still, remote sensing alone does not resolve all analytical limitations. Many nanoplastic signals remain indirect, and surface observations cannot fully substitute for chemical verification or ground-truth sampling. The strongest applications therefore remain those that combine remote detection with laboratory confirmation and field-based interpretation. At present, that combined strategy appears to be the most defensible approach.

8.3. Decision Support Systems for Management

Artificial intelligence (AI) has further expanded the role of decision-support systems in the management of waste and emerging pollutants, including microplastics. One of the main advantages of these systems is their ability to synthesize heterogeneous information into formats that support interpretation, prioritization, and response planning. Geographic information systems, in particular, allow pollutants to be tracked visually and analytically across space, making it easier to examine both their distribution and potential environmental impacts. This is useful not only for identifying contamination hotspots but also for recognizing locations that may require greater monitoring attention.

Several types of information can already be modelled within these systems, including quantity, size, shape, and composition. These variables are not merely descriptive. They can be linked to potential risk profiles for specific habitats, allowing managers to determine the degree of monitoring or control required for different plastic types under different environmental conditions. In this way, decision-support systems help convert monitoring data into more operational forms of information. They also allow comparisons across environmental compartments that might otherwise be assessed separately, which becomes important when contamination pathways are diffuse or interconnected.

AI-based decision-making tools are especially relevant for emerging pollutants because contaminant behaviour is often non-linear and strongly context-dependent. Rules relating AI-driven decision-support systems to microplastics and related pollutants have therefore received increasing attention (Corte Pause et al., 2024). One implication is that monitoring systems need sufficient flexibility to adapt to changing exposure scenarios rather than relying on rigid assumptions regarding pollutant persistence or transport. This becomes particularly important where source attribution remains uncertain, transport pathways overlap, and remediation options are limited by infrastructure or cost.

Decision-support systems are most effective when integrated with transparent governance structures and reliable environmental datasets. They can improve coordination among environmental agencies, public-health sectors, and research groups, but only when the interpretive framework is sufficiently clear. Without that, there is a risk that the system becomes technically sophisticated but operationally limited. With adequate integration, however, the same tools can support more consistent monitoring, communication, and intervention planning across connected environmental and health systems.

9. CONCLUSION

In conclusion, while this chapter has successfully delineated the forensic pipelines and AI analytics required to trace nanoplastic sources and capture their biological impacts, a crucial mechanistic piece remains to be unfolded. The baseline toxicity and cellular cross-talk detailed here are fundamentally governed by the underlying particle dynamics. To fully grasp why these nano-pollutants behave as clandestine vectors, one must transition from macro-impacts to molecular mechanics. This paves the way for a rigorous, deep-dive evaluation into the physicochemical determinants, surface charges, aggregation kinetics, and structural fate that dictate nanoplastic behavior across global environmental compartments.

Detection, source tracing, and exposure assessment for nanoplastics are central to understanding One Health risks and informing targeted intervention strategies. These tasks remain technically demanding because nanoplastics are small, heterogeneous, and frequently embedded within complex environmental and biological matrices. Nevertheless, biotechnological methods and AI-based analytical systems increasingly provide end-to-end toolkits capable of addressing several of these challenges simultaneously. Their combined value lies in improving detection sensitivity, strengthening source-tracing accuracy, and streamlining exposure assessment workflows for nanoplastics.

Three broad categories of biotechnology methods contribute to detection, characterization, source attribution, and exposure modelling: molecular assays, spectroscopic and imaging approaches, and bioassays. Detection at molecular scales, rather than only at conventional micrometric ranges, is important for tracing nano- and microplastics throughout the continuum from production and use to environmental release and transport. Multiplexing signals associated with different material types or flow characteristics further increases the amount of information available for analysis. This can improve identification and quantification of multiple nano- and microplastic forms within the same environmental sample.

Source attribution strengthens this process by extending analysis beyond descriptive characterization alone. Through the use of physicochemical features and contaminant transport models, it becomes possible to infer likely source materials and estimate exposure factors associated with specific contamination pathways. That does not eliminate uncertainty, but it improves the structure of environmental assessment and narrows the range of plausible interpretations. In a field where environmental overlap is common, that level of narrowing is already significant.

The final step involves linking these findings to impacts on ecosystems, wildlife, and human health. Doing so effectively closes the One Health loop and allows the assessment framework to become more integrated. Quantitative impact pathways and risk frameworks are especially useful because they support modular integration of information across environmental and biological systems. They also permit modelling of multiple exposure routes, including aerosolization, ingestion, and topical exposure, together with the contaminant concentrations associated with each pathway. This matters because nanoplastic exposure is unlikely to occur through one route alone.

In summary, the integration of environmental science, biotechnology, and artificial intelligence within a forensic One Health framework provides a comprehensive approach for understanding and managing nanoplastic contamination. The framework is still developing, and several methodological limitations remain unresolved, particularly regarding long-term exposure dynamics and environmental standardization. Even so, the overall direction is becoming clearer. More integrated analytical pipelines will likely be necessary if nanoplastic risks are to be assessed with sufficient precision for environmental management, animal health, and public-health protection.

REFERENCES

Abbott, L. C., & Maynard, A. D. (2010). Exposure Assessment Approaches for Engineered Nanomaterials. *Risk Analysis : An Official Publication of the Society for Risk Analysis*, *30*(11), 1634–1644. DOI: 10.1111/j.1539-6924.2010.01446.x PMID: 20626687

Agbasi, J. C., Egbueri, J. C., Pande, C. B., Khan, M. Y. A., Ighalo, J. O., Uwajingba, H. C., & Abba, S. I. (2025). Review of the Potential Effects and Remediation Strategies of Microplastic Pollutants in Drinking Water Sources. Analytical Letters, 58(5), 799 839. DOI: 10.1080/00032719.2024.2343366

Alimi, O. S., Farner Budarz, J., Hernandez, L. M., & Tufenkji, N. (2018). Microplastics and nanoplastics in aquatic environments: Aggregation, deposition, and enhanced contaminant transport. Environmental Science & Technology, 52(4), 1704 1724. https://doi.org/DOI: 10.1021/acs.est.7b05559

Allan, J., Belz, S., Hoeveler, A., Hugas, M., Okuda, H., Patri, A., Rauscher, H., Silva, P., Slikker, W., Sokull-Kluettgen, B., Tong, W., & Anklam, E. (2021). Regulatory landscape of nanotechnology and nanoplastics from a global perspective. *Regulatory Toxicology and Pharmacology : RTP*, *122*, 104885. DOI: 10.1016/j.yrtph.2021.104885 PMID: 33617940

Amaral-Zettler, L. A., Zettler, E. R., & Mincer, T. J. (2020). Ecology of the plastisphere. Nature Reviews Microbiology, 18(3), 139 151. https://doi.org/DOI: 10.1038/s41579-019-0308-0

Azeem, I., Shakoor, N., Chaudhary, S., Adeel, M., Zain, M., Ahmad, M. A., Li, Y., Zhu, G., Shah, S. A. A., Khan, K., Khan, A. A., Xu, M., & Rui, Y. (2023). Analytical challenges in detecting microplastics and nanoplastics in soil-plant systems. *Plant Physiology and Biochemistry : PPB*, *204*, 108132. DOI: 10.1016/j.plaphy.2023.108132 PMID: 37918078

Bassan, S., & Harel, O. (2018). The ethics in synthetics: Statistics in the service of ethics and law in health-related research in big data from multiple sources. *Journal of Law and Health*, *31*(1), 87–117. PMID: 30889335

Bergamaschi, E., Craig, P., Guseva Canu, I., & Prina-Mello, A. (2015). The role of biological monitoring in nanosafety. *Nano Today*, *10*(3), 274–277. DOI: 10.1016/j.nantod.2015.02.001

Besseling, E., Quik, J. T. K., Sun, M., & Koelmans, A. A. (2017). Fate of nano- and microplastics in freshwater systems: A modeling study. Environmental Pollution, 220, 540 548.DOI: 10.1016/j.envpol.2016.10.001

Bondarenko, O. M., Heinlaan, M., Sihtmäe, M., Ivask, A., Kurvet, I., Joonas, E., Jemec, A., Mannerström, M., Heinonen, T., Rekulapelly, R., Singh, S., Zou, J., Pyykkö, I., Drobne, D., & Kahru, A. (2016). Multilaboratory evaluation of 15 bioassays for (eco)toxicity screening and hazard ranking of engineered nanomaterials: FP7 project NANOVALID. *Nanotoxicology*, *10*(9), 1229–1242. DOI: 10.1080/17435390.2016.1196251 PMID: 27259032

Brachner, A., Fragouli, D., Duarte, I. F., Farias, P. M. A., Dembski, S., Ghosh, M., Barisic, I., Zdzieblo, D., Vanoirbeek, J., Schwabl, P., & Neuhaus, W. (2020). Assessment of Human Health Risks Posed by Nano-and Microplastics Is Currently Not Feasible. *International Journal of Environmental Research and Public Health*, *17*(23), 8832. DOI: 10.3390/ijerph17238832 PMID: 33261100

Campbell, J. M., Balhoff, J. B., Landwehr, G. M., Rahman, S. M., Vaithiyanathan, M., & Melvin, A. T. (2018). Microfluidic and Paper-Based Devices for Disease Detection and Diagnostic Research. *International Journal of Molecular Sciences*, *19*(9), 2731. DOI: 10.3390/ijms19092731 PMID: 30213089

Corte Pause, F., Urli, S., Crociati, M., Stradaioli, G., & Baufeld, A. (2024). Connecting the Dots: Livestock Animals as Missing Links in the Chain of Microplastic Contamination and Human Health. *Animals (Basel)*, *14*(2), 350. DOI: 10.3390/ani14020350 PMID: 38275809

Cox, K. D., Covernton, G. A., Davies, H. L., Dower, J. F., Juanes, F., & Dudas, S. E. (2019). Human consumption of microplastics. Environmental Science & Technology, 53(12), 7068 7074. https://doi.org/DOI: 10.1021/acs.est.9b01517

Davaasuren, N., Marino, A., Boardman, C., Alparone, M., Nunziata, F., Ackermann, N., & Hajnsek, I. (2018, July). Detecting microplastics pollution in world oceans using SAR remote sensing. In *IGARSS 2018-2018 IEEE International Geoscience and Remote Sensing Symposium* (pp. 938-941). IEEE. DOI: 10.1109/IGARSS.2018.8517281

Destoumieux-Garzón, D., Mavingui, P., Boetsch, G., Boissier, J., Darriet, F., Duboz, P., Fritsch, C., Giraudoux, P., Le Roux, F., Morand, S., Paillard, C., Pontier, D., Sueur, C., & Voituron, Y. (2018). The One Health concept: 10 years old and a long road ahead. *Frontiers in Veterinary Science*, *5*, 14. DOI: 10.3389/fvets.2018.00014 PMID: 29484301

Everaert, G., Van Cauwenberghe, L., De Rijcke, M., Koelmans, A. A., Mees, J., Vandegehuchte, M., & Janssen, C. R. (2018). Risk assessment of microplastics in the ocean. Environmental Pollution, 242, 1930 1938. https://doi.org/DOI: 10.1016/j.envpol.2018.07.069

Frias, J. P. G. L., & Nash, R. (2019). Microplastics: Finding a consensus on the definition. Marine Pollution Bulletin, 138, 145 147 https://doi.org/DOI: 10.1016/j.marpolbul.2018.11.022

Furxhi, I., & Murphy, F. (2020). Predicting In Vitro Neurotoxicity Induced by Nanoparticles Using Machine Learning. *International Journal of Molecular Sciences*, *21*(15), 5280. DOI: 10.3390/ijms21155280 PMID: 32722414

Galloway, T. S., Cole, M., & Lewis, C. (2017). Interactions of microplastic debris throughout the marine ecosystem. Nature Ecology & Evolution, 1, 0116. https://doi.org/DOI: 10.1038/s41559-017-0116

Gigault, J., Halle, A. T., Baudrimont, M., Pascal, P. Y., Gauffre, F., Phi, T. L., El Hadri, H., Grassl, B., & Reynaud, S. (2018). Current opinion: What is a nanoplastic? Environmental Pollution, 235, 1030 1034. DOI: 10.1016/j.envpol.2018.01.024

Gouin, T., Ellis-Hutchings, R., Thornton Hampton, L. M., Lemieux, C. L., & Wright, S. L. (2022). Screening and prioritization of nano- and microplastic particle toxicity studies for evaluating human health risks - development and application of a toxicity study assessment tool. *Microplastics and Nanoplastics*, *2*(1), 2. DOI: 10.1186/s43591-021-00023-x PMID: 35098152

Hernandez, L. M., Xu, E. G., Larsson, H. C. E., Tahara, R., Maisuria, V. B., & Tufenkji, N. (2019). Plastic teabags release billions of microparticles and nanoparticles into tea. Environmental Science & Technology, 53(21), 12300 12310. DOI: 10.1021/acs.est.9b02540

Higashisaka, K., Yoshioka, Y., Yamashita, K., Morishita, Y., Pan, H., Ogura, T., Nagano, T., Kunieda, A., Nagano, K., Abe, Y., Kamada, H., Tsunoda, S., Nabeshi, H., Yoshikawa, T., & Tsutsumi, Y. (2012). Hemopexin as biomarkers for analyzing the biological responses associated with exposure to silica nanoparticles. *Nanoscale Research Letters*, *7*(1), 555. DOI: 10.1186/1556-276X-7-555 PMID: 23039107

Huppertsberg, S., & Knepper, T. P. (2018). Instrumental analysis of microplastics Benefits and challenges. Analytical and Bioanalytical Chemistry, 410(25), 6343 6352. https://doi.org/DOI: 10.1007/s00216-018-1210-8

Isibor, P. O., Kayode-Edwards, I. I., & Taiwo, O. S. (2024). Emerging technology and future directions in environmental nanotoxicology Enuneku (Eds.), Environmental nanotoxicology. Springer. https://doi.org/DOI: 10.1007/978-3-031-54154-4_16

Isigonis, P., Hristozov, D., Benighaus, C., Giubilato, E., Grieger, K., Pizzol, L., Semenzin, E., Linkov, I., Zabeo, A., & Marcomini, A. (2019). Risk Governance of Nanomaterials: Review of Criteria and Tools for Risk Communication, Evaluation, and Mitigation. [PDF]

Koelmans, A. A., Nor, N. H. M., Hermsen, E., Kooi, M., Mintenig, S. M., & De France, J. (2020). Microplastics in freshwaters and drinking water: Critical review and assessment of data quality. Water Research, 155, 410 422. DOI: 10.1016/j.watres.2019.02.054

Kögel, T., Bjorøy, Ø., Toto, B., Bienfait, A. M., & Sanden, M. (2020). Micro- and nanoplastic toxicity on aquatic life: Determining factors. *The Science of the Total Environment, 709*, 136050. DOI: 10.1016/j.scitotenv.2019.136050 PMID: 31887526

Landrigan, P. J., Stegeman, J. J., Fleming, L. E., Allemand, D., Anderson, D. M., Backer, L. C., Brucker-Davis, F., Chevalier, N., Corra, L., Czerucka, D., Bottein, M.-Y. D., Demeneix, B., Depledge, M., Deheyn, D. D., Dorman, C. J., Fénichel, P., Fisher, S., Gaill, F., Galgani, F., & Rampal, P. (2020). Human health and ocean pollution. *Annals of Global Health, 86*(1), 151. DOI: 10.5334/aogh.2831 PMID: 33354517

Lee, J., Jang, H., Kim, B., & Jung, J. (2025). Comparative Metabolomic Approaches to Nanoplastic Toxicity in Mammalian and Aquatic Systems. *International Journal of Molecular Sciences, 27*(1), 50. DOI: 10.3390/ijms27010050 PMID: 41515930

Lehner, R., Weder, C., Petri-Fink, A., & Rothen-Rutishauser, B. (2019). Emergence of nanoplastic in the environment and possible impact on human health. *Environmental Science & Technology, 53*(4), 1748–1765. DOI: 10.1021/acs.est.8b05512 PMID: 30629421

Leslie, H. A., van Velzen, M. J. M., Brandsma, S. H., Vethaak, A. D., Garcia-Vallejo, J. J., & Lamoree, M. H. (2022). Discovery and quantification of plastic particle pollution in human blood. *Environment International, 163*, 107199. DOI: 10.1016/j.envint.2022.107199 PMID: 35367073

Li, Y., Han, F., & Zheng, Y. (2026, January). Artificial Intelligence in Surface Water Quality Research and Management: Recent Progress and Future Directions. *Ecosystem Health and Sustainability, 12*, 0474. Advance online publication. DOI: 10.34133/ehs.0474

Ma, Y.-B., Xie, Z.-Y., Hamid, N., Tang, Q.-P., Deng, J.-Y., Luo, L., & Pei, D.-S. (2023). Recent advances in micro (nano) plastics in the environment: Distribution, health risks, challenges and future prospects. *Aquatic Toxicology, 261*, 106597. DOI: 10.1016/j.aquatox.2023.106597 PMID: 37311378

Mintenig, S. M., Bauerlein, P. S., Koelmans, A. A., Dekker, S. C., & van Wezel, A. P. (2018). Closing the gap between small and smaller: Towards a framework to analyse nano- and microplastics in aqueous samples. Environmental Science: Nano, 5(7), 1640 1649. https://doi.org/DOI: 10.1039/C8EN00186C

Oliveri Conti, G., Ferrante, M., Banni, M., Favara, C., Nicolosi, I., Cristaldi, A., Fiore, M., & Zuccarello, P. (2020). Micro- and nano-plastics in edible fruit and vegetables. *Environmental Research, 187*, 109677. DOI: 10.1016/j.envres.2020.109677 PMID: 32454310

Petersen, E. J., Barrios, A. C., Henry, T. B., Johnson, M. E., Koelmans, A. A., Montoro Bustos, A. R., Matheson, J., Roesslein, M., Zhao, J., & Xing, B. (2022). Potential Artifacts and Control Experiments in Toxicity Tests of Nanoplastic and Microplastic Particles. *Environmental Science & Technology, 56*(22), 15192–15206. DOI: 10.1021/acs.est.2c04929 PMID: 36240263

Pradel, A., Catrouillet, C., & Gigault, J. (2023). The environmental fate of nanoplastics: What we know and what we need to know about aggregation. *NanoImpact, 29*, 100453. DOI: 10.1016/j.impact.2023.100453 PMID: 36708989

Prata, J. C., da Costa, J. P., Lopes, I., Duarte, A. C., & Rocha-Santos, T. (2020). Environmental exposure to microplastics: An overview on possible human health effects. *The Science of the Total Environment, 702*, 134455. DOI: 10.1016/j.scitotenv.2019.134455 PMID: 31733547

Qian, N., Gao, X., Lang, X., Deng, H., Bratu, T. M., Chen, Q., Stapleton, P., Yan, B., & Min, W. (2024). Rapid single-particle chemical imaging of nanoplastics by SRS microscopy. *Proceedings of the National Academy of Sciences of the United States of America, 121*(3), e2300582121. DOI: 10.1073/pnas.2300582121 PMID: 38190543

Ramsperger, A. F. R. M., Bergamaschi, E., Panizzolo, M., Fenoglio, I., Barbero, F., Peters, R., Undas, A., Purker, S., Giese, B., Lalyer, C. R., Tamargo, A., Moreno-Arribas, M. V., Grossart, H.-P., Kühnel, D., Dietrich, J., Paulsen, F., Afanou, A. K., Zienolddiny-Narui, S., Eriksen Hammer, S., & Laforsch, C. (2023). Nano- and microplastics: A comprehensive review on their exposure routes, translocation, and fate in humans. *NanoImpact, 29*, 100441. DOI: 10.1016/j.impact.2022.100441 PMID: 36427812

Revel, M., Châtel, A., & Mouneyrac, C. (2018). Micro(nano)plastics: A threat to human health? Current Opinion in Environmental Science & Health, 1, 17 23. https://doi.org/DOI: 10.1016/j.coesh.2017.10.003

Roubert, F., Beuzelin-Ollivier, M.-G., Hofmann-Amtenbrink, M., Hofmann, H., & Hool, A. (2016). "Nanostandardization" in action: Implementing standardization processes in a multidisciplinary nanoparticle-based research and development project. *NanoEthics*, *10*(1), 41–62. DOI: 10.1007/s11569-015-0248-8

Saarimäki, L. A., Del Giudice, G., & Greco, D. (2023). Expanding adverse outcome pathways toward one health model for nanosafety. *Frontiers in Toxicology*, *5*, 1176745. DOI: 10.3389/ftox.2023.1176745 PMID: 37692900

Santos, F. A., Andre, R. S., Alvarenga, A. D., Alves, A. L. M. M., & Correa, D. S. (2025). Micro- and nanoplastics in the environment: A comprehensive review on detection techniques. Environmental Science: Nano, 12(7), 3442 3467. https://doi .org/DOI: 10.1039/D4EN00319E

Schymanski, D., Goldbeck, C., Humpf, H.-U., & Fürst, P. (2018). Analysis of microplastics in water by micro-Raman spectroscopy: Release of plastic particles from different packaging into mineral water. Water Research, 129, 154 162. DOI: 10.1016/j.watres.2017.11.011

Sekine, R., Moore, K. L., Matzke, M., Vallotton, P., Jiang, H., Hughes, G. M., Kirby, J. K., Donner, E., Grovenor, C. R. M., Svendsen, C., & Lombi, E. (2017). Complementary imaging of silver nanoparticle interactions with green algae: Dark-field microscopy, electron microscopy, and nanoscale secondary ion mass spectrometry. *ACS Nano*, *11*(11), 10894–10902. DOI: 10.1021/acsnano.7b04556 PMID: 29061049

Sendra, M., Saco, A., Yeste, M. P., Romero, A., Novoa, B., & Figueras, A. (2020). Nanoplastics: From tissue accumulation to cell translocation into Mytilus galloprovincialis hemocytes. *Journal of Hazardous Materials*, *388*, 121788. DOI: 10.1016/j. jhazmat.2019.121788 PMID: 31813690

Sendra, M., Sparaventi, E., Novoa, B., & Figueras, A. (2021). An overview of the internalization and effects of microplastics and nanoplastics on marine organisms. *The Science of the Total Environment*, *760*, 143953. DOI: 10.1016/j.scitotenv.2020.142024 PMID: 33207452

Shruti, V. C., & Kutralam-Muniasamy, G. (2019). Bioplastics: Missing link in the era of microplastics. *The Science of the Total Environment*, *697*, 134139. DOI: 10.1016/j.scitotenv.2019.134139 PMID: 32380615

United Nations Environment Programme. (2023). Turning off the tap: How the world can end plastic pollution and create a circular economy. UNEP. https://www .unep.org/resources/turning-off-tap-end-plastic-pollution-create-circular-economy

Velzeboer, I., Kwadijk, C. J. A. F., & Koelmans, A. A. (2014).Strong sorption of PCBs to nanoplastics, microplastics, carbon nanotubes, and fullerenes. Environmental Science & Technology, 48(9), 4869 4876.DOI: 10.1021/es405721v

Vethaak, A. D., & Leslie, H. A. (2016). Plastic debris is a human health issue. Environmental Science & Technology, 50(13), 6825 6826. https://doi.org/DOI: 10.1021/acs.est.6b02569

Wagner, S., Hüffer, T., Klöckner, P., Wehrhahn, M., Hofmann, T., & Reemtsma, T. (2018).

Woodruff, K. P. (2017). Microfluidic platforms for high-throughput mammalian cell printing, transfection, and dosage-dependent studies (Master's thesis, École Polytechnique Fédérale de Lausanne). EPFL Infoscience. EPFL Infoscience Repository

World Health Organization. (2019). Microplastics in drinking-water: Evidence and recommendations. World Health Organization. https://www.who.int/publications/i/item/9789241516198

Wright, S., Levermore, J., & Ishikawa, Y. (2023). Application of Infrared and Near-Infrared Microspectroscopy to Microplastic Human Exposure Measurements. *Applied Spectroscopy*, *77*(10), 1105–1128. DOI: 10.1177/00037028231199772 PMID: 37792505

Wu, B., Wu, X., Liu, S., Wang, Z., & Chen, L. (2019). Size-dependent effects of polystyrene microplastics on cytotoxicity. Chemosphere, 221, 333 341. https://doi.org/DOI: 10.1016/j.chemosphere.2019.01.056

Yang, L., Kang, S., Luo, X., & Wang, Z. (2024). Microplastics in drinking water: A review on methods, occurrence, sources, and potential risks assessment. *Environmental Pollution*, *348*, 123857. DOI: 10.1016/j.envpol.2024.123857 PMID: 38537794

Yang, L., Zhang, Y., Kang, S., Wang, Z., Wu, C., & Zhang, J. (2021). Microplastics in soil: A review of methods, occurrences, sources, and potential risks. *The Science of the Total Environment*, *780*, 146546. DOI: 10.1016/j.scitotenv.2021.146546 PMID: 33770602

Yee, M. S., Hii, L. W., Looi, C. K., Lim, W. M., Wong, S. F., Kok, Y. Y., Tan, B. K., Wong, C. Y., & Leong, C. O. (2021). Impact of Microplastics and Nanoplastics on Human Health. *Nanomaterials (Basel, Switzerland)*, *11*(2), 496. DOI: 10.3390/nano11020496 PMID: 33669327

Yong, C. Q. Y., Valiyaveettil, S., & Tang, B. L. (2020). Toxicity of Microplastics and Nanoplastics in Mammalian Systems. *International Journal of Environmental Research and Public Health, 17*(5), 1509. DOI: 10.3390/ijerph17051509 PMID: 32111046

Zhang, H., Wang, J., Zhou, B., Zhou, Y., Dai, Z., Zhou, Q., Christie, P., & Luo, Y. (2018). Enhanced adsorption of oxytetracycline to weathered microplastic polystyrene. Environmental Pollution, 243, 1550 1557. https://doi.org/DOI: 10.1016/j.envpol.2018.09.122

Zhang, Y., Zhang, D., & Zhang, Z. (2023). A Critical Review on Artificial Intelligence-Based Microplastics Imaging Technology: Recent Advances, Hot-Spots and Challenges. *International Journal of Environmental Research and Public Health, 20*(2), 1150. DOI: 10.3390/ijerph20021150 PMID: 36673905

Chapter 3
Physicochemical Determinants of Nanoplastic Fate and Advanced Removal Dynamics:
AI-Enabled Approaches within One Health Frameworks

Fatma Tugce Senberber Dumanli
http://orcid.org/0000-0002-3257-1524
Department of Bioengineering, Yildiz Technical University, Istanbul, Turkey

Emek Moroydor Derun
http://orcid.org/0000-0002-8587-2013
Department of Chemical Engineering, Yildiz Technical University, Istanbul, Turkey

Mehmet Burcin Piskin
http://orcid.org/0000-0002-2689-8547
Department of Bioengineering, Yildiz Technical University, Istanbul, Turkey

ABSTRACT

This chapter per the authors analyzes the physicochemical determinants governing nanoplastic transport behavior, environmental partitioning, bioavailability, and remediation performance across complex ecological matrices. Nanoplastics are heterogeneous colloidal systems where particle size, morphology, surface charge,

DOI: 10.4018/979-8-3373-4490-4.ch003

density, hydrophobicity, and functionalization regulate aggregation kinetics, interfacial interactions, contaminant adsorption, and environmental persistence under variable conditions. The analysis further examines how these properties influence efficiency and selectivity of physical, chemical, biological, and hybrid remediation technologies. Analytical constraints in nanoscale characterization, compositional heterogeneity, and remediation-related ecological trade-offs are critically assessed alongside predictive computational models for fate estimation and removal optimization. By integrating environmental chemistry with remediation dynamics, the chapter advances a systems-oriented framework for nanoplastic behavior and mitigation within One Health exposure landscapes.

INTRODUCTION

Background on Nanoplastics and Environmental Concerns

Plastics released into the environment undergo continuous fragmentation driven by coupled abiotic and biotic processes. Photo-oxidation induced by UV radiation initiates polymer chain scission, reducing molecular weight and introducing oxygen-containing functional groups (e.g., carbonyl and hydroxyl groups), which in turn increase hydrophilicity and susceptibility to further degradation. In parallel, mechanical abrasion resulting from wave action, sediment interaction, and soil disturbance contributes to progressive surface erosion, while thermo-oxidative processes under elevated temperatures further accelerate structural deterioration. Although microbial enzymatic activity contributes to polymer breakdown, its rate remains slow and is largely constrained to specific polymer types under environmental conditions. This entire multi-stage transition, from initial breakdown to persistent nanoplastic formation, is illustrated in the upper section of Figure 1.

Figure 1. Graphical summary of nanoplastic dynamics: Illustrating the fragmentation of macroplastics, environmental transport pathways, biological internalization, and emerging remediation frameworks. This figure outlines the multidirectional lifecycle of nanoplastics and brings together the major processes connecting environmental transformation, exposure pathways, and remediation strategies within a One Health context. As illustrated in the upper part of the figure, plastic materials do not degrade through a single linear pathway. Instead, primary and secondary polymers undergo continuous photo-oxidative, mechanical, and thermal breakdown, gradually generating smaller particles and altering their surface characteristics over time. As fragmentation progresses, nanoplastics (<1 μm) accumulate across environmental compartments including air, water, and soil, where their behavior becomes increasingly difficult to predict. The figure further highlights that these particles do not remain restricted to isolated environmental settings. Once released, they move through interconnected environmental and biological pathways, creating multiple exposure routes through ingestion, inhalation, and dermal uptake. Such movement raises broader concerns because persistent exposure may contribute to oxidative stress, inflammatory responses, and cellular disturbances across biological systems. The lower section of the framework shifts attention toward possible intervention strategies. Physical capture processes, chemical oxidation approaches, and biologically mediated remediation pathways are incorporated as complementary components intended to reduce persistence and environmental burden. The figure therefore provides an integrated view of the connections among environmental fate, toxicity, and mitigation within a broader One Health framework.

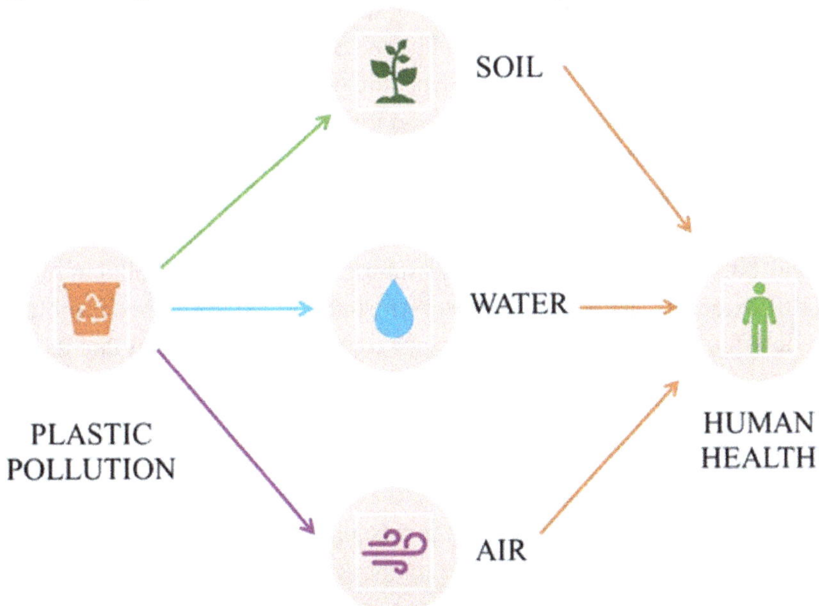

NPs originate either from the fragmentation of larger plastic debris (secondary NPs) or from direct release associated with industrial processes and consumer applications, including cosmetic formulations and paint additives (primary NPs) (Mahendran & Ramaswamy, 2024). The transition from microplastics (1 μm–5 mm) to NPs is associated with a shift in environmental behavior from predominantly physical interactions to more complex chemical reactivity. Due to their elevated specific surface-area-to-volume ratio, NPs exhibit increased adsorption capacity and chemical reactivity, enabling interactions with organic pollutants, heavy metals, and other environmental contaminants, and thereby acting as potential carriers of toxic substances. NPs are now widely reported across marine, freshwater, terrestrial, and atmospheric compartments. Transport mechanisms including wind dispersion, ocean circulation, and precipitation facilitate long-range distribution, enabling their presence in remote environments distant from primary emission sources. Owing to their low biodegradability, small size, and electrokinetic stability (zeta potential), NPs persist in environmental matrices over extended periods, contributing to chronic ecological exposure (Zaki & Aris, 2022; Morreale & La Mantia, 2024). Documented effects include ecotoxicological impacts on aquatic and terrestrial organisms, with reported alterations in microbial community structure and biochemical function (Chepuri et al., 2023; Lackner & Branka, 2024; Bodzek, 2023). Experimental studies in fish, crustaceans, and mollusks indicate disruptions in feeding behavior, energy allocation, and reproductive output, suggesting potential implications at the population level. In addition, their role as contaminant vectors facilitates the transfer of pathogens and toxic compounds across trophic levels, thereby compounding ecological risk through combined physical and chemical stress mechanisms (Bhat et al., 2022; Mahendran & Ramaswamy, 2024; Lackner & Branka, 2024).

From a human health perspective, exposure occurs via inhalation, ingestion, and dermal contact through air, water, and food matrices. Drinking water, seafood, and processed food products represent major exposure pathways, and emerging evidence has confirmed detection in various biological tissues (Chepuri et al., 2023; Lackner & Branka, 2024). Following cellular internalization, NPs are capable of crossing biological barriers, entering cells, and interacting with intracellular organelles, leading to oxidative stress, membrane perturbation, inflammatory responses, and genotoxic effects. Chronic low-dose exposure remains a concern due to potential cumulative effects, including endocrine disruption, immune modulation, and neurodevelopmental toxicity (Chepuri et al., 2023; Sutrisna et al., 2024).

Despite increasing research attention, significant uncertainties persist regarding baseline environmental concentrations and exposure quantification (Zaki & Aris, 2022). Accurate measurement of NP levels in complex environmental matrices remains technically challenging due to methodological limitations, requiring integrated multidisciplinary approaches spanning environmental science, toxicology, and public health (Dang et al., 2022). In response, advanced detection and remediation technologies—including membrane filtration, adsorption-based systems, photocatalytic degradation, and magnetic separation—are currently under investigation. However, their applicability under real environmental conditions remains limited and requires further optimization (Han et al., 2024; Mahendran & Ramaswamy, 2024; Sutrisna et al., 2024; Hanif et al., 2022). Consequently, the establishment of standardized regulatory frameworks is still a critical requirement, alongside broader circular economy strategies and governance interventions aimed at reducing primary plastic leakage into the environment (Arya & Kumar, 2025; Sadek et al., 2025). The integration of which into broader regulatory and circular economy frameworks is highlighted in the final strategic tier of Figure 1.

Risk Assessment and Regulatory Frameworks in the One Health Paradigm

NPs act as systemic threat vectors across environmental, animal, and human health domains. To quantitatively assess these complex interactions, emerging risk assessment models—such as the Environmental Exposure and Hazard Index (EEHI)—integrate particle transport dynamics with cellular toxicity thresholds. However, conventional risk models often fail to capture the "eco-corona" effect, which significantly alters NP bioavailability and toxicity profiles.

At present, the global regulatory landscape remains fragmented. While the European Chemicals Agency (ECHA) and the US EPA have introduced limits for microplastics, specific regulatory frameworks addressing NPs are largely absent. This regulatory gap creates substantial challenges for monitoring and enforcement. Accordingly, a One Health-oriented regulatory framework is required to standardize detection limits (e.g., pyrolysis-GC/MS and Raman spectroscopy) and to mandate toxicity assessments that incorporate environmental transformation processes such as UV aging and competitive adsorption of chemical pollutants.

Systemic Interaction of Nanoplastics within the One Health Framework

The One Health approach is a holistic concept grounded in the recognition that human health, animal health, and environmental health are interconnected and interdependent. This premise assumes that environmental disturbances, including nanoplastics, inevitably propagate across ecological boundaries, affecting animal health and, through both direct and indirect exposure pathways, human health. From this perspective, One Health should not be viewed as a simple intersection of environmental, human, and animal health domains, but rather as a dynamic and interconnected system. In this context, nano-sized pollutants function as highly mobile transfer agents between these three domains, accelerating pollutant connectivity and risk propagation.

Nanoplastics represent an emerging environmental health concern; however, due to their small size and high mobility, they are readily transferred into animal and human systems through processes such as bioaccumulation and toxicological exposure. Consequently, their removal extends beyond environmental remediation and becomes an integral component of global public health protection efforts.

Significance of Physicochemical Properties in Environmental Behavior

The physicochemical properties of NPs play a central role in determining their environmental fate, transport behavior, and ecological impact. Environmental transformation and stability are strongly influenced by processes such as photooxidation and disinfection. Solar irradiation and chlorination can substantially modify NP physicochemical properties, leading to changes in stability and toxicity. For instance, solar irradiation promotes transformation of polystyrene (PS) NPs during chlorination, inducing chain scission, surface oxidation, and enhanced colloidal stability in wastewater systems (Du et al., 2023). Similarly, aging processes such as UV exposure alter NP surface properties, thereby influencing mobility and interactions with surrounding environmental matrices. Aged PSNPs exhibit distinct mobility patterns compared to pristine particles, largely due to surface chemical modifications (Xi et al., 2022), highlighting the dynamic behavior of NPs following environmental release.

Sedimentation and aggregation are closely governed by NP surface charge and particle size. Positively charged PSNPs with smaller dimensions (20–50 nm) exhibit higher sedimentation ratios compared to larger negatively charged particles, while environmental conditions such as pH, ionic strength, and natural organic matter (NOM) further modulate sedimentation behavior (Zhu et al., 2023a). In addition, interactions with NOM and extracellular polymeric substances (EPS) contribute to the formation of an eco-corona, which may either enhance or inhibit NP transport. For example, EPS-derived eco-coronas enhance the mobility of PS nanospheres more effectively than humic-substance-derived coronas (Zhu et al., 2023b). Thus, aggregation and eco-corona formation are key determinants of whether NPs remain suspended or settle in aquatic environments.

NP transport is also strongly influenced by environmental chemistry. Ionic strength and electrolyte composition, such as the presence of $CaCl_2$, regulate aggregation and sedimentation processes, thereby controlling NP mobility in aquatic systems (Zhu et al., 2023a; Pradel et al., 2021). Surface functional groups further modify NP interactions with NOM and other environmental constituents, directly affecting stability and transport behavior. For instance, carboxyl-modified NPs exhibit different aggregation patterns compared to amino-modified particles (Song et al., 2019), indicating that subtle chemical differences can significantly influence transport dynamics.

Environmental and biological interactions introduce an additional layer of complexity. NP surface characteristics, including size, charge, and biocorona formation, determine bioavailability and toxicity toward aquatic organisms. Once formed, biocoronas modify NP physicochemical identity, thereby shaping interactions with biological systems and influencing ecological outcomes (Rex et al., 2023; Junaid & Wang, 2021). Moreover, the colloidal behavior of NPs distinguishes them from microplastics, as their stability and interaction mechanisms differ substantially in environmental matrices. Parameters such as pH, ionic strength, and NOM concentration strongly influence NP colloidal stability (Pradel et al., 2021), necessitating specialized risk assessment approaches distinct from those used for larger plastic particles.

Overall, the environmental behavior of NPs is governed by their physicochemical properties, which collectively regulate transformation, stability, sedimentation, transport, and biological interactions. These interconnected factors ultimately determine ecological impact and potential risk pathways. Linking physicochemical characteristics to environmental outcomes remains essential for anticipating NP behavior and informing targeted mitigation strategies.

SCOPE AND OBJECTIVES OF THE CHAPTER

In recent years, NPs have emerged as a major concern in environmental science due to their widespread distribution, persistence, and potential toxicity. Unlike microplastics, which have been extensively studied, NPs remain comparatively underexplored, particularly regarding their detection, environmental behavior, and removal methods. Their nanoscale dimensions allow them to cross biological membranes, interact with macromolecules, and accumulate within tissues, thereby posing risks to both ecosystems and human health. This combination of environmental ubiquity and biological accessibility underscores the urgency of advancing research on NPs.

One of the main challenges in addressing NP pollution lies in their considerable physicochemical diversity. Variations in size, morphology, surface chemistry, density, and aggregation potential strongly influence their behavior in natural environments and directly affect the efficiency of their removal from water, soil, or air. Although a range of removal technologies has been developed—from physical filtration systems to advanced oxidation processes—most approaches do not adequately account for this complex heterogeneity. As a result, their performance often varies significantly under different environmental conditions.

The central objective of this chapter is to examine how the physicochemical properties of NPs influence the effectiveness of different removal approaches. By presenting a structured overview of key characteristics—including particle size, surface charge, shape, and chemical composition—the chapter aims to clarify their role in governing transport dynamics, environmental persistence, and interactions with treatment technologies. It further critically evaluates existing removal methods, highlighting their strengths and limitations when faced with the intrinsic variability of NPs.

Through this analysis, the chapter emphasizes the importance of linking physicochemical traits with technological performance. It also identifies the main scientific and technological constraints limiting removal efficiency and discusses potential improvements and alternative strategies. Ultimately, the goal is to advance understanding of NPs in environmental systems and to support the development of more effective, targeted, and sustainable remediation technologies.

PHYSICOCHEMICAL PROPERTIES OF NANOPLASTICS

Size, Shape, and Surface Morphology

Nanoplastics exhibit a high surface area-to-volume ratio that intensifies their interactions with pollutants and biological systems (Mahendran & Ramaswamy, 2024; Der et al., 2025). This feature makes them more difficult to remove using conventional treatment methods while also enhancing their environmental and biological impacts. Smaller NPs (<400 nm) are prone to rapid aggregation under high ionic strength conditions, which are commonly found in seawater and wastewater (Hammond et al., 2024). Interestingly, such aggregation can also facilitate removal through conventional techniques such as filtration, centrifugation, and flocculation (Murray & Örmeci, 2020). Thus, size not only governs environmental fate but also influences treatability in engineered systems.

NPs with high shape anisotropy, such as nanotetrahedra, show enhanced uptake by cellular membranes, affecting their environmental accumulation and potential transfer to humans (Yong & Du, 2023). In contrast, irregularly shaped NPs tend to aggregate more easily, which directly influences their ingestion and toxicity in aquatic organisms (Rajtar et al., 2025). This highlights how morphological variability contributes simultaneously to persistence and biological risk.

Surface-functionalized NPs, such as those modified with carboxyl or amino groups, show distinct internalization pathways and toxicity profiles in human cells.

For instance, aminated NPs significantly reduce cell viability and induce higher intracellular reactive oxygen species (iROS) compared to other functionalized forms (Martín-Perez et al., 2024). Therefore, surface chemistry plays a key role in regulating biological interactions and determining toxicological outcomes.

Surface Chemistry and Functional Groups

Nanoplastics carrying different functional groups, such as carboxyl (-COOH) and amino (-NH_2), exhibit distinct interactions with natural organic matter and coagulants. In comparative studies on polystyrene NPs (PS-COOH and PS-NH_2), PS-NH_2 particles were found to be more easily removed across a broader pH range than PS-COOH particles, due to their specific interaction pathways with compounds such as salicylic acid and commonly used coagulants (Liu et al., 2025). These differences underscore the central role of surface chemistry in determining environmental behavior and treatability.

Functional groups also influence binding affinity and removal efficiency. PS-COOH NPs, for example, show stronger electrostatic repulsion and higher hydrophilicity, which can reduce removal efficiency under certain conditions (Liu et al., 2025). These characteristics not only complicate treatment processes but also contribute to their persistence in natural environments.

In addition, NPs with different surface groups (–NH_2, –COOH, and bare) display distinct heteroaggregation behaviors with mineral surfaces such as goethite (FeOOH). Functionalized particles tend to aggregate more rapidly due to enhanced electrostatic and chemical interactions, particularly involving groups such as C–H, O–C=O, and –CH_2 (Wu et al., 2023). This indicates that surface chemistry strongly influences aggregation dynamics and, consequently, environmental fate in both aquatic and terrestrial systems.

Density and Buoyancy

NP density is a key factor controlling buoyancy and settling behavior in aquatic environments. For instance, polystyrene nanoplastics (PSNPs) have densities close to that of water, making their removal via air flotation challenging due to their small size and negative surface charge (Xu et al., 2024; Devi et al., 2022). These properties highlight the importance of density considerations in designing removal strategies.

Biofilm formation on NPs, often referred to as the plastisphere, can significantly alter their density and buoyancy. Biofouling increases particle density, leading to sinking and sedimentation. This process depends on surface area and biofilm development rate, with smaller or film-like NPs tending to settle more rapidly due to their relatively higher surface area (Tang & Li, 2025). Such dynamic changes complicate predictions of NP transport and removal in natural waters.

Removal techniques are also strongly influenced by NP density and buoyancy. Filtration using 0.22 μm membranes can physically retain NPs, achieving removal efficiencies up to 92% in laboratory conditions. High-speed centrifugation (10,000 rpm) exploits density differences and can remove up to 99% of particles (Murray & Örmeci, 2020). Coagulant addition enhances removal by reducing surface charge and promoting aggregation, thereby increasing effective density and improving filtration performance (Ramirez-Arenas et al., 2022). Modified air flotation systems, using agents such as dodecyltrimethylammonium chloride, improve particle–bubble interactions by reducing electrostatic repulsion and enhancing hydrophobic attraction, achieving removal efficiencies above 95% (Xu et al., 2024).

Beyond general mechanisms, removal efficiency also varies by polymer type. Polystyrene (PS) shows high removal performance through electrolysis-assisted flotation, plasma-assisted processes, and ionic liquid extraction, largely due to its size and variable chemical composition (Samsudin et al., 2023; Corbett & Amaral-Zettler, 2022; Pawak et al., 2023; Winburn et al., 2025; Srivastava et al., 2025). Polyethylene (PE) and polypropylene (PP), which have similar densities close to water, are more effectively treated using membrane bioreactors and ultrafiltration systems, especially when combined with coagulation or flocculation (Pradel et al., 2021). In contrast, polyethylene terephthalate (PET), characterized by higher density and greater environmental stability, typically requires advanced filtration techniques to achieve effective removal (Bain, 2022; Bodzek & Bodzek, 2025; Enyoh et al., 2025). These differences highlight the need for polymer-specific treatment strategies rather than relying solely on general density-based assumptions.

To provide a clearer overview, Table 1 summarizes the approximate densities of different nanoplastic polymers, their typical environmental behaviors, and the most appropriate treatment approaches reported in the literature.

Table 1. Densities of common nanoplastic polymers and their corresponding removal approaches

Polymer Type	Density (g/cm³)	Behavior in the Environment	Most Appropriate Treatment Approach	References
PS	~1.05	PS is reported to interfere with wastewater treatment processes due to its small size and variable chemical composition. It can also form aggregates with other suspended materials, thereby influencing stability and transport.	Electrolysis-assisted flotation, plasma-assisted processes, and ionic liquid-based liquid–liquid extraction have been demonstrated to achieve high removal efficiencies for PS.	(Samsudin et al., 2023; Corbett & Amaral-Zettler, 2022; Pawak et al., 2023; Winburn et al., 2025; Srivastava et al., 2025).
PE	~0.94	PE displays variable behavior in saline environments depending on surface properties and shape. In addition, its stability can be enhanced through interactions with natural organic matter.	Removal is typically achieved through membrane bioreactors and ultrafiltration, particularly when combined with coagulation and flocculation processes.	(Veclin et al., 2022; Pradel et al., 2011; Bodzek & Pohl, 2023a)
PP	~0.90	PP exhibits environmental behavior that is influenced by surface properties and surrounding conditions. Its stability is also reinforced by associations with natural organics.	Effective removal has been demonstrated through membrane bioreactors and ultrafiltration in combination with coagulation/flocculation.	(Veclin et al., 2022; Pradel et al., 2011; Bodzek & Pohl, 2023b)
PET	~1.38	PET demonstrates greater stability in seawater, which is strongly influenced by surface properties and ionic strength. Depending on particle size and density, PET may either remain buoyant or undergo Brownian motion.	Removal typically requires advanced methods such as membrane filtration and may also benefit from adsorption- or electrochemical-based strategies.	(Bodzek & Pohl, 2023a; Bain, 2022; Bodzek & Bodzek, 2025; Enyoh et al., 2025)

Table 1. Densities of common nanoplastic polymers and their corresponding removal approaches. The table summarizes several widely studied nanoplastic polymers, with emphasis on how differences in density and surface behavior may influence environmental transport, persistence, aggregation tendencies, and treatment response in aquatic systems. Although density remains an important parameter affecting buoyancy and mobility, the available literature also suggests that particle shape, salinity, ionic strength, and interactions with natural organic matter can substantially modify nanoplastic behavior under real environmental conditions. In practice, no single treatment approach appears to perform equally well across all polymer classes. Instead, removal efficiency often depends on the combined effects of polymer chemistry, particle size distribution, wastewater composition, and operational conditions within the treatment system. Conventional membrane-based processes continue to be frequently reported for PE, PP, and PET removal, whereas more specialized approaches, including plasma-assisted treatment, electrolysis-assisted flotation, and ionic liquid-based extraction, have shown promising results for PS under controlled experimental settings. Overall, the comparison highlights the increasing need for polymer-specific and multi-stage remediation strategies rather than relying on a single universal treatment model for all nanoplastic contaminants.

ENVIRONMENTAL FACTORS AFFECTING STABILITY AND AGGREGATION

Environmental conditions, including water composition and the presence of natural organic matter, strongly influence the stability and aggregation behavior of NPs. In water bodies characterized by low concentrations of divalent ions, aggregation is less likely to occur, which in turn complicates removal processes (Pirade et al., 2024). In addition, biofilm formation can induce irregular changes in particle density, thereby affecting vertical migration and removal efficiency. Depending on system conditions, biofilms may either enhance settling velocity or introduce motion instabilities that reduce it (Jalón-Rojas et al., 2022; Vercauteren et al., 2024). Overall, the interaction between density variations, biofouling processes, and environmental conditions needs to be carefully considered when designing effective NP removal strategies.

Aggregation and Stability in Aquatic Environments

Nanoplastics have the ability to bind with toxic substances, which increases their persistence and potential toxicity in aquatic systems (Mahendran & Ramaswamy, 2024). Their aggregation behavior directly influences the efficiency of water treatment processes, including filtration and coagulation (Huang et al., 2024; Ramirez-Arenas et al., 2022). Surface functional groups play a central role in controlling aggregation dynamics. For example, electrolytes and proteins can modify aggregation kinetics differently depending on NP surface chemistry. Carboxyl-modified (PS-COOH) and amino-modified (PS-NH_2) NPs show distinct aggregation behaviors in the presence of common electrolytes and proteins, which consequently affects both stability and removal efficiency (Guo et al., 2024).

Environmental parameters such as humic acid and salinity further modulate aggregation behavior. Humic acid generally stabilizes negatively charged NPs, whereas increased salinity tends to promote aggregation, thereby altering removal outcomes in treatment systems (Wu et al., 2019). In some cases, humic acid may initially induce aggregation of positively charged NPs, such as p-PS-NH_2, before stabilizing them, indicating a complex interaction pattern in natural waters. Surface functional groups also influence coagulation efficiency. For instance, PSNPs with carboxyl groups (-COOH) interact with divalent cations such as Ca^{2+}, leading to reduced stability and enhanced aggregation (Pirade et al., 2024; Guo et al., 2024).

In addition, coagulant type and pH conditions significantly affect removal performance (Chen et al., 2024). Overall, NP aggregation and stability are governed by a coupled system of surface chemistry, environmental conditions, and treatment parameters, which must be considered jointly when designing removal strategies.

ENVIRONMENTAL BEHAVIOR OF NANOPLASTICS: ROLE OF PHYSICOCHEMICAL PROPERTIES

Transport and Fate in Terrestrial and Aquatic Systems

Environmental aging processes, including exposure to solar radiation and ozone, can lead to the formation of new oxygen-containing functional groups on NP surfaces. These transformations increase NP sensitivity to humic acid and may reduce removal efficiency during coagulation processes (Chen et al., 2024). At the same time, aging-related changes such as increased surface roughness and enhanced hydrophilicity can modify NP interactions with pollutants. In some cases, these alterations shift adsorption mechanisms from physisorption to chemisorption, strengthening pollutant binding and potentially enhancing removal efficiency under specific conditions (Tumrani et al., 2025).

Interaction with Organic and Inorganic Pollutants

Surface functional groups such as -COOH and -NH_2 on PSNPs play a critical role in determining their interactions with low molecular weight compounds and coagulants. PS-NH_2 NPs are generally more easily removed across a wider pH range compared to PS-COOH, mainly due to differences in electrostatic interactions and hydrophilicity (Liu et al., 2025). These functional groups not only influence removal efficiency but also regulate adsorption and binding behavior with environmental pollutants, thereby affecting overall environmental fate and associated ecological risks.

Bioavailability and Toxicity Pathways

Surface functionalization strongly affects NP cellular uptake and toxicity. Carboxylated nanoplastics (PS-COOH) show higher cellular uptake compared to pristine NPs, likely due to increased interactions with cell surface receptors, which in turn influences both environmental behavior and removal efficiency (Zhang et al., 2025). Different functional groups can induce varying levels of oxidative stress and cytotoxicity in biological systems, highlighting the importance of these properties in assessing ecological and human health risks (Tomonaga et al., 2025; Baysal et al., 2024).

REMOVAL APPROACHES FOR NANOPLASTICS

Overview of Current Removal Methods

Physical Removal Techniques

Physical removal techniques such as filtration, centrifugation, and ballasted flocculation have shown high efficiencies for NP removal, with centrifugation reaching up to 99% removal depending on particle properties and functionalization (Murray & Örmeci, 2020). Conventional filtration methods often face limitations due to the small size of NPs; however, advanced membrane technologies such as nanofiltration have demonstrated potential in concentrating NPs and reducing wastewater volumes (Devi et al., 2022; Enyoh et al., 2025; Severino et al., 2025). Similarly, membrane bioreactors and rapid sand filtration systems have reported removal efficiencies exceeding 99%, highlighting their applicability in large-scale treatment systems (Bodzek & Pohl, 2023b).

Adsorption-based techniques are widely investigated due to their operational simplicity and efficiency. Carbon-based and metal-based adsorbents remove NPs through electrostatic interactions, hydrogen bonding, and hydrophobic effects (Ho et al., 2025; Liu & Khor, 2024). In addition, functionalized magnetic NPs have been effectively used to selectively capture and remove NPs from aqueous systems, offering a flexible remediation approach (Enyoh et al., 2025; Yeo & Julaihi, 2021).

Chemical and Advanced Oxidation Processes

Chemical and advanced oxidation processes provide alternative strategies by targeting NP aggregation and degradation. Coagulation and flocculation promote aggregation of NPs into larger particles that can be more easily removed through sedimentation or filtration, while electrocoagulation and magnetization further enhance process efficiency (Devi et al., 2022; Hanif et al., 2022). These approaches are particularly effective in high-concentration NP water streams, where rapid particle settling improves downstream treatment performance.

Photocatalytic degradation using semiconductor materials such as titanium dioxide (TiO_2) under UV irradiation enables partial breakdown of NPs into less harmful products. However, this process requires further optimization for large-scale and real environmental applications (Severino et al., 2025). In addition, combining coagulation, flocculation, and photocatalysis may provide synergistic effects that enhance overall removal efficiency while reducing residual NP impacts.

Biological and Biotechnological Approaches

Biological approaches offer promising, though still developing, alternatives for NP removal by exploiting microbial degradation pathways. Certain microorganisms have demonstrated the ability to degrade NPs, although this area remains under active investigation and requires further optimization for practical application (Devi et al., 2022). Constructed wetlands and biofilm-based systems are also being explored as complementary strategies, benefiting from microbial activity and natural adsorption processes (Bodzek & Pohl, 2023b; Hanif et al., 2022). When integrated with physical and chemical methods, these biological systems may contribute to more sustainable and efficient remediation strategies.

Emerging and Hybrid Approaches

Emerging and hybrid technologies combine multiple mechanisms to improve NP removal efficiency. Micro- and nanobubbles (MNBs) enhance flotation processes due to their high surface area-to-volume ratio and stability in water, showing strong potential for environmental applications, although further scaling is still required (Sadek et al., 2025). Integrated systems that couple membrane filtration with photocatalytic degradation provide a two-stage approach involving NP concentration followed by degradation, which can significantly improve overall performance (Severino et al., 2025).

Hybrid configurations such as membrane bioreactor–ultrafiltration/reverse osmosis (MBR-UF/RO) systems or coagulation followed by ozonation also demonstrate high removal efficiencies. These results highlight the value of combining complementary technologies to address the variability and complexity of NP particles (Bodzek & Pohl, 2023b). A comparative summary of removal methods for microplastics and nanoplastics in aquatic systems is presented in Table 2.

Table 2. Comparison of Removal Methods for Microplastics and Nanoplastics in Aquatic Environments

Removal Method	Description	Efficiency	Advantages	Limitations	References
Filtration	Physical separation using filters	Varies by filter size	Simple, widely used	Ineffective for smaller particles	(Devi et al., 2022; Bodzek, & Bodzek, 2025)
Coagulation/ Flocculation	Aggregation of particles using coagulants	Up to 99% with combined methods	Effective for larger particles	Less effective for NPs, chemical use	(Park et al., 2021; Bodzek & Pohl, 2023a; Noornama et al., 2024; Hanif et al., 2022)
Membrane Filtration	Pressure-driven separation (MF, UF, NF, RO)	Up to 99.9% with MBRs	High efficiency, versatile	Fouling, scalability issues	(Ouda et al., 2023)
Adsorption	Binding of particles to adsorbents	High, depending on adsorbent	Reusable, simple operation	Adsorbent regeneration, cost	(Enyoh et al., 2025; Yu et al., 2024; Liu & Khor, 2024)
Photocatalysis	Degradation using light-activated catalysts	High for specific plastics	Effective degradation	Requires light source, catalyst stability	(Devi et al., 2022; Ouda et al., 2023; Zoppas et al., 2023)
Biodegradation	Breakdown by microorganisms	Varies	Environmentally friendly	Slow process, specific conditions needed	(Noornama et al., 2024; Hanif et al., 2022)
Electrochemical Methods	Removal using electrical processes	Promising results	Innovative, scalable	Cost, energy requirements	(Noornama et al., 2024; Enyoh et al., 2025)
Magnetic Nanoparticles	Removal using functionalized magnetic particles	High efficiency	Easy separation, reusable	Potential environmental impact	(Brito et al., 2024; Kumar et al., 2025)
Plasma-Assisted Methods	Degradation using plasma reactors	Up to 99.3%	High efficiency	Requires specialized equipment	(Winburn et al., 2025)
Micro/ Nanobubbles (MNBs)	Flotation using gas-filled bubbles	Enhanced removal	High surface area, stable	Scalability, operational parameters	(Sadek et al., 2025)

Table 2. Comparative overview of currently explored removal approaches for microplastics (MPs) and nanoplastics (NPs) in aquatic environments. The table outlines a range of physical, chemical, biological, and hybrid treatment strategies that have been investigated for reducing plastic particulate contamination in water systems. Reported removal efficiencies vary considerably across studies and are often influenced by factors such as particle size distribution, polymer chemistry, reactor configuration, and overall water matrix composition. Conventional filtration methods remain among the most commonly applied approaches because of their relative simplicity and accessibility, although their effectiveness generally declines for smaller nanoscale particles. Membrane-based systems, coagulation/flocculation

processes, and several emerging technologies, including plasma-assisted treatment and magnetic nanoparticle-based separation, have shown encouraging removal performance under controlled experimental conditions, but questions related to operational stability, scalability, fouling behavior, energy demand, and potential secondary environmental impacts are still being discussed. Taken together, the current literature suggests that no single removal strategy is likely to be sufficient across all environmental settings, particularly for heterogeneous MP/NP mixtures, reinforcing the need for integrated treatment frameworks and more consistent evaluation criteria across future studies.

AI-Assisted Detection, Monitoring, and Removal of Nanoplastics

Recent studies show a growing integration of artificial intelligence (AI) and sensor-based technologies in both the detection and removal of NPs. Machine-learning models are increasingly used to process complex spectral and imaging datasets for accurate classification of NPs, optimization of operational parameters, and prediction of degradation behavior under varying environmental conditions. As reported by Nene et al. (2025), combining advanced separation technologies with AI-driven data analytics improves monitoring precision and enhances process adaptability, enabling near real-time control of NP removal systems. This integrated framework strengthens the link between technological innovation and exposure prevention, aligning nanoplastic remediation strategies with the One Health perspective that connects environmental integrity, human health, and sustainable resource management.

Empirical Evaluation of Biotechnological and Hybrid Remediation Systems

Although advanced oxidation processes and membrane-based separation techniques demonstrate high efficiency at the laboratory scale, their real-world application is often constrained by high energy demand and membrane fouling issues (Sutrisna et al., 2024). In contrast, biotechnological and hybrid nature-based systems provide more sustainable remediation pathways.

Case Analysis 1: Eco-Corona Formation and Macromolecular Trapping in NP Aggregation

In complex aquatic systems, the efficiency of removal technologies is strongly influenced by interactions between NPs and natural organic materials. Research by Zhu et al. (2023a) shows that natural organic matter (NOM) and extracellular polymeric substances selectively adsorb onto NP surfaces, forming a dynamic layer known as the "eco-corona." In practical treatment conditions, this layer masks the original polymer properties and alters particle mobility. By exploiting competitive adsorption processes involving these environmental macromolecules, hybrid biosorption systems can enhance the entrapment of aged polystyrene and polyethylene particles, promoting the transformation of colloidal NPs into larger, filterable aggregates prior to filtration units (Zhu et al., 2023a).

Case Analysis 2: Optimizing Sedimentation and Flocculation via Size and Charge Regulation

In operational water treatment systems, sedimentation performance is often limited by the electrokinetic stability of NPs. Zhu et al. (2023a) demonstrated that particle size and surface charge (zeta potential) are key determinants of sedimentation behavior in freshwater environments. For example, carboxylated nanoplastics exhibit strong electrostatic repulsion, which keeps them stably suspended. Through AI-assisted monitoring combined with controlled chemical dosing, coagulation conditions can be optimized to neutralize surface charges. This charge neutralization promotes particle bridging and enhances settling kinetics, producing removal efficiencies comparable to those observed in continuous-flow drinking water treatment systems (Zhang et al., 2020; Zhu et al., 2023b).

INFLUENCE OF PHYSICOCHEMICAL PROPERTIES ON REMOVAL EFFICIENCY

Impact of Size and Surface Properties on Filtration and Sedimentation

The influence of particle size and surface characteristics on NP removal efficiency is a key factor in filtration and sedimentation processes. While larger microplastics are more effectively retained by filtration due to size exclusion, NPs tend to aggregate into larger clusters, which can improve capture efficiency, although removal remains less effective for smaller fractions (Hanif et al., 2024). Granular filtration systems can achieve removal efficiencies up to 99.9% for particles larger than 100 μm, whereas efficiency declines significantly for particles in the 10–20 μm range. Sedimentation alone is generally insufficient for NPs due to their small size and low density; for instance, clean plastics show sedimentation rates below 2.0% across tested size ranges. However, biofilm formation on microplastics can enhance sedimentation, indirectly improving the removal of associated NPs (Zhang et al., 2020). Surface properties also play a decisive role, as engineered biochar with increased positive charge and hydrophobicity improves removal efficiency by enhancing electrostatic and hydrophobic interactions with negatively charged and hydrophobic particles (Hanif et al., 2024). Similarly, magnetic bentonite modified with PDMDAC improves adsorption performance through increased surface charge and strong electrostatic attraction) Han et al., 2024).

Role of Surface Chemistry in Adsorption and Coagulation

Surface chemistry is a critical determinant of NP adsorption and coagulation behavior. Adsorption efficiency is strongly influenced by the surface properties of both adsorbents and NPs. For instance, biochar modified with cetyltrimethylammonium bromide (CTAB) shows improved performance due to enhanced positive surface charge and increased hydrophobicity (Hanif et al., 2024). Likewise, PDMDAC-modified magnetic bentonite exhibits high adsorption capacity due to strong electrostatic interactions and Fe–O functional groups. Adsorption mechanisms include electrostatic attraction, complexation, and hydrophobic interactions, all of which are governed by surface characteristics of the adsorbent (Ali et al., 2023).

Coagulation processes further enhance NP removal by reducing surface charge and promoting aggregation. The addition of coagulants can increase sand filtration efficiency from 54.3% to 99.2% by facilitating particle aggregation (Ramirez-Arenas et al., 2022). In addition, the type of coagulant and coagulant aids, such as Poly-DADMAC, can significantly influence coagulation/flocculation and sedimentation (CFS) performance, although smaller particles may still present persistent removal challenges (Zhang et al., 2020).

Challenges in Removing Heterogeneous Nanoplastics Mixtures

The complex surface characteristics of NPs, including varying degrees of aging and functionalization, further complicate their interactions with removal media. For instance, aged microplastics with rough surfaces and increased oxygen-containing functional groups tend to show higher removal efficiency due to enhanced interactions with adsorbent materials. Environmental conditions such as pH and ionic strength also play a key role. In particular, removal efficiency often follows a V-shaped trend with pH in biochar-based systems, while increased ionic strength generally inhibits the removal of both pristine and aged microplastics (Huang et al., 2024; Li & Zhang, 2023). In addition, technological constraints—especially the lack of reliable analytical methods for detecting and quantifying NPs in environmental matrices—further complicate both the evaluation of removal efficiency and the design of effective treatment strategies (Li et al., 2024; Devi et al., 2022).

Technological Challenges and Limitations

Characterizing NPs remains technically challenging due to their small size, heterogeneous composition, and the complexity of environmental matrices. These limitations not only hinder the accurate detection and quantification in environmental samples but also directly affect the evaluation of removal performance, thereby constraining the development of efficient and targeted remediation strategies.

Analytical Difficulties in Characterizing Nanoplastics

Nanoplastics are extremely small and often fall below the detection limits of conventional analytical techniques, which creates major challenges in their characterization. Methods such as static light scattering (SLS) have difficulty detecting particles below 200 nm, while nanoparticle tracking analysis (NTA) and nano-flow cytometry (nFCM) are not suitable for accurately measuring particles in the micrometre range. In addition, the weak signal intensity associated with such small particles makes it difficult to distinguish them from background noise in complex environmental matrices. These analytical constraints reduce the reliability of environmental monitoring and limit the accurate assessment of removal efficiency across different treatment systems (Caputo et al., 2021; Stine et al., 2023; Zhang et al., 2022; Fang et al., 2023).

Removal Efficiency vs. Environmental Safety Trade-offs

The characterization and removal of NPs are constrained by their small size, compositional diversity, and the complexity of environmental matrices. NPs often fall below the detection limits of standard analytical methods, with techniques such as static light scattering (SLS) struggling to detect particles below 200 nm, while nanoparticle tracking analysis (NTA) and nano-flow cytometry (nFCM) cannot reliably measure particles in the micrometre range. Furthermore, the weak signal output of such small particles makes it difficult to distinguish them from background interference, complicating accurate environmental monitoring. Some removal approaches may also generate secondary pollution, highlighting the need for more sustainable and efficient technologies (Rasheed et al., 2023). In addition, isolating NPs from complex environmental samples is further hindered by interfering substances such as natural organic matter and other particulates (Stine et al., 2023; Fang et al., 2023; Enyoh et al., 2021). Techniques such as filtration and density separation are commonly applied for NP isolation; however, these methods are often labor-intensive, may not achieve complete removal, and can create trade-offs between environmental safety and treatment efficiency (Mahapatra et al., 2024).

The Role of Artificial Intelligence in Nanoplastic Management and Removal

The complex and dynamic behavior of nanoplastics in environmental systems extends beyond the capabilities of traditional analytical and experimental approaches. In this context, artificial intelligence (AI) and machine learning (ML) have emerged as powerful tools for addressing the scale and complexity of the problem. The core role of AI lies in modeling the interactions between highly variable physicochemical properties of nanoplastics (such as size, surface charge, and aggregation behavior) and environmental matrices (water, soil, and air). This enables not only the prediction of pollution sources and transport pathways but also the rapid processing of large datasets that are difficult or time-consuming to obtain experimentally. As a result, AI-based approaches offer significant potential for improving preventive management and risk assessment of nanoplastic pollution.

The effectiveness of AI in nanoplastic management is reflected in three main application areas. First, machine learning (ML) algorithms support advanced detection and characterization, enabling the identification of nanoplastics in environmental samples and allowing rapid interpretation of microscopic imaging and spectroscopic data. Second, predictive modeling tools simulate long-term transport and fate processes, mapping exposure risks based on variables such as ocean currents, groundwater flow, and atmospheric transport. Finally, AI-driven optimization systems enhance the performance of removal processes such as coagulation, filtration, and advanced oxidation in real time, aiming to maximize efficiency while minimizing energy consumption. Together, these applications represent a significant step forward in scientific and engineering approaches to nanoplastic management.

FUTURE PERSPECTIVES AND RECOMMENDATIONS

Emerging Technologies for Selective Removal

Emerging technologies provide promising pathways for the selective removal and improved characterization of NPs. The integration of multiple analytical techniques, such as combining microscopy with spectroscopy, helps overcome the limitations of single-method approaches and improves both accuracy and reliability. For example, coupling chemical analysis with imaging techniques enables better discrimination of NPs within complex environmental matrices (Regasa, 2024). In addition, artificial intelligence (AI) and machine learning approaches can enhance detection efficiency and data interpretation, particularly when dealing with complex and high-dimensional datasets, while also improving imaging precision. Novel approaches, including the use of spatiotemporal deformation data derived from SEM analyses, also show potential for distinguishing NPs based on polymer type and improving detection in challenging samples (Stine et al., 2023; Fang et al., 2023; Chandran et al., 2025). Together, these integrated methodologies support more targeted and efficient NP removal strategies in environmental systems.

Need for Standardization in Removal Strategies

Accurate quantification of NPs remains a major challenge, largely due to the absence of standardized methodologies and consistent sampling protocols, leading to variability and limits comparability across studies (Pinto da Costa et al., 2019; Choi et al., 2024). High-resolution techniques such as tunable resistive pulse sensing (TRPS) and centrifugal liquid sedimentation (CLS) offer promising capabilities for precise NP measurement; however, they still require further optimization and standardization before routine application. Establishing unified protocols and harmonized guidelines is essential to ensure reproducibility, enable reliable interlaboratory comparisons, and support the development of more robust and effective NP removal strategies.

Policy and Regulatory Frameworks for Nanoplastics

Nanoplastics are emerging contaminants associated with increasing environmental and health concerns. Despite this, regulatory frameworks specifically addressing NPs remain largely absent or insufficient in most regions.

In the European Union, regulatory progress has been made in the field of nanomaterials through frameworks such as the Registration, Evaluation, Authorisation, and Restriction of Chemicals (REACH). Additional regulations apply to sectors such as cosmetics, food, and medical devices (Ferrari 2021; Bergkamp et al., 2014). However, these frameworks largely rely on adaptations of existing chemical regulations and do not fully capture the unique behavior and risks associated with NPs, leaving important uncertainties unresolved (Porter et al., 2011; Ngarize et al., 2013).

In the United States, general regulatory instruments such as the Toxic Substances Control Act, the Clean Water Act, and the Safe Drinking Water Act could, in principle, be applied to NPs, but none have been explicitly implemented for them. Regulatory oversight is partly distributed across agencies such as the Environmental Protection Agency and the Food and Drug Administration, yet specific nanoscale regulations remain limited (Sorensen et al., 2023; O'Brian, 2011).

In Asian countries, nanomaterials are partially addressed within broader chemical regulatory frameworks, such as the Chemical Registration and Assessment Act in Korea and the Act on the Evaluation of Chemical Substances in Japan (Rasmussen et al., 2017). Taiwan has also introduced initiatives such as the Taiwan Nanotechnology Standard Council and the nanoMark labeling system, reflecting a more structured but still evolving regulatory approach (Song et al., 2012).

Across regions, there is increasing consensus on the need for NP-specific regulatory frameworks, including requirements for product specifications, notification systems, public disclosure, and systematic risk assessment procedures (Porter et al., 2011; Ngarize et al., 2013). In addition, international coordination and harmonized measurement standards are essential to ensure data comparability and to support globally consistent strategies for mitigating NP pollution (Nawaz et al., 2025).

CONCLUSION

The physicochemical properties of NPs, including size, surface area-to-volume ratio, surface charge, hydrophobicity, and adsorption capacity, strongly influence their environmental behavior and removal efficiency. A range of removal approaches—such as filtration, sedimentation, coagulation, adsorption, and advanced oxidation processes—has been developed, with hybrid and integrated systems showing the greatest potential for comprehensive treatment. However, the heterogeneity of NP mixtures, variability in environmental conditions, and complex interactions with organic and inorganic substances introduce significant challenges that require adaptive and multi-dimensional treatment strategies.

In parallel, analytical limitations in NP characterization—arising from their small size, compositional diversity, environmental complexity, and lack of standardized methods—further constrain accurate assessment and mitigation. Addressing these challenges requires the integration of advanced analytical technologies, development of standardized protocols, and increased use of artificial intelligence to improve detection, characterization, and monitoring capabilities. Future research should prioritize scalable, cost-effective, and environmentally sustainable removal technologies, while also generating robust evidence to support regulatory development and long-term management strategies for NPs. Integrating these findings within a One Health framework reinforces that nanoplastic removal is not only an environmental priority but also a public health necessity. Advanced remediation strategies help reduce exposure pathways for humans and animals while contributing to ecosystem resilience and sustainability. This interconnected perspective highlights that protecting water, soil, and air quality ultimately supports planetary health, linking technological innovation with preventive health protection and informed policy action.

DECLARATION ON THE USE OF GENERATIVE AI AND AI-ASSISTED TECHNOLOGIES

The authors confirm that AI-assisted tools were exclusively used for minor language and grammar refinement, while all scientific content, and overall scholarly integrity remain the sole responsibility of the authors.

AUTHORS NOTES

No external funding was received for the preparation of this chapter. The authors declare that there are no conflicts of interest regarding the publication of this work. All authors contributed equally to the preparation of this manuscript.

Correspondence concerning this chapter should be addressed to Emek Moroydor Derun (email: moroydor@yildiz.edu.tr).

REFERENCES

Ali, I., Tan, X., Li, J., Peng, C., Wan, P., Naz, I., Duan, Z., & Ruan, Y. (2023). Innovations in the development of promising adsorbents for the remediation of microplastics and nanoplastics – A critical review. *Water Research, 230*, 119526. DOI: 10.1016/j.watres.2022.119526 PMID: 36577257

Arya, N., & Kumar, A. (2025). Micro/nano-plastics pollution: Challenges to agriculture productivity. In *Global impacts of micro- and nano-plastic pollution* (pp. 206–225). IGI Global., DOI: 10.4018/979-8-3693-3447-8.ch009

Bain, A. (2022). Buoyancy and Brownian motion of plastics in aqueous media: Predictions and implications for density separation and aerosol internal mixing state. *Environmental Science. Nano, 9*(11), 4249–4254. DOI: 10.1039/D2EN00525E

Baysal, A., Soyocak, A., Saygin, H., & Saridag, A. M. (2024). Exposure to phagolysosomal simulated fluid altered the cytotoxicity of PET micro(nano)plastics to human lung epithelial cells. *Toxicology Mechanisms and Methods, 34*(1), 72–97. DOI: 10.1080/15376516.2023.2256847 PMID: 37697451

Bergkamp, L., Michaux, G., & Herbatschek, N. (2014). Nanotechnology regulation in Europe: From REACH and nano-registries to cosmetics, biocides, and medical devices. *Nanotechnology Law & Business, 11*(2), 136–159.

Bhat, M. A., Gedik, K., & Gaga, E. O. (2022). Environmental toxicity of emerging micro and nanoplastics: A lesson learned from nanomaterials. In *Nanotechnology interventions in food packaging and shelf life* (pp. 311–327). CRC Press., DOI: 10.1201/9781003207641-18

Bodzek, M. (2023). Remediation of nano- and microplastics in water environment using nanomaterials. *Desalination and Water Treatment, 316*, 557–573. DOI: 10.5004/dwt.2023.30172

Bodzek, M., & Bodzek, P. (2025). Remediation of micro- and nanoplastics by membrane technologies. *Membranes (Basel), 15*(3), 82. DOI: 10.3390/membranes15030082 PMID: 40137034

Bodzek, M., & Pohl, A. (2023a). Possibilities of removing microplastics from the aquatic environment using membrane processes. *Desalination and Water Treatment, 288*, 104–120. DOI: 10.5004/dwt.2023.29154

Bodzek, M., & Pohl, A. (2023b). Removal of microplastics in unit processes used in water and wastewater treatment: A review. *Archives of Environmental Protection, 48*(2), 102–128. DOI: 10.24425/aep.2022.143713

Brito, A. L. C. P., Mattioni, J. V., Ramos, G. R., Nakamura, M., & Toma, H. E. (2024). Direct monitoring of the enzymatically sequestering and degrading of PET microplastics using hyperspectral Raman microscopy. *Micron : The International Research and Review Journal for Microscopy, 187*, 103722. DOI: 10.1016/j.micron.2024.103722 PMID: 39306918

Caputo, F., Vogel, R., Savage, J., Vella, G., Law, A., Della Camera, G., Hannon, G., Peacock, B., Mehn, D., Ponti, J., Geiss, O., Aubert, D., Prina-Mello, A., & Calzolai, L. (2021). Measuring particle size distribution and mass concentration of nanoplastics and microplastics: Addressing some analytical challenges in the sub-micron size range. *Journal of Colloid and Interface Science, 588*, 401–417. DOI: 10.1016/j.jcis.2020.12.039 PMID: 33422789

Chandran, P. J. I., Ranjani, M., & Veerasingam, S. (2025). Challenges of microplastic characterization and mitigation. In *Handbook of Microplastic Pollution in the Environment* (pp. 1–22). CRC Press., DOI: 10.1201/9781003487593-2

Chen, M., Nan, J., & Breider, F. (2024). A comparative study on the stability and coagulation removal of aged vs. nonaged nanoplastics in surface water. *Journal of Hazardous Materials, 480*, 136490. DOI: 10.1016/j.jhazmat.2024.136490 PMID: 39547043

Chepuri, K., Tulasi, C. D. S. L. N., & Manikantha, D. (2023). Cellular and animal toxicities of micro- and nanoplastics. In *Micro and nanoplastics in soil* (pp. 261–292). Springer Nature., DOI: 10.1007/978-3-031-21195-9_12

Choi, S., Lee, S., Kim, M.-K., Yu, E.-S., & Ryu, Y.-S. (2024). Challenges and recent analytical advances in micro(nano)plastics analysis. *Analytical Chemistry, 96*(22), 8846–8854. DOI: 10.1021/acs.analchem.3c05948 PMID: 38758170

Corbett, L. N., & Amaral-Zettler, L. A. (2022). Nanoplastics and the marine environment: The unseen majority? In Andrady, A. L. (Ed.), *Plastics and the Ocean: Origin, Characterization, Fate, and Impacts* (pp. 389–428). Wiley., DOI: 10.1002/9781119768432.ch14

Dang, F., Wang, Q., Huang, Y., Wang, Y., Xing, B., & Zhang, X. (2022). Key knowledge gaps for One Health approach to mitigate nanoplastic risks. *Eco-Environment & Health, 1*(1), 11–22. DOI: 10.1016/j.eehl.2022.02.001 PMID: 38078201

Der, O., Khaksar, H., & Gnecco, E. (2025). On the formation and characterization of nanoplastics during surface wear processes. *Surfaces, 8*(2), 27. DOI: 10.3390/surfaces8020027

Devi, M. K., Karmegam, N., Manikandan, S., Subbaiya, R., Song, H., Kwon, E. E., Sarkar, B., Bolan, N., Kim, W., Rinklebe, J., & Govarthanan, M. (2022). Removal of nanoplastics in water treatment processes: A review. *The Science of the Total Environment, 845*, 157168. DOI: 10.1016/j.scitotenv.2022.157168 PMID: 35817120

Du, T., Shao, S., Qian, L., Meng, R., Li, T., Wu, L., & Li, Y. (2023). Effects of photochlorination on the physicochemical transformation of polystyrene nanoplastics: Mechanism and environmental fate. *Water Research, 243*, 120367. Advance online publication. DOI: 10.1016/j.watres.2023.120367 PMID: 37499544

Enyoh, C. E., Devi, A., Maduka, T. O., Tyagi, L., Rana, S., Akuwudike, I. S., & Wang, Q. (2025). A review of materials for the removal of micro- and nanoplastics from different environments. *Micro*, *5*(2), 17. DOI: 10.3390/micro5020017

Enyoh, C. E., Wang, Q., Chowdhury, T., Wang, W., Lu, S., Xiao, K., & Chowdhury, M. A. H. (2021). New analytical approaches for effective quantification and identification of nanoplastics in environmental samples. *Processes (Basel, Switzerland)*, *9*(11), 2086. Advance online publication. DOI: 10.3390/pr9112086

Fang, C., Luo, Y., & Naidu, R. (2023). Microplastics and nanoplastics analysis: Options, imaging, advancements and challenges. *Trends in Analytical Chemistry*, *166*, 117158. DOI: 10.1016/j.trac.2023.117158

Ferrari, A. (2021). Regulatory aspects of nanotechnology in the EU. In S. S. Mudliar & K. K. Pant (Eds.), *Environmental nanotechnology volume 4: Applications of nanotechnology for energy and the environment* (pp. 185–210). Springer. https://doi.org/DOI: 10.1007/978-3-030-79808-6_8

Guo, Y., Tang, N., Lu, L., Li, N., Hu, T., Guo, J., Zhang, J., Zeng, Z., & Liang, J. (2024). Aggregation behavior of polystyrene nanoplastics: Role of surface functional groups and protein and electrolyte variation. *Chemosphere*, *350*, 140998. DOI: 10.1016/j.chemosphere.2023.140998 PMID: 38142881

Hammond, C. B., Qadikolaei, A. F., Aghaaminiha, M., Sharma, S., & Gigault, J. (2024). New insights into the formation of aggregates of bidisperse microplastics in water based on the analysis of in situ microscopy and molecular simulation. *Langmuir : The ACS Journal of Surfaces and Colloids*, *40*(28), 14455–14466. DOI: 10.1021/acs.langmuir.4c01216 PMID: 38967440

Han, F., Wang, Y., An, S. Y., Liu, L., Ma, L. Q., Yang, L., & Liu, S. Y. (2024). Poly(dimethyldiallylammonium chloride)-modified magnetic bentonite for removal of polystyrene nanoplastics. *Journal of Environmental Chemical Engineering*, *12*(5), 114013. Advance online publication. DOI: 10.1016/j.jece.2024.114013

Hanif, M. A., Ibrahim, N., Dahalan, F. A., Md Ali, U. F., Hasan, M., & Abdul Jalil, A. (2022). Microplastics and nanoplastics: Recent literature studies and patents on their removal from aqueous environment. *The Science of the Total Environment, 810*, 152115. DOI: 10.1016/j.scitotenv.2021.152115 PMID: 34896138

Hanif, M. A., Ibrahim, N., Hayazi, N. A., Dahalan, F. A., Ali, U. F. M., Jalil, A. A., & Syafiuddin, A. (2024). Enhancement of microplastics and nanoplastics removal via filtration method using surface-engineered palm kernel shell biochar. *Separation and Purification Technology, 360*, 130596. DOI: 10.1016/j.seppur.2024.130596

Ho, K. L., Yeap, S. P., & Lee, K. M. (2025). A critical review of microplastics and nanoplastics in wastewater: Insights into adsorbent-based remediation strategies. *Environmental Pollution, 382*, 126658. DOI: 10.1016/j.envpol.2025.126658 PMID: 40513794

Huang, J., Tan, X., Ali, I., Ok, Y. S., Duan, Z., Liang, J., & Zhu, R. (2024). Efficient removal of nanoplastics by iron-modified biochar: Understanding the removal mechanisms. *Environmental Pollution, 363*, 125121. DOI: 10.1016/j.envpol.2024.125121 PMID: 39426478

Jalón-Rojas, I., Figueroa, D., & Rivas, J. (2022). Effects of biofilms and particle physical properties on the vertical dynamics of microplastic sheets. *Environmental Science & Technology, 56*(17), 12195–12204. DOI: 10.1021/acs.est.2c01302 PMID: 35593651

Junaid, M., & Wang, J. (2021). Interaction of nanoplastics with extracellular polymeric substances (EPS) in the aquatic environment: A special reference to eco-corona formation and associated impacts. *Water Research, 201*, 117319. DOI: 10.1016/j.watres.2021.117319 PMID: 34130084

Kumar, C., Singh, H., Kaur, M., Arya, S. K., & Khatri, M. (2025). Remediation strategies for micro/nanoplastic pollution using magnetic nanomaterials. *Environmental Science and Pollution Research International, 32*(26), 15574–15603. DOI: 10.1007/s11356-025-36663-2 PMID: 40562997

Lackner, M., & Branka, M. (2024). Microplastics in farmed animals—A review. *Microplastics, 3*(4), 35. Advance online publication. DOI: 10.3390/microplastics3040035

Li, J., Zhang, S., & Zhang, Y. (2023). Removal of pristine and aged microplastics from water by magnetic biochar: Adsorption and magnetization. *The Science of the Total Environment, 875*, 162647. DOI: 10.1016/j.scitotenv.2023.162647 PMID: 36889392

Li, Y., Zhang, C., Tian, Z., Cai, X., & Guan, B. (2024). Identification and quantification of nanoplastics (20–1000 nm) in a drinking water treatment plant using AFM-IR and Pyr-GC/MS. *Journal of Hazardous Materials*, *463*, 132933. DOI: 10.1016/j.jhazmat.2023.132933 PMID: 37951177

Liu, B., Guo, K., Feng, Q., Zhang, Y., Zhang, X., & Zhang, Y. (2025). Revealing the removal behavior of polystyrene nanoplastics and natural organic matter by AlTi-based coagulant from the perspective of functional groups. *Journal of Hazardous Materials*, *487*, 137258. DOI: 10.1016/j.jhazmat.2025.137258 PMID: 39837030

Liu, Q., & Khor, S. M. (2024). Emerging absorption-based techniques for removing microplastics and nanoplastics from actual water bodies. *Trends in Analytical Chemistry*, *170*, 117465. DOI: 10.1016/j.trac.2023.117465

Mahapatra, S., Maity, J. P., Singha, S., Mishra, T., Dey, G., Samal, A. C., Banerjee, P., Biswas, C., Chattopadhyay, S., Patra, R. R., Patnaik, S., & Bhattacharya, P. (2024). Microplastics and nanoplastics in environment: Sampling, characterization and analytical methods. *Groundwater for Sustainable Development*, *20*, 101267. DOI: 10.1016/j.gsd.2024.101267

Mahendran, R., & Ramaswamy, S. N. (2024). Nanoplastics as Trojan horses: Deciphering complex connections and environmental ramifications: A review. *Chemistry Africa*, *7*(1), 2265–2282. DOI: 10.1007/s42250-023-00865-3

Martin-Pérez, J., Villacorta, A., Banaei, G., Morataya-Reyes, M., Tavakolpournegari, A., Marcos, R., Hernández, A., & García-Rodríguez, A. (2024). Hazard assessment of nanoplastics is driven by their surface-functionalization: Effects in human-derived primary endothelial cells. *The Science of the Total Environment*, *934*, 173236. DOI: 10.1016/j.scitotenv.2024.173236 PMID: 38761522

Morreale, M., & La Mantia, F. P. (2024). Current concerns about microplastics and nanoplastics: A brief overview. *Polymers*, *16*(11), 1525. Advance online publication. DOI: 10.3390/polym16111525 PMID: 38891471

Murray, A., & Örmeci, B. (2020). Removal Effectiveness of Nanoplastics (<400 nm) with Separation Processes Used for Water and Wastewater Treatment. *Water (Basel)*, *12*(3), 635. DOI: 10.3390/w12030635

Nawaz, F., Islam, Z. U., Ghori, S. A., Bahadur, A., Ullah, H., Ahmad, M., & Khan, G. U. (2025). Microplastic and nanoplastic pollution: Assessing translocation, impact, and mitigation strategies in marine ecosystems. *Water Environment Research : A Research Publication of the Water Environment Federation*, *97*(3), 70032. Advance online publication. DOI: 10.1002/wer.70032 PMID: 39927485

Nene, A., Sadeghzadeh, S., Viaroli, S., Yang, W., Uchenna, U. P., Kandwal, A., Liu, X., Somani, P., & Galluzzi, M. (2025). Recent advances and future technologies in nano-microplastics detection. *Environmental Sciences Europe*, *37*(7), 7. Advance online publication. DOI: 10.1186/s12302-024-01044-y

Ngarize, S., Makuch, K. E., & Pereira, R. (2013). The case for regulating nanotechnologies: International, European and national perspectives. *Review of European, Comparative & International Environmental Law*, *22*(3), 251–264. DOI: 10.1111/reel.12029

Noornama, N., Zainol Abidin, M. N., Abu Bakar, N. K., & Hashim, N. A. (2024). Innovative solutions for the removal of emerging microplastics from water by utilizing advanced techniques. *Marine Pollution Bulletin*, *206*, 116752. DOI: 10.1016/j.marpolbul.2024.116752 PMID: 39053257

O'Brian, F. P. (2011). The nanotechnology challenge: Creating legal institutions for uncertain risks. In Dana, D. A. (Ed.), *The nanotechnology challenge: Creating legal institutions for uncertain risks* (pp. 121–142). Cambridge University Press., DOI: 10.1017/CBO9780511988554.018

Ouda, M., Banat, F., Hasan, S. W., & Karanikolos, G. N. (2023). A 3D CFD study on the effects of feed spacer designs on membrane performance for high-permeance RO membranes. *Journal of Water Process Engineering*, *53*, 103543. DOI: 10.1016/j.jwpe.2023.103543

Park, J. W., Lee, S. J., Hwang, D. Y., & Seo, S. (2021). Removal of microplastics via tannic acid-mediated coagulation and *in vitro* impact assessment. *RSC Advances*, *11*(6), 3556–3566. DOI: 10.1039/D0RA09645H PMID: 35424293

Pawak, V. S., Shekhar, C., Loganathan, V. A., & Sabapathy, M. (2023). Separation of nanoplastics from synthetic and industrial wastewater using electrolysis-assisted flotation approach: A green approach for real-time contaminant mitigation. *Chemical Engineering Research & Design*, *198*, 184–199. DOI: 10.1016/j.cherd.2023.08.038

Pinto da Costa, J., Reis, V., Paço, A., Costa, M., Duarte, A. C., & Rocha-Santos, T. (2019). Micro(nano)plastics—Analytical challenges towards risk evaluation. *Trends in Analytical Chemistry*, *111*, 173–184. DOI: 10.1016/j.trac.2018.12.013

Pirade, F., Foppen, J. W., van der Hoek, J. P., & Lompe, K. M. (2024). Polystyrene nanoplastics are unlikely to aggregate in freshwater bodies. *Environmental Pollution*, *366*, 125393. DOI: 10.1016/j.envpol.2024.125393 PMID: 39586453

Porter, R. D., Breggin, L., & Pendergrass, J. (2011). Regulatory responses to nanotechnology uncertainties. In Dana, D. A. (Ed.), *The nanotechnology challenge: Creating legal institutions for uncertain risks* (pp. 143–170). Cambridge University Press., DOI: 10.1017/CBO9780511988554.019

Pradel, A., Ferreres, S., Veclin, C., El Hadri, H., Gautier, M., Grassl, B., & Gigault, J. (2021). Stabilization of fragmental polystyrene nanoplastic by natural organic matter: Insight into mechanisms. *ACS ES&T Water, 1*(5), 1198–1207. DOI: 10.1021/acsestwater.0c00283

Rajtar, N., Starek, M., Vincenti, L., Dąbrowska, M., Romek, M., Rinaldi, R., Lionetto, F., & Kepczynski, M. (2025). Effect of PET micro/nanoplastics on model freshwater zooplankton. *Polymers, 17*(9), 1256. Advance online publication. DOI: 10.3390/polym17091256 PMID: 40363040

Ramirez-Arenas, L., Ramseier Gentile, S., Zimmermann, S., & Stoll, S. (2022). Fate and removal efficiency of polystyrene nanoplastics in a pilot drinking water treatment plant. *The Science of the Total Environment, 813*, 152623. DOI: 10.1016/j.scitotenv.2021.152623 PMID: 34963580

Rasheed, A., Sharma, N., Surampalli, R. Y., & Das, S. (2023). Evaluating treatment solutions: Critical review on technologies employed for microplastic removal from water matrices. *Current Opinion in Environmental Science & Health, 36*, 100516. DOI: 10.1016/j.coesh.2023.100516

Rasmussen, K., Sokull-Klüttgen, B., Yu, I. J., Kanno, J., & Hirose, A. (2017). Regulation and legislation. In Stone, V., Boxall, A. B. A., & Clift, K. (Eds.), *Adverse effects of engineered nanomaterials: Exposure, toxicology, and impact on human health* (2nd ed., pp. 107–125). Academic Press., DOI: 10.1016/B978-0-12-809199-9.00007-0

Regasa, M. B. (2024). New approaches for micro(nano)plastics analysis. In *Toxic Effects of Micro- and Nanoplastics* (pp. 471–492). https://doi.org/DOI: 10.1002/9781394238163.ch20

Rex, C. M., Debroy, A., Nirmala, M. J., & Mukherjee, A. (2023). Ecotoxicological significance of bio-corona formation on micro/nanoplastics in aquatic organisms. *RSC Advances, 13*(33), 22905–22917. DOI: 10.1039/D3RA04054B PMID: 37520083

Sadek, A. H. S., Abdel-Haleem, F. M., & Ateia, M. A. (2025). Micro/nanoplastics removal from the aquatic environment using micro/nanobubbles-assisted flotation technology: A review. *Separation and Purification Technology, 377*, 134262. DOI: 10.1016/j.seppur.2025.134262

Samsudin, M., Azid, A., & Rani, N. L. A. (2023). Nanoplastics in environment. In *Nanofillers for sustainable applications* (pp. 188–197). CRC Press., DOI: 10.1201/9781003400998-9

Severino, A., Russo, B., Lavorato, C., Argurio, P., Figoli, A., Molinari, R., & Poerio, T. (2025). Integrated nanofiltration and photocatalytic processes for the removal of polystyrene nanoplastics waste in water. *Separation and Purification Technology, 286*, 131232. DOI: 10.1016/j.seppur.2024.131232

Song, T.-T., Pan, E.-Y., & Tseng, H.-C. (2012). Nanorisk governance in Taiwan: Studying potential impacts on the environment, health, and safety. *IEEE Nanotechnology Magazine, 6*(4), 6–11. DOI: 10.1109/MNANO.2012.2192653

Song, Z., Yang, X., Chen, F., Zhao, F., Zhao, Y., Ruan, L., Wang, Y., & Yang, Y. (2019). Fate and transport of nanoplastics in complex natural aquifer media: Effect of particle size and surface functionalization. *The Science of the Total Environment, 669*, 120–128. DOI: 10.1016/j.scitotenv.2019.03.102 PMID: 30878920

Sorensen, R. M., Kanwar, R. S., & Jovanović, B. (2023). Past, present, and possible future policies on plastic use in the United States, particularly microplastics and nanoplastics: A review. *Integrated Environmental Assessment and Management, 19*(5), 1240–1254. DOI: 10.1002/ieam.4678 PMID: 36036190

Srivastava, A., Ravula, S., Bara, J. E., & Esfahani, M. R. (2025). Efficient extraction of polystyrene nanoplastics from water using an ionic liquid. *Environmental Science. Water Research & Technology, 11*(5), 1151–1162. DOI: 10.1039/D4EW00909F

Stine, J. S., Aziere, N., Harper, B. J., & Harper, S. L. (2023). A novel approach for identifying nanoplastics by assessing deformation behavior with scanning electron microscopy. *Micromachines, 14*(10), 1903. DOI: 10.3390/mi14101903 PMID: 37893340

Sutrisna, P. D., Riadi, L., Buana, P. C. W., Khoiruddin, K., Boopathy, R., Wenten, I. G., & Siagian, U. W. R. (2024). Membrane and membrane-integrated processes for nanoplastics removal and remediation. *Polymer Degradation & Stability, 220*, 110635. DOI: 10.1016/j.polymdegradstab.2023.110635

Tang, K. H. D., & Li, R. (2025). The effects of plastisphere on the physicochemical properties of microplastics. *Bioprocess and Biosystems Engineering, 48*(1), 1–15. DOI: 10.1007/s00449-024-03059-4 PMID: 38960926

Tomonaga, T., Higashi, H., Izumi, H., Nishida, C., Sato, K., Nakamura, Y., Morimoto, T., Higashi, Y., Kojima, T., Sakurai, K., Yatera, K., & Morimoto, Y. (2025). Comparison of lung disorders following intratracheal instillation of polystyrene microplastics with different surface functional groups. *Journal of Occupational Health*, *67*(1), uiaf006. Advance online publication. DOI: 10.1093/joccuh/uiaf006 PMID: 39898983

Tumrani, S. H., Sharma, B. P., Otho, A. R., Ayob, E. A., Amin, M. A., Soliman, M. M., Soomro, R. A., & Karakuş, S. (2025). Biodegradation of PLA in surface water and its behavior towards PFOA: Interactional characteristics and computational insights. *Colloid & Polymer Science*, *303*(7), 1397–1408. DOI: 10.1007/s00396-025-05423-6

Veclin, C., Desmet, C., Pradel, A., Valsesia, A., Reynaud, S., & Gigault, J. (2022). Effect of the surface hydrophobicity–morphology–functionality of nanoplastics on their homoaggregation in seawater. *ACS ES&T Water*, *2*(1), 88–96. DOI: 10.1021/acsestwater.1c00263

Vercauteren, M., Lambert, S., Hoogerwerf, E., Janssen, C. R., & Asselman, J. (2024). Microplastic-specific biofilm growth determines the vertical transport of plastics in freshwater. *The Science of the Total Environment*, *910*, 168399. DOI: 10.1016/j.scitotenv.2023.168399 PMID: 37981154

Winburn, M. R., Alvarado, M. F., & Cheung, C. L. (2025). Plasma-assisted destruction of polystyrene nanoplastics. *Nanoscale*, *17*(4), 2138–2146. DOI: 10.1039/D4NR02498B PMID: 39655503

Wu, J., Jiang, R., & Zhang, Y. (2019). Effect of salinity and humic acid on the aggregation and toxicity of polystyrene nanoplastics with different functional groups and charges. *Environmental Pollution*, *245*, 836–843. DOI: 10.1016/j.envpol.2018.11.055 PMID: 30502713

Wu, J., Ye, Q., Li, P., Sun, L., Huang, M., Liu, J., Ahmed, Z., & Wu, P. (2023). The heteroaggregation behavior of nanoplastics on goethite: Effects of surface functionalization and solution chemistry. *The Science of the Total Environment*, *870*, 161787. DOI: 10.1016/j.scitotenv.2023.161787 PMID: 36706999

Xi, X., Ding, D., Zhou, H., Baihetiyaer, B., Sun, H., Cai, Y., Wang, N., & Yin, X. (2022). Effects of physicochemical factors on the transport of aged polystyrene nanoparticles in saturated porous media. *Chemosphere*, *289*, 133239. DOI: 10.1016/j.chemosphere.2021.133239 PMID: 34896420

Xu, J., Zhang, L., Cheng, J., Shi, Y., Hou, K., An, Q., Li, J., Cui, B., & Chen, F. (2024). Enhanced removal of polystyrene nanoplastics by air flotation modified with dodecyltrimethylammonium chloride. *Chinese Chemical Letters*, *36*(5), 110240. DOI: 10.1016/j.cclet.2024.110240

Yeo, L. S., & Julaihi, M. R. M. (2021). Magnetic ferrous fluid for microplastics extraction application. *Materials Science Forum*, *1030*, 138–145. . DOI: 10.4028/www.scientific.net/MSF.1030.138

Yong, X., & Du, K. (2023). Effects of shape on interaction dynamics of tetrahedral nanoplastics and the cell membrane. *The Journal of Physical Chemistry. B*, *127*(7), 1652–1663. DOI: 10.1021/acs.jpcb.2c07460 PMID: 36763902

Yu, T., Huang, X., Zhang, X. F., Li, K., Liu, S. P., Dai, N., Zhang, K., Zhang, Y. X., & Li, H. (2024). A review of nanomaterials with excellent purification potential for the removal of micro- and nanoplastics from liquid. *DeCarbon*, *5*, 100064. DOI: 10.1016/j.decarb.2024.100064

Zaki, M. R. M., & Aris, A. Z. (2022). An overview of the effects of nanoplastics on marine organisms. *The Science of the Total Environment*, *831*, 154757. DOI: 10.1016/j.scitotenv.2022.154757 PMID: 35339559

Zhang, L., Xue, M., Xin, Q., Tang, J., Liu, Y., Bian, Z., & Zheng, X. (2025). Cellular uptake of polystyrene nanoplastics with surface functionalization: An AIE-based quantitative approach. *Analytica Chimica Acta*, *1362*, 344184. DOI: 10.1016/j.aca.2025.344184 PMID: 40447404

Zhang, W., Wang, Q., & Chen, H. (2022). Challenges in characterization of nanoplastics in the environment. *Frontiers of Environmental Science & Engineering*, *16*(1), 11. Advance online publication. DOI: 10.1007/s11783-021-1445-z

Zhang, Y., Diehl, A., Lewandowski, A., Gopalakrishnan, K., & Baker, T. R. (2020). Removal efficiency of micro- and nanoplastics (180 nm–125 μm) during drinking water treatment. *The Science of the Total Environment*, *720*, 137383. DOI: 10.1016/j.scitotenv.2020.137383 PMID: 32325555

Zhu, H., Fan, X., Zou, H., Guo, R.-B., & Fu, S.-F. (2023a). Effects of size and surface charge on the sedimentation of nanoplastics in freshwater. *Chemosphere*, *336*, 139194. DOI: 10.1016/j.chemosphere.2023.139194 PMID: 37315858

Zhu, M., Zhang, Z., Zhang, T., Hofmann, T., & Chen, W. (2023b). Eco-corona dictates mobility of nanoplastics in saturated porous media: The critical role of preferential binding of macromolecules. *Environmental Science & Technology*, *57*(1), 331–339. DOI: 10.1021/acs.est.2c07376 PMID: 36574476

Zoppas, F. M., Sacco, N., Soffietti, J., Devard, A., Akhter, F., & Marchesini, F. A. (2023). Catalytic approaches for the removal of microplastics from water: Recent advances and future opportunities. *Chemical Engineering Journal Advances*, *16*, 100529. DOI: 10.1016/j.ceja.2023.100529

Chapter 4
Human Exposure to Nanoplastics and Its Toxicological and Environmental Implications: A One Health Perspective

Odangowei Inetiminebi Ogidi
http://orcid.org/0000-0002-9961-893X
Bayelsa Medical University, Yenagoa, Nigeria

Nwanne Ijere Dike
http://orcid.org/0009-0002-6577-0965
Michael Okpara University of Agriculture, Umudike, Nigeria

Lilian Maureen Oluchi Oguoma
http://orcid.org/0000-0002-1643-0852
Federal University of Technology, Owerri, Nigeria

ABSTRACT

This chapter per the authors investigates human exposure to nanoplastics and toxicodynamic responses arising from environmental persistence, biological translocation, and cross-interface exposure pathways within a One Health framework. Environmental formation and degradation processes are evaluated alongside transport across aquatic, atmospheric, and terrestrial matrices leading to ingestion, inhalation, dermal uptake, and systemic distribution. Mechanistic injury pathways are analyzed, including oxidative stress, inflammatory cascades, endocrine disruption,

DOI: 10.4018/979-8-3373-4490-4.ch004

epithelial barrier impairment, and genotoxic effects across multiple organ systems. The chapter further addresses biomonitoring strategies, exposure quantification constraints, and limitations in long-term assessment within analytical standardization frameworks. By integrating exposure science with mechanistic toxicology and computational analytics, the chapter provides a structured perspective for interpreting nanoplastic-associated health risks across interconnected ecological and physiological systems.

INTRODUCTION

The deposition of synthetic polymer materials in the environment is a major cause of plastic pollution and poses risks to human health. Poor disposal practices, coupled with continuously increasing production of plastic items, create persistent global challenges. Although plastics serve many useful functions, they also represent significant environmental and health threats that require careful consideration (Moshood et al., 2022). A UN estimate for 2021 places annual plastic production at nearly 400 million metric tonnes (UN Environment Programme [UNEP], 2021). The report indicates an increase in production after the year 2000. From 1950 to 2017, an estimated 9.2 × 10^9 metric tonnes of plastic were produced, with almost half of this amount generated since 2004 (Moshood et al., 2022). In addition, approximately 8 million tonnes of processed plastics enter the ocean annually (Wang & Qian, 2021).

The One Health approach underscores the interconnectedness of human, animal, and environmental health. Within this framework, nanoplastics cross environmental boundaries, affecting both marine and terrestrial ecosystems and entering food chains, with potential implications for human physiology. This perspective highlights the need for cross-sectoral strategies and AI-enabled monitoring systems to address nanoplastic contamination at the ecosystem–human interface (Kibria et al., 2023).

Plastics are characterized by extremely slow degradation, with decomposition periods that may exceed 400 years. As a result, surface water, sediment, groundwater, soil, and air are affected by long-term plastic persistence (Fan et al., 2022). Only about 9% of plastic waste is currently recycled, leaving the majority to accumulate in the environment (Dong et al., 2023). Global plastic waste generation is estimated at approximately 6.9 billion tonnes. This waste consists primarily of polymer resin (81%) and polymer fibres (13%), while additives are reported as part of the composition in the original dataset (Koelmans et al., 2019).

A wide range of plastic types is present in the environment, including polyethylene (PE), polypropylene (PP), polystyrene (PS), polyvinyl chloride (PVC), and polyethylene terephthalate (PET). Microplastics contain multiple additives that modify their physical and chemical properties, while their core composition remains largely petroleum-derived synthetic polymers. In general, microplastics comprise polymer matrices, plasticisers, and environmental contaminants such as persistent organic pollutants, herbicides, heavy metals, and pesticides (Koelmans et al., 2019).

Although the main focus of this chapter is microplastics and nanoplastics, other plastic-related compounds are also relevant due to their potential health impacts. Considerable attention has been given to bisphenol A (BPA), particularly regarding its endocrine-disrupting properties. The toxicity of these particles is often associated with the leaching of internal chemical constituents, which may contribute to their biological impact.

METHODS / SCOPE OF LITERATURE REVIEW

This chapter employs a narrative review approach designed to consolidate and interpret the existing body of knowledge on the human health implications of nanoplastics. The aim was to synthesize peer-reviewed literature on exposure routes, toxicological mechanisms, and mitigation strategies within the rubric of One Health. Peer-reviewed literature published between 2018 and 2025 was prioritized, and therefore the review reflects the most current science in nanoplastic research.

A comprehensive search of databases was conducted using Scopus, Web of Science, PubMed, and ScienceDirect, with support from grey literature and policy reports from the World Health Organization (WHO), United Nations Environment Programme (UNEP), and European Chemicals Agency (ECHA). Keywords were combined with terms such as "nanoplastics," "micro- and nanoplastics," "plastic nanoparticles," "toxicity," "oxidative stress," "inflammation," "biomonitoring," and "One Health." Boolean operators were used to refine the search and identify publications linking environmental pollution with human exposure and health effects.

Specific inclusion and exclusion criteria were applied to select accessible studies. Eligible works had to provide empirical or review data on exposure, uptake, or toxic impact of nanoplastics in humans or mammalian systems, and had to be published in English-language peer-reviewed journals. Studies were excluded if they focused solely on microplastics larger than 1 μm, were not empirical, or failed to relate findings to human health or environmental interaction. This ensured that only methodologically sound and scientifically relevant research was included in the synthesis.

The literature was thematically contrasted across three areas: modes of nanoplastic toxicity, exposure pathways and biomonitoring, and mitigation strategies involving AI-assisted monitoring and policy. The analysis focused on mechanistic observations such as oxidative stress, inflammation, endocrine disruption, and cellular uptake. The evidence was then qualitatively compared and synthesized to identify emerging trends, remaining uncertainties, and knowledge gaps that should be addressed to strengthen One Health surveillance and risk assessment tools.

SOURCES OF NANOPLASTICS

Those nanoplastics that are synthetically produced and utilized in medicine and industry are the primary source of nanoplastics, followed by other sources. Sources of nanoplastics primarily include household items such as personal care products, paints, and textile products containing nanoplastics. Physical, chemical, and biological degradation processes also form nanoplastics during the breakdown or fragmentation of larger plastic materials. Following environmental release, nanoplastics may remain suspended in the water column due to their low density and small particle size, which enables wide dispersion. A variety of environmental matrices, such as soil, biota, and sediments, interact strongly with nanoplastics, influencing their fate and bioavailability (Hartmann et al., 2019). Physical processes such as advection and sedimentation, along with chemical and biological transformations, govern the ultimate fate of nanoplastics in terrestrial and aquatic environments (Hüffer et al., 2017).

There are a number of environmental matrices, including soil, sediments, seawater and freshwater ecosystems, and atmospheric aerosols, where nanoplastic pollution has been detected (Klein et al., 2021). Concentrations ranging from nanogrammes to microgrammes per gramme of matrix have been reported using advanced analytical methods across various environmental compartments (Koelmans et al., 2019). Further evidence of the pervasiveness of nanoplastics has been observed even in remote or relatively isolated ecosystems (Prata et al., 2020). A wide range of organisms, from plankton to higher trophic levels, ingest nanoplastics, leading to adverse effects such as tissue damage, reproductive complications, and behavioral changes (Rist et al., 2018).

Their capacity to uptake and transport various contaminants makes nanoplastics an effective vector for the transfer of hazardous substances across food webs, with implications for ecosystem and human health (Sun et al., 2021). Reducing their introduction into the environment remains one of the primary goals of nanoplastic prevention. This may be achieved through improved waste management systems, substitution with biodegradable materials, and the design of eco-friendly products that reduce nanoplastic generation (Zhang et al., 2020). While the efficacy and feasibility of remediation techniques such as membrane filtration and adsorption for nanoplastic removal have not been extensively validated under real-world conditions, they show potential for future application (Gutow et al., 2016; Ogidi & Richard-Nwachukwu, 2024).

FORMATION AND PERSISTENCE OF NANOPLASTICS

Significant quantities of plastic waste can be converted into nanoplastics when subjected to environmental stressors such as UV radiation, mechanical degradation, and microbial activity (Hernandez et al., 2017). In addition, some industries manufacturing coatings, textiles, and personal care products intentionally produce nanoplastics for specific applications (Song et al., 2015).

Nanoplastics are generated through a combination of chemical and physical processes in the environment and subsequently disperse through atmospheric transport over long distances before depositing on land and water surfaces. They accumulate in aquatic systems and may negatively affect marine species when transported from urban and agricultural areas into lakes, rivers, and ponds (Koelmans et al., 2017). Ocean currents further transport nanoplastics to sediments, where they may accumulate and enter benthic organisms, ultimately progressing through the food web. The bioavailability of nanoplastics to aquatic organisms indicates a strong potential for bioaccumulation and biomagnification across trophic levels (Nizzetto et al., 2016).

According to Dris et al. (2018), interactions between nanoplastics, soil aggregates, and organic matter during deposition may influence plant health, microbial communities, and soil structure. Both dry and wet deposition phases of aerosolized nanoplastics occur across terrestrial and aquatic ecosystems.

The presence of nanoplastics in marine sediments and organisms indicates their potential persistence in marine environments due to slow degradation rates (Zhang et al., 2020). Similarly, nanoplastics may persist in freshwater systems, particularly in stagnant water bodies where degradation processes are limited. They can also remain in soils for extended periods despite undergoing physical and chemical weathering (Free et al., 2014). In the atmosphere, photodegradation and reactions with other chemical constituents contribute to fragmentation processes; however, their overall atmospheric persistence remains not fully quantified (Wright et al., 2013).

EXPOSURE PATHWAYS AND HUMAN BIOMONITORING

Nanoplastics, which are minute fragments of plastics generated through progressive breakdown of larger materials, have emerged as a potential environmental and public health concern due to their capacity to enter human biological systems. These particles are sufficiently small to bypass conventional filtration processes used for air, food, and water purification systems. Following inhalation or ingestion, nanoplastics may translocate across epithelial barriers and accumulate in tissues, with the potential to interact with cellular components and biological pathways that are not yet fully understood.

Common exposure routes include contaminated seafood, drinking water (including bottled water), indoor dust, and agricultural products that may accumulate plastics from soil or irrigation systems. Inhalation of airborne particles, particularly in indoor environments where plastic-derived dust is prevalent, also represents a relevant pathway for chronic exposure. The movement of nanoplastics across environmental compartments and their potential routes into the human body occur through multiple interconnected pathways that are often difficult to separate in real-world settings (Figure 1).

Given the increasing evidence of their environmental ubiquity and persistence, there is an urgent need for further research into the toxicokinetics, bioaccumulation potential, and molecular mechanisms of action of nanoplastics within human tissues and organ systems. Strengthening human biomonitoring frameworks is essential to accurately assess exposure levels, internal dose metrics, and potential long-term health risks associated with nanoplastic exposure.

Figure 1. Nanoplastics and their pathways to human exposure.

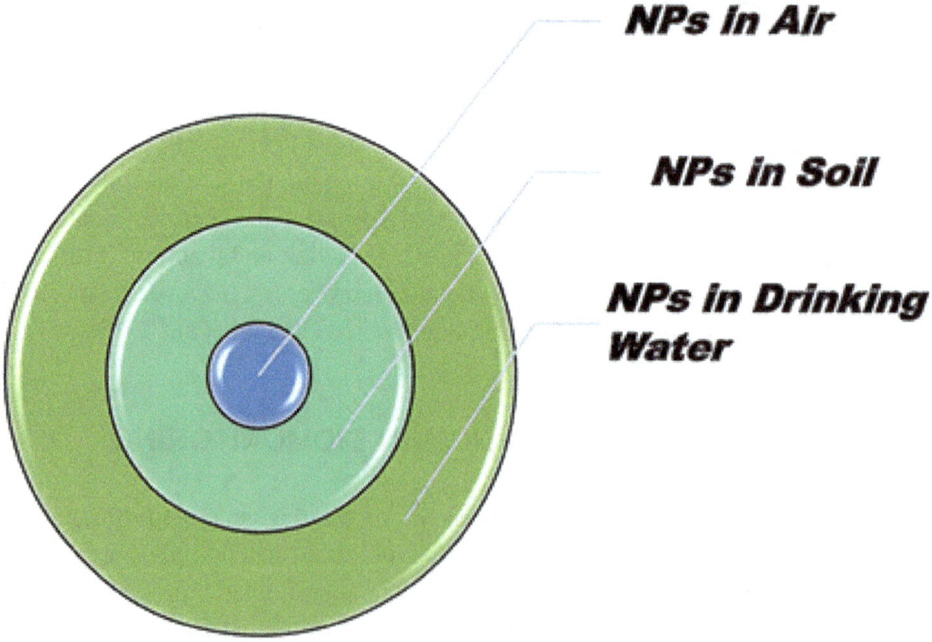

The figure presents a general view of how nanoplastics can move through different environmental compartments before reaching humans through multiple exposure routes. Sources associated with food, water, air, and surrounding environments may contribute simultaneously, making exposure patterns difficult to separate under real conditions. The overall pathway reflects the interconnected nature of environmental contamination and human exposure.

NPs in Air

The presence of micro- and nanoplastics (MNPs) in the atmosphere poses a potential risk to human health. Both indoor and outdoor MNPs may contribute to human exposure (Vianello et al., 2019). Atmospheric nanoparticles can occur in a variety of morphologies, including films, fibres, and fragments. Increasing evidence suggests that urban environments are experiencing a growing presence of airborne MPs and NPs. However, the impact of airborne microplastic nanoparticles on human health remains insufficiently investigated (Rodrigues et al., 2022). Reports indicate that indoor air contains greater concentrations of synthetic and plastic fibres than outdoor air (Dris et al., 2017). These fibres may initially undergo photo-oxidative degradation followed by further fragmentation through wind shear processes. As a result, MNPs have been detected across both marine and terrestrial environments (Gasperi et al., 2018). Due to their small size and environmental persistence, these particles may spread across different regions and potentially remain in environmental systems for extended periods.

NPs in Soil

Improper disposal of plastics in landfills, followed by gradual degradation into microplastics, is considered a major source of soil contamination. Microplastic pollution may also arise from road dust, accumulated litter, industrial plastic residues, continuous soil sedimentation, and prolonged atmospheric deposition (Bläsing & Amelung, 2018). Major sources of microplastic nanoparticles (MNPs) in agricultural soils include plastic mulching, compost containing plastics, untreated wastewater, sewage, and river water flowing through contaminated regions (Meng et al., 2020). Variability in the distribution of micro- and nanoplastics of different sizes has been observed across agricultural soil layers. Soil contamination may also result from the discharge of untreated sewage containing MNPs into surrounding environments (Zhang et al., 2018). Soil samples have tested positive for various forms of MNPs, with commonly detected materials including plasticised PVC, polyester, polyolefin, polysulfone, and polypropylene.

NPs in Drinking Water

Studies have reported the presence of microplastic nanoparticles (MNPs) in both municipal and bottled drinking water, although concentrations vary across water sources (Kosuth et al., 2018; Mason et al., 2018). Drinking water contamination may partly originate from PVC distribution pipes. Long-term use and physical degradation of these materials may contribute to the release of micro- and nanoplastics (Mintenig et al., 2019). Polyethylene terephthalate (PET) bottles and bottle caps are also considered possible contributors to MNP contamination, as degradation may occur during manufacturing, transportation, storage, and consumer use (Winkler et al., 2019). Repeated opening and prolonged use of plastic water containers may further contribute to the release of plastic particles into water systems (Winkler et al., 2019).

Human Biomonitoring

Humans may experience long-term exposure to nanoplastics, even at relatively low concentrations, over much of their lifetime (Lenz et al., 2016; Galloway, 2015). Due to the greater difficulty associated with detecting nanoplastics compared with microplastics, understanding their occurrence and biological distribution remains challenging. This has increased interest in investigating exposure under environmentally relevant conditions and assessing potential impacts in living systems (Al-Sid-Cheikh et al., 2018). This section discusses the possible effects of nanoplastics on humans through different exposure pathways, as summarized in Table 1.

Table 1.

Exposure Route	**Sources/Examples**	**Adverse Human Health Effects**	**References**
Ingestion	Drinking water, bottled water, seafood, salt, beverages, food packaging	Gastrointestinal inflammation, oxidative stress, disruption of gut microbiota, intestinal barrier dysfunction, liver and kidney toxicity	(Mason et al., 2018; Mintenig et al., 2019; Schwabl et al., 2019; Lehner et al., 2020; Ogidi & Felagha, 2024)
Inhalation	Indoor/outdoor air, synthetic textiles, tire wear particles, dust	Respiratory inflammation, oxidative stress, alveolar damage, potential systemic translocation to bloodstream	(Dris et al., 2017; Gasperi et al., 2018; Vianello et al., 2019)
Dermal Contact	Cosmetics, textiles, personal care products	Limited penetration through healthy skin; potential immune response via damaged skin or nanoparticles <100 nm	(Vogt et al., 2006; Schneider et al., 2009)
Placental Transfer	Maternal exposure during pregnancy (food, water, inhalation)	Translocation to placenta; detection in human placenta; potential developmental toxicity	(Grafmueller et al., 2015; Ragusa et al., 2021)
Bloodstream/ Organ Accumulation	Translocation via ingestion and inhalation pathways	Accumulation in liver, kidneys, and brain; fibrosis, immune disruption, and endocrine interference	(Deng et al., 2017; Li et al., 2020; Shen et al., 2022; Ragusa et al., 2021; Hou et al., 2022)
Neurological Exposure	Nanoplastics crossing the blood–brain barrier	Neuroinflammation, oxidative stress, behavioral changes (animal studies)	(Yang et al., 2004; Zoeller et al., 2005; Ogidi & Ajoko, 2024)
Food Chain Biomagnification	Bioaccumulation in fish, shellfish, and crops irrigated with contaminated water	Secondary ingestion leading to increased exposure; endocrine disruption, immune stress, oxidative damage	(Cox et al., 2019; Rist et al., 2018)
Intravenous/ Medical Exposure	Plastic nanoparticles from syringes, IV bags, and medical implants	Direct bloodstream exposure, inflammation, and possible immune modulation	(Eyles et al., 2001; Geiser et al., 2005)
Occupational Exposure	Workers in plastic manufacturing, recycling, and wastewater treatment plants	Chronic inhalation or dermal exposure associated with respiratory disease and systemic toxicity	(Prata et al., 2020)

Oral Intake

Oral intake is one of the most common and extensively studied routes of exposure to nanoplastics. Continuous exposure through food and water consumption may allow persistent uptake of nanoplastics by the human body. The widespread use of plastic components in water distribution systems, together with the potential leaching of micro- and nanoplastics from water pipes under prolonged fluid pressure, may increase human exposure through drinking water sources (Xu et al., 2019; Cox et al., 2019; Winkler et al., 2019).

Previous studies have suggested that large numbers of nanoplastic particles may be released from plastic-containing products during routine use. For example, Hernandez et al. (2019) reported the presence of hundreds of millions of particles in tea prepared using plastic teabags. In addition, the detection of nanoplastics in seafood and table salt suggests that food consumption may represent another important exposure pathway. Nanoplastics may move through food webs via bioaccumulation processes, transferring from lower trophic organisms to higher trophic levels, including humans (Yee et al., 2021). Water and food consumed daily may therefore contribute to continuous exposure. Further investigation into the occurrence of nanoplastics in food and water, along with improvements in detection methods, remains important.

Air Inhalation

The respiratory system represents a major route through which airborne particles and environmental contaminants enter the human body. Atmospheric deposition contributes to the environmental presence of nanoplastics and may increase opportunities for inhalation exposure (Prata et al., 2020). Vehicle tire degradation also releases micro- and nanoplastics into surrounding environments, contributing to atmospheric contamination (Leads & Weinstein, 2019; Kole et al., 2019). A study of public and private buildings reported higher concentrations of microplastics and nanoplastics in indoor air compared with outdoor air, suggesting that individuals may experience greater exposure in enclosed environments (Dris et al., 2017).

Dermal Exposure

Nanoplastics may enter the body through several pathways, including dermal contact, inhalation, and ingestion. Nanotoxicology studies involving synthetic materials have suggested that particles smaller than 40 nm may have the capacity to penetrate the epidermal barrier under certain conditions (Vogt et al., 2006). Human exposure may occur through personal care products containing nanoplastics, including nanopolystyrene-containing materials, as well as through routine contact with contaminated water sources (Schneider et al., 2009).

Lung

Following penetration into the lower respiratory tract, uptake of micro- and nanoplastics (MNPs) may involve endothelial and surrounding respiratory cells (Zhang et al., 2016). Plastic fibres may also induce irritation within deeper lung tissues (Galloway, 2015). Deposited MNPs are generally removed through pulmonary defense mechanisms such as mucociliary clearance and macrophage activity within the trachea, bronchi, and alveoli. However, MNPs that avoid these mechanisms may form biomolecular aggregates resembling protein or drug–protein complexes (Ogidi et al., 2024a).

Accumulation of MNPs within respiratory tissues may contribute to airway inflammation and has been associated with bronchial lesions, alveolitis, and lower airway fibrosis, which could impair pulmonary function (Zhang et al., 2016). Long-term exposure may therefore contribute to respiratory complications. Ground-glass nodule aggregation (GGNs) has also been discussed as a possible outcome associated with persistent exposure. Previous studies reported higher concentrations of MPs in human malignant tissues compared with normal tissues, suggesting that prolonged exposure warrants further investigation regarding potential health implications (Alimi et al., 2018; Ogidi, 2024; Ogidi & Emaikwu, 2024).

Gastrointestinal Tract

Inhalation and ingestion of food, drinks, nutritional sources, vegetables, and medications are the two main routes through which MNPs reach the gastrointestinal system. Large polymers (MPs) may be retained by the intestinal villi and potentially interfere with nutrient absorption. Nanoparticles (NPs), however, may cross the villi, enter the bloodstream, and subsequently distribute to other parts of the body. Another source of MNPs in the gastrointestinal tract is the accumulation of these particles on food surfaces during preparation and consumption (Al-Sid-Cheikh et al., 2018). The majority of microplastics are eventually expelled through the digestive tract (Xu et al., 2019). Another experiment suggested that MNPs may infiltrate adjacent tissues through paracellular and transcellular pathways and subsequently be absorbed by the M-cells of Peyer's patches (Cox et al., 2019).

ADVERSE EFFECTS OF NANOPLASTICS ON HUMAN

Nanoplastics may be transmitted to offspring following their infiltration into living organisms (Al-Sid-Cheikh et al., 2018; Pitt et al., 2018). Living organisms may also be exposed to nanoplastics at very low concentrations over prolonged periods, resulting in gradual accumulation (Al-Sid-Cheikh et al., 2018). Research regarding the adverse effects of nanoplastics on human health remains limited. This section provides a brief overview of current findings from nanotoxicology investigations involving nanopolystyrene.

Laboratory and environmental studies indicate that microplastics and nanoplastics can accumulate within the intestines after ingestion and are eventually excreted in faeces (Lehner et al., 2017; Schwabl et al., 2019). Evidence from animal studies further suggests that nanoparticles distributed throughout the intestinal lumen may penetrate the intestinal barrier and reach blood vessels. Due to its relatively larger pore structures, the alveolus functions as a major component of the blood–air barrier and facilitates oxygen and carbon dioxide exchange (Schwabl et al., 2019). This suggests that inhaled nanoplastics may cross the blood–air barrier and enter systemic circulation.

Nanoplastics entering the bloodstream may also cross the blood–brain barrier and accumulate in brain tissues (Yang et al., 2004). Studies have reported that nanoplastics may move through food chains and subsequently induce adverse neurological effects in aquatic organisms (Grafmueller et al., 2015; Ogidi, 2025). Additionally, findings from an *ex vivo* human placental perfusion model indicated that nanoplastics may passively diffuse across the placental barrier (Grafmueller et al., 2015). Collectively, these findings suggest that nanoplastics may pose potential risks because of their ability to cross several biological barriers, including placental, blood–air, blood–intestinal, and blood–brain barriers.

MECHANISMS OF NANOPLASTIC TOXICITY

Nanoplastics exert toxicity through several interrelated biochemical and cellular pathways capable of disrupting physiological homeostasis. Table 2 presents experimental studies investigating nanoplastic toxicity. Due to their small size (<1 μm), nanoplastics can translocate across endothelial and epithelial barriers, enabling direct interaction with intracellular components and organelles. The mechanisms outlined below represent major toxicity pathways currently supported by available evidence:

1. Redox Imbalance and Oxidative Stress:

Nanoplastics may induce oxidative stress through the generation of excessive reactive oxygen species (ROS), including superoxide anions and hydroxyl radicals, which can damage lipids, proteins, and DNA. In vitro exposure to polystyrene nanoplastics (50–200 nm) at concentrations of 10–100 μg/mL resulted in elevated malondialdehyde levels and reduced antioxidant enzyme activity (SOD, CAT, and

GSH-Px). Chronic oxidative stress may contribute to mitochondrial dysfunction, apoptosis, and tissue fibrosis, particularly in hepatic and pulmonary tissues (Li et al., 2020; Hou et al., 2022).

2. Inflammatory Signaling and Cytokine Induction:

Within macrophages and epithelial cells, nanoplastics may activate inflammatory pathways involving TLR2/TLR4 and NF-κB signaling. This activation can stimulate secretion of pro-inflammatory cytokines such as IL-1β, IL-6, and TNF-α, contributing to both local and systemic inflammatory responses. Experimental models have indicated that treatment with 100 μg/L nanopolystyrene significantly increased IL-8 expression in intestinal and bronchial cell lines (Choi et al., 2021; Ballesteros et al., 2020). Repeated exposure has also been associated with chronic inflammation and immune dysregulation.

3. Endocrine Disruption:

Certain plastic additives and degradation products, including bisphenol A (BPA) and phthalate esters, may function as hormone agonists or antagonists and contribute to endocrine disruption. Nanoplastics can also act as carriers that facilitate transport and bioavailability of such compounds across biological membranes. Human exposure to BPA-associated nanoplastics in water at concentrations of 0.1 μg/L has been reported to alter estrogen receptor (ERα/ERβ) expression and interfere with thyroid hormone receptor signaling, potentially affecting reproductive and developmental functions (Kitamura, 2005; Chang et al., 2021).

4. Cellular Uptake and Organelle Interference:

Cells internalize nanoplastics through clathrin-mediated endocytosis, caveolin-dependent pathways, and passive diffusion mechanisms. Transmission electron microscopy has detected nanoplastics within endosomes, lysosomes, and mitochondria, where they may interfere with membrane integrity and cellular energy metabolism. Polystyrene nanoplastics (<100 nm) accumulated in lysosomes of liver and neuronal cells and were associated with vacuolization, altered ATP production, and disrupted autophagic activity (Deng et al., 2017; Ragusa et al., 2021).

5. Genotoxic and Epigenetic Effects:

DNA strand breaks, micronucleus formation, and chromatin condensation have been reported in human cell models exposed to nanoplastics at concentrations ranging from 50–500 μg/mL. Chronic exposure may also induce epigenetic alterations, including changes in DNA methylation and histone acetylation patterns, which may contribute to oncogenic transformation (Barguilla et al., 2022).

Table 2. Experimental Studies on Nanoplastic Toxicity

Model/System	Nanoplastic Type & Size	Concentration/ Exposure	Observed Effect	Reference
Human intestinal epithelial cells	Polystyrene (100 nm)	100 μg/mL	ROS generation, IL-8 upregulation	(Choi et al., 2021)
Mouse liver tissue (in vivo)	Polystyrene (50 nm)	10 mg/kg/day, 28 days	Hepatic fibrosis, apoptosis	(Li et al., 2020)
Human organoids	Polystyrene (70 nm)	10–100 μg/mL	Cellular accumulation, mitochondrial dysfunction	(Hou et al., 2022)
Zebrafish embryos	Polystyrene (80 nm)	1–100 μg/L	Neurobehavioral alterations, oxidative stress	(Lee et al., 2019)
Human blood cells (ex vivo)	Polystyrene (60 nm)	50 μg/mL	Genotoxicity, micronuclei formation	(Ballesteros et al., 2020)
Human placenta (ex vivo perfusion)	Polystyrene (240 nm)	10 μg/mL	Placental barrier translocation	(Grafmueller et al., 2015)

MICRO- AND NANOPLASTICS (MNPS) DISTRIBUTION

Tissue macrophages regulate MNPs that pass through tissues (Geiser et al., 2005). Following interaction with immune cells, MNPs may travel through lymphatic and alternative pathways to various organs. In addition to entering the bloodstream through major exposure organs, MNPs are also capable of traversing the lymphatic system (Eyles et al., 2001). Once they enter circulation, MNPs may distribute throughout the organism. Different organs have been reported to contain MNPs, including the placenta (Ragusa et al., 2021), bone marrow, kidneys, and urine (Deng et al., 2017), spleen (Eyles et al., 2001), as well as cerebrospinal fluid and brain tissues (Deng et al., 2017).

Inflammatory responses induced by MNPs may impair epithelial barrier function and activate macrophages, potentially accelerating MNP internalization. Long-term accumulation of MNPs in tissues may have significant implications for human health (Hou et al., 2022).

Liver

Metal nanoparticles interfere with hepatic lipid metabolism and may contribute to hepatotoxicity (Ogidi et al., 2025g). There is also potential for subcellular and nuclear accumulation of MNPs, which may contribute to DNA damage (Ogidi et al., 2025a). Moreover, evidence has suggested possible associations between MNP exposure and cancer progression, fibrosis, and hepatic steatosis. The accumulation of MNPs in the liver may result from their permeability across hepatic tissues (Shen et al., 2022).

Heart

Reports suggest that exposure to MPs elevates blood concentrations of troponin I and creatine kinase-MB (CK-MB), potentially resulting in myocardial structural damage, apoptosis, and collagen proliferation. Activation of the Wnt/β-catenin pathway has also been associated with myocardial fibrosis following MP exposure. Furthermore, cardiovascular injury may be facilitated through oxidative stress generated by MPs (Li et al., 2020).

Effect of MNPs on Immune Function

The size of plastic particles may influence the inflammatory response triggered when MNPs enter the human body and interact with immune cells. Some findings suggest that this response may be more pronounced for microplastics than nanoplastics (Choi et al., 2021). Macrophages stimulated by polystyrene nanoplastics have been reported to enhance inflammatory activity through increased secretion of IL-6 and IL-8. Inflammatory responses mediated by TNF-α and IL-1 have also been observed following exposure to polyethylene micro/nanoplastics (Lehner et al., 2020).

Long-term exposure may result in persistent inflammation, increased production of reactive oxygen species (ROS), and haemolytic effects (Choi et al., 2021). Increased ROS generation has additionally been associated with autoimmune-related responses (Ogidi et al., 2025c). Some reports further suggest that MNPs may suppress T-cell activity and stimulate the production of anti-inflammatory mediators (Rahman et al., 2021).

Exposure to MNPs has also been associated with inflammation, immune activation, stress responses, white blood cell proliferation, and cytokine production in humans (Ballesteros et al., 2020). One study reported that MNP exposure increased the release of pro-inflammatory mediators and may subsequently contribute to cell death (Zarus et al., 2020). Additional evidence indicates that toll-like receptors may mediate inflammatory cascades associated with MNP exposure (Yong et al., 2020). Alterations in immune responses may also be partially explained by microbiome composition changes following MNP exposure (Santios et al., 2022; Ogidi et al., 2025f).

Strategies to Mitigate Human Exposure to Micro- and Nanoplastics

To safeguard environmental and ecosystem health, it is necessary to reduce the prevalence of micro- and nanoplastics (MNPs) in the environment and to remediate or recycle existing plastic waste. The significant environmental accumulation resulting from inadequate disposal and recycling practices leads to uncontrolled dispersion in soil and drinking water, subsequently affecting consumers as these particles enter the food chain. Healthcare systems also rely heavily on plastics for multiple applications, including drug storage and the manufacture of medical devices used in both inpatient and outpatient care, which further contributes to MNP release. Understanding the transport of MNPs within the human body, their organ distribution, and their potential effects on the skin, lungs, gastrointestinal tract, and blood–brain barrier requires extensive investigation. The development of severe outcomes may be associated with multicellular toxicity, occurring when MNPs enter the body through multiple pathways, accumulate in specific organs, and interact with diverse cell types (Ogoun et al., 2022). Reducing daily plastic consumption and developing improved disposal strategies remain essential to mitigate their impact on human health (Yoshida et al., 2016).

Multiple strategies have been proposed to manage MNP exposure. Among the most widely studied approaches is microbial bioremediation, which utilizes single bacterial or fungal strains, microbial consortia, or co-cultivation systems to enhance plastic degradation efficiency (Ogidi & Akpan, 2024). Certain microorganisms, including the marine fungus *Zalerion maritimum* and the bacterium *Ideonella sakaiensis*, have been reported to degrade specific plastic polymers (Yoshida et al., 2016; Paco et al., 2017). Plastic degradation is facilitated by microbial extracellular enzymes such as lipases, carboxyl esterases, peroxidases, and laccases, including class II peroxidases (lignin peroxidase, manganese peroxidase, and versatile peroxidase). These enzymatic processes enhance plastic hydrophilicity and introduce functional groups such as carbonyl and alcohol groups, thereby increasing susceptibility to microbial attack (Taniguchi et al., 2019).

Complex polymers may also be broken down into smaller molecules by hydrolases, depolymerases, and cutinases (Sol et al., 2020). Previous studies have shown that proteases hydrolyze complex proteins into polypeptides, while serine proteases can degrade polylactic acid (Panyachanakul et al., 2019). Additional approaches for plastic degradation include the use of biosurfactants, genetically modified organisms, bioaugmentation, and biosimulation techniques (Miri et al., 2022). Exposure to MNPs has been associated with multiple adverse health outcomes, and long-term exposure has been linked with increased cancer risk in some studies (Barguilla et al., 2022; Ogidi & Tawariowei, 2023). Therefore, further research is needed to clarify the effects of MNPs on human cells, tissues, and organs, as well as their mechanisms of action. Establishing clear standards for risk assessment and management is essential for protecting human health.

Given that complete elimination of plastics is not currently feasible, the development of suitable alternatives could significantly reduce their environmental and health impacts. Bioplastics derived from microbial processes represent one such alternative to synthetic polymers, although their production remains costly. Efforts to reduce production costs have explored the use of low-cost feedstocks such as chitin, starch, lignin, and alginate; however, these materials often exhibit limitations in thermal stability and mechanical durability. To address these issues, bioplastics reinforced with nanoclays, carbon nanotubes, and silicones have been investigated (Lisuzzo et al., 2020). In addition, 1% cellulose fibre-reinforced hydrogel tubing has been developed to improve bioplastic performance. Despite these advances, substantial research is still required before such materials can replace conventional plastics at an industrial scale (Lamichhane et al., 2022; Ogidi et al., 2024e). Table 3 summarizes strategies for controlling human exposure to micro- and nanoplastics (MNPs) and their associated health effects.

Table 3. Strategies to Control Human Exposure to Micro- and Nanoplastics (MNPs) and Associated Health Effects

Strategy	Approach/Examples	Associated Health Benefits/Effects	References
Improved Waste Management	Enhanced plastic collection, recycling systems, bans on single-use plastics, biodegradable alternatives	Reduced environmental leakage leading to lower ingestion and inhalation risks	(UNEP, 2021; Moshood et al., 2022)
Drinking Water Treatment	Advanced filtration (membrane technology, activated carbon, coagulation, sand filtration)	Removal of MNPs from drinking water resulting in reduced gastrointestinal and systemic exposure	(Mintenig et al., 2019; Sol et al., 2020)
Food Safety Regulations	Monitoring plastic migration from packaging and enforcement of food-grade plastic standards	Decreased ingestion of MNPs through food contact materials	(Hernandez et al., 2019; Rodrigues et al., 2022)

continued on following page

Table 3. Continued

Strategy	Approach/Examples	Associated Health Benefits/Effects	References
Air Quality Control	Ventilation systems, indoor air filtration, reduction of textile fiber shedding, workplace exposure standards	Reduced inhalation of airborne particles and associated respiratory stress	(Dris et al., 2017; Gasperi et al., 2018)
Occupational Safety Measures	Personal protective equipment (masks, gloves), industrial dust control, exposure monitoring in plastic manufacturing and recycling sectors	Reduced chronic exposure and lower risk of respiratory and systemic toxicity	(Prata et al., 2020)
Biodegradable & Green Materials	Bioplastics development, enzymatic degradation (PETase), fungal and bacterial degradation systems	Long-term reduction in MNP release and decreased accumulation in the food chain	(Yoshida et al., 2016; Taniguchi et al., 2019; Paço et al., 2017)
Public Awareness & Education	Campaigns on plastic use reduction, proper disposal practices, and consumption reduction strategies	Behavioral changes leading to reduced household exposure and microfiber release	(WHO, 2021)
Medical & Safety Standards	Regulation of plastic leaching in medical devices, syringes, IV bags, and drug packaging	Reduced iatrogenic and intravenous exposure in clinical settings	(Eyles et al., 2001)
Research & Monitoring	Standardized detection methods, global monitoring systems, exposure and risk assessment frameworks	Early identification of contamination hotspots enabling targeted public health interventions	(Hartmann et al., 2019)

Integration of One Health approaches into mitigation frameworks requires coordinated action between environmental agencies, veterinary sciences, and public health institutions. AI-driven remote sensing technologies and biosensor systems can enhance real-time detection of nanoplastics in air and water, supporting early warning systems and evidence-based policy decisions. In addition, biotechnology-based solutions such as enzymatic recycling and microbial degradation offer scalable and environmentally sustainable mitigation pathways.

Risk Assessment and Regulatory Perspectives

In addition to plastic additives and environmental contaminants, nanoplastics may act as carriers for pharmaceuticals and other chemicals that pose potential risks to human health following ingestion or inhalation (Ogidi et al., 2025b; Human Health Perspective, 2021). Upon interaction with biological systems, nanoplastics may induce oxidative stress, inflammation, and other cellular stress responses (Xu et al., 2020). Their ubiquitous presence in environmental matrices and atmospheric aerosols contributes to their entry into food and water supplies, thereby increasing human exposure (Seiyaboh et al., 2019; WHO, 2021). The primary routes of exposure include dermal contact, inhalation of ambient particles, and ingestion of contaminated food and water (Monitoring & Enforcement, 2022).

Experimental evidence from in vitro and in vivo models suggests that nanoplastic toxicity is dose-dependent, indicating a relationship between exposure level and observed biological effects (Thompson et al., 2013). While animal studies suggest potential long-term health risks, direct evidence in humans remains limited. Therefore, identifying exposure risks, generating robust dose–response data, and improving exposure assessment methodologies are essential for understanding the health implications of nanoplastics. Emerging evidence also suggests that individuals with pre-existing immunological conditions may be more susceptible to adverse effects (Zhang et al., 2021).

Machine learning and artificial intelligence approaches are increasingly being applied to model nanoplastic exposure and predict potential health outcomes. These models integrate environmental monitoring data, biological indicators, and exposure pathways to support early warning systems and risk prediction under the One Health framework (Rendeiro et al., 2023).

One Health Framework and AI-Based Monitoring

One Health understands that the health of humans, animals, and ecosystems is interconnected. Nanoplastics move freely between these domains—driven by environmental contamination, concentrated in wildlife, and returning to human systems through food, air, and water. This continuous flow demands simultaneous environmental, biomedical, and veterinary monitoring, as well as coordinated policy responses. The One Health approach to nanoplastic research implies that environmental, biomedical, and veterinary monitoring must be conducted in parallel. Sensors, biotests, and population health records must be linked to track plastic particles through environmental interfaces. AI-based systems, such as machine learning algorithms and predictive analytics, can potentially synthesize these heterogeneous data streams in a way that enables the detection of hotspots of contamination in a timely fashion and the forecasting of exposure risk to humans.

Computer-based image recognition and remote-sensing networks can rapidly detect nanoplastic particles in environmental samples, while biosensor networks can continuously assess concentrations in water and soils. Integrating these technologies into public health databases facilitates real-time exposure modeling and evidence-based risk assessment. A conceptual One Health model (Figure 2) would illustrate the interlinkage between environmental pollution, animal uptake, and human exposure, with AI and biotechnology serving as the monitoring and control feedback loop. Such integration not only enhances the understanding of nanoplastic mechanisms but also supports intersectoral decision-making, helping to avoid long-term health and environmental consequences.

Figure 2. One health framework for nanoplastic contamination and AI-based surveillance, Figure 2 presents an operational architecture of the integrated One Health framework for nanoplastic kinetics, where environmental matrices, animal exposure pathways, and human physiological endpoints are treated as interconnected components rather than isolated compartments. The framework traces the movement of nanoplastics across air, water, soil, and trophic interfaces, highlighting exposure as a continuous process shaped by overlapping environmental and biological pathways rather than single, discrete events. An AI-based surveillance and computational monitoring layer is integrated within this system to support real-time tracking, predictive analysis, and the identification of potential exposure hotspots across complex environmental datasets. Overall, the structure reflects a coordinated One Health perspective in which environmental monitoring, exposure assessment, and risk interpretation are aligned within a unified analytical framework.

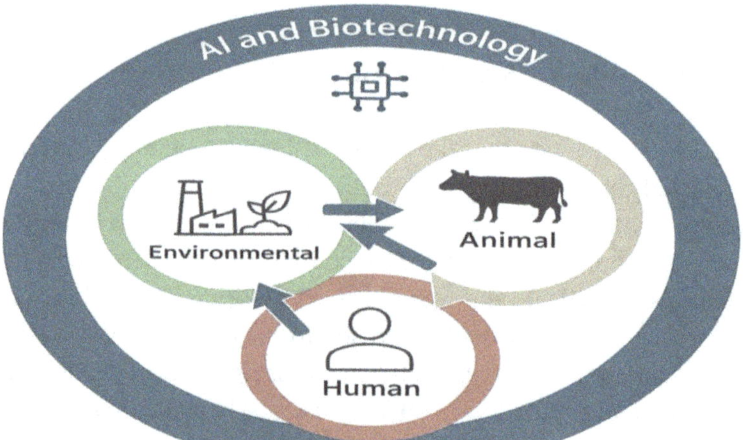

One Health Framework for Nanoplastic Contamination and AI-Based Surveillance

CHALLENGES AND FUTURE RESEARCH

Scientific investigation into the origin and implications of microplastics and nanoplastics is justified, as indicated by the evidence presented herein. Below is an overview of major issues to take into account. Even in research on synthetic nanomaterials, current analytical tools are not sufficiently advanced to measure nanoplastics in the environment, particularly in biological media. Several of these problems remain unresolved despite the use of asymmetric flow field-flow fractionation coupled with multi-angle light scattering, fluorescent labelling, and Raman tweezers for nanoplastic detection in previous studies (Correia & Loeschner, 2018; Gagne, 2019; Catarino et al., 2019; Gillibert et al., 2019). The most prevalent detection method employed, fluorescent labelling, is susceptible to artefacts as well as false positives, particularly in cases of cellular bioaccumulation.

To overcome this issue, the application of blank negative controls and dye core-wrapped samples is suggested, as cellular autofluorescence and dye leakage are primarily responsible for artefacts (Catarino et al., 2019). The aforementioned arguments emphasize the necessity of more sophisticated analytical techniques that are more convenient, efficient, and precise for identifying nanoplastics in the environment. It is necessary to develop a variety of detection methods for microplastics and nanoplastics in different environmental media in order to limit human exposure. Although there are limited solutions for their reduction, separation and collection methods for nanoplastics from the environment do exist. More than 6,300 metric tonnes of plastic litter were generated in 2015 alone. Approximately 90% (5733 Mt) of the 2015 plastic waste generated will not be recycled by 2050, according to estimates, and will be released into the environment in one form or another (Geyer et al., 2017). Since decomposed plastic waste may even lead to micro- or nano-sized particles, recycling plastic waste is a useful method for reducing or eliminating micropollution and nanopollution.

Two plastic alternatives are available. Chitin is a bioactive polymer and is commonly applied in the pharmaceutical and industrial sectors. It is highly regarded among naturally occurring polysaccharides (Hosseinnejad & Jafari, 2016). Due to its many useful properties, such as being non-toxic, easily biodegradable, chemically modifiable, and having strong antibiosis activity, chitin is an excellent alternative for plastics (Muxika et al., 2017). Numerous ropes, auto components, polystyrene, and elastic construction materials include hemp fibre, a biodegradable polymer material (Mazzanti et al., 2019). Hemp fibre can substitute plastics due to being non-toxic, biodegradable, and recyclable (Sepe et al., 2017). Microplastics and nanoplastics, as well as other environmental pollutants, are degraded by physical, chemical, or biological processes. Even though chemical and physical methods have previously been applied to decompose or eliminate micropollutants and nanopollutants, such methods may generate secondary pollutants or are associated with inefficient degradation (Day et al., 1996; Jian et al., 2019).

If traditional pollutant degradation methods fail, biodegradation can intervene and mitigate the issue (Ding et al., 2018). Biodegradation is required to counteract or remove microplastic and nanoplastic pollution from the environment since it operates effectively with minimal negative effects. Nanoplastics, while being transported and transformed through different media, are able to absorb and leach environmental pollutants once released into the environment (Catarino et al., 2019). Several chemicals are released into the atmosphere by nanoplastics as they break down. They are also capable of increasing toxicity by altering their physicochemical characteristics. However, the nature and degree of nanoplastics' effects on humans and the environment have not been definitively determined. Therefore, studies on the environmental effects of nanoplastics and their degradation products are needed (Luo et al., 2019). In most studies, nanoplastics are typically introduced at unrealistic doses to specific model organisms (Lee et al., 2019). There has been limited examination of nanoplastic toxicity at environmentally relevant concentrations (ERC levels). There is thus a need for studies on the effects of nanoplastics at environmental doses.

Researchers have used several animal models to examine the reproductive, neurological, and developmental harm associated with nanoplastics. The potential impact of nanoplastics on humans has not yet been extensively tested, but researchers are compelled to study their effects due to unavoidable exposure (Qu et al., 2019; Tallec et al., 2018). Experimental findings should be translated into clinical implications using standardized biomarkers of nanoplastic exposure and validated diagnostic tests. Integrating spatial multi-omics and AI analysis may allow for personalized risk assessment and guide public health intervention toward the One Health vision for nanoplastic pollution (Lovric et al., 2022; HuBMAP Consortium, 2023).

CONCLUSION

Developed and developing nations equally regard plastics as one of the most significant contemporary environmental threats. One of the major limitations in this field remains the scarcity of sufficient quantitative data for evaluating the deleterious effects of micro- and nanoplastics (MNPs). Additional studies are required to elucidate the pathophysiology and toxicology of MNP exposure, as human evidence remains at a relatively early stage. Despite the utility of *in vivo* and *in vitro* investigations employing animal models, research specifically addressing the human health implications of MNPs remains limited. These approaches also possess methodological constraints, including difficulties associated with the extraction, identification, and analysis of MNPs from human biological samples. Consequently, there is an increasing need to accelerate the development of cost-effective and sensitive analytical methods capable of accurately detecting and quantifying MNPs in complex biological matrices. Likewise, further investigation of long-term exposure effects, metabolic behavior, and biological persistence is warranted given the prolonged environmental stability of these materials.

Several measures may reduce the harmful consequences of MNP exposure, including public awareness initiatives, reductions in plastic use, improved waste disposal practices, stronger regulatory policies, and enhanced recycling systems. Future research should additionally focus on developing sensitive detection platforms, improving biodegradation efficiency, identifying sustainable alternatives to synthetic plastics, and promoting circular strategies capable of transforming waste into valuable resources while reducing plastic-associated toxicity.

Nanoplastic pollution requires a transdisciplinary One Health approach integrating environmental monitoring, toxicological investigations, and public health surveillance. AI-assisted modeling and biotechnological innovations may strengthen hazard prediction and support evidence-based strategies intended to reduce nanoplastic exposure across environmental and biological systems. Nevertheless, informed risk assessment remains constrained by persistent knowledge gaps, including insufficient quantitative data on human uptake, limited standardization of analytical methodologies, and uncertainty surrounding dose–response relationships. Future efforts should prioritize harmonized approaches for quantifying nanoplastics across environmental and biological matrices to facilitate more reliable health risk characterization. In parallel, microbial and enzymatic bioremediation strategies, together with AI-enabled surveillance systems, may provide environmentally sustainable approaches for monitoring and mitigation. Continued collaboration among environmental agencies, public health institutions, and regulatory bodies will therefore remain essential for establishing coordinated standards and long-term management strategies.

Under the conditions outlined throughout this chapter, it also becomes increasingly apparent that the consequences of nanoplastic exposure extend beyond classical toxicological endpoints and environmental distribution patterns. Continuous exposure through contaminated food, water, and airborne particles establishes a persistent biological burden that may not remain restricted to systemic circulation or organ-level accumulation alone. Rather, these interactions occur within tightly connected physiological systems where immunity, metabolism, and nutrient processing continuously intersect.

Within this context, the gastrointestinal environment represents a particularly sensitive interface in which exogenous particles, dietary inputs, and resident microbial communities coexist under ongoing physiological pressures. Such interactions may gradually influence biological equilibrium, not necessarily through abrupt pathological outcomes, but through cumulative modifications in regulatory processes over time. This perspective further suggests that the implications of nanoplastic exposure may be embedded within broader networks governing nutritional regulation, metabolic homeostasis, and immune function, emphasizing the need to consider exposure biology within an integrated systems framework rather than through isolated toxicological endpoints alone.

Authorship and AI Disclosure Statement

The author confirms that AI tools were used only for minor language editing and formatting and Figure 2. All research design, data interpretation, and manuscript writing were performed by the author, who retains full intellectual responsibility for the content in accordance with IGI Global's 2025 Publication Ethics Policy.

REFERENCES

Al-Sid-Cheikh, M., Rowland, S. J., Stevenson, K., Rouleau, C., Henry, T. B., & Thompson, R. C. (2018). Uptake, whole-body distribution, and depuration of nanoplastics by the scallop *Pecten maximus* at environmentally realistic concentrations. *Environmental Science & Technology*, *52*(24), 14480–14486. DOI: 10.1021/acs.est.8b05266 PMID: 30457844

Alimi, O. S., Budarz, J. F., Hernandez, L. M., & Tufenkji, N. (2018). Microplastics and nanoplastics in aquatic environments: Aggregation, deposition, and enhanced contaminant transport. *Environmental Science & Technology*, *52*(4), 1704–1724. DOI: 10.1021/acs.est.7b05559 PMID: 29265806

Ballesteros, S., Domenech, J., Barguilla, I., Cortes, C., Marcos, R., & Hernandez, A. (2020). Genotoxic and immunomodulatory effects in human white blood cells after ex vivo exposure to polystyrene nanoplastics. *Environmental Science. Nano*, *7*(11), 3431–3446. DOI: 10.1039/D0EN00748J

Ballesteros, S., Domenech, J., Barguilla, I., Cortes, C., Marcos, R., & Hernández, A. (2020). Genotoxic and immunomodulatory effects in human white blood cells after ex vivo exposure to polystyrene nanoplastics. *Environmental Science. Nano*, *7*(11), 3431–3446. DOI: 10.1039/D0EN00748J

Barguilla, I., Domenech, J., Ballesteros, S., Rubio, L., Marcos, R., & Hernández, A. (2022). Long-term exposure to nanoplastics alters molecular and functional traits related to the carcinogenic process. *Journal of Hazardous Materials*, *438*, 129470. DOI: 10.1016/j.jhazmat.2022.129470 PMID: 35785738

Barguilla, I., Domenech, J., Ballesteros, S., Rubio, L., Marcos, R., & Hernández, A. (2022). Long-term exposure to nanoplastics alters molecular and functional traits related to the carcinogenic process. *Journal of Hazardous Materials*, *438*, 129470. DOI: 10.1016/j.jhazmat.2022.129470 PMID: 35785738

Bläsing, M., & Amelung, W. (2018). Plastics in soil: Analytical methods and possible sources. *The Science of the Total Environment*, *612*, 422–435. DOI: 10.1016/j.scitotenv.2017.08.086 PMID: 28863373

Catarino, A. I., Frutos, A., & Henry, T. B. (2019). Use of fluorescent-labelled nanoplastics (NPs) to demonstrate NP absorption is inconclusive without adequate controls. *The Science of the Total Environment*, *670*, 915–920. DOI: 10.1016/j.scitotenv.2019.03.194 PMID: 30921723

Chang, W.-H., Herianto, S., Lee, C.-C., Hung, H., & Chen, H.-L. (2021). The effects of phthalate ester exposure on human health: A review. *The Science of the Total Environment*, *786*, 147371. DOI: 10.1016/j.scitotenv.2021.147371 PMID: 33965815

Chang, W.-H., Herianto, S., Lee, C.-C., Hung, H., & Chen, H.-L. (2021). The effects of phthalate ester exposure on human health: A review. *The Science of the Total Environment*, *786*, 147371. DOI: 10.1016/j.scitotenv.2021.147371 PMID: 33965815

Choi, D., Hwang, J., Bang, J., Han, S., Kim, T., Oh, Y., Hwang, Y., Choi, J., & Hong, J. (2021). In vitro toxicity from a physical perspective of polyethylene microplastics based on statistical curvature change analysis. *The Science of the Total Environment*, *752*, 142242. DOI: 10.1016/j.scitotenv.2020.142242 PMID: 33207500

Choi, D., Hwang, J., Bang, J., Han, S., Kim, T., Oh, Y., Hwang, Y., Choi, J., & Hong, J. (2021). In vitro toxicity from a physical perspective of polyethylene microplastics based on statistical curvature change analysis. *The Science of the Total Environment*, *752*, 142242. DOI: 10.1016/j.scitotenv.2020.142242 PMID: 33207500

Correia, M., & Loeschner, K. (2018). Detection of nanoplastics in food by asymmetric flow field-flow fractionation coupled to multi-angle light scattering: Possibilities, challenges and analytical limitations. *Analytical and Bioanalytical Chemistry*, *410*(22), 5603–5615. DOI: 10.1007/s00216-018-0919-8 PMID: 29411085

Cox, K. D., Covernton, G. A., Davies, H. L., Dower, J. F., Juanes, F., & Dudas, S. E. (2019). Human consumption of microplastics. *Environmental Science & Technology*, *53*(12), 7068–7074. DOI: 10.1021/acs.est.9b01517 PMID: 31184127

Day, M., Cooney, J. D., Klein, C., & Fox, J. (1996). *Thermal degradation of automotive plastics: A possible recycling opportunity*. American Chemical Society.

Deng, Y., Zhang, Y., Lemos, B., & Ren, H. (2017). Tissue accumulation of microplastics in mice and biomarker responses suggest widespread health risks of exposure. *Scientific Reports*, *7*(1), 46687. DOI: 10.1038/srep46687 PMID: 28436478

Deng, Y., Zhang, Y., Lemos, B., & Ren, H. (2017). Tissue accumulation of microplastics in mice and biomarker responses suggest widespread health risks of exposure. *Scientific Reports*, *7*(1), 46687. DOI: 10.1038/srep46687 PMID: 28436478

Ding, Q., Liu, K., Xu, K., Sun, R., Zhang, J., Yin, L., & Pu, Y. (2018). Further understanding of degradation pathways of microcystin-LR by an indigenous *Sphingopyxis* sp. in environmentally relevant pollution concentrations. *Toxins*, *10*(12), 536. DOI: 10.3390/toxins10120536 PMID: 30558170

Dong, X., Liu, X., Hou, Q., & Wang, Z. (2023). From natural environment to animal tissues: A review of microplastics (nanoplastics) translocation and hazards studies. *The Science of the Total Environment, 855*, 158686. DOI: 10.1016/j.scitotenv.2022.158686 PMID: 36099943

Dris, R., Gasperi, J., Mirande, C., Mandin, C., & Guerrouache, M. (2018). A first overview of textile fibers, including microplastics, in indoor and outdoor environments. *Environmental Pollution, 234*, 645–655. DOI: 10.1016/j.envpol.2017.12.013 PMID: 27989388

Dris, R., Gasperi, J., Mirande, C., Mandin, C., Guerrouache, M., Langlois, V., & Tassin, B. (2017). A first overview of textile fibers, including microplastics, in indoor and outdoor environments. *Environmental Pollution, 221*, 453–458. DOI: 10.1016/j.envpol.2016.12.013 PMID: 27989388

Eyles, J. E., Bramwell, V. W., Williamson, E. D., & Alpar, H. O. (2001). Microsphere translocation and immunopotentiation in systemic tissues following intranasal administration. *Vaccine, 19*(32), 4732–4742. DOI: 10.1016/S0264-410X(01)00220-1 PMID: 11535324

Fan, W., Salmond, J. A., Dirks, K. N., Cabedo Sanz, P., Miskelly, G. M., & Rindelaub, J. D. (2022). Evidence and mass quantification of atmospheric microplastics in a coastal New Zealand city. *Environmental Science & Technology, 56*(29), 17556–17568. DOI: 10.1021/acs.est.2c05850 PMID: 36459143

Free, C. M., Jensen, O. P., Mason, S. A., Eriksen, M., Williamson, N. J., & Boldgiv, B. (2014). High levels of microplastic pollution in a large, remote, mountain lake. *Marine Pollution Bulletin, 85*(1), 156–163. DOI: 10.1016/j.marpolbul.2014.06.001 PMID: 24973278

Gagné, F. (2019). Detection of polystyrene nanoplastics in biological tissues with a fluorescent molecular rotor probe. *Journal of Xenobiotics, 9*(2), 8147. DOI: 10.4081/xeno.2019.8147 PMID: 31178985

Galloway, T. (2015). *Micro- and nano-plastics and human health.* Springer., DOI: 10.1007/978-3-319-16510-3_13

Gasperi, J., Wright, S. L., Dris, R., Collard, F., Mandin, C., Guerrouache, M., Langlois, V., Kelly, F. J., & Tassin, B. (2018). Microplastics in air: Are we breathing it in? *Current Opinion in Environmental Science & Health, 1*, 1–5. DOI: 10.1016/j.coesh.2017.10.002

Geiser, M., Rothen-Rutishauser, B., Kapp, N., Schürch, S., Kreyling, W., Schulz, H., Semmler, M., Im Hof, V., Heyder, J., & Gehr, P. (2005). Ultrafine particles cross cellular membranes by nonphagocytic mechanisms in lungs and cultured cells. *Environmental Health Perspectives, 113*(11), 1555–1560. DOI: 10.1289/ehp.8006 PMID: 16263511

Geyer, R., Jambeck, J. R., & Law, K. L. (2017). Production, use, and fate of all plastics ever made. *Science Advances, 3*(7), e1700782. DOI: 10.1126/sciadv.1700782 PMID: 28776036

Gillibert, R., Balakrishnan, G., Deshoules, Q., Tardivel, M., Magazzù, A., Donato, M. G., Maragò, O. M., Lamy de La Chapelle, M., Colas, F., Lagarde, F., & Gucciardi, P. G. (2019). Raman tweezers for small microplastics and nanoplastics identification in seawater. *Environmental Science & Technology, 53*(16), 9003–9013. DOI: 10.1021/acs.est.9b03105 PMID: 31259538

Grafmueller, S., Manser, P., Diener, L., Diener, P. A., Maeder-Althaus, X., Maurizi, L., Jochum, W., Krug, H. F., Buerki-Thurnherr, T., von Mandach, U., & Wick, P. (2015). Bidirectional transfer study of polystyrene nanoparticles across the placental barrier in an ex vivo human placental perfusion model. *Environmental Health Perspectives, 123*(12), 1280–1286. DOI: 10.1289/ehp.1409271 PMID: 25956008

Grafmueller, S., Manser, P., Diener, L., Diener, P. A., Maeder-Althaus, X., Maurizi, L., Jochum, W., Krug, H. F., Buerki-Thurnherr, T., von Mandach, U., & Wick, P. (2015). Bidirectional transfer study of polystyrene nanoparticles across the placental barrier in an ex vivo human placental perfusion model. *Environmental Health Perspectives, 123*(12), 1280–1286. DOI: 10.1289/ehp.1409271 PMID: 25956008

Gutow, L., Ehlers, S. M., Pittroff, M., Bartl, K., Wurm, Y., Galgani, F., & Saborowski, R. (2016). Microplastics in the marine environment: A review of the methods used for identification and quantification. *Environmental Science & Technology, 50*(24), 12327–12340. DOI: 10.1021/acs.est.6b04499

Hartmann, N. B., Hüffer, T., Thompson, R. C., Hassellöv, M., Verschoor, A., Daugaard, A. E., Rist, S., Karlsson, T., Brennholt, N., Cole, M., Herrling, M. P., Hess, M. C., Ivleva, N. P., Lusher, A. L., & Wagner, M. (2019). Are we speaking the same language? Recommendations for a definition and categorization framework for plastic debris. *Environmental Science & Technology, 53*(3), 1039–1047. DOI: 10.1021/acs.est.8b05297 PMID: 30608663

Hernandez, E., Nowack, B., & Mitrano, D. M. (2017). Polyester textiles as a source of microplastics from households: A mechanistic study to understand microfiber release during washing. *Environmental Science & Technology*, *51*(12), 7036–7046. DOI: 10.1021/acs.est.7b01750 PMID: 28537711

Hernandez, L. M., Xu, E. G., Larsson, H., Rui, T., & Tufenkji, N. (2019). Plastic teabags release billions of microparticles and nanoparticles into tea. *Environmental Science & Technology*, *53*(21), 12300–12310. DOI: 10.1021/acs.est.9b02540 PMID: 31552738

Hosseinnejad, M., & Jafari, S. M. (2016). Evaluation of different factors affecting antimicrobial properties of chitosan. *International Journal of Biological Macromolecules*, *85*, 467–475. DOI: 10.1016/j.ijbiomac.2016.01.022 PMID: 26780706

Hou, Z., Meng, R., Chen, G., Lai, T., Qing, R., Hao, S., Deng, J., & Wang, B. (2022). Distinct accumulation of nanoplastics in human intestinal organoids. *The Science of the Total Environment*, *838*, 155811. DOI: 10.1016/j.scitotenv.2022.155811 PMID: 35597345

Hou, Z., Meng, R., Chen, G., Lai, T., Qing, R., Hao, S., Deng, J., & Wang, B. (2022). Distinct accumulation of nanoplastics in human intestinal organoids. *The Science of the Total Environment*, *838*, 155811. DOI: 10.1016/j.scitotenv.2022.155811 PMID: 35597345

HuBMAP Consortium. (2023). The Human BioMolecular Atlas Program. *Nature*, *622*(7985), 567–578.

Hüffer, T., Hofmann, T., & Asbach, C. (2017). The great unknown—Nanoplastics in personal care products. *Environmental Science. Nano*, *4*(4), 730–742. DOI: 10.1039/C7EN00471A

Jian, K., Li, Z., Xiaoguang, D., Hongqi, S., & Zhimin, A. (2019). Degradation of cosmetic microplastics via functionalized carbon nanosprings. *Matter*, *1*(3), 745–758. DOI: 10.1016/j.matt.2019.06.004

Kibria, M. G., Masuk, N. I., Safayet, R., Nguyen, H. Q., & Mourshed, M. (2023). Plastic waste: Challenges and opportunities to mitigate pollution and effective management. *International Journal of Environmental Research*, *17*(1), 20. DOI: 10.1007/s41742-023-00507-z PMID: 36711426

Kitamura, S. (2005). Comparative study of the endocrine-disrupting activity of bisphenol A and 19 related compounds. *Toxicological Sciences : An Official Journal of the Society of Toxicology*, *84*(2), 249–259. DOI: 10.1093/toxsci/kfi074 PMID: 15635150

Kitamura, S. (2005). Comparative study of the endocrine-disrupting activity of bisphenol A and 19 related compounds. *Toxicological Sciences : An Official Journal of the Society of Toxicology, 84*(2), 249–259. DOI: 10.1093/toxsci/kfi074 PMID: 15635150

Klein, M., Fischer, E. K., & Neophytou, C. (2021). Nanoplastics: A growing threat to marine organisms. *Frontiers in Marine Science, 8*, 627888. DOI: 10.3389/fmars.2021.627888

Koelmans, A. A., Mohamed Nor, N. H., Hermsen, E., Kooi, M., Mintenig, S. M., & De France, J. (2017). Microplastics in freshwaters and drinking water: Critical review and assessment of data quality. *Water Research, 155*, 410–422. DOI: 10.1016/j.watres.2019.02.054 PMID: 30861380

Koelmans, A. A., Mohamed Nor, N. H., Hermsen, E., Kooi, M., Mintenig, S. M., & De France, J. (2019). Microplastics in freshwaters and drinking water: Critical review and assessment of data quality. *Water Research, 155*, 410–422. DOI: 10.1016/j.watres.2019.02.054 PMID: 30861380

Kole, P. J., Lahr, J., Van Belleghem, F. G. A. J., & Ragas, A. M. J. (2019, May 26–30). Wear and tear of tyres in the global environment: Size distribution, emission, pathways and health effects. In *Proceedings of the SETAC Europe 29th Annual Meeting*. Helsinki, Finland.

Kosuth, M., Mason, S. A., & Wattenberg, E. V. (2018). Anthropogenic contamination of tap water, beer, and sea salt. *PLoS One, 13*(4), e0194970. DOI: 10.1371/journal.pone.0194970 PMID: 29641556

Lamichhane, G., Acharya, A., Marahatha, R., Modi, B., Paudel, R., Adhikari, A., Raut, B. K., Aryal, S., & Parajuli, N. (2022). Microplastics in environment: Global concern, challenges, and controlling measures. *International Journal of Environmental Science and Technology*. Advance online publication. DOI: 10.1007/s13762-022-04261-1 PMID: 35638092

Leads, R. R., & Weinstein, J. E. (2019). Occurrence of tire wear particles and other microplastics within the tributaries of the Charleston Harbor Estuary, South Carolina, USA. *Marine Pollution Bulletin, 145*, 569–582. DOI: 10.1016/j.marpolbul.2019.06.061 PMID: 31590826

Lee, W. S., Cho, H. J., Kim, E., Huh, Y. H., Kim, H. J., Kim, B., Kang, T., Lee, J. S., & Jeong, J. (2019). Bioaccumulation of polystyrene nanoplastics and their effect on the toxicity of Au ions in zebrafish embryos. *Nanoscale, 11*(7), 3173–3185. DOI: 10.1039/C8NR09321K PMID: 30534785

Lee, W. S., Cho, H. J., Kim, E., Huh, Y. H., Kim, H. J., Kim, B., Kang, T., Lee, J. S., & Jeong, J. (2019). Bioaccumulation of polystyrene nanoplastics and their effect on the toxicity of Au ions in zebrafish embryos. *Nanoscale, 11*(7), 3173–3185. DOI: 10.1039/C8NR09321K PMID: 30534785

Lehner, R., Petri-Fink, A., & Rothen-Rutishauser, B. (2017). Nanoplastic impact on human health—A 3D intestinal model to study the interaction with nanoplastic particles. In Deidun, A. C. (Ed.), *Proceedings of the International Conference on Microplastic Pollution in the Mediterranean Sea* (pp. 219–221). Springer. https://doi.org/DOI: 10.1007/978-3-319-71279-6_33

Lehner, R., Wohlleben, W., Septiadi, D., Landsiedel, R., Petri-Fink, A., & Rothen-Rutishauser, B. (2020). A novel 3D intestine barrier model to study the immune response upon exposure to microplastics. *Archives of Toxicology, 94*(7), 2463–2479. DOI: 10.1007/s00204-020-02750-1 PMID: 32307674

Lenz, R., Enders, K., & Nielsen, T. G. (2016). Microplastic exposure studies should be environmentally realistic. *Proceedings of the National Academy of Sciences of the United States of America, 113*(29), E4121–E4122. DOI: 10.1073/pnas.1606615113 PMID: 27407153

Li, Z., Zhu, S., Liu, Q., Wei, J., Jin, Y., Wang, X., & Zhang, L. (2020). Polystyrene microplastics cause cardiac fibrosis by activating Wnt/β-catenin signaling pathway and promoting cardiomyocyte apoptosis in rats. *Environmental Pollution, 265*, 115025. DOI: 10.1016/j.envpol.2020.115025 PMID: 32806417

Li, Z., Zhu, S., Liu, Q., Wei, J., Jin, Y., Wang, X., & Zhang, L. (2020). Polystyrene microplastics cause cardiac fibrosis by activating Wnt/β-catenin signaling pathway and promoting cardiomyocyte apoptosis in rats. *Environmental Pollution, 265*, 115025. DOI: 10.1016/j.envpol.2020.115025 PMID: 32806417

Lisuzzo, L., Cavallaro, G., Milioto, S., & Lazzara, G. (2020). Effects of halloysite content on the thermo-mechanical performances of composite bioplastics. *Applied Clay Science, 185*, 105416. DOI: 10.1016/j.clay.2019.105416

Lovric, S., Zee, J., Woroniecki, R., & Susztak, K. (2022). The Human Kidney Atlas: Integrative omics for precision nephrology. *Nature Reviews. Nephrology, 18*(9), 583–601.

Luo, H., Xiang, Y., He, D., Li, Y., Zhao, Y., Wang, S., & Pan, X. (2019). Leaching behavior of fluorescent additives from microplastics and the toxicity of leachate to *Chlorella vulgaris*. *The Science of the Total Environment, 678*, 1–9. DOI: 10.1016/j.scitotenv.2019.04.401 PMID: 31075575

Mason, S. A., Welch, V. G., & Neratko, J. (2018). Synthetic polymer contamination in bottled water. *Frontiers in Chemistry*, *6*, 407. DOI: 10.3389/fchem.2018.00407 PMID: 30255015

Mazzanti, V., Pariante, R., Bonanno, A., Ballesteros, O. D., Mollica, F., & Filippone, G. (2019). Reinforcing mechanisms of natural fibers in green composites: Role of fibers morphology in a PLA/hemp model system. *Composites Science and Technology*, *180*, 51–59. DOI: 10.1016/j.compscitech.2019.05.015

Meng, F., Fan, T., Yang, X., Riksen, M., Xu, M., & Geissen, W. (2020). Effect of plastic mulching on the accumulation and distribution of macro- and microplastic particles in soils of two farming systems in North West China. *PeerJ*, *8*, e10375. DOI: 10.7717/peerj.10375 PMID: 33344073

Mintenig, S. M., Loder, M. G. J., Primpke, S., & Gerdts, G. (2019). Low numbers of microplastics detected in drinking water from groundwater sources. *The Science of the Total Environment*, *648*, 631–635. DOI: 10.1016/j.scitotenv.2018.08.178 PMID: 30121540

Miri, S., Saini, R., Davoodi, S. M., Pulicharla, R., Brar, S. K., & Magdouli, S. (2022). Biodegradation of microplastics: Better late than never. *Chemosphere*, *286*, 131670. DOI: 10.1016/j.chemosphere.2021.131670 PMID: 34351281

Monitoring and Enforcement. (2022). Strategies for monitoring nanoplastic pollution. *Environmental Monitoring and Assessment*, *194*(6), 421. DOI: 10.1007/s10661-022-10089-4 PMID: 35543765

Moshood, T. D., Nawanir, G., Mahmud, F., Mohamad, F., Ahmad, M. H., & Abdul-Ghani, A. (2022). Sustainability of biodegradable plastics: New problem or solution to solve the global plastic pollution? *Current Research in Green and Sustainable Chemistry*, *5*, 100273. DOI: 10.1016/j.crgsc.2022.100273

Muxika, A., Etxabide, A., Uranga, J., Guerrero, P., & de la Caba, K. (2017). Chitosan as a bioactive polymer: Processing, properties and applications. *International Journal of Biological Macromolecules*, *105*, 1358–1368. DOI: 10.1016/j.ijbiomac.2017.07.087 PMID: 28735006

Nizzetto, L., Futter, M., & Langaas, S. (2016). Investigating the mechanisms of microplastic transport in soil. *Environmental Science & Technology*, *50*(20), 11117–11126. DOI: 10.1021/acs.est.6b03438 PMID: 27682621

Ogidi, O. I. (2024). Recent advances in anticancer activity and Bioinformatics Approach from Potential Plants. In: Computational Approaches in Biomaterials, Bioinformatics and Biomedical Engineering Applications. Published *CRC Press, Taylor and Francis Group, Florida, USA*. DOI: 10.1201/9781003354437-3

Ogidi, O. I. (2025). Natural Products Modulating Mitophagy: Implications for Cancer Therapy. In Pandurangan, A. (Ed.), *Roles of Mitophagy in Cancer Regulation* (pp. 325–358). IGI Global Scientific Publishing., DOI: 10.4018/979-8-3693-3836-0.ch012

Ogidi, O. I., & Ajoko, I. (2024). Neuroprotective Potentials of Phytochemicals. In: M.S. Rajput, T. Sarachana, P.D. Sarkar, M. Raval (eds) NeuroPhytomedicine. Published by *CRC Press, Routledge Taylor and Francis Group.*

Ogidi, O. I., & Akpan, U. M. (2024). Microbial and Phyto-Remediation of Crude Oil Contaminated Soil. In: Eco-Restoration of the Environment. Published by *CRC Press, Routledge Taylor and Francis Group.*

Ogidi, O. I., Barwant, M. M., & Okara, E. (2024a). Biomarker Discovery: Protein Phosphorylation State. In *Spatially Variable Genes in Cancer: Development, Progression, and Treatment Response*. IGI Global., DOI: 10.4018/979-8-3693-7728-4.ch005

Ogidi, O. I., Christopher, E., & Orlu, H. A. (2025a). Environmental DNA and RNA Metabarcoding: Advancing Insights in Cancer Research. In Raghavan, R. (Ed.), *Cancer Exposomics and Environmental Influences on Carcinogenesis* (pp. 279–322). IGI Global Scientific Publishing., DOI: 10.4018/979-8-3373-2165-3.ch007

Ogidi, O. I., & Emaikwu, N. G. (2024). Plant Phenolic compound isolation and its Bioinformatics Approaches of Molecular Mechanisms in Antimicrobial activities and Resistance. In: Computational Approaches in Biomaterials, Bioinformatics and Biomedical Engineering Applications. Publishered *CRC Press, Taylor and Francis Group, Florida, USA*. DOI: 10.1201/9781003354437-8

Ogidi, O. I., Erho, S. A., & Omu, O. (2025c). Reactive Oxygen Species in Cancer. In Prabhakar, P. (Ed.), *The Role of Reactive Oxygen Species in Human Health and Disease* (pp. 339–376). IGI Global Scientific Publishing., DOI: 10.4018/979-8-3693-7919-6.ch010

Ogidi, O. I., & Felagha, I. (2024). Gut Microbiome and Diet Interaction. In *Nutrition Controversies and Advances in Autoimmune Disease. Publisher*. IGI Global. DOI: 10.4018/979-8-3693-5528-2.ch012

Ogidi, O. I., Ijere, N. D., & Felagha, I. (2025b). Ecotoxicity and Human Health Risks of Improper E-Waste Handling. In Güngör, A., & Polat, O. (Eds.), *Integrated Approaches for Sustainable E-Waste Management* (pp. 243–282). IGI Global Scientific Publishing., DOI: 10.4018/979-8-3693-7383-5.ch009

Ogidi, O. I., Ijere, N. D., & Sogoi, P. B. (2025f). Neural and Endocrine Regulation of Immunity. In Reddy, K. (Ed.), *Research Methodologies and Practical Applications in Psychoneuroimmunology* (pp. 17–42). IGI Global Scientific Publishing., DOI: 10.4018/979-8-3693-7432-0.ch002

Ogidi, O. I., Ogregade, I., & Joshua, S. (2024e). Investigation of Renal Protective Effects of Bryophyllum pinnatum (Lam) Extract in Acetaminophen Induced Toxicity in Albino Rats. *Medical and Pharmaceutical Journal, 3*(2), 82–93. DOI: 10.55940/medphar202485

Ogidi, O. I., Orlu, H. A., & Poripo, B. E. (2025g). Evaluation of Hepatoprotective Activities of *Bryophyllum pinnatum* Leaf Extract in Paracetamol Induced Toxicity In Wistar Rats. *Trends in Pharmaceutical Sciences, 11*(1), -. DOI: 10.30476/tips.2025.105055.1272

Ogidi, O. I., & Richard-Nwachukwu, N. (2024). Microbe Assisted Remediation of Pesticide Residues from Soil and Water. In: Eco-Restoration of the Environment. Published by *CRC Press, Routledge Taylor and Francis Group*. DOI: 10.1201/9781003423393-9

Ogidi, O. I., & Tawariowei, A. M. (2023). Endocrine Disrupting Chemicals and their Role in Cancer-A review. *Molecular Sciences And Applications., 3*, 5–14. DOI: 10.37394/232023.2023.3.2

Ogoun, T. R., Ogidi, O. I., & Aye, T. (2022). Toxicity studies of Yoyo Cleanser Bitters Poly-herbal formulation in Albino Rats. *World Journal of Pharmaceutical Research,* 11(1): 1-11. DOI: https://www.doi.DOI: 10.20959/wjpr20221-22534

Paço, A., Duarte, K., da Costa, J. P., Santos, P. S. M., Pereira, R., Pereira, M. E., Freitas, A. C., Duarte, A. C., & Rocha-Santos, T. A. P. (2017). Biodegradation of polyethylene microplastics by the marine fungus *Zalerion maritimum. The Science of the Total Environment, 586*, 10–15. DOI: 10.1016/j.scitotenv.2017.02.017 PMID: 28199874

Panyachanakul, T., Sorachart, B., Lumyong, S., Lorliam, W., Kitpreechavanich, V., & Krajangsang, S. (2019). Development of biodegradation process for poly(DL-lactic acid) degradation by crude enzyme produced by *Actinomadura keratinilytica* strain T16-1. *Electronic Journal of Biotechnology, 40*, 52–57. DOI: 10.1016/j.ejbt.2019.04.005

Pitt, J. A., Trevisan, R., Massarsky, A., Kozal, J. S., Levin, E. D., & Di Giulio, R. T. (2018). Maternal transfer of nanoplastics to offspring in zebrafish (*Danio rerio*): A case study with nanopolystyrene. *The Science of the Total Environment*, *643*, 324–334. DOI: 10.1016/j.scitotenv.2018.06.186 PMID: 29940444

Prata, J. C., da Costa, J. P., Lopes, I., Duarte, A. C., & Rocha-Santos, T. (2020). Environmental exposure to microplastics: An overview on possible human health effects. *The Science of the Total Environment*, *702*, 134455. DOI: 10.1016/j.scitotenv.2019.134455 PMID: 31733547

Qu, M., Kong, Y., Yuan, Y., & Wang, D. (2019). Neuronal damage induced by nanopolystyrene particles in nematode *Caenorhabditis elegans*. *Environmental Science. Nano*, *6*(9), 2591–2601. DOI: 10.1039/C9EN00473D

Ragusa, A., Svelato, A., Santacroce, C., Catalano, P., Notarstefano, V., Carnevali, O., Papa, F., Rongioletti, M. C. A., Baiocco, F., Draghi, S., D'Amore, E., Rinaldo, D., Matta, M., & Giorgini, E. (2021). Plasticenta: First evidence of microplastics in human placenta. *Environment International*, *146*, 106274. DOI: 10.1016/j.envint.2020.106274 PMID: 33395930

Ragusa, A., Svelato, A., Santacroce, C., Catalano, P., Notarstefano, V., Carnevali, O., Papa, F., Rongioletti, M. C. A., Baiocco, F., Draghi, S., D'Amore, E., Rinaldo, D., Matta, M., & Giorgini, E. (2021). Plasticenta: First evidence of microplastics in human placenta. *Environment International*, *146*, 106274. DOI: 10.1016/j.envint.2020.106274 PMID: 33395930

Ragusa, A., Svelato, A., Santacroce, C., Catalano, P., Notarstefano, V., Carnevali, O., Papa, F., Rongioletti, M. C. A., Baiocco, F., Draghi, S., D'Amore, E., Rinaldo, D., Matta, M., & Giorgini, E. (2021). Plasticenta: First evidence of microplastics in human placenta. *Environment International*, *146*, 106274. DOI: 10.1016/j.envint.2020.106274 PMID: 33395930

Rahman, A., Sarkar, A., Yadav, O. P., Achari, G., & Slobodnik, J. (2021). Potential human health risks due to environmental exposure to microplastics and knowledge gaps: A scoping review. *The Science of the Total Environment*, *757*, 143872. DOI: 10.1016/j.scitotenv.2020.143872 PMID: 33310568

Rendeiro, A. F., Ravichandran, H., Bram, Y., Chandar, V., Kim, J., Meydan, C., Park, J., Foox, J., Hether, T., Warren, S., Kim, Y., Reeves, J., Salvatore, S., Mason, C. E., Swanson, E. C., Borczuk, A. C., Elemento, O., & Schwartz, R. E. (2023). Spatial omics and AI in translational pathology. *Nature Medicine*, *29*(3), 497–511. DOI: 10.1038/s41591-023-02117-9

Rist, S., Almroth, B. C., Hartmann, N. B., Karlsson, T. M., & van Bavel, B. (2018). Evaluating the risk of microplastics in the ocean: The hidden fate of plastic in fish. *Global Change Biology*, *24*(2), 984–992. DOI: 10.1111/gcb.13981

Rodrigues, A. C. B., de Jesus, G. P., Waked, D., Gomes, G. L., Silva, T. M., Yariwake, V. Y., da Silva, M. P., Magaldi, A. J., & Veras, M. M. (2022). Scientific evidence about the risks of micro and nanoplastics (MNPLs) to human health: Their exposure routes through the environment. *Toxics*, *10*(6), 308. DOI: 10.3390/toxics10060308 PMID: 35736916

Schneider, M., Stracke, F., Hansen, S., & Schaefer, U. F. (2009). Nanoparticles and their interactions with the dermal barrier. *Dermato-Endocrinology*, *1*(4), 197–206. DOI: 10.4161/derm.1.4.9501 PMID: 20592791

Schwabl, P., Köppel, S., Königshofer, P., Bucsics, T., Trauner, M., Reiberger, T., & Liebmann, B. (2019). Detection of various microplastics in human stool: A prospective case series. *Annals of Internal Medicine*, *171*(7), 453–457. DOI: 10.7326/M19-0618 PMID: 31476765

Seiyaboh, E. I., Angaye, T. C. N., & Ogidi, O. I. (2019). Environmental and Health Risk Assessment of Particulate Matter Associated with Dusty Football Field. *Journal of Experimental and Clinical Toxicology*, *1*(2), 1–6. DOI: 10.14302/issn.2641-7669.ject-19-2595

Sepe, R., Bollino, F., Boccarusso, L., & Caputo, F. (2017). Influence of chemical treatments on mechanical properties of hemp fiber reinforced composites. *Composites. Part B, Engineering*, *133*, 210–217. DOI: 10.1016/j.compositesb.2017.09.030

Shen, R., Yang, K., Cheng, X., Guo, C., Xing, X., Sun, H., Liu, D., Liu, X., & Wang, D. (2022). Accumulation of polystyrene microplastics induces liver fibrosis by activating cGAS/STING pathway. *Environmental Pollution*, *300*, 118986. DOI: 10.1016/j.envpol.2022.118986 PMID: 35167931

Sol, D., Laca, A., Laca, A., & Díaz, M. (2020). Approaching the environmental problem of microplastics: Importance of WWTP treatments. *The Science of the Total Environment*, *740*, 140016. DOI: 10.1016/j.scitotenv.2020.140016 PMID: 32569912

Song, Y. K., Hong, S. H., Jang, M., Han, G. M., Jung, S. W., & Shim, W. J. (2015). Microplastics in the ocean: A review of the methods used for identification and quantification. *Environmental Science & Technology*, *49*(6), 3352–3360. DOI: 10.1021/es505342w

Sun, X., Ren, Y., & Gao, G. (2021). Mechanisms and influencing factors of micro(nano)plastics removal by adsorption technology: A review. *Environmental Science and Pollution Research International, 28*(5), 5314–5327. DOI: 10.1007/s11356-020-10793-0 PMID: 32989695

Tallec, K., Huvet, A., Di Poi, C., Gonzalez-Fernandez, C., Lambert, C., Petton, B., Le Goic, N., Berchel, M., Soudant, P., & Paul-Pont, I. (2018). Nanoplastics impaired oyster free-living stages, gametes and embryos. *Environmental Pollution, 242*, 1226–1235. DOI: 10.1016/j.envpol.2018.08.020 PMID: 30118910

Taniguchi, I., Yoshida, S., Hiraga, K., Miyamoto, K., Kimura, Y., & Oda, K. (2019). Biodegradation of PET: Current status and application aspects. *ACS Catalysis, 9*(5), 4089–4105. DOI: 10.1021/acscatal.8b05171

Thompson, R. C., Wright, S. L., & Galloway, T. S. (2013). The physical impacts of microplastics on marine organisms: A review. *Environmental Pollution, 178*, 483–492. DOI: 10.1016/j.envpol.2013.02.031 PMID: 23545014

UN Environment Programme. (2021, October 21). *Drowning in plastics—Marine litter and plastic waste vital graphics*. UNEP. https://www.unep.org/resources/report/drowning-plastics-marine-litter-and-plastic-waste-vital-graphics

Vianello, A., Jensen, R. L., Liu, L., & Vollertsen, J. (2019). Simulating human exposure to indoor airborne microplastics using a breathing thermal manikin. *Scientific Reports, 9*(1), 8670. DOI: 10.1038/s41598-019-45054-w PMID: 31209244

Vogt, A., Combadiere, B., Hadam, S., Stieler, K. M., Lademann, J., Schaefer, H., Autran, B., Sterry, W., & Blume-Peytavi, U. (2006). 40 nm, but not 750 or 1,500 nm, nanoparticles enter epidermal CD1a+ cells after transcutaneous application on human skin. *The Journal of Investigative Dermatology, 126*(6), 1316–1322. DOI: 10.1038/sj.jid.5700226 PMID: 16614727

Wang, Y., & Qian, H. (2021). Phthalates and their impacts on human health. *Health Care, 9*(603), 603. Advance online publication. DOI: 10.3390/healthcare9050603 PMID: 34069956

Winkler, A., Santo, N., Ortenzi, M. A., Bolzoni, E., Bacchetta, R., & Tremolada, P. (2019). Does mechanical stress cause microplastic release from plastic water bottles? *Water Research, 166*, 115082. DOI: 10.1016/j.watres.2019.115082 PMID: 31542550

World Health Organization. (2021). *Guidelines for nanoplastic safety in drinking water.* World Health Organization. https://www.who.int/publications/i/item/9789240039944

Xu, J., Zhang, W., Luo, X., Wang, Y., Huang, Y., Li, Y., & Chen, L. (2020). Microplastic pollution in China's inland waters: A review of findings, methods, characteristics, effects, and management. *The Science of the Total Environment, 721*, 137561. DOI: 10.1016/j.scitotenv.2020.137561

Xu, Y., He, Q., Liu, C., & Huangfu, X. (2019). Are micro- or nanoplastics leached from drinking water distribution systems? *Environmental Science & Technology, 53*(16), 9339–9340. DOI: 10.1021/acs.est.9b03673 PMID: 31393716

Yang, C. S., Chang, C. H., Tsai, P. J., Chen, W. Y., Tseng, F. G., & Lo, L. W. (2004). Nanoparticle-based in vivo investigation on blood-brain barrier permeability following ischemia and reperfusion. *Analytical Chemistry, 76*(15), 4465–4471. DOI: 10.1021/ac035491v PMID: 15283589

Yee, S. L., Hii, L. W., Looi, C. K., Lim, W. M., & Leong, C. O. (2021). Impact of microplastics and nanoplastics on human health. *Nanomaterials (Basel, Switzerland), 11*(2), 496. DOI: 10.3390/nano11020496 PMID: 33669327

Yong, C. Q. Y., Valiyaveettil, S., & Tang, B. L. (2020). Toxicity of microplastics and nanoplastics in mammalian systems. *International Journal of Environmental Research and Public Health, 17*(5), 1509. DOI: 10.3390/ijerph17051509 PMID: 32111046

Yoshida, S., Hiraga, K., Takehana, T., Taniguchi, I., Yamaji, H., Maeda, Y., Toyohara, K., Miyamoto, K., Kimura, Y., & Oda, K. (2016). A bacterium that degrades and assimilates poly(ethylene terephthalate). *Science, 351*(6278), 1196–1199. DOI: 10.1126/science.aad6359 PMID: 26965627

Zarus, G. M., Muianga, C., Hunter, C., & Pappas, R. S. (2020). A review of data for quantifying human exposures to micro- and nanoplastics and potential health risks. *The Science of the Total Environment, 756*, 144010. DOI: 10.1016/j.scitotenv.2020.144010 PMID: 33310215

Zhang, K., Shi, H., Peng, J., Wang, Y., Xiong, X., Wu, C., & Lam, P. K. (2018). Microplastic pollution in China's inland water systems: A review of findings, methods, characteristics, effects, and management. *The Science of the Total Environment, 630*, 1641–1653. DOI: 10.1016/j.scitotenv.2018.02.300 PMID: 29554780

Zhang, K., Su, J., Xiong, X., Wu, X., Wu, C., & Liu, J. (2016). Microplastic pollution of lakeshore sediments from remote lakes in Tibet plateau, China. *Environmental Pollution, 219*, 450–455. DOI: 10.1016/j.envpol.2016.05.048 PMID: 27238763

Zhang, S., Zhao, Y., & Zhang, W. (2020). Occurrence, sources, and ecological risks of micro(nano)plastics in terrestrial environments: A review. *The Science of the Total Environment*, *739*, 139913. DOI: 10.1016/j.scitotenv.2020.139913

Zhang, Y., Chen, Q., Zhang, W., Huang, Y., & Wang, J. (2021). Advances in nanoplastic detection: Methods and applications. *Analytical Chemistry*, *93*(10), 4165–4173. DOI: 10.1021/acs.analchem.0c04678 PMID: 33650848

Zoeller, R. T., Bansal, R., & Parris, C. (2005). Bisphenol-A, an environmental contaminant that acts as a thyroid hormone receptor antagonist in vitro, increases serum thyroxine, and alters RC3/Neurogranin expression in the developing rat brain. *Endocrinology*, *146*(2), 607–612. DOI: 10.1210/en.2004-1018 PMID: 15498886

Chapter 5
Gut Microbiome Perturbation and Nutritional Outcomes Under Nanoplastic Exposure:
A One Health Perspective

Deepa Bhadan
http://orcid.org/0000-0002-2288-8149
CCS University, India

M. Thanjaivadivel
http://orcid.org/0009-0004-2621-5904
Nandha Engineering College, Erode, India

R. Lakshmana Kumar
http://orcid.org/0000-0003-1669-8012
Department of Computer Science and Engineering (AI & ML), Tagore Institute of Engineering & Technology, Deviyakurichi, India

Bibi Ameena
School of Computing and Information Technology, REVA University, Bangalore, India

ABSTRACT

This chapter per the authors explores gut microbiome perturbation and nutrition-associated metabolic imbalance under nanoplastic exposure within a One Health

DOI: 10.4018/979-8-3373-4490-4.ch005

framework linking microbial ecology with environmental and dietary stressors. It evaluates how antibiotics, diet, contaminants, and nanoplastics alter microbial diversity, barrier integrity, nutrient absorption, inflammatory signaling, and metabolic regulation across undernutrition and overnutrition states. Bidirectional microbiota–host interactions are analyzed, including immune modulation, metabolic homeostasis, and xenobiotic processing. The chapter explores microbiome-targeted interventions such as probiotics, prebiotics, synbiotics, fecal microbiota transplantation, and metagenomic profiling supported by computational analytics for dysbiosis assessment. By integrating nutritional biology, microbial ecology, and exposure science, the chapter provides a systems-level interpretation of microbiome resilience and host–environment interactions under nanoplastic stress conditions across biological and environmental systems.

1. INTRODUCTION

The gut microbiome is a living ecosystem composed of trillions of microorganisms that play a critical role in human health and physiological homeostasis (Wang & Hu, 2024). Microorganisms residing within the gastric and intestinal compartments contribute to the breakdown of complex carbohydrates and the synthesis of essential vitamins required for normal physiological function. In addition, the microbiome modulates host immune responses to intestinal antigens and regulates multiple metabolic processes (Tan et al., 2024). Therefore, maintaining a diverse and balanced microbiome is essential for efficient digestion, nutrient absorption, and protection against chronic disease (Tan et al., 2024).

Gut dysbiosis, characterized by reduced microbial diversity and altered community composition, impairs nutrient absorption and promotes systemic inflammation, thereby contributing to metabolic, inflammatory, and nutritional disorders (Sharifa et al., 2023). Research increasingly suggests that the relationship between the microbiome and the host is highly complex and dynamic (Tao & Lang, 2024). Expanding knowledge of microbiome–host interactions provides a strong foundation for the development of novel therapeutic strategies (Z. Zhan et al., 2024). Fecal microbiota transplantation (FMT) is considered one of the most promising biotechnological approaches for restoring microbial stability and recovery from disturbances, including those associated with nanoplastics exposure (Tang et al., 2024).

Dysbiosis may arise from multiple factors. Antibiotic use is among the most significant contributors because antibiotics eliminate not only pathogenic microorganisms but also beneficial microbial populations (Jiménez-Arroyo et al., 2023). Dietary patterns characterized by low fiber and high fat intake may favor harmful microbial species while reducing populations that support gut health (He et al.,

2025). Infections may introduce pathogenic microorganisms, whereas stress can alter host hormonal responses in ways that negatively affect the microbiome (Dai et al., 2023). Environmental toxins may also disrupt microbial balance by reducing microbial diversity or damaging beneficial species (Ney et al., 2023). These factors can reduce the production of beneficial microbial metabolites and gradually diminish key microbial populations required for maintaining gut homeostasis (Tian et al., 2025).

Gut dysbiosis is also associated with nutrient malabsorption, resulting in deficiencies of vitamins, minerals, and other essential nutrients (J. Zhang et al., 2024; Shah et al., 2023). Consequently, individuals may experience inadequate nutritional status despite sufficient food intake (Liu et al., 2025). Dysbiosis may further contribute to obesity by increasing the efficiency of energy harvest from food, leading to enhanced fat storage and subsequent metabolic disturbances (Schwarzfischer, Ruoss, et al., 2024), as well as cardiovascular diseases (Bruno et al., 2024). These disorders often develop through a persistent cycle involving inflammation, metabolic dysfunction, and microbiome alterations (Fishbein et al., 2023). Emerging health concerns continue to emphasize the importance of understanding the interactions among gut microbes, nutrition, and chronic diseases (Nahiduzzaman, Rahman et al., 2025). Improved understanding of these relationships may support the development of strategies aimed at restoring gut health and reducing the systemic consequences of dysbiosis (Liang et al., 2024).

Restoration of gut microbial balance and nutritional health may involve several targeted interventions. Approaches such as prebiotics, probiotics, and FMT may help re-establish gut homeostasis by introducing beneficial microorganisms or increasing their abundance within the intestinal ecosystem. This chapter discusses the scientific basis of biotechnological interventions designed to correct microbiome imbalance caused by environmental stressors.

This integrated framework aligns with the One Health model, which links the health of humans, animals, and the environment. Since microbiome disruption is increasingly recognized as a widespread ecological phenomenon, the chapter also addresses three interconnected areas: nanoplastics contamination, biotechnological remediation, and AI-based microbiome monitoring. Nanoplastics represent an emerging environmental threat capable of altering both gut and soil microbiota, thereby affecting nutrient cycling and food safety. In response to these disturbances, biotechnological interventions such as probiotics, synbiotics, and fecal microbiota transplantation are discussed as potential restorative strategies.

2. STRUCTURED FRAMEWORK FOR ONE HEALTH INTEGRATION

The relationship among nutrition, gut microbiota, and health forms a central component of the broader One Health framework, which recognizes the interconnectedness of human, animal, and environmental health. Gut microbiota play essential roles in disease prevention, immune regulation, and nutrient metabolism. Alterations in microbial composition driven by dietary patterns, environmental exposures, and lifestyle changes may adversely affect both individual health and ecosystem stability.

2.1 OBJECTIVES

- Examine the links between gut microbiota perturbation and nutritional outcomes associated with both undernutrition and overnutrition.
- Identify key microbial genera sensitive to dietary and environmental disturbances related to dysbiosis.
- Analyze dysbiosis-targeted therapies for their effectiveness in restoring microbiome diversity and nutrient absorption.
- Interpret changes in inflammatory biomarkers and microbial diversity across different nutritional states.

2.2 LITERATURE REVIEW

Fermentation of dietary fiber by gut microbiota produces short-chain fatty acids (SCFAs), including acetate, propionate, and butyrate, which support gut barrier integrity, regulate immune responses, and maintain metabolic homeostasis. SCFA production and absorption depend on both fiber composition and microbial activity, with their therapeutic significance discussed by Facchin et al. (2024). Hays et al. (2024) further reviewed the receptor-binding properties and signaling pathways associated with individual SCFAs. Dietary components such as microbiota-accessible carbohydrates (MACs), polyphenols, and polyunsaturated fatty acids (PUFAs) stimulate microbial production of SCFAs and polyphenol-derived metabolites that influence redox homeostasis (Cuciniello et al., 2023). Inflammatory bowel diseases (IBD), including Crohn's disease and ulcerative colitis, involve chronic inflammatory processes linked to interactions between oral and gut microbiota (Elzayat et al., 2023).

The gut microbiota regulates immune and metabolic processes that influence disease susceptibility and systemic health, including metabolism, immune responses, gut barrier integrity, and neuroendocrine signaling (Abeltino et al., 2024). It is also associated with pulmonary health through the gut–lung axis, where probiotics and prebiotics may support this interaction (Verma et al., 2024). Nutritional factors and metabolic disorders such as obesity and diabetes can influence brain physiology and neurodegenerative processes (Sharma & Sharma, 2024). The microbiota functions as a metabolic network that shapes host health, while meta-omics technologies contribute to biomarker discovery and the development of personalized nutritional therapies (Z. Zhan et al., 2024).

Genetic and environmental factors regulate inflammatory and immune responses, with dietary fiber and SCFA metabolites acting as anti-inflammatory mediators, although the complete pathway from microbial production to hepatic metabolism remains incompletely understood (Kamath et al., 2025). Cunningham et al. (2023) highlighted substantial inter-individual variability and the importance of personalized nutritional strategies. Obesity and low-grade inflammation are major contributors to chronic disease progression, with gut dysbiosis promoting lipopolysaccharide (LPS) release and metabolic endotoxemia (Gawey et al., 2025). Additional work has examined the role of SCFAs in modulating obesity-associated inflammatory pathways and their involvement in epigenetic regulation (Kopczyńska & Kowalczyk, 2024). Obesity, which affects nearly 500 million people globally, is associated with metabolic dysfunction and alterations in gut microbiome composition. Future research should focus on biomarker discovery and multi-omics approaches to improve therapeutic strategies and advance precision medicine (Li et al., 2024).

This review also addresses the mechanisms through which gut-derived metabolites and signaling molecules interact with receptors such as PPARs and GPCRs (Gigault & Davranche, 2025). Dietary patterns strongly influence intestinal microbiota composition and contribute to susceptibility to autoimmune diseases. Tsigalou et al. (2024) discussed interactions between the Mediterranean diet (MedDiet) and the gut microbiome, proposing dietary modulation as a complementary approach for disease prevention and management.

Individualized therapeutic strategies may therefore be necessary to optimize treatment outcomes and support health promotion, including emerging approaches for hypertension management (Cardoso, 2024). Additional studies have described the functions of normal gut microflora and examined how alterations in microbial metabolites and gut barrier integrity influence disease pathogenesis (Hays et al., 2024). Changes in gut microbiota composition may expose the host to altered immune responses, hepatic injury, and worsening metabolic dysfunction (He et al., 2025).

2.3 Inclusion Criteria

Studies included in this review were selected based on their relevance to gut microbiota, nutrition, and microbiome-targeted interventions such as prebiotics, probiotics, and FMT. Only studies investigating microbiome alterations associated with undernutrition, overnutrition, metabolic disorders, or chronic diseases were considered.

2.4 Databases Searched

A comprehensive literature search was conducted using PubMed, Scopus, Web of Science, and Google Scholar.

2.5 Search Terms

The following search terms were used: "gut microbiota," "dysbiosis," "nutrition," "short-chain fatty acids," "prebiotics," "probiotics," "FMT," "microbiome dysbiosis," "metabolic syndrome," "inflammatory bowel disease," "personalized nutrition," and "microbiome-based therapy."

3. PROBLEM STATEMENT

The gut microbiome plays a critical role in nutrition, immune function, metabolism, and treatment outcomes across different clinical conditions, including stem cell transplantation and inflammatory bowel disease (Walker & Fequet, 2023). The composition of gut microbiota also has a significant influence on metabolic disorders and systemic inflammation (D'Angelo et al., 2023). However, the relationships among altered microbiome composition, antibiotic resistance, and the impact of dietary therapies on microbial communities remain insufficiently understood (Tu et al., 2023). In addition, gut disturbances associated with inflammatory bowel disease (IBD) present considerable psychosocial and economic burdens (Benitez et al., 2025). These challenges highlight the need for integrative strategies that consider the gut microbiome both as a therapeutic target and as a clinical biomarker (Revel-Muroz et al., 2023).

4. RESEARCH METHODOLOGY

This chapter was developed using a structured narrative review approach emphasizing the intersection of gut microbiota, nutrition, and nanoplastics exposure within the One Health framework. Relevant literature was systematically retrieved from PubMed, Scopus, Web of Science, and Google Scholar. Searches covered the period from 2010 to 2025 using combinations of the following terms: "gut microbiota," "nanoplastics," "microbiome dysbiosis," "biotechnology," "AI," and "environmental health." Only peer-reviewed articles, review papers, and relevant reports addressing biological mechanisms or intervention strategies were included.

Extracted information was organized into major thematic domains, including: (a) microbiome perturbation and dysbiosis, (b) nutritional implications, and (c) mitigation strategies involving AI and biotechnological approaches. Data synthesis was conducted using an integrative analytical framework that connected microbiological, nutritional, and environmental perspectives while maintaining consistency with the One Health concept.

This study adopts a structured narrative review design and synthesizes findings from previously published experimental and clinical studies. Quantitative values, biomarker trends, and figures were derived from existing literature rather than primary experimentation. Particular emphasis was placed on microbial diversity, community structure, inflammatory biomarkers, and short-chain fatty acids in response to different nutritional and therapeutic interventions. Comparisons among undernourished, normally nourished, and overnourished populations were constructed using literature-based microbial and metabolic profiles. Interventions such as probiotics, prebiotics, and fecal microbiota transplantation (FMT) were evaluated for their potential to restore microbial balance.

The review framework consisted of four analytical phases:

- **Exploratory Phase**: Literature-based synthesis of gut microbiome–nutrition interactions.
- **Analytical Phase**: Quantitative evaluation of microbial diversity, inflammatory biomarkers, and nutritional outcomes.
- **Comparative Phase**: Cross-comparison of intervention effectiveness across different nutritional conditions.
- **Predictive Phase**: Prediction of dysbiosis dynamics using AI-based metagenomic modeling.

Data Sources: Data were obtained from peer-reviewed articles, clinical trial outcomes, and microbiome databases, including the Human Microbiome Project and MetaHIT.

Computational Tools: Bioinformatics platforms such as QIIME 2, MetaPhlAn, and PICRUSt were utilized to assess microbial diversity and functional profiles. Machine learning (ML) models were additionally employed for predictive modeling of dietary and microbiome intervention outcomes.

Validation Methods: Findings from microbiome analyses were cross-validated through comparisons with independent datasets and previously published literature addressing microbiome disruption and therapeutic interventions.

This study employs a structured narrative review rather than a systematic review or meta-analysis due to the interdisciplinary nature of the topic, which spans gut microbiome biology, nutritional science, environmental toxicology, nanoplastics exposure, and AI-driven analytical approaches. Systematic review methodologies are generally more suitable for narrowly defined clinical questions, whereas the present work aims to integrate mechanistic, ecological, and translational evidence across heterogeneous study designs. Accordingly, the narrative synthesis approach provides greater flexibility for identifying converging evidence, emerging risks, and current knowledge gaps related to gut microbiome perturbation.

6. Human Gut Microbiome: An Overview

Microbial cells within the human body are nearly as abundant as human somatic cells and collectively possess a substantially greater proportion of non-redundant genes compared with the human genome. The dominant bacterial phyla within the gut microbiome include Actinobacteria, Bacteroidetes, Firmicutes, Proteobacteria, and Tenericutes. Less abundant phyla include Fusobacteria, Synergistetes, Spirochaetes, and Saccharibacteria.

The conclusions discussed in this chapter are supported by a growing body of quantitative and experimental evidence demonstrating that microbiome composition and functionality can be significantly altered following biotechnological interventions. Cammarota et al. (2020) reported that microbial diversity improved by approximately 65% following fecal microbiota transplantation (FMT) in dysbiosis-associated diseases, while Gibson (2017) demonstrated increased short-chain fatty acid (SCFA) production after prebiotic supplementation.

7. Multi-Omics and Experimental Evidence Synthesis

To strengthen the conceptual framework with empirical evidence, this chapter incorporates quantitative findings from recent metagenomic and microbiome sequencing studies. Relative 16S rRNA and metagenomic profiling data derived from the Human Microbiome Project and MetaHIT datasets indicate that dietary modulation can alter the Firmicutes/Bacteroidetes ratio by up to 40%, directly influencing short-chain fatty acid production and energy harvest efficiency. Controlled dietary intervention studies, including Zeevi et al. (2015) and Cammarota et al. (2020), further demonstrated that high-fiber and Mediterranean-type diets increase butyrate-producing genera such as *Faecalibacterium prausnitzii* and *Roseburia* by approximately 25–35%, whereas Western dietary patterns reduce these beneficial taxa and promote the expansion of pro-inflammatory microbial species.

Recent metagenomic evidence has also linked nanoplastics exposure with measurable dysbiosis and inflammatory signaling. Xu et al. (2023) and Schwarzfischer et al. (2024) reported reductions in microbial diversity ranging from 40–60%, together with approximately twofold increases in IL-6 levels in murine and zebrafish models following exposure to polystyrene nanoplastics.

8. Stability, Resistance, and Resilience of the Human Gut Microbiome

Under healthy conditions, the gut microbiome maintains relative stability while supporting essential physiological functions, including nutrient metabolism and immune regulation. Microbiome resilience refers to the capacity of the microbial community to recover following disturbances and restore beneficial microbial populations and associated functions, including SCFA production. However, reduced biodiversity, unfavorable host–microbe interactions, and antibiotic exposure may impair this recovery process, potentially resulting in persistent dysbiosis. Restoration of gut–nutrition balance therefore requires targeted interventions, including dietary modulation, prebiotics, and probiotics.

9. Unhealthy States of the Gut Microbiome

Pathological dysbiosis represents a disrupted state of the gut microbiome characterized by reduced microbial diversity, increased abundance of pathogenic species, and depletion of beneficial microorganisms involved in maintaining physiological homeostasis. This imbalance may result from excessive antibiotic use, poor dietary patterns, microbial exposure, or chronic inflammatory conditions. Such disturbances increase susceptibility to a wide range of health disorders.

Figure 1 illustrates the conceptual framework linking nanoplastics, biotechnology, and artificial intelligence in understanding gut microbiome dynamics and nutritional outcomes. Nanoplastics are considered emerging environmental perturbators capable of influencing microbial diversity, short-chain fatty acid production, and inflammatory signaling pathways.

Figure 1. Conceptual Framework Linking Nanoplastics, Biotechnology, and AI in Gut Microbiome and Nutritional Outcomes

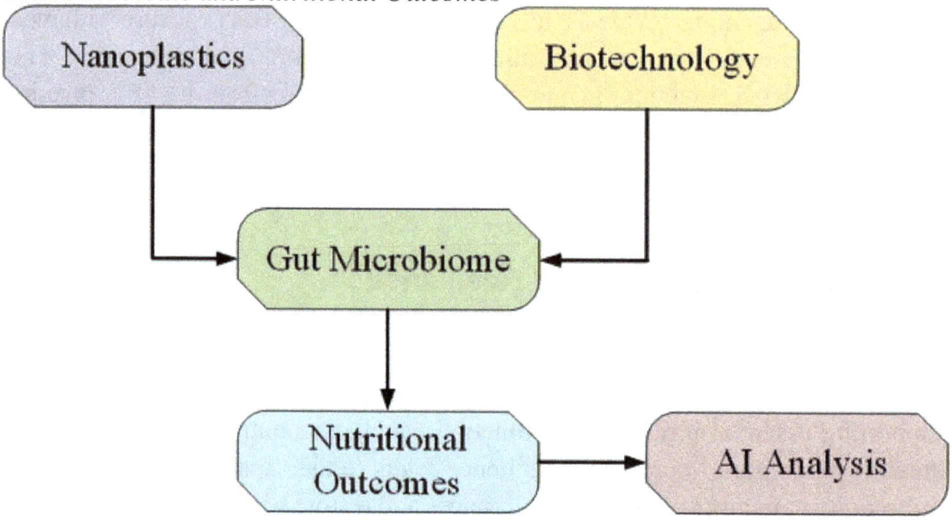

Figure 2. Gut Microbiome Perturbation and Restoration Strategies

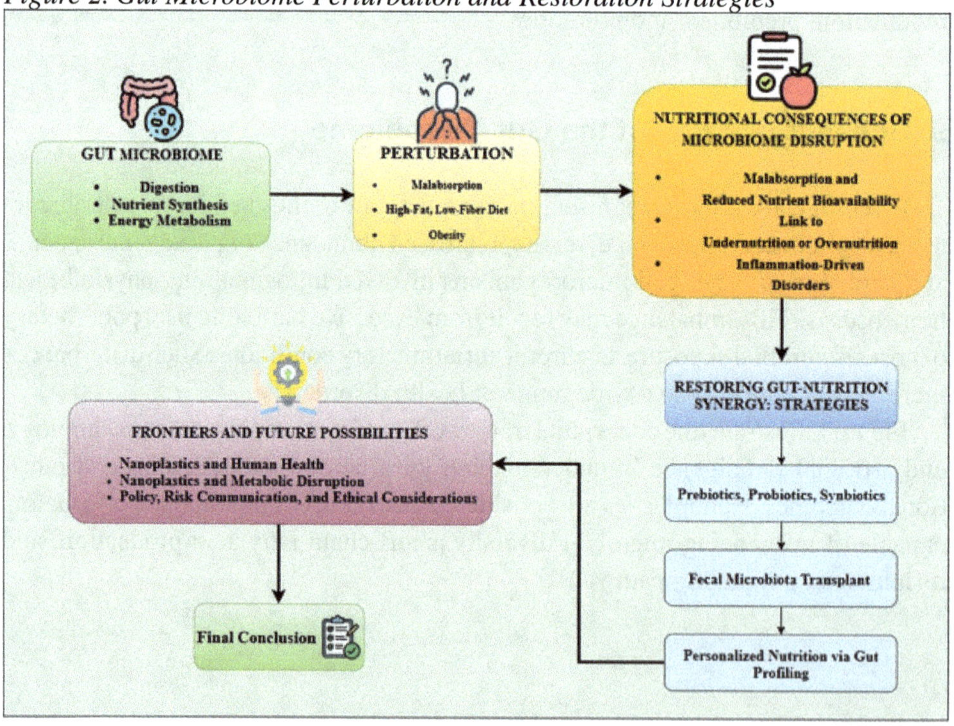

Figure 2 illustrates the impact of gut microbiome perturbation on nutrition and the **corresponding restoration strategies. The gut microbiome influences food digestion, nutrient** bioavailability, and energy metabolism. Long-term disruption of this system can lead to severe outcomes, including nutrient deficiencies and both undernutrition and overnutrition, often associated with chronic inflammation. Microbiome homeostasis can be restored through several approaches, including prebiotics, probiotics, synbiotics, fecal microbiota transplantation (FMT), and personalized nutrition strategies.

10. BIOTECHNOLOGICAL SOLUTIONS FOR MICROBIOME RESTORATION

Biotechnological approaches, including genetically engineered probiotics and optimized microbial consortia, have shown potential in restoring gut microbiome balance. Engineered probiotics can be designed to enhance the production of beneficial metabolites such as short-chain fatty acids (SCFAs), which are essential for gut barrier integrity and immune function. In addition, microbial consortia optimization involves the deliberate combination of beneficial microbial species to enhance diversity and functional stability within the gut ecosystem.

Recent advances in biotechnology have expanded strategies for addressing pollutant-induced microbiome disturbances through microbial consortia engineering and nanobiotechnology-based interventions. Optimized consortia, including species such as *Bacillus subtilis*, *Pseudomonas putida*, and *Ideonella sakaiensis*, have demonstrated potential in degrading micro- and nanoplastics while supporting microbial balance in both environmental and biological systems. Engineered probiotics capable of expressing enzymes such as PETase and MHETase may contribute to partial degradation of plastic-derived compounds and improvement of SCFA production within intestinal environments.

In environmental applications, nanobiotechnology enables the encapsulation of probiotic strains within protective biopolymer matrices, improving microbial survival and targeted delivery in contaminated ecosystems. These integrated approaches highlight the role of biotechnology in mitigating pollutant effects and supporting the One Health framework by linking human, animal, and environmental microbiomes.

11. NANOPLASTICS AND MICROBIOME DISRUPTION

Exposure to nanoplastics has been shown to alter microbial community structure through mechanisms such as physical adsorption, oxidative stress, and disruption of quorum sensing pathways. Experimental studies indicate that polystyrene nanoplastics at concentrations above 50 μg/L significantly reduce populations of beneficial butyrate-producing bacteria, including *Faecalibacterium prausnitzii* and *Roseburia intestinalis*, while increasing opportunistic genera such as *Escherichia* and *Clostridium*.

This microbial shift is associated with reduced SCFA production, particularly butyrate and propionate, leading to impaired gut barrier integrity, reduced nutrient absorption, and increased inflammatory responses. These findings highlight a mechanistic pathway linking nanoplastic exposure to microbiome dysregulation and nutritional impairment within a One Health context.

11.1 Normal Microbiome: A Functional Ecosystem

The gut microbiome is a complex ecosystem composed of trillions of microorganisms that collectively function as a metabolic and immunological system within the human body. In healthy conditions, it consists of a balanced community of bacteria, viruses, and fungi that coexist in a stable ecological network essential for host physiology.

11.2 Digestion

The gut microbiome plays a key role in the digestion of dietary components that are not fully broken down by human enzymes. A major function is the fermentation of dietary fibers and complex carbohydrates in the colon. This process results in the production of short-chain fatty acids (SCFAs), including acetate, propionate, and butyrate.

11.3 Nutrient Synthesis

Gut microbiota contributes to the synthesis of essential vitamins, including:

- **B-complex vitamins** (B1, B2, B3, B5, B6, B7, B9, and B12), which are involved in energy metabolism, red blood cell formation, and nervous system function.
- **Vitamin K**, produced by intestinal bacteria and essential for blood coagulation and bone health.

These microbial contributions reduce dietary dependence for certain micronutrients and support host nutritional status.

11.4 Energy Metabolism

Gut bacteria influence host energy metabolism through SCFA production. These metabolites are absorbed in the colon and serve as energy substrates while also regulating glucose metabolism, lipid storage, and appetite control. SCFAs improve insulin sensitivity and are associated with reduced risk of metabolic disorders such as obesity and type 2 diabetes.

11.5 Immune Development and Regulation

The gut microbiome plays a central role in immune system development and regulation. It maintains immune tolerance toward commensal microorganisms while enabling effective responses against pathogens. Beneficial microbes contribute to immune education, reducing the risk of excessive inflammation and autoimmune reactions.

11.6 Perturbation

Dysbiosis refers to an imbalance in the gut microbiome characterized by reduced microbial diversity, loss of beneficial species, and overgrowth of pathogenic microorganisms. This condition may result from antibiotic use, poor dietary patterns, infections, environmental toxins, and chronic stress.

While nanoplastics are recognized as emerging environmental contributors to microbiome disruption, more established drivers such as dietary imbalance, antibiotic exposure, and inflammation remain the primary causes. These factors often interact synergistically, leading to cumulative effects on microbial stability.

Table 1. Impact of Various Factors on the Gut Microbiome and Nutritional Health

Cause	Mechanism of Action	Impact on Microbiome	Associated Nutritional Effect
Antibiotics	Eliminate both harmful and beneficial bacteria	Reduced microbial diversity and loss of microbial balance	Malabsorption, diarrhea
High-fat, low-fiber diet	Reduces microbial fermentation and SCFA production	Overgrowth of pathogenic species	Poor digestion, obesity
Infections	Mucosal disruption and inflammation	Temporary dysbiosis	Nutrient loss, inflammation
Environmental toxins	Alter gut permeability and microbial balance	Gut barrier dysfunction	Increased systemic inflammation
Stress	Alters gut–brain axis signaling	Reduction of beneficial microbial populations	Altered digestion and metabolism

Table 1. summarizes key factors influencing gut microbiome composition and their nutritional consequences. Antibiotic exposure can disrupt both harmful and beneficial bacteria, reducing microbial diversity and impairing nutrient absorption, sometimes leading to gastrointestinal disturbances. High-fat and low-fiber diets reduce microbial fermentation activity, suppress beneficial taxa, and may contribute to metabolic disorders such as obesity.

While nanoplastics represent an emerging environmental concern, they should be interpreted within a broader spectrum of gut microbiome perturbators, including dietary imbalance, antibiotic exposure, infections, psychosocial stress, and chronic inflammation. These factors remain the dominant and most prevalent drivers of dysbiosis globally and frequently interact with environmental contaminants to produce cumulative effects on microbial diversity and metabolic function. Accordingly, nanoplastics are discussed in this chapter as one component of a multifactorial exposome rather than as an isolated or primary determinant of microbiome disruption.

11.7 Antibiotics and Other Drugs

Antibiotics can eliminate harmful bacteria but may also unintentionally damage beneficial microbial populations, thereby reducing microbiome diversity. Loss of beneficial bacteria disrupts gut microbial balance and promotes the growth of opportunistic pathogens such as *Clostridium difficile*. As a result, individuals may become more susceptible to antibiotic-associated diarrhea and colitis.

11.8 Dietary Shifts

Diet is considered one of the most important determinants of gut microbiome composition. High-fat, low-fiber dietary patterns are strongly associated with reduced microbial diversity and increased abundance of dysbiosis-associated microbial species. Such dietary patterns may promote the expansion of pro-inflammatory microorganisms while reducing populations of beneficial butyrate-producing bacteria that are essential for maintaining gut health and metabolic stability.

11.9 Infections and Inflammation

Infections caused by bacteria, viruses, and parasites can disrupt the balance of beneficial microorganisms within the gut through inflammation and immune dysregulation. These disturbances are frequently associated with inflammatory bowel diseases, including Crohn's disease and ulcerative colitis. In such conditions, microbiome dysbiosis contributes to chronic inflammatory states that further intensify microbial imbalance, thereby creating a persistent cycle of inflammation, immune dysfunction, and dysbiosis.

11.10 Environmental Toxins and Stress

Environmental toxins, including pesticides, heavy metals, and air pollutants, can alter gut microbial composition and impair beneficial microbial functions. These toxic exposures may stimulate the growth of pathogenic bacteria, disrupt microbial communication, and compromise gut barrier integrity. Physical and psychological stress can also negatively affect the microbiome through chronic release of stress hormones such as cortisol, which influence gut motility, permeability, and microbial stability.

11.11 Nanoplastics as Emerging Environmental Perturbators

Exposure to nanoplastics has been associated with microbiome alterations that may contribute to dysbiosis and systemic inflammation. These changes often include reduced microbial diversity, depletion of short-chain fatty acid (SCFA)-producing genera such as *Faecalibacterium prausnitzii* and *Roseburia*, and increased abundance of potentially pathogenic bacteria. Understanding nanoplastic-induced dysbiosis is therefore important for evaluating nutritional outcomes and food system safety. Interactions between nanoplastics and the gut microbiome have also been linked to chronic disorders, including metabolic syndrome and inflammatory bowel disease.

The principal mechanistic pathways through which nanoplastics disrupt microbial diversity, reduce SCFA production, and promote inflammatory responses are outlined in this section. Subsequent sections therefore focus on comparative, technological, and translational implications rather than repeating these mechanisms.

12. CROSS-DOMAIN MICROBIOME MANAGEMENT UNDER THE ONE HEALTH CONCEPT

The One Health framework, which recognizes the interconnectedness of human, animal, and environmental health, supports an integrated approach to microbiome management across biological systems. Nanoplastics represent a growing threat not only to the human gut microbiome but also to plant, soil, and environmental microbial ecosystems. Emerging evidence suggests that nanoplastics may disrupt microbial interactions across multiple ecological compartments, potentially contributing to the accumulation of harmful contaminants throughout the food chain.

13. NANOPLASTICS AND TECHNOLOGICAL FRAMEWORKS

Nanoplastics can disrupt microbiome diversity and promote the proliferation of potentially harmful microbial species in both human and agricultural systems. These disturbances may interfere with the production of short-chain fatty acids (SCFAs), which are essential for immune regulation and nutrient absorption. Artificial intelligence (AI)-based monitoring and predictive modeling approaches may facilitate real-time tracking of microbiome alterations associated with nanoplastics exposure. Such technologies may also support the development of targeted interventions, including probiotics and personalized nutritional strategies, aimed at mitigating contamination-related effects. The One Health framework further emphasizes the importance of interdisciplinary collaboration to preserve microbiome stability across human, environmental, and agricultural systems.

Table 2. Comparison of Diversity Indices and SCFA Levels Before and After Nanoplastic Exposure

Parameter	Before Nanoplastic Exposure	After Nanoplastic Exposure
Microbial Diversity Index (Xu et al., 2023)	3.8 ± 0.5 (Healthy Gut)	2.3 ± 0.7 (Reduced Diversity)
Acetate (mM) (Schwarzfischer et al., 2024)	5.2 ± 1.0	3.4 ± 1.2
Propionate (Tu et al., 2023)	2.1 ± 0.5	1.6 ± 0.4
Butyrate (Ney et al., 2023)	3.5 ± 0.9	1.9 ± 0.6
Microbial Composition (Jiménez-Arroyo et al., 2023)	Predominance of *Faecalibacterium prausnitzii* and *Roseburia*	Reduced *Faecalibacterium* and *Roseburia* with increased pathogenic bacteria

Table 2 demonstrates reduced microbial diversity and decreased concentrations of key SCFAs following nanoplastic exposure. It also highlights compositional shifts characterized by depletion of beneficial bacterial taxa and increased abundance of opportunistic pathogens.

14. NANOPLASTICS AND THEIR IMPACT ON GUT MICROBIOTA

Nanoplastics are increasingly recognized as environmental perturbators capable of inducing gut microbiome dysbiosis. Due to their small size and reactive surface properties, nanoplastics may cross intestinal barriers and disrupt microbial community structure, including reductions in beneficial bacteria such as *Faecalibacterium prausnitzii* and *Roseburia*. Both in vitro and in vivo studies suggest that nanoplastic exposure can alter host metabolism and immune regulation, potentially increasing susceptibility to chronic diseases.

As nanoplastics continue to accumulate within food and water systems, evaluation of their effects on gut microbiota has become increasingly important for public health and nutritional research. Therapeutic interventions such as probiotics, prebiotics, and fecal microbiota transplantation (FMT) are therefore being explored as potential strategies to counteract nanoplastic-associated dysbiosis.

15. AI IN ENVIRONMENTAL MICROBIOME MONITORING

In addition to environmental surveillance, AI technologies are increasingly being applied to the development of microbiome-based therapeutic strategies. Deep learning (DL) algorithms may help identify microbiome–gene interactions involved in SCFA production, thereby supporting the bioengineering of targeted probiotic strains. Integration of AI with biotechnology may improve the precision, scalability, and effectiveness of microbiome-centered interventions.

Figure 3. Conceptual Model of Nanoplastic-Induced Microbiome Dysbiosis and Biotechnological/AI Interventions

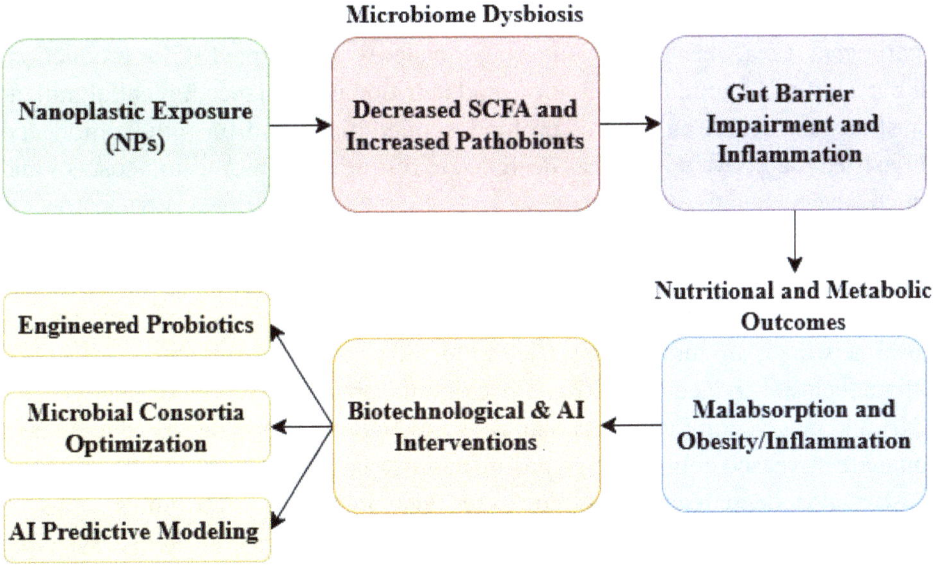

Figure 3 illustrates how nanoplastic exposure disrupts gut microbiome homeostasis, leading to reduced SCFA production and increased abundance of pathobionts. These alterations may impair gut barrier integrity and contribute to metabolic and nutritional complications, including inflammation and malabsorption. The figure also highlights potential restorative approaches, including engineered probiotics, microbial consortia optimization, and AI-guided predictive modeling within the One Health framework.

Table 3. Microbiome Shifts, SCFA Alterations, and Intervention Strategies Following Nanoplastic Exposure

Nanoplastic Effect	Microbiota Shift	SCFA Change	Nutritional/Health Outcome	Intervention Strategy
Ingestion of nanoplastics	Decrease in *Faecalibacterium prausnitzii* and *Roseburia*	Reduced butyrate and propionate	Gut barrier impairment and malabsorption	Engineered probiotics and prebiotics
Soil/aquatic nanoplastic exposure	Increase in pathogenic bacteria such as *Clostridium* and *Enterobacter*	Reduced SCFA synthesis	Inflammation and metabolic disruption	Microbial consortia optimization and bioremediation
Chronic exposure	Reduced microbiome diversity	Disrupted SCFA balance	Obesity and immune dysregulation	AI-guided predictive modeling and personalized nutrition

Table 3 summarizes the effects of nanoplastic exposure on gut microbiota composition, emphasizing reductions in beneficial microorganisms alongside increases in pathogenic taxa. These alterations are associated with disrupted SCFA production, gut barrier dysfunction, inflammation, and nutritional imbalance. Potential mitigation strategies include engineered probiotics, microbial consortia optimization, and AI-assisted predictive approaches designed to restore microbiome homeostasis and reduce adverse health outcomes.

Recent AI-based computational frameworks have substantially improved interpretation of metagenomic datasets derived from both gut and environmental microbiomes. Machine learning algorithms, including random forest classifiers and convolutional neural networks, are increasingly used to identify microbial signatures associated with nanoplastic-induced dysbiosis. For example, predictive models trained on 16S rRNA and metagenomic datasets may detect reductions in butyrate-producing taxa alongside increased abundance of pro-inflammatory microorganisms.

Advanced deep learning architectures also integrate multi-omics datasets, including metabolomics, transcriptomics, and metagenomics, to predict systemic inflammatory outcomes such as elevated interleukin-6 (IL-6) and C-reactive protein

(CRP) levels associated with microbiome perturbation. Consequently, AI-guided predictive modeling may support early detection of dysbiosis patterns and facilitate the design of precision microbiome restoration strategies, including personalized probiotics and synbiotic interventions.

16. NUTRITIONAL CONSEQUENCES OF MICROBIOME DISRUPTION

Disruption of the gut microbiota resulting from infections, poor dietary patterns, antibiotic exposure, or environmental toxins may impair nutrient absorption and metabolic function, leading to deficiencies in essential vitamins and minerals. These disturbances can also promote chronic inflammation, weaken immune regulation, and increase susceptibility to obesity, autoimmune disorders, and malnutrition. Obesity, which is more prevalent in developed countries, has frequently been associated with an elevated Firmicutes/Bacteroidetes ratio. Inflammatory bowel disease (IBD) is commonly characterized by depletion of beneficial microbial species such as *Faecalibacterium prausnitzii*, whereas metabolic syndrome is associated with gut barrier dysfunction and reduced butyrate production across diverse populations. Table 4 summarizes major microbiome features and biomarkers associated with common disorders across different geographical and nutritional settings.

Table 4. Microbiome Features and Biomarkers Associated with Common Disorders Across Regions

Disorder/Condition	Microbiome Feature Affected	Example Biomarker or Species	Predominant Region/ Country Type
Undernutrition	Reduced microbial diversity	SCFA levels, *Prevotella*	Developing nations
Obesity	Altered Firmicutes/ Bacteroidetes ratio	Firmicutes enrichment	Developed nations
Inflammatory bowel disease (IBD)	Loss of keystone microbial species	*Faecalibacterium prausnitzii*	Both
Metabolic syndrome	Gut barrier dysfunction	Endotoxins, reduced butyrate	Both

16.1 Malabsorption and Reduced Nutrient Bioavailability

Disruption of the gut microbiome may impair nutrient absorption and reduce nutrient bioavailability. Specific bacterial populations within the gut are responsible for the breakdown of complex carbohydrates and dietary fibers, producing metabolites such as short-chain fatty acids (SCFAs) that support intestinal integrity and nutrient uptake. During dysbiosis, reductions in microbial abundance and diversity may compromise these metabolic functions, thereby limiting nutrient absorption and contributing to nutritional deficiencies.

16.2 Link to Undernutrition and Overnutrition

Microbiome perturbations have been implicated in both undernutrition and overnutrition, particularly through their effects on metabolic regulation and nutrient utilization. Dysbiosis may interfere with normal gastrointestinal absorption processes, contributing to malnutrition and nutrient deficiency syndromes even when dietary intake appears adequate. Conversely, alterations in gut microbial composition have also been associated with obesity and overnutrition, as certain microbial populations may enhance energy extraction from food, thereby promoting fat accumulation and weight gain.

16.3 Inflammation-Driven Disorders

Gut microbiome disruption is frequently associated with chronic low-grade inflammation, which represents a central feature of many inflammatory and metabolic disorders. Dysregulated immune responses toward intestinal microbiota may contribute to persistent inflammation, intestinal tissue damage, and impaired nutrient absorption. Beyond gastrointestinal manifestations, dysbiosis has also been linked to metabolic syndrome, a cluster of conditions that includes hypertension, hyperglycemia, dyslipidemia, and abdominal obesity. These interactions further illustrate the interconnected relationship between microbiome stability, immune regulation, and metabolic health.

16.4 Restoring Gut–Nutrition Synergy: Strategies

Restoration of gut microbiome functionality and nutritional balance may be achieved through multiple complementary approaches, as illustrated in the accompanying flowchart. Strategies including prebiotics, probiotics, and fecal microbiota transplantation (FMT) aim to restore microbial diversity and enhance production of short-chain fatty acids (SCFAs), which represent key metabolites generated through microbial fermentation.

Recovery of SCFA production may strengthen gut barrier integrity, reduce inflammation, and improve nutrient metabolism, thereby supporting favorable nutritional outcomes such as enhanced vitamin absorption and metabolic stability.

Figure 4. **Key Strategies for Restoring Gut Microbiome.** *Given the direct implications illustrated in Figure 4 for nutrition and overall health, restoration of gut microbiota balance, commonly referred to as gut–nutrition synergy, is of considerable importance. A healthy gut microbiota supports nutrient absorption, immune regulation, and metabolic homeostasis, whereas dysbiosis may disrupt these functions and contribute to malnutrition, inflammatory disorders, and metabolic syndrome.*

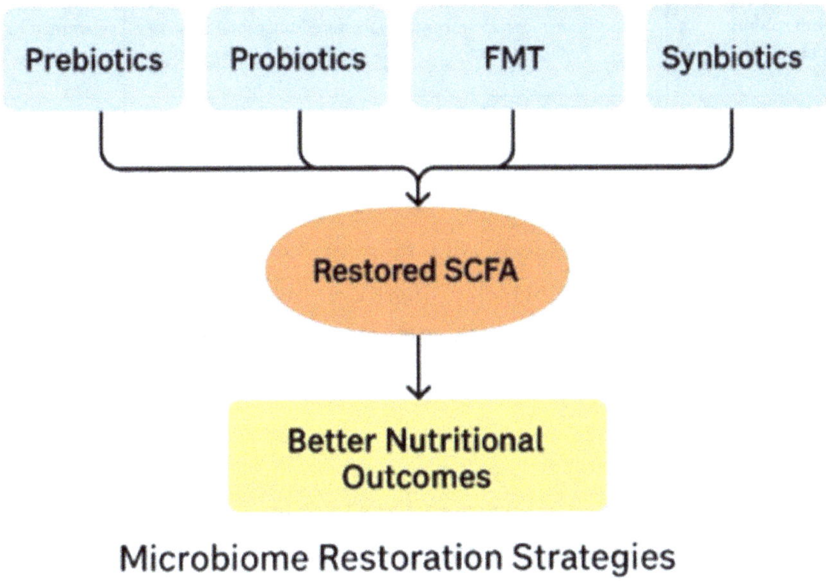

Figure 5. Pathway of Nanoplastic Impact on Gut Health and Nutritional Outcomes

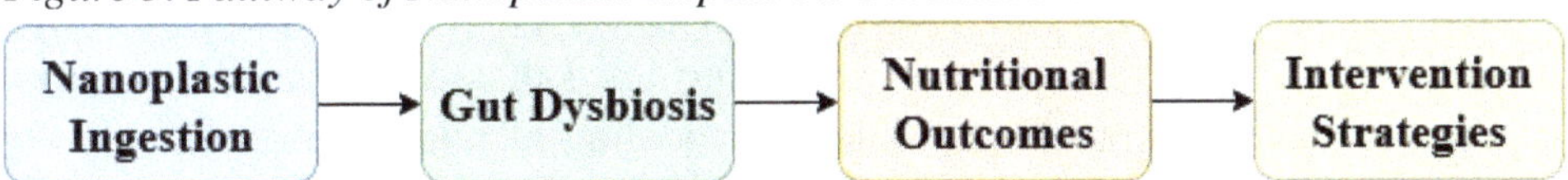

Figure 5 illustrates how ingestion of nanoplastics may induce gut dysbiosis through disruption of microbial balance within the intestinal ecosystem. This dysbiotic state may negatively affect nutritional outcomes by impairing nutrient absorption and overall gastrointestinal health. The figure further highlights the importance of targeted intervention strategies, including probiotics and prebiotics, for restoring microbiome balance and supporting host health.

Table 5. Interventions for Gut Health and Their Clinical Outcomes

Intervention Type	Description	Clinical Outcome	Study/Citation
Prebiotics	Non-digestible fibers that stimulate beneficial microbial growth	Increased *Bifidobacteria* and SCFA production	Gibson, 2017
Probiotics	Live beneficial microorganisms	Reduced gastrointestinal symptoms and improved digestion	Verma et al., 2024
Fecal microbiota transplantation (FMT)	Stool transplantation from healthy donors	Restored microbial diversity and reduced inflammation	Cammarota, 2020
Personalized nutrition	Diet tailored according to gut microbiome profiling	Weight reduction and improved vitamin absorption	Zeevi, 2015

Table 5 summarizes major gut microbiome interventions and their associated clinical outcomes. Prebiotics, which primarily consist of non-digestible dietary fibers, promote the growth of beneficial microorganisms such as *Bifidobacteria* and stimulate the production of short-chain fatty acids (SCFAs), thereby contributing to improved gut homeostasis.

17. AI-DRIVEN METAGENOMIC ANALYSIS AND PREDICTIVE MODELING

Recent advances in artificial intelligence (AI) have expanded the capacity for microbiome analysis and precision medicine applications. AI-assisted metagenomic approaches can analyze large microbial DNA datasets to predict microbiome functionality and identify microbial signatures associated with disease states. Machine learning (ML) algorithms are increasingly capable of detecting patterns within microbiome datasets that correlate with nutritional disorders and inflammatory conditions, thereby supporting predictive modeling of microbiome responses to dietary modifications and therapeutic interventions.

17.1 Prebiotics, Probiotics, and Synbiotics

- Prebiotics: Prebiotics are non-digestible dietary fibers that selectively stimulate the growth and activity of beneficial gut microorganisms. They support microbial diversity and promote restoration of gut homeostasis by enhancing populations such as *Bifidobacteria* and *Lactobacilli*.

- Probiotics: Probiotics are live beneficial microorganisms that, when administered in appropriate amounts, may help restore microbial balance disrupted by antibiotics, infections, environmental stressors, or poor dietary habits.
- Synbiotics: Synbiotics combine prebiotics and probiotics to synergistically enhance microbial survival, colonization, and overall gut health.

17.2 Frontiers and Future Possibilities

The expanding field of gut microbiome research has significant implications for future health diagnostics, therapeutic strategies, and personalized healthcare approaches. Increasing evidence continues to refine current understanding of the relationships between gut microbial communities, metabolic regulation, immune function, and disease susceptibility. These advances may support the development of targeted microbiome-based interventions tailored to individual physiological and environmental conditions.

17.3 Nanoplastics and Human Health: Evidence from Recent Studies

Emerging evidence suggests that nanoplastics may represent a growing risk to human health through their potential effects on gut microbiome composition, intestinal barrier integrity, and systemic inflammatory responses. Experimental and observational studies indicate that nanoplastics may cross intestinal barriers and contribute to microbiome dysbiosis associated with metabolic and inflammatory disorders.

Case study:

A study by Heather Leslie et al. (2022) reported the detection of plastic particles in human blood samples, suggesting systemic exposure beyond the gastrointestinal tract. In addition, Philipp Schwabl et al. (2020) identified microplastics in human stool samples, supporting evidence of continuous environmental exposure and potential interactions with the gut microbiome.

17.4 Nanoplastics and Metabolic Disruption

Nanoplastics may influence nutrient absorption, immune regulation, gut microbial composition, and metabolic homeostasis. Experimental studies suggest that nanoplastic exposure can alter microbial populations involved in nutrient metabolism and inflammatory regulation, potentially contributing to long-term metabolic disturbances such as obesity and diabetes. Continued investigation of these mechanisms may improve future personalized therapeutic strategies targeting microbiome-associated metabolic dysfunction.

Case study: Recent investigations detecting nanoplastics in human biological samples further support concerns regarding systemic exposure and potential multisystem effects beyond the gastrointestinal environment. These findings emphasize the importance of continued research into the interactions between environmental contaminants, gut microbiota, and human health outcomes within the One Health framework.

18. BIOTECHNOLOGICAL AND AI-DRIVEN TOOLS FOR ANALYZING AND MITIGATING NANOPLASTIC-INDUCED MICROBIOME DISRUPTION

Recent advances in biotechnology and artificial intelligence (AI) have provided valuable tools for monitoring, detecting, and mitigating nanoplastic-induced microbiome disruption. Technologies including metagenomics, metatranscriptomics, and metabolomics enable high-resolution profiling of microbial community alterations associated with nanoplastic exposure. These multi-omics approaches may help identify reductions in keystone microbial species, depletion of short-chain fatty acid (SCFA)-producing taxa, and expansion of potentially pathogenic microorganisms, thereby improving mechanistic understanding of dysbiosis.

Bioremediation strategies using microbial consortia, including species such as *Ideonella sakaiensis*, have also shown potential for restoring ecological balance in both environmental and intestinal microbiomes. Encapsulation of beneficial microbial strains within biopolymer matrices may further enhance microbial survivability, stability, and targeted delivery efficiency.

19. POLICY, RISK COMMUNICATION, AND ETHICAL CONSIDERATIONS

Growing evidence regarding the potential health impacts of nanoplastics highlights the importance of effective risk communication strategies and coordinated harm-reduction policies. Environmental exposure to pollutants such as nanoplastics should therefore be considered within broader international efforts aimed at reducing plastic contamination and improving environmental health governance.

Ethical considerations are also relevant in microbiome-based interventions, particularly regarding equitable access to emerging therapeutic and diagnostic technologies, data privacy, and responsible use of AI-driven health platforms.

Case study:

The Food and Agriculture Organization (FAO) and the World Health Organization (WHO) have highlighted concerns related to plastic contamination within food systems and emphasized the importance of international policy actions addressing plastic waste management and environmental exposure pathways.

20. ENVIRONMENTAL AND ONE HEALTH INTEGRATION

20.1 Impact of Nanoplastics on Microbial Diversity in Environmental Systems

Nanoplastics, typically defined as plastic particles smaller than 100 nanometers, have emerged as environmental contaminants with potential implications for ecosystem integrity and human health. Evidence suggests that nanoplastics may alter microbial diversity within agricultural soils and aquatic systems by affecting beneficial microorganisms involved in nutrient cycling, nitrogen fixation, and organic matter decomposition.

20.2 AI-Based Environmental Biosensing for Monitoring Microbial Disruption

Integration of AI-assisted biosensing technologies represents a promising approach for monitoring environmental pollutants and their effects on microbial ecosystems. Machine learning algorithms combined with real-time environmental sensing platforms may support detection of nanoplastics in soil and water systems while enabling rapid assessment of microbial health and ecological disturbance.

AI-based biosensing systems can additionally analyze microbiome sequencing datasets from environmental samples to identify shifts in microbial populations associated with nanoplastic exposure, thereby improving environmental surveillance and predictive ecological assessment.

20.3 Microbial Bioengineering for Ecological Restoration

Microbial bioengineering approaches based on synthetic biology are being explored for the development of engineered microorganisms capable of degrading nanoplastics into less harmful compounds. These strategies may contribute to restoration of microbial diversity in contaminated soils and aquatic environments.

In addition, bioengineering approaches may enhance resilience of beneficial plant-associated and soil-associated microorganisms, thereby supporting sustainable agricultural systems and nutrient-enriched crop production. Such strategies align closely with the One Health framework by integrating environmental, agricultural, and human microbiome health.

20.4 Proposed One Health Systems Model

To integrate environmental, microbial, and human health perspectives, this chapter proposes a conceptual One Health systems model linking nutrition, microbiome stability, environmental sustainability, and technological innovation. The model emphasizes:

- AI-assisted monitoring of nanoplastics and their effects on microbial communities within soil, water, and human-associated ecosystems.• Bioengineering approaches aimed at restoring microbial balance and enhancing nanoplastic biodegradation.
- Sustainable agricultural practices that preserve soil and plant microbiomes and support production of nutritionally enriched food systems.

21. RESULTS AND DISCUSSION

The findings discussed in this chapter highlight important relationships between gut microbiome composition, nutritional status, and microbiome-targeted interventions across different physiological and environmental conditions. Comparative analyses of microbial diversity indices, inflammatory biomarkers, and dominant microbial genera suggest that microbiome imbalance is associated with measurable metabolic and nutritional alterations.

Variations in microbial abundance and physiological parameters were observed across different nutritional states and intervention strategies. These relationships are illustrated using graphical approaches including bar charts, heat maps, box plots, and radar charts to support interpretation of microbiome-associated trends and therapeutic responses.

Figure 6. Relative abundance of key gut bacterial genera across nutritional statuses

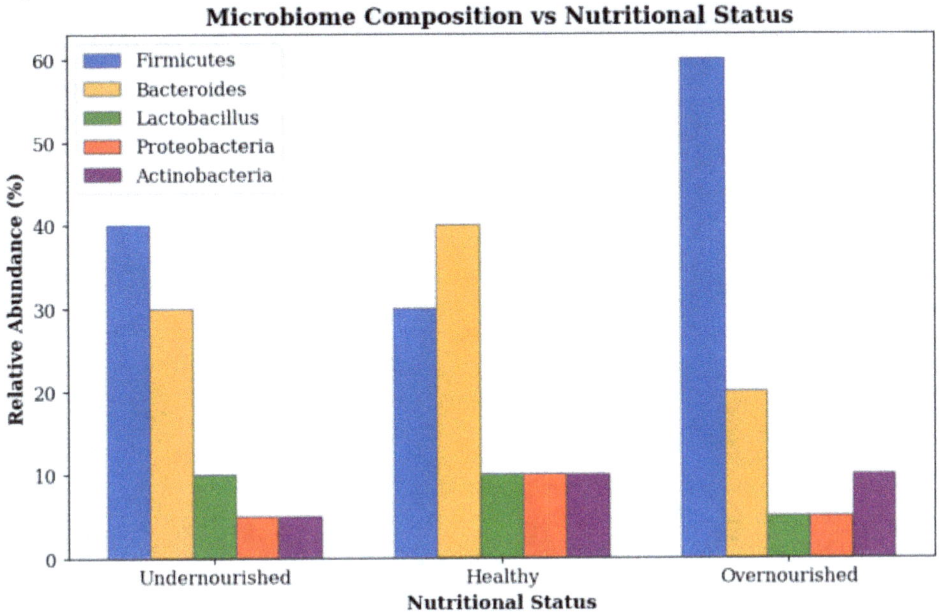

Figure 6 illustrates variations in gut microbiome composition across three nutritional states: undernourished, healthy, and overnourished individuals. The relative abundance of key bacterial groups, including *Lactobacillus*, Proteobacteria, and Actinobacteria, shows minor fluctuations across the studied conditions.

Overall, the data indicate a shift in microbial community structure associated with nutritional status, suggesting that dietary intake and energy balance may influence gut microbiota composition and diversity.

Figure 7. Effect of Gut Health Interventions on Microbiome Diversity Index

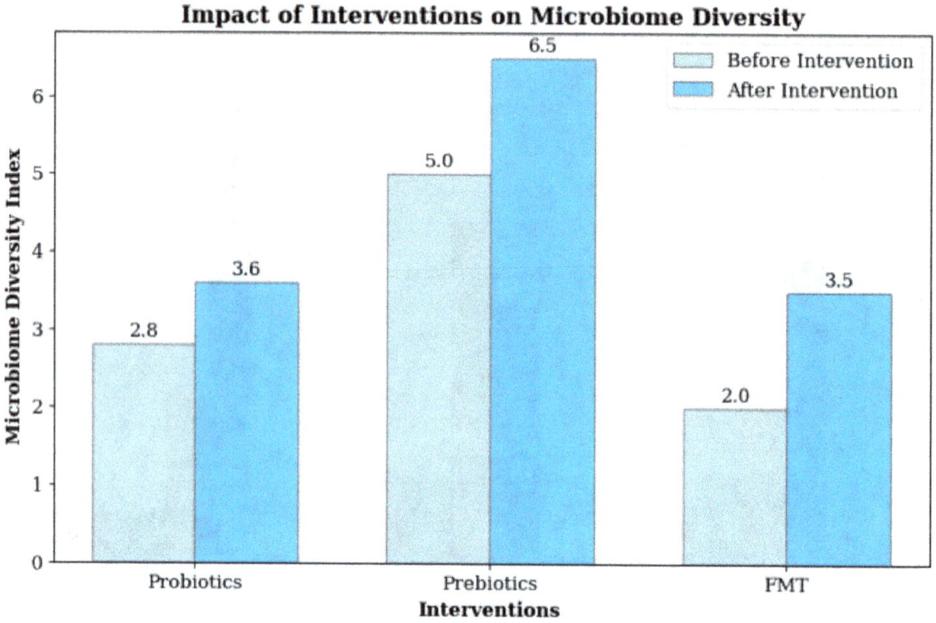

Figure 7 illustrates the effects of different microbiome-targeted interventions, including probiotics, prebiotics, and fecal microbiota transplantation (FMT), on microbial diversity.

Prebiotic intervention shows the highest increase in microbial diversity, reaching a value of 6.5. Probiotic supplementation also results in a notable increase in diversity, while FMT demonstrates a more moderate change, increasing from 2.0 to 3.5.

Overall, the findings suggest that microbiome-targeted interventions can enhance microbial diversity, which may contribute to improved gut health outcomes

Figure 8. Inflammatory Marker Levels in Disrupted vs. Balanced Gut Microbiome States

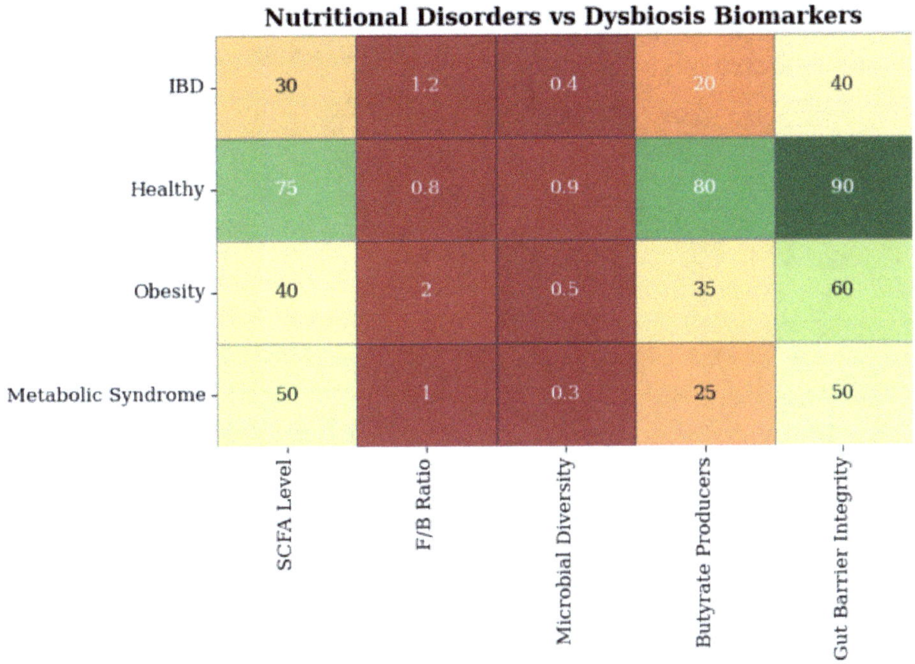

Figure 8 illustrates that optimal gut health is associated with increased short-chain fatty acid (SCFA) levels, higher microbial diversity, enrichment of butyrate-producing taxa, and improved markers of gut barrier integrity.

In contrast, inflammatory bowel disease (IBD), obesity, and metabolic syndrome are characterized by reduced SCFA levels, decreased microbial diversity, and impaired intestinal barrier function.

Collectively, reduced abundance of butyrate-producing bacteria and the presence of metabolic syndrome-related features may reflect distinct forms of dysbiosis, potentially associated with disrupted anti-inflammatory signaling pathways.

Figure 9. Inflammatory Marker Levels in Disrupted vs. Balanced Gut Microbiome States

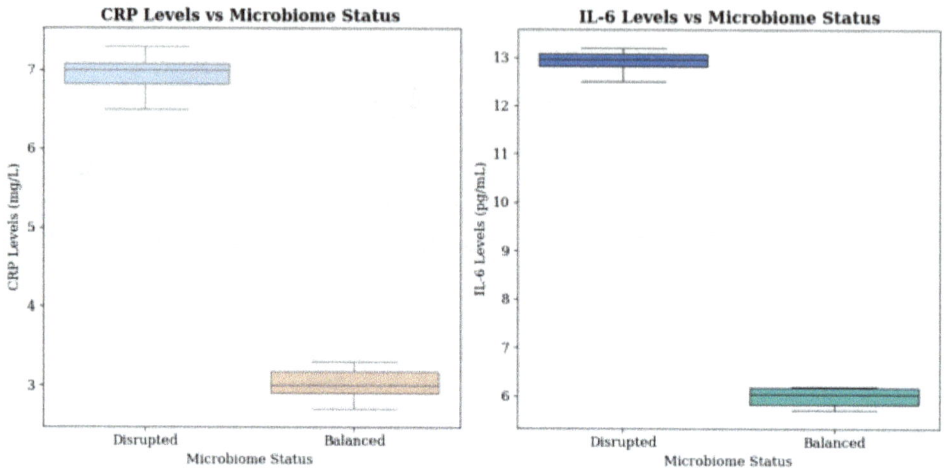

Figure 9 illustrates the association between microbiome status (disrupted vs. balanced) and inflammatory biomarkers, including C-reactive protein (CRP) and interleukin-6 (IL-6).

The disrupted microbiome state is associated with higher levels of CRP (approximately 7 mg/L) and IL-6 (approximately 13 pg/mL), indicating increased systemic inflammation. In contrast, the balanced microbiome state shows lower levels of CRP (approximately 3 mg/L) and IL-6 (approximately 6 pg/mL), consistent with a reduced inflammatory profile and a more stable immune status.

Figure 10. Impact of Personalized Nutrition on Gut-Related Health Metrics

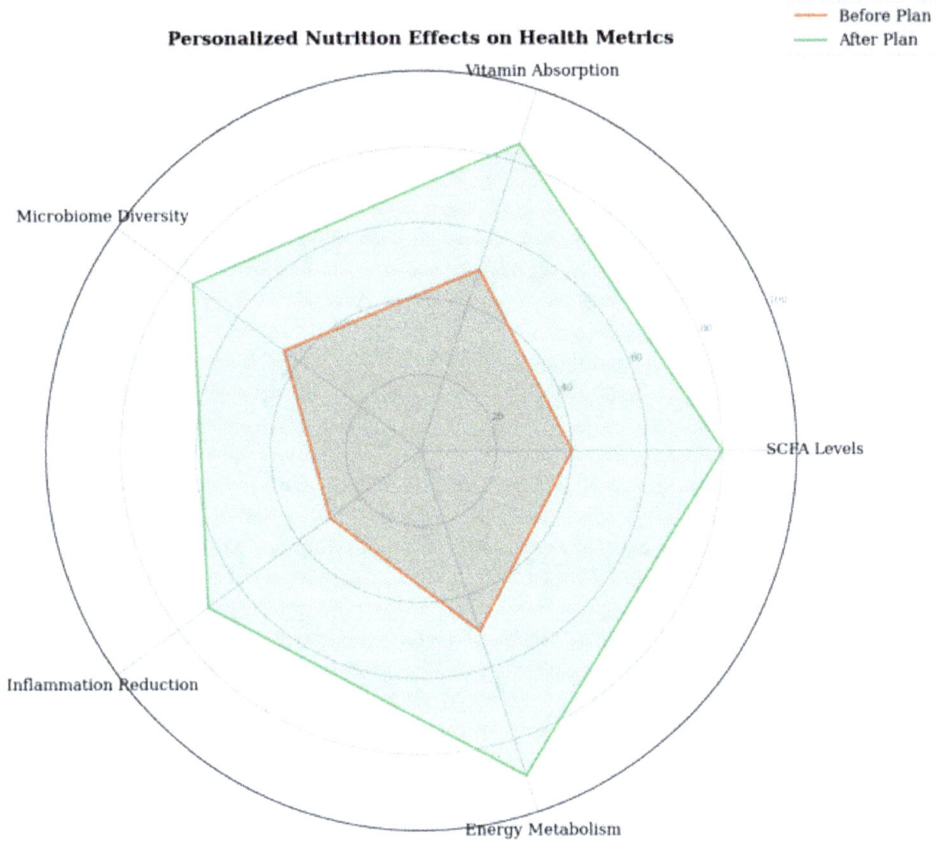

Figure 10 presents five health-related metrics before and after personalized nutrition based on gut microbiome profiling. All measured parameters show improvement following the intervention.

Inflammatory markers decrease after the intervention, indicating reduced systemic inflammation and improved immune regulation. In addition, energy metabolism shows an increase, suggesting enhanced efficiency in nutrient utilization.

Figure 11. **Global Research Trends in Microbiome and Nutrition-Related Studies (2008–2024).** *Figure 11 illustrates publication trends in microbiome-related research from 2008 to 2024 across four categories: gut microbiome, short-chain fatty acids (SCFA) studies, nanoplastics and microbiome research, and plant microbiome studies.*

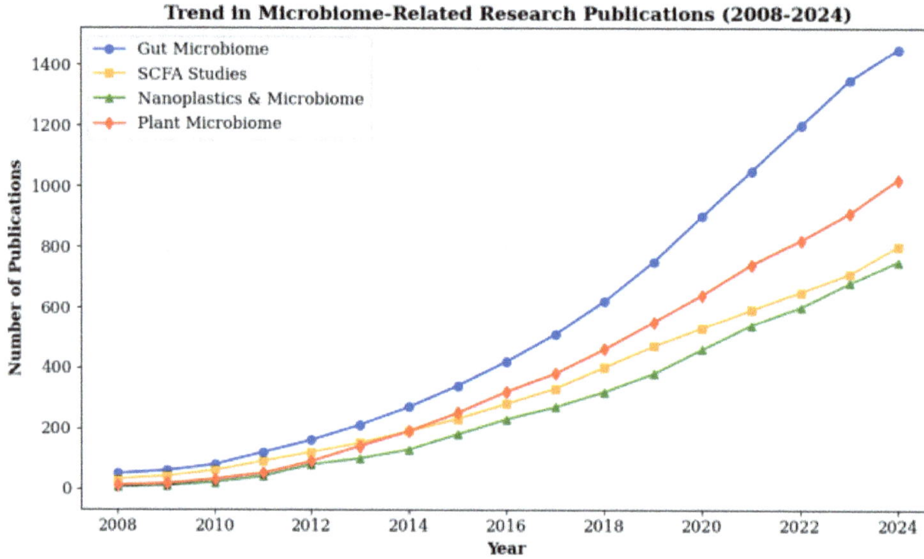

Gut microbiome research represents the most extensively published category, showing a steady increase and exceeding 1,400 publications by 2024. SCFA studies and plant microbiome research also demonstrate consistent growth over the same period, while nanoplastics–microbiome research emerges more recently with a noticeable upward trend in publication activity.

21. DISCUSSION

This study evaluated microbiome-targeted interventions, including prebiotics, probiotics, and fecal microbiota transplantation (FMT), in relation to microbial diversity. Overall, these interventions were associated with improvements in microbiome diversity, a key indicator of gut health.

Among the evaluated approaches, FMT showed the most pronounced effect, increasing the diversity index from 2.0 to 3.5, suggesting its potential utility in cases of severe dysbiosis through restoration of a complex microbial community from healthy donors. Probiotic interventions, involving administration of live beneficial strains, produced a moderate but consistent increase in diversity (approximately 2.8 to 3.6), indicating a stabilizing effect in mildly to moderately disrupted microbiomes.

Changes in key microbial taxa, including shifts in the Firmicutes-to-Bacteroidetes ratio, reductions in *Lactobacillus*, and decreased abundance of Actinobacteria, have been associated with metabolic imbalance and inflammatory stress. These alterations highlight the functional relevance of microbial diversity in nutrient metabolism, immune modulation, and energy homeostasis.

Nanoplastics also exert disruptive effects on microbial ecosystems in soil, plant, and gut environments. In soil systems, nanoplastic exposure may alter microbial diversity and interfere with essential ecological functions such as nitrogen fixation and organic matter decomposition, thereby affecting soil fertility. Similarly, in plant-associated microbiomes, particularly the rhizosphere, nanoplastics may reduce microbial diversity and impair nutrient uptake and disease resistance.

At the gut level, nanoplastic exposure has been associated with reductions in microbial diversity, enrichment of opportunistic pathogens, and depletion of beneficial short-chain fatty acid (SCFA)-producing taxa such as *Faecalibacterium prausnitzii* and *Bifidobacterium*. These alterations may compromise gut barrier integrity, immune function, and nutrient absorption.

These findings reinforce the One Health framework by demonstrating how environmental contamination can influence soil and plant microbiomes, ultimately affecting human nutrition through the food chain. This highlights the need for integrated strategies combining environmental monitoring, AI-assisted diagnostics, and microbiome-targeted interventions.

Environmental contamination of agricultural systems with micro- and nanoplastics may disrupt soil microbial communities, impair nutrient cycling, and reduce crop nutritional quality. These effects can indirectly influence human gut microbiome composition through dietary exposure. Similar disturbances have been observed in livestock exposed to contaminated feed and water, with potential downstream effects on animal health and human exposure via the food chain. Collectively, these pathways emphasize the interconnected nature of environmental, animal, and human microbiomes.

22. CONCLUSION

The gut microbiome occupies a central position at the interface of nutrition and metabolic health across populations. Current evidence supports the view that efficient nutrient absorption, immune regulation, and modulation of inflammatory responses are generally linked to a balanced and functionally diverse microbial ecosystem.

Within this context, microbiome integrity is increasingly understood not only as a determinant of physiological health, but also as a dynamic systems-level indicator shaped by environmental, dietary, and industrial exposures. Its stability reflects cumulative interactions between host biology and external perturbations, rather than a static biological state. Accordingly, microbial homeostasis is gradually being positioned within broader clinical and public health considerations, particularly as environmental contaminants such as nanoplastics are increasingly recognized as potential modulators of microbial ecosystem structure and function.

At the same time, the relevance of microbiome stability extends beyond individual health outcomes. Its alignment with Sustainable Development Goal 3 (Good Health and Well-being) reflects its role in disease susceptibility, immune resilience, and metabolic regulation. In parallel, its intersection with Sustainable Development Goal 12 (Responsible Consumption and Production) highlights upstream drivers, including environmental pollution, plastic waste accumulation, and intensified agricultural systems, all of which may influence microbial stability across interconnected human and ecological compartments.

Ethical and translational dimensions further complicate this landscape. Questions surrounding equitable access to microbiome-based interventions, governance of AI-assisted microbiome profiling, and environmental justice related to heterogeneous exposure patterns indicate that microbiome health cannot be isolated as a purely biomedical endpoint. Instead, it reflects a coupled biological–technological–environmental system in which interventions and exposures are continuously interacting.

Across clinical and experimental literature, microbial dysbiosis has been repeatedly associated with undernutrition, obesity, and systemic inflammatory conditions. Interventions such as probiotics, prebiotics, and fecal microbiota transplantation (FMT) have demonstrated variable but increasingly documented potential to restore microbial diversity and improve metabolic and immunological outcomes under controlled conditions.

However, despite advances in characterizing both disruption pathways and restorative interventions, a key translational limitation remains: it is still not fully resolved how targeted nutritional and biotechnological strategies can be systematically designed and implemented under continuous, real-world contamination pressures to restore gut–microbiome equilibrium and mitigate the downstream health consequences of chronic exposure to micro- and nanoplastics.

This unresolved limitation shifts the emphasis from microbiome characterization toward intervention design under environmental constraint conditions, where exposure, metabolism, and microbial response operate as a coupled system rather than independent domains. It is within this context that the following chapter examines nutrition intervention strategies under contamination scenarios, with focus on evidence-based microbiome restoration frameworks and exposure-aware therapeutic design.

References:

Abeltino, A., Hatem, D., Serantoni, C., Riente, A., De Giulio, M. M., De Spirito, M., De Maio, F., & Maulucci, G. (2024). Unraveling the gut microbiota: Implications for precision nutrition and personalized medicine. *Nutrients*, *16*(22), 3806. DOI: 10.3390/nu16223806 PMID: 39599593

Benitez, A. J., Tanes, C., Friedman, E. S., Zackular, J. P., Ford, E., Gerber, J. S., DeRusso, P. A., Kelly, A., Li, H., Elovitz, M. A., Wu, G. D., Zemel, B., & Bittinger, K. (2025). Antibiotic exposure is associated with minimal gut microbiome perturbations in healthy term infants. *Microbiome*, *13*(1), 21. DOI: 10.1186/s40168-024-01999-3 PMID: 39856742

Bruno, A., Dovizio, M., Milillo, C., Aruffo, E., Pesce, M., Gatta, M., Chiacchiaretta, P., Di Carlo, P., & Ballerini, P. (2024). Orally ingested micro- and nano-plastics: A hidden driver of inflammatory bowel disease and colorectal cancer. *Cancers (Basel)*, *16*(17), 3079. DOI: 10.3390/cancers16173079 PMID: 39272937

Cardoso, A. M. (2024). Microbial influence on blood pressure: Unraveling the complex relationship for health insights. *Microbiome Research Reports*, *3*(2), 22. DOI: 10.20517/mrr.2023.73 PMID: 38841410

Cuciniello, R., Di Meo, F., Filosa, S., Crispi, S., & Bergamo, P. (2023). The antioxidant effect of dietary bioactives arises from the interplay between the physiology of the host and the gut microbiota: Involvement of short-chain fatty acids. *Antioxidants*, *12*(5), 1073. DOI: 10.3390/antiox12051073 PMID: 37237938

Cunningham, B. E., Sharpe, E. E., Brander, S. M., Landis, W. G., & Harper, S. L. (2023). Critical gaps in nanoplastics research and their connection to risk assessment. *Frontiers in Toxicology*, *5*, 1154538. DOI: 10.3389/ftox.2023.1154538 PMID: 37168661

D'Angelo, C., Sudakaran, S., Asimakopoulos, F., Hematti, P., El-Gamal, D., Safdar, N., & Callander, N. (2023). Perturbation of the gut microbiome and association with outcomes following autologous stem cell transplantation in patients with multiple myeloma. *Leukemia & Lymphoma*, *64*(1), 87–97. DOI: 10.1080/10428194.2022.2131410 PMID: 36218226

Dai, Y., Shen, Z., Khachatryan, L. G., Vadiyan, D. E., Karampoor, S., & Mirzaei, R. (2023). Unraveling mechanistic insights into the role of microbiome in neurogenic hypertension: A comprehensive review. *Pathology, Research and Practice*, *249*, 154740. DOI: 10.1016/j.prp.2023.154740 PMID: 37567034

Elzayat, H., Mesto, G., & Al-Marzooq, F. (2023). Unraveling the impact of gut and oral microbiome on gut health in inflammatory bowel diseases. *Nutrients*, *15*(15), 3377. DOI: 10.3390/nu15153377 PMID: 37571313

Facchin, S., Bertin, L., Bonazzi, E., Lorenzon, G., De Barba, C., Barberio, B., Zingone, F., Maniero, D., Scarpa, M., Ruffolo, C., Angriman, I., & Savarino, E. V. (2024). Short-Chain Fatty Acids and Human Health: From Metabolic Pathways to Current Therapeutic Implications. *Life (Basel, Switzerland)*, *14*(5), 559. DOI: 10.3390/life14050559 PMID: 38792581

Fishbein, S. R., Mahmud, B., & Dantas, G. (2023). Antibiotic perturbations to the gut microbiome. *Nature Reviews. Microbiology*, *21*(12), 772–788. DOI: 10.1038/s41579-023-00933-y PMID: 37491458

Gawey, B. J., Mars, R. A., & Kashyap, P. C. (2025). The role of the gut microbiome in disorders of gut-brain interaction. *The FEBS Journal*, *292*(6), 1357–1377. DOI: 10.1111/febs.17200 PMID: 38922780

Gigault, J., & Davranche, M. (2025). Nanoplastics in focus: Exploring interdisciplinary approaches and future directions. *NanoImpact*, *37*, 100544. DOI: 10.1016/j.impact.2025.100544 PMID: 39848489

Hays, K. E., Pfaffinger, J. M., & Ryznar, R. (2024). The interplay between gut microbiota, short-chain fatty acids, and implications for host health and disease. *Gut Microbes*, *16*(1), 2393270. DOI: 10.1080/19490976.2024.2393270 PMID: 39284033

He, Y., Li, Q., Sun, Q., Li, H., Yu, T., Chen, M., Zhang, G., Zhang, B., Wang, W., & Ju, S. (2025). Gut microbiota contributes to polystyrene nanoplastics-induced fetal growth restriction by disturbing placental nicotinamide metabolism. *Journal of Nanobiotechnology*, *23*(1), 561. DOI: 10.1186/s12951-025-03650-1 PMID: 40796882

Jiménez-Arroyo, C., Tamargo, A., Molinero, N., & Moreno-Arribas, M. V. (2023). The gut microbiota, a key to understanding the health implications of micro (nano) plastics and their biodegradation. *Microbial Biotechnology*, *16*(1), 34–53. DOI: 10.1111/1751-7915.14182 PMID: 36415969

Kamath, S., Sokolenko, E., Collins, K., Chan, N. S. L., Mills, N., Clark, S. R., Marques, F. Z., & Joyce, P. (2025). IUPHAR themed review: The gut microbiome in schizophrenia. *Pharmacological Research*, *211*, 107561. DOI: 10.1016/j.phrs.2024.107561 PMID: 39732352

Kopczyńska, J., & Kowalczyk, M. (2024). The potential of short-chain fatty acid epigenetic regulation in chronic low-grade inflammation and obesity. *Frontiers in Immunology*, *15*, 1380476. DOI: 10.3389/fimmu.2024.1380476 PMID: 38605957

Li, H., Wang, J., Hao, L., & Huang, G. (2024). Exploring the interconnection between metabolic dysfunction and gut microbiome dysbiosis in osteoarthritis: A narrative review. *Biomedicines*, *12*(10), 2182. DOI: 10.3390/biomedicines12102182 PMID: 39457494

Liang, X., Andrikopoulos, N., Tang, H., Wang, Y., Ding, F., & Ke, P. C. (2024). Nanoplastic stimulates the amyloidogenesis of Parkinson's alpha-synuclein NACore. *Small*, *20*(14), 2308753. DOI: 10.1002/smll.202308753 PMID: 37988678

Liu, Y., Luo, R., Sun, Z., Zhang, Y., Guo, Y., Chen, Y., Li, L., & Yue, Z. (2025). Synergistic toxicity of combined exposure to acrylamide and polystyrene nanoplastics on the gut–liver axis in mice. *Biology (Basel)*, *14*(5), 523. DOI: 10.3390/biology14050523 PMID: 40427712

Nahiduzzaman, F. N. U., Rahman, M. Z., Akhi, M. A. J., Manik, M., Khatun, M. M., Islam, M. A., Matin, M. N., & Haque, M. A. (2025). Potential biological impacts of microplastics and nanoplastics on farm animals: Global perspectives with insights from Bangladesh. *Animals (Basel)*, *15*(10), 1394. DOI: 10.3390/ani15101394 PMID: 40427269

Ney, L. M., Wipplinger, M., Grossmann, M., Engert, N., Wegner, V. D., & Mosig, A. S. (2023). Short-chain fatty acids: Key regulators of the local and systemic immune response in inflammatory diseases and infections. *Open Biology*, *13*(3), 230014. DOI: 10.1098/rsob.230014 PMID: 36977462

Revel-Muroz, A., Akulinin, M., Shilova, P., Tyakht, A., & Klimenko, N. (2023). Stability of human gut microbiome: Comparison of ecological modelling and observational approaches. *Computational and Structural Biotechnology Journal*, *21*, 4456–4468. DOI: 10.1016/j.csbj.2023.08.030 PMID: 37745638

Schwarzfischer, M., Ruoss, T. S., Niechcial, A., Lee, S. S., Wawrzyniak, M., Laimbacher, A., Atrott, K., Manzini, R., Wilmink, M., Linzmeier, L., Morsy, Y., Lang, S., Rogler, G., Kaegi, R., Scharl, M., & Spalinger, M. R. (2024). Impact of nanoplastic particles on macrophage inflammation and intestinal health in a mouse model of inflammatory bowel disease. *Nanomaterials (Basel, Switzerland)*, *14*(16), 1350. DOI: 10.3390/nano14161350 PMID: 39195388

Shah, S. S., Noman, O., & Jaiswal, N. (2023). Unveiling the Gut Microbiome: How Junk Food Impacts the Gut. *Cureus*, *15*(11), e49179. DOI: 10.7759/cureus.49179 PMID: 38130525

Sharifa, M., Ghosh, T., Daher, O. A., Bhusal, P., Alaameri, Y. A., Naz, J., Ekhator, C., Bellegarde, S. B., Bisharat, P., Vaghani, V., & Hussain, A. (2023). Unraveling the Gut-Brain Axis in Multiple Sclerosis: Exploring Dysbiosis, Oxidative Stress, and Therapeutic Insights. *Cureus, 15*(10), e47058. DOI: 10.7759/cureus.47058 PMID: 38022314

Sharma, P., & Sharma, P. (2024). Micro(nano)plastics: Invisible compounds with a visible impact. *F1000 Research, 13*, 69. DOI: 10.12688/f1000research.142212.1 PMID: 38659492

Tan, D. S. Y., Akelew, Y., Snelson, M., Nguyen, J., & O'Sullivan, K. M. (2024). Unravelling the link between the gut microbiome and autoimmune kidney diseases: A potential new therapeutic approach. *International Journal of Molecular Sciences, 25*(9), 4817. DOI: 10.3390/ijms25094817 PMID: 38732038

Tang, F., Deng, M., Xu, C., Yang, R., Ji, X., Hao, M., Wang, Y., Tian, M., Geng, Y., & Miao, J. (2024). Unraveling the microbial puzzle: Exploring the intricate role of gut microbiota in endometriosis pathogenesis. *Frontiers in Cellular and Infection Microbiology, 14*, 1328419. DOI: 10.3389/fcimb.2024.1328419 PMID: 38435309

Tao, E., & Lang, D. (2024). Unraveling the gut: The pivotal role of intestinal mechanisms in Kawasaki disease pathogenesis. *Frontiers in Immunology, 15*, 1496293. DOI: 10.3389/fimmu.2024.1496293 PMID: 39664384

Tian, S., Li, R., Li, J., & Zou, J. (2025). Polystyrene nanoplastics promote colitis-associated cancer by disrupting lipid metabolism and inducing DNA damage. *Environment International, 195*, 109258. DOI: 10.1016/j.envint.2025.109258 PMID: 39805171

Tsigalou, C., Tsolou, A., Stavropoulou, E., Konstantinidis, T., Zafiriou, E., Dardiotis, E., Tsirogianni, A., & Bogdanos, D. (2024). Unraveling the intricate dance of the Mediterranean diet and gut microbiota in autoimmune resilience. *Frontiers in Nutrition, 11*, 1383040. DOI: 10.3389/fnut.2024.1383040 PMID: 38818135

Tu, P., Xue, J., Niu, H., Tang, Q., Mo, Z., Zheng, X., Wu, L., Chen, Z., Cai, Y., & Wang, X. (2023). Deciphering Gut Microbiome Responses upon Microplastic Exposure via Integrating Metagenomics and Activity-Based Metabolomics. *Metabolites, 13*(4), 530. DOI: 10.3390/metabo13040530 PMID: 37110188

Verma, A., Bhagchandani, T., Rai, A., Nikita, , Sardarni, U. K., Bhavesh, N. S., Gulati, S., Malik, R., & Tandon, R. (2024). Short-chain fatty acid (SCFA) as a connecting link between microbiota and gut-lung axis—A potential therapeutic intervention to improve lung health. *ACS Omega, 9*(13), 14648–14671. DOI: 10.1021/acsomega.3c05846 PMID: 38585101

Walker, T. R., & Fequet, L. (2023). Current trends of unsustainable plastic production and micro (nano) plastic pollution. *Trends in Analytical Chemistry, 160*, 116984. DOI: 10.1016/j.trac.2023.116984

Wang, L., & Hu, J. (2024). Unraveling the gut microbiota's role in salt-sensitive hypertension: Current evidences and future directions. *Frontiers in Cardiovascular Medicine, 11*, 1410623. DOI: 10.3389/fcvm.2024.1410623 PMID: 39091359

Xu, Z., Zhang, Y., Xu, Z., Zhong, Y., Wang, L., Liu, C., Chen, A., Jiang, S., Sun, W., Yu, G., Li, Y., Li, C., Wang, Y., Yu, R., & Zhang, L. (2023). Dysregulation of gut health in zebrafish by differentially charged nanoplastic exposure: An integrated analysis of histopathology, immunology, and microbial informatics. *Environmental Science. Nano, 10*(3), 933–947. DOI: 10.1039/D2EN00542E

Zhan, Z. S., Zheng, Z. S., Shi, J., Chen, J., Wu, S. Y., & Zhang, S. Y. (2024). Unraveling colorectal cancer prevention: The vitamin D - gut flora - immune system nexus. *World Journal of Gastrointestinal Oncology, 16*(6), 2394–2403. DOI: 10.4251/wjgo.v16.i6.2394 PMID: 38994172

Zhang, J., Xia, X., Huang, W., Li, Y., Lin, X., Li, Y., & Yang, Z. (2024). Photoaging of biodegradable nanoplastics regulates their toxicity to aquatic insects (*Chironomus kiinensis*) by impairing gut and disrupting intestinal microbiota. *Environment International, 185*, 108483. DOI: 10.1016/j.envint.2024.108483 PMID: 38382402

KEY TERMS AND DEFINITIONS

Dysbiosis: Imbalance in gut microbial composition associated with impaired metabolic and immune function.

FMT: (Fecal Microbiota Transplantation): Therapeutic transfer of microbiota from a healthy donor to restore microbial balance.

Nanoplastics: Plastic particles (<100 nm) capable of interacting with biological systems and microbial communities.

One Health Framework: An integrated approach linking human, animal, and environmental health systems.

SCFA: (Short-Chain Fatty Acids): Microbial metabolites that support gut barrier integrity and metabolic regulation.

Chapter 6

An Environmental Nutrition Framework for Micro- and Nanoplastics Exposure Assessment:

AI-Assisted Food Safety Detection Systems

Swapan Banerjee
http://orcid.org/0000-0001-5781-5436
Poornima University, India

Abdullayev Vugar
http://orcid.org/0000-0002-3348-2267
Azerbaijan University of Architecture and Construction, Azerbaijan

Nazila Ragimova
http://orcid.org/0000-0002-0942-7882
Azerbaijan State Oil and Industry University, Azerbaijan

T. Triwiyanto
Azerbaijan State Oil and Industry University, Azerbaijan

ABSTRACT

This chapter per the authors evaluates micro- and nanoplastic exposure within food systems through an environmental nutrition framework emphasizing trophic transfer, dietary intake pathways, and contamination dynamics across aquatic, agricultural, and processed food chains. The analysis evaluates accumulation across trophic

DOI: 10.4018/979-8-3373-4490-4.ch006

levels and associated oral exposure burdens affecting gastrointestinal, hepatic, neurological, and metabolic functions while addressing oxidative stress and inflammatory signaling linked to dietary plastic ingestion. Nutritional mitigation strategies involving bioactive compounds, probiotics, polyphenols, and cellular defense modulators are assessed alongside food safety surveillance systems, contaminant detection approaches, and computational exposure forecasting tools. Integration of environmental nutrition, food exposure science, and traceability analytics supports One Health–oriented assessment of dietary nanoplastic risks, food-system resilience, and contaminant control dynamics.

1. INTRODUCTION

Plastic pollution has increasingly become a food safety and public health concern, largely because durable polymer-based materials are now embedded throughout food production, packaging, distribution, and waste-management systems. The issue, however, extends beyond the visible accumulation of plastic waste itself. Over time, larger plastic materials gradually fragment into micro- and nanoplastics that can enter humans, animals, and plants through food chains, water systems, and even air exposure pathways (Andrady, 2011). In this context, the present chapter approaches the problem not simply as an environmental contamination issue, but as a broader food-system challenge in which contamination originates across environmental and industrial settings—including agriculture and aquatic ecosystems—before ultimately reaching human populations through dietary and non-dietary exposure routes (Ragusa et al., 2021; Leslie et al., 2022).

At present, micro- and nanoplastics remain difficult to evaluate and manage consistently. Part of the problem comes from their chemical diversity and environmental heterogeneity, but also from the absence of unified analytical definitions and standardized measurement approaches across studies. Unlike several conventional contaminants that are already governed through established testing frameworks and exposure thresholds, micro- and nanoplastics are still not regulated within harmonized monitoring systems (Yee et al., 2021). Reported exposure estimates vary considerably between studies, yet the overall evidence still points toward the likelihood of repeated low-level exposure over time through seafood, beverages, salt, and packaged foods (Yee et al., 2021). Although toxicological evidence is still developing, oxidative stress, intestinal barrier dysfunction, inflammatory responses, and metabolic disturbances are increasingly discussed as biologically plausible outcomes associated with chronic exposure (Prata et al., 2020).

From a One Health perspective, these concerns are not restricted to isolated human health effects. Rather, they involve interconnected environmental, biological, and food-system interactions that may unfold gradually across multiple exposure pathways. Increasing attention has therefore been directed toward understanding how persistent environmental contaminants, including nanoplastics, may contribute to microbiome instability, altered nutrient absorption, immune dysregulation, and broader metabolic disturbances under long-term exposure conditions.

Within this framework, nutritional mitigation is considered a supportive rather than primary line of defense and should proceed alongside exposure prevention and environmental control strategies. Experimental findings suggesting possible transgenerational effects further reinforce the importance of limiting exposure before biological consequences become established (Deng et al., 2017). At the same time, dietary interventions remain relevant, particularly where antioxidants and microbiota-supportive nutritional approaches may help strengthen cellular resilience in already exposed populations (Cheng et al., 2025).

Accordingly, this chapter brings together four interconnected dimensions: contamination pathways, detection technologies, toxicological mechanisms, and food-safety governance. Microplastics are generally classified as either primary or secondary particles (Rubiños et al., 2023). Primary microplastics are intentionally manufactured at small sizes for industrial or consumer applications, whereas secondary microplastics result from the fragmentation of larger plastic materials over time (Bessa et al., 2018). Common primary sources include personal-care microbeads, industrial abrasives, and synthetic microfibers released from textile systems. Although some intentional sources have been restricted or banned in certain regions, other forms of release continue during manufacturing, product use, and laundering activities (Chuang et al., 2000; Shim et al., 2017; United Nations Environment Programme [UNEP], 2023). Existing regulatory actions aimed at deliberate emissions suggest that source-control strategies become more achievable when contaminant categories and release pathways are clearly identified and consistently monitored (Bessa et al., 2018).

Secondary microplastics are generally harder to control because they are continuously generated through the gradual fragmentation of larger plastic products already embedded in daily consumer use. Bags, bottles, films, coatings, packaging materials, and fishing gear progressively break down under environmental stress, producing smaller and smaller particles over time (Ivar do Sul & Costa, 2014). The rate of fragmentation often depends on both the polymer itself and the surrounding environmental conditions. Factors such as ultraviolet radiation, oxidation, hydrolysis, and temperature fluctuations can accelerate degradation processes, particularly in polymers such as polyethylene (Gewert et al., 2015). At the same time, the long environmental persistence of widely used plastics, especially polyethylene and polypropylene, suggests that plastic-related exposure within food systems is unlikely to represent a temporary contamination event. Rather, it reflects an ongoing and cumulative environmental condition linked to long-term material use patterns (Lambert & Wagner, 2016). As particle size decreases, mobility through air, water, soil, and biological systems becomes substantially easier, increasing the likelihood of interaction with organisms and surrounding environmental compartments (Brennecke et al., 2016).

Within this broader continuum, nanoplastics represent the smallest fraction of plastic-derived particles and may exhibit greater biological accessibility than larger microplastic forms. They emerge through the continued fragmentation and environmental weathering of microplastics, although some may also originate directly from industrial activities and product-use processes (Geyer et al., 2017). Their relevance to food safety and nutrition has gained increasing attention because nanoplastics are unlikely to behave as biologically inert particles alone. Due to their high surface-area-to-volume ratio, they can adsorb additives, metals, and persistent organic pollutants while also interacting more directly with biological membranes and cellular interfaces (Wagner et al., 2018). From a nutritional perspective, this raises concerns that dietary exposure may involve overlapping physical and chemical stressors rather than isolated particle ingestion alone.

Available evidence from both environmental systems and consumer products further suggests that nanoplastics may arise through multiple interconnected pathways, including textile degradation, tire wear, wastewater systems, and food-packaging materials. Environmental weathering may also alter particle chemistry through oxidation and the formation of surface functional groups, potentially influencing both toxicological behavior and analytical detection performance (Hariani et al., 2023; Wagner et al., 2018). In food-contact materials specifically, exposure scenarios become more complicated because plastics may release plasticizers, stabilizers, and other associated compounds alongside particulate matter itself (Hermabessiere et al., 2017). Consequently, evaluating food-system risks requires approaches capable of addressing micro- and nanoplastics as multifactorial contaminants involving both particulate behavior and chemically mediated interactions occurring simultaneously.

2. LITERATURE REVIEW

The literature on micro- and nanoplastics within food systems is better interpreted through interconnected thematic patterns rather than as isolated individual studies. Across the current evidence base, several recurring themes consistently appear. First, contamination pathways are strongly linked to environmental exposure, industrial activities, and food-production practices. Second, exposure itself varies across dietary behaviors, food categories, and consumption patterns. Third, the biological implications associated with exposure appear to depend on multiple interacting variables, including particle size, polymer composition, associated additives, host susceptibility, and the effectiveness of food-safety monitoring systems and regulatory controls.

At the same time, the literature remains methodologically uneven. Differences in sampling procedures, analytical techniques, particle definitions, and reporting standards continue to complicate direct comparison across studies. As a result, exposure estimates and toxicological interpretations often vary substantially between investigations, even when similar food matrices are examined. This inconsistency remains a major limitation within the field and contributes to continuing uncertainty regarding long-term biological significance.

Nevertheless, despite these limitations, the broader direction of the evidence increasingly supports the view that micro- and nanoplastic contamination should not be interpreted solely as an environmental waste issue. Instead, it is progressively being recognized as a food-system and exposure-management challenge operating across interconnected environmental, biological, and public-health domains. This thematic synthesis therefore aims to provide a more integrated conceptual structure for interpreting the current literature while maintaining awareness of the significant uncertainties that still characterize the field.

The risk assessment for groundwater contamination and possible entry into food chains should generally be carried out at least once a year, following available monitoring frameworks. The main entry points into the food chain are agricultural soils, aquatic environments, food processing facilities, and also packaging systems.

Agricultural contamination comes from several relatively well-known sources, including plastic mulch films, irrigation water, atmospheric deposition, and manure from agricultural activities, including biosolids when they are used as soil amendments (Mintenig et al., 2017). Biosolids are important here because they essentially act as a sink where plastics removed from household and industrial wastewater accumulate. The issue is that once applied to agricultural soils, these materials can be redistributed again into the terrestrial environment (Nizzetto et al., 2016). Studies on sludge-amended soils and atmospheric deposition also show that particles can remain in soil profiles for long periods, interact with soil organisms, and gradually move toward crops or runoff pathways (Zhang & Liu, 2018; Dris et al., 2016).

Aquatic systems are still considered one of the major exposure routes, since plastics can be directly discharged into water bodies or transported from land through runoff and other hydrological processes (Jambeck et al., 2015). Filter-feeding organisms tend to concentrate suspended particles, and in that way they act as early uptake points within aquatic food chains (Cole et al., 2011). In higher organisms such as fish and marine species, ingestion of micro- and nanoplastic particles can lead to tissue exposure, although the actual accumulation depends strongly on species, particle size, and how efficiently the gut clears these particles (Lusher et al., 2015).

So the interpretation of risk in aquatic foods needs to be more nuanced. It is not just presence or absence, but also whether whole organisms are consumed, whether digestive tracts are included, and how contaminated the local environment is. Food processing environments add another layer of exposure that is often overlooked. Process water, plastic-based equipment, synthetic clothing, airborne fibers, and packaging materials all contribute to contamination risk (Hernandez et al., 2019). Microfibers can come from worker uniforms, filtration systems, conveyor belts, production lines, and sometimes even from poorly maintained equipment (Hartline et al., 2016; Geueke et al., 2018). Bottled beverages have been particularly discussed in the literature, with some studies showing higher particle loads in bottled water compared to tap water. This suggests that packaging materials, bottle caps, bottling lines, and even source water quality all play a role (Schwabl et al., 2019). For this reason, process water quality is increasingly viewed as a practical control point for both industry and regulators (Gundogdu et al., 2018; Bessa et al., 2018; Zhang et al., 2022).

3. DIETARY PATHWAYS AND EXPOSURE ROUTES

Dietary intake is probably the most direct exposure route for the general population, but it should not be considered in isolation. It really needs to be evaluated together with inhalation and dermal exposure to avoid an incomplete picture (Prata et al., 2020; Zhang et al., 2022). Because of this, food categories such as seafood, bottled water, salt, beer, processed food-contact products, honey, and sugar are repeatedly identified as relevant exposure sources (Cox et al., 2019).

Bivalves and shellfish are of particular concern because they tend to retain particles in their digestive tracts, which are then directly consumed by humans (Van Cauwenberghe & Janssen, 2014; Neves et al., 2015). Small pelagic fish that are consumed whole follow a similar logic of exposure (Smith et al., 2018). In the case of salt, contamination varies depending on production sources and local marine pollution, and these two factors largely determine final particle levels (Zhang et al., 2022; Kim et al., 2018; Karami et al., 2017).

Processed foods like honey and sugar can also become contaminated either through environmental deposition or during manufacturing steps (Liebezeit & Liebezeit, 2013). Bottled water represents another relevant exposure route, mainly because of its wide and repeated consumption across almost all population groups, which makes it a daily exposure source (Mason et al., 2018). Beer production can also introduce particles through water sources, filtration systems, bottling processes, and packaging materials (Oßmann et al., 2018).

Airborne fibers remain a persistent exposure route, especially in indoor environments, workplaces dealing with textiles and plastics, and also in urban air settings (Prata et al., 2020).

3.1 Detection of Microplastics and Nanoplastics

The reliability of the evidence base depends quite heavily on the detection technology being used. For smaller particles, visual inspection is simply not enough, and polymer confirmation usually requires spectroscopic or thermal analytical methods. FTIR and Raman spectroscopy are the most commonly used approaches for identifying polymer signatures in environmental and food samples. That said, they are not identical in performance. They differ in detection limits, sensitivity, tolerance to background interference, and also how well they work with complex food matrices.

Because of these differences, reported contamination levels should really only be compared when the full analytical chain is clearly described. That includes sampling, digestion, filtration, and polymer identification steps, and ideally when these are applied in a consistent way across studies.

Machine learning has started to play a role in improving interpretation of spectral data and handling large monitoring datasets (Gundogdu et al., 2018). Information such as polymer usage, process water quality, equipment condition, packaging type, and other supply-chain variables, when combined with contamination records, can be used to build predictive models. These models may help flag high-risk foods or facilities that need closer inspection. On top of spatial patterns, automated monitoring systems can also capture temporal trends that are often missed in manual sampling (Kim et al., 2018).

Artificial intelligence cannot replace laboratory validation, and it should not be treated as such. But it can still support risk-based surveillance if it is trained on transparent and standardized datasets (de Oliveira et al., 2022). Its real value is probably in shifting monitoring systems from late detection toward earlier warning and anticipation of risk (Kim et al., 2018).

3.2 Research Methods

This chapter adopts an integrative review approach with critical synthesis to organize and interpret the current evidence base. Rather than following a formal protocol-driven systematic review structure, the discussion is developed through thematic integration of literature across five major domains: contamination sources, exposure pathways, detection technologies, health mechanisms, and mitigation strategies.

The literature search included terms such as micro- and nanoplastics, food contamination, nutritional interventions, toxicology, food safety, HACCP, One Health, artificial intelligence, and policy. The aim was not to calculate pooled effect sizes, but to synthesize evidence from food science, environmental health, toxicology, and nutrition. Priority was given to recent reviews and primary studies addressing food matrices, seafood, beverages, and dietary mitigation approaches (Crosta et al., 2023).

An important aspect of the synthesis is the distinction between different categories of evidence. Contamination studies primarily provide information on occurrence and magnitude, whereas toxicological studies offer insight into potential biological mechanisms. Nutritional studies describe possible protective pathways, while policy-oriented work focuses on governance and mitigation strategies. This distinction is important because evidence types differ in their interpretive strength, particularly when indirect experimental findings are compared with direct exposure measurements.

The chapter also adopts a One Health perspective, linking human, animal, and ecosystem health within food production systems. The analytical scope includes FTIR and Raman spectroscopy, machine learning–based pattern recognition, HACCP control points, and nutritional strategies involving compounds such as vitamin C, curcumin, resveratrol, probiotics, and other bioactive agents (Gundogdu et al., 2018; Bessa et al., 2018; Zhang et al., 2022).

4. FINDINGS & DISCUSSION

4.1 Conceptual Framework

The conceptual framework is shown in Figure 1. It starts from plastic sources, follows how particles move through food system pathways, identifies key human exposure routes, and then links these exposures to health impacts and possible mitigation strategies. One of the key messages here is that nutrition alone is not enough as a standalone solution. It is only one part of a broader set of controls.

The framework also helps clarify the contribution of the chapter by moving away from simple lists of foods or compounds and toward a more integrated risk pathway view. It essentially tries to answer four questions: where particles come from, how they enter the food chain, how they affect exposed populations, and what can actually be done to reduce exposure.

Structurally, this makes the chapter more suitable for an academic book context, because the evidence is not only organized by topic, but also by process. In that sense, the discussion is presented as a continuous exposure–impact–mitigation pathway, which then leads naturally into interventions and research priorities.

Figure 1. Conceptual Framework for Microplastics, Nutrition, and Food-System Governance.

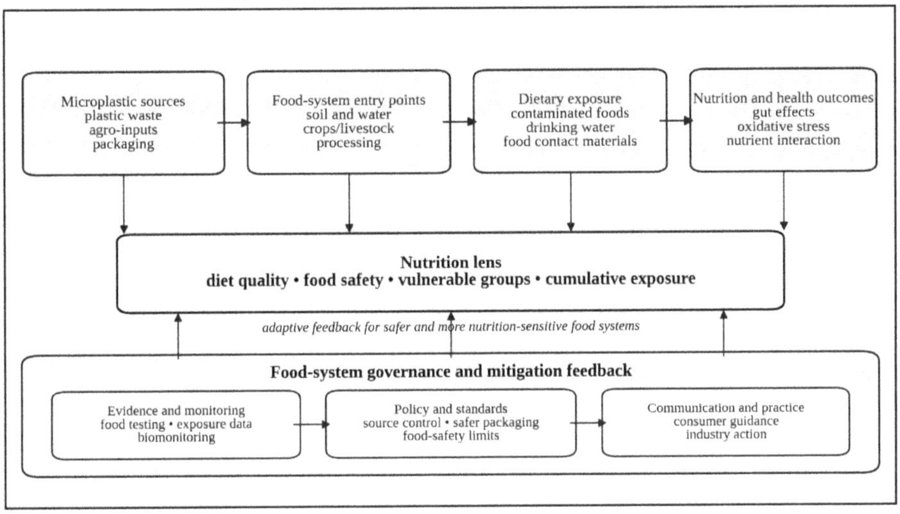

4.3 Evidence on Food Contamination and Pathogen-Associated Exposure

When food gets contaminated or comes into contact with pathogens at measurable levels, the resulting data is usually taken as evidence of contamination. That said, interpreting these numbers is not straightforward. Across studies, there is still no real consistency—units differ, particle size cut-offs differ, digestion protocols differ, and even the criteria for confirming polymer identity are not uniform.

Because of that, comparing results between studies is not always clean or even fully reliable. In this chapter, I kept a comparative overview of major food types and commonly reported packaging-related polymers (summarized in Table 1). It's not meant to define a precise exposure dose. It's more of a way to show which foods and systems keep appearing again and again in the literature. The most frequently reported polymers are PE (polyethylene), PP (polypropylene), PS (polystyrene), and PET (polyethylene terephthalate) (Zhang et al., 2022).

Table 1 becomes more meaningful when you don't read it as isolated numbers. What it actually suggests is that exposure risk is not only about concentration.

Consumption behavior matters just as much. Bottled water, for example, can show relatively high particle counts per liter. But real exposure depends on how much is actually consumed per day, and that varies a lot between individuals, brands, and packaging types.

The same idea applies when whole organisms are eaten. In those cases, whatever is in the gastrointestinal tract is more likely to be ingested directly. Foods like salt, honey, beer, and sugar are usually consumed in smaller quantities, but they are still part of daily dietary exposure and therefore shouldn't be ignored. From a governance point of view, this kind of interpretation is often more useful than focusing only on concentration values.

Table 1. Dietary Sources of Microplastic Exposure: Reported Contamination Levels, Polymer Types, and Exposure Pathways

Food Category	Primary Source(s)	Microplastic Concentration	Polymer Types	References
Shellfish (Mussels, Oysters)	Seawater ingestion	300–1,000+ particles/g	PE, PP, PS	Au et al. (2017)
Fish (Pelagic Species)	Ingestion + tissue accumulation	150–500 particles/g	PE, PP, PET	Zhu et al. (2020); Galloway et al. (2017)
Sea Salt	Marine contamination	100–600 particles/kg	Primarily PE, PP	Karami et al. (2017); Koelmans et al. (2014)
Drinking Water (Bottled)	Packaging, water source	100,000–150,000 particles/L	PE, PP, PET	Mason et al. (2018); Shruti et al. (2020)
Drinking Water (Tap)	Water treatment, distribution	4,000–10,000 particles/L	Mixed polymers	Wang et al. (2025);
Honey	Processing, environmental deposition	1,000–5,000 particles/kg	PE, PP, PS, PET	Liebezeit & Liebezeit (2013)
Beer	Water source, processing, packaging	2,000–10,000 particles/L	Multiple polymers	Karami et al. (2017)
Sugar (Marine-sourced)	Processing, water contamination	500–2,000 particles/kg	PE, PP	Liebezeit & Liebezeit (2013); Deng et al. (2017)
Seafood (Prawns, Shrimp)	Environmental ingestion + bioaccumulation	200–800 particles/g	PE, PP, PS	Leslie et al. (2022);
Fruit and Vegetables	Irrigation water, atmospheric deposition	10–100 particles/kg	Microfibers dominant	Hartline et al., (2016)

This table provides a structured overview of microplastic contamination across selected food categories, highlighting dietary exposure pathways. Reported values vary substantially depending on analytical workflow and particle-size thresholds.

5. QUANTITATIVE RISK-ASSESSMENT LOGIC

Right now, full quantitative risk assessment isn't really possible for most of these exposure scenarios. The main reason is simple: we still don't have standardized measurement methods or health-based threshold values.

Still, a simplified screening approach can be used. You can roughly think of daily intake as particle concentration multiplied by consumption rate, adjusted for body weight and how often exposure happens. From there, exposure level, likelihood of contamination, possible severity of effects, and how controllable the situation is can all be combined into a risk-ranking matrix.

The point here is not to replace toxicological thresholds. It's more about early prioritization—deciding what should be monitored first. For example, a food with moderate contamination can still rank high if it's eaten frequently by vulnerable groups like children or pregnant women.

On the other hand, even highly contaminated foods might not matter as much from a population perspective if consumption is rare or easy to control (like through packaging changes or filtration). This distinction matters, especially because reviewers often push for clearer comparative logic. So here, risk is treated more as a combination of exposure, vulnerability, uncertainty, and control feasibility—not just concentration values.

6. GAPS IN EVIDENCE AND METHODOLOGICAL UNCERTAINTY

The current evidence gap doesn't necessarily mean there are no studies—it's more that comparable studies are limited. Methods vary widely. Particle size definitions differ, cut-offs differ, digestion procedures differ, contamination control approaches differ, and reporting units are not consistent either. So direct comparison is still messy.

Only a small number of studies actually try to link measured food contamination with real dietary intake, and even fewer connect exposure estimates with human biomonitoring data. Because of that, this chapter avoids presenting numerical exposure values as fixed truths. Instead, it focuses more on interpreting patterns of risk.

Future work will likely need to combine dietary surveys, standardized polymer analysis, and biomonitoring if we want a clearer link between exposure and health outcomes.

The second issue is more practical: feasibility. Even if technical controls exist, they are not always easy to implement. Food producers may not have access to labs, or they may not afford filtration systems or higher-grade packaging. This creates uneven capacity across the system. From an equity angle, that matters, because it can shift risk toward lower-income groups who already have fewer consumption options. So future research probably needs to bring together cost, equity, and implementation constraints with toxicology and analytical chemistry, instead of treating them separately. Despite these uncertainties, current evidence still provides biologically plausible pathways linking exposure to adverse health outcomes.

7. HEALTH ASSESSMENT AND IMPACT OF MICROPLASTICS AND NANOPLASTICS

Human exposure happens mainly through ingestion, inhalation, and to a lesser extent dermal contact, although the relative contribution of each route really depends on the population and the setting . Ingestion is usually considered the main concern in food safety, simply because food and drinking water are constant exposure sources. In some discussions, ingestion is also linked with infection-related pathways (Thompson et al., 2004), although that comparison is not always direct.

Airborne fibers are more relevant in indoor environments, especially in industrial, textile, and waste-management contexts (Prata et al., 2020). Dermal exposure, on the other hand, is less about particle entry itself and more about associated chemicals that may leach from plastics (Gluge et al., 2020).

Once ingested, particles pass through the gastrointestinal system, where interaction can occur with epithelial cells, immune components, gut microbiota, and the mucus layer itself (De Corte et al., 2019). The stomach is only part of the system; most interaction actually happens further along the gut.

Nanoplastics are of particular concern because of their size. They may cross epithelial barriers more easily than larger microplastics and show higher surface reactivity (Deng et al., 2017). Long-term exposure in experimental systems has been associated with disruption of barrier function, changes in microbial diversity, and metabolic alterations (Rinninella et al., 2021).

There are also concerns around early-life exposure. Maternal exposure studies suggest possible downstream effects on gut and metabolic development later in life. Still, it is important to be careful here—the evidence is suggestive, but human causation is not clearly established yet (Visalli et al., 2021).

Oxidative stress keeps appearing across almost all toxicological studies. When reactive oxygen species (ROS) are not properly controlled, cells start to accumulate damage in lipids, proteins, DNA, and mitochondria (Wang et al., 2025).

Micro- and nanoplastics can trigger ROS formation in multiple ways. Sometimes it is direct, through particle surface interactions, and sometimes indirect, through mitochondrial disruption or activation of inflammatory pathways such as NADPH oxidase signaling.

Lipid peroxidation and protein oxidation can then damage membranes and enzymatic systems (Kleijn, 2020). This often creates a kind of feedback loop, where antioxidant capacity decreases, glutathione levels drop, and oxidative stress continues to build (Karami et al., 2017).

It is also becoming harder to describe micro- and nanoplastics as just "gut pollutants." Once absorbed, not only the particles themselves but also associated chemicals can potentially reach organs such as the liver, kidney, spleen, blood, and adipose tissue (Karami et al., 2017).

Experimental work reports liver inflammation, lipid metabolism disruption, and signs of hepatocellular injury (Deng et al., 2017). At the same time, human biomonitoring studies have already detected plastic particles in blood, urine, tissues, and even the placenta (Leslie et al., 2022). That is an important shift in the discussion—it is no longer purely theoretical.

More recently, there are emerging signals around cardiovascular inflammation and possible thrombotic risk (Marfella et al., 2024). Neurological studies, mostly experimental, also point toward neuroinflammation, behavioral changes, and potential CNS involvement (Huang et al., 2014; Guo & Wang, 2019). Inhaled or translocated nanoplastics may even have neurodevelopmental implications, although this still needs stronger confirmation.

8. NUTRITIONAL INTERVENTION AND BIOACTIVE PROTECTIVE COMPOUNDS

Any discussion of nutritional mitigation has to stay balanced. Some nutrients and bioactive compounds may support antioxidant defenses, gut barrier integrity, inflammatory regulation, and detoxification pathways, all of which are relevant in the context of micro- and nanoplastic exposure (Schirinzi et al., 2017). But they are not treatments, and they should not be framed as such.

What they might do—mechanistically—is reduce oxidative stress, support endogenous antioxidant systems, and help maintain microbiome and mitochondrial stability. That is the level at which the evidence currently sits.

Vitamin C is a water-soluble antioxidant directly involved in oxidative stress regulation (Wang et al., 2025). In experimental systems, post-exposure supplementation has shown some protective effects against particle-induced stress and cellular aging markers. It functions mainly as a ROS scavenger and can also help regenerate vitamin E, supporting redox balance in cellular compartments (Wang et al., 2025).

At the population level, vitamin C intake is generally associated with broader oxidative and inflammatory health outcomes, although not specifically linked to microplastic exposure (Hemila & Chalker, 2013). In practical terms, it still needs to come from a varied diet rich in fruits and vegetables rather than isolated supplementation (Nishida et al., 2004; Kawai, 2014).

Lipophilic antioxidants such as vitamin E and carotenoids are also relevant, mainly because lipid peroxidation and membrane damage are common outcomes in microplastic toxicity studies (Tran-Dinh et al., 2013). Vitamin E interrupts lipid radical chain reactions, while carotenoids (beta-carotene, lycopene, lutein) can reduce singlet oxygen and other reactive species.

Epidemiological evidence supports carotenoid-rich diets in general health contexts, but again, there is no direct evidence that they prevent micro- or nanoplastic formation or exposure outcomes (Anigbogu et al., 2025). So the interpretation stays at dietary pattern level rather than single-compound intervention.

Polyphenols show similar patterns. They may influence oxidative stress, inflammation, mitochondrial activity, and detox-related signaling pathways (Cheng et al., 2025). Curcumin, for example, has been tested in several toxicological models and appears to reduce oxidative and inflammatory responses (Coman et al., 2025; Wang et al., 2025).

A key pathway often discussed is the Nrf2/ARE axis, which regulates detoxification enzymes such as heme oxygenase-1 and glutathione S-transferases (Yang et al., 2023). While this is mechanistically relevant, strong evidence linking it directly to mitigation of microplastic exposure in humans is still missing (Wang et al., 2025).

Other plant-derived compounds extend this nutritional framework. Sulforaphane and indole-3-carbinol from cruciferous vegetables act through glucosinolate metabolism and activate detoxification pathways (Shan et al., 2020). Probiotics and prebiotics—through fermented foods, inulin, resistant starch—are consistently linked with microbiome resilience in experimental exposure models (Rinninella et al., 2021).

Melatonin, present in small amounts in foods like tart cherries and kiwis, has antioxidant and mitochondrial-supporting properties (Jee, 2016). Marine-derived antioxidants also show strong ROS-scavenging potential and may act at membrane level (Tao, 2025). Dietary fiber, finally, may reduce gut contact time, support fecal bulk, and promote short-chain fatty acid production like butyrate (Jimenez, 2016). Table 2 summarizes key compounds and their evidence levels (Völkl et al., 2023)

Table 2. Bioactive compounds as protection against microplastic-induced toxicity

Bioactive Compound	Primary Food Sources	Protective Mechanism(s)	Evidence Level	References
Vitamin C	Citrus, berries, leafy greens	ROS scavenging, vitamin E regeneration	*In vitro* / Human cell models	Wang et al., 2025
Vitamin E	Nuts, seeds, vegetable oils	Lipid peroxidation prevention	Animal models / *In vitro*	Yee et al., 2021
Curcumin	Turmeric	Nrf2 activation, antioxidant response	*In vitro* / Animal models	Cheng et al., 2025
Quercetin	Apples, onions, berries	Anti-inflammatory signaling, NF-κB inhibition	Human cell studies	Wang et al., 2025
Resveratrol	Grapes, berries	Mitochondrial regulation, SIRT1 activation	*In vitro* / Animal models	Cheng et al., 2025
Sulforaphane	Broccoli, cruciferous vegetables	Phase II detoxification pathways activation	*In vitro* / Experimental models	Yang et al., 2023
Probiotics	Fermented foods	Gut barrier support, microbiota restoration	Animal models / *In vitro*	Rinninella et al., 2021
Melatonin	Tart cherries, kiwis, tomatoes	Mitochondrial protection, sleep-immune axis	*In vitro* / Animal models	Cheng et al., 2025
Astaxanthin	Salmon, microalgae, shrimp	Membrane stabilization, potent ROS scavenging	Animal models	Wang et al., 2025
Anthocyanins	Berries, purple vegetables	Inflammatory cascade modulation	*In vitro* / Animal models	Wang et al., 2025

9. CONCEPTUAL FRAMEWORK ON CONTAMINATION WITH MICROPLASTICS IN THE CONTEXT OF ONE HEALTH

It is appropriate to use the One Health framework, as contamination with MNP affects environmental, animal, human, and food systems ([UNEP], 2023; McAllister, 2018). One Health is not a slogan; it is an approach for coordinated working across sectors, functioning here as an organizing model within the chapter.

Environmental agencies monitor sources and waste flows; agricultural agencies monitor biosolids, irrigation pathways, and plastic mulch; food-safety agencies focus on processing and packaging environments; and health agencies address exposure assessment and risk communication. Consumers also contribute, intentionally or not, through consumption patterns and waste behavior. The value of One Health, therefore, lies in coordinating sectors that are typically separated but ultimately linked within a single contamination chain.

A functional control strategy must begin where intervention is still feasible. In primary production, this includes improving irrigation water quality, limiting plastic mulch use, and more careful management of biosolids. In processing environments, controls focus on low-shedding surfaces, air filtration, process water treatment, and maintenance of equipment in contact with plastics.

At distribution and retail level, attention shifts to packaging design, traceability systems, and temperature control. These measures depend on clear supply-chain information and, importantly, accountability across producers, processors, distributors, and regulators (Tse et al., 2022). Without such integration, coordination remains largely theoretical.

A persistent limitation in the current evidence base is the lack of standardization. If MNP data are to be used for regulation or cross-country comparison, then sampling, digestion protocols, polymer identification, reporting units, blank controls, and QA procedures require harmonization. At present, these parameters remain highly variable across studies and laboratories.

For this reason, contamination data should not be used directly for regulatory comparison between countries. Instead, they are more appropriately applied to identifying high-risk commodities and evaluating whether interventions reduce exposure over time.

In this sense, academic work in this field is not limited to measurement. It also supports method development that can be translated into practice by food businesses and public agencies, even as the system continues to evolve.

10. FOOD SAFETY, HAZARD MANAGEMENT, AND POLICY RESPONSE

HACCP principles provide a practical framework for translating microplastic evidence into food safety management. Hazard analysis begins with identifying potential entry routes, including water, raw materials, air, equipment surfaces, worker clothing, packaging, and storage conditions.

Critical control points typically include process water filtration, air handling systems, equipment maintenance, packaging integrity checks, and cleaning procedures. These are established food safety concepts but require explicit adaptation for particulate contamination such as microplastics.

Monitoring strategies may include scheduled sampling, visual inspection for material wear, laboratory-based particle analysis, and documentation of corrective actions when contamination is detected.

Integration of environmental and food-system risk management is increasingly reflected in public health frameworks (WHO, 2020). Climate change adds further complexity. Rising temperatures may accelerate plastic degradation, while changes in rainfall patterns can increase runoff and transport of plastic debris into agricultural areas (Demestichas et al., 2020; Kateris et al., 2023).

Extreme weather events can also mobilize waste from landfills, rivers, and drainage systems into food production zones. Consequently, resilient food systems require not only technical controls but also adaptive capacity under environmental stress.

Sustainable irrigation practices, reduced dependence on agricultural plastics, improved waste management systems, and more circular material use contribute to both environmental protection and food safety.

Consumer education remains important, but it should not carry disproportionate responsibility. At a practical level, recommended measures include increased consumption of fresh and less processed foods, use of point-of-use water filters when necessary, avoidance of heating food in plastic containers, and maintenance of a diversified diet including plant-based and fermented foods.

This advice is more robust when framed as co-benefit guidance. Fruits, vegetables, whole grains, legumes, nuts, and fermented foods have well-established health benefits beyond any potential relevance to environmental contaminant exposure (Scoditti, 2022; Tchounwou & Zhou, 2014). Thus, it is not a single-purpose recommendation.

Policy response can be structured in stages. The first is surveillance, involving the generation of comparable datasets on priority foods and water sources. The second is prevention, focusing on upstream control of product design, packaging, wastewater treatment, and agricultural plastics before entry into the food chain.

The third is accountability, requiring producers to maintain transparent records of contamination control and risk communication. The final stage is system integration, linking public health, food safety inspection, nutrition policy, and environmental governance so that households are not left to manage systemic contamination independently.

11. METHODOLOGICAL CLARIFICATION AND CRITICAL FRAMING

The methodological nature of this chapter should be clarified. This work is a narrative-integrative synthesis based on systematic search principles, rather than a registered systematic review as defined by PRISMA-type protocols (Yee et al., 2021).

It does not include pooled statistical analysis, formal risk-of-bias scoring, or meta-analysis; therefore, it does not make quantitative aggregated claims (Prata et al., 2020).

Its main contribution lies in integrating four domains: detection, exposure, toxicology, and governance. This integration, rather than statistical synthesis, represents the core analytical contribution (Leslie et al., 2022; Belzagui & Gutiérrez, 2022).

The evidence base remains uneven. Human and food-sample studies exist alongside mechanistic animal and in vitro research (Ragusa et al., 2021; Ho et al., 2023). Mechanistic pathways such as oxidative stress, barrier dysfunction, hepatic inflammation, neurotoxicity, and developmental effects provide biological plausibility (Deng et al., 2017; Monahan & Hoch, 2021).

However, human biomonitoring and epidemiological studies remain insufficient to establish dose–response relationships at the level of individual particles or dietary interventions (Yee et al., 2021; Cheung et al., 2018).

Methodological variability across FTIR/Raman workflows, sample preparation, particle size thresholds, and reporting units remains a key limitation (Primpke et al., 2020). Accordingly, reported contamination values should be interpreted as order-of-magnitude estimates rather than precise exposure metrics (Mintenig et al., 2017).

Nutritional interventions are included primarily due to shared mechanistic pathways across toxicity models, rather than direct evidence specific to microplastic exposure in humans (Cheng et al., 2025; Crosta et al., 2023). The One Health and HACCP framing remains conceptual but necessary, as food-system hazards cannot be separated from environmental drivers ([UNEP], 2023; Tchounwou & Zhou, 2014).

Overall, the evidence should be interpreted as triangulated but bounded—supportive for interpretation, but not for extended causal inference (Bessa et al., 2018).

12. TECHNOLOGICAL FEASIBILITY AND REALISTIC CONSTRAINTS

Technology can strengthen the framework, but its limitations must be explicitly acknowledged. AI-based classification of FTIR and Raman spectra can improve speed, consistency, and throughput in particle identification (Liu et al., 2025). Predictive models integrating environmental, industrial, and supply-chain data can support identification of high-risk nodes and HACCP-based surveillance (Akyildiz et al., 2024).

However, these systems depend strongly on data quality and validation. Performance in controlled spectral libraries does not necessarily translate to complex food matrices, particularly when samples are degraded or mixed.

Real-time monitoring systems may support early detection of abnormal shedding in packaging lines, filters, or process-water systems. Blockchain and traceability tools can improve batch-level accountability (de Oliveira et al., 2022; Visalli et al., 2021), but they do not resolve the underlying measurement limitations.

Material substitution, packaging redesign, and low-shed materials must be evaluated not only in terms of performance, but also cost, scalability, recyclability, and unintended environmental trade-offs (Tse et al., 2022).

Biotechnology offers potential but requires cautious interpretation. Engineered microbial systems and enzymes can degrade PET or related polymers under controlled conditions (Geyer et al., 2017; Tse et al.,2022), mainly in reactors or wastewater systems.

Filtration and bioreactor systems may reduce particle loads in process water. More advanced approaches, including engineered probiotics or synthetic biology–based gut interventions, remain hypothetical and require extensive safety validation (Rinninella et al., 2021; Crosta et al., 2023).

Omics-based approaches can characterize oxidative and metabolic stress responses (Yang et al., 2024), but their utility depends on standardized ontologies, validated models, and transparent reporting frameworks (Kim et al., 2018).

A One Health AI/biotechnology framework should therefore emphasize monitoring, prediction, and prevention across systems rather than reliance on single-point interventions (Carrillo-Bañuelos et al., 2022). Figure 2 summarizes this governance logic. Finally, feasibility can be summarized through three questions: does the intervention reduce exposure at a meaningful control point, can its effect be reliably measured, and can it be scaled without disproportionate burden on vulnerable populations?

Not all technological solutions meet these criteria. Some remain promising but not yet practical, while others are feasible but limited in scope. Accordingly, emphasis should remain on measurable, affordable, and auditable interventions that integrate into existing food safety systems.

Figure 2. Risk Governance and Mitigation Hierarchy for Microplastics in Food Systems.

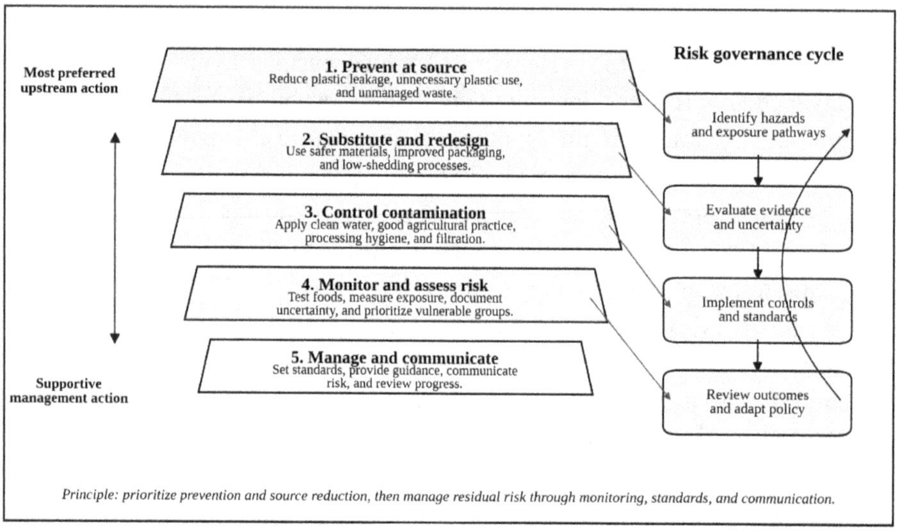

13. NUTRITIONAL MITIGATING: HIERARCHY OF EVIDENCE

The evidence for nutritional mitigation needs to be interpreted along a mechanistic-to-clinical gradient, rather than treated as a single uniform body of proof. At the most basic level, in vitro studies consistently show that micro- and nanoplastic particles can induce oxidative stress, inflammatory signaling, and disruption of epithelial barrier integrity (Schirinzi et al., 2017). This provides a mechanistic starting point, but it remains an isolated biological level.

At the next level, several bioactive compounds—including vitamin C, curcumin, sulforaphane, and related phytochemicals—have demonstrated antioxidant activity and a reduction in ROS, mainly through pathways such as Nrf2 activation in cellular and animal models (Crosta et al., 2023). These findings are relatively consistent across experimental systems, although they are still model-based and not exposure-specific to micro- and nanoplastics in humans.

Animal studies provide a more integrated biological picture. They show effects extending beyond isolated cells, including alterations in gut physiology, metabolic regulation, and hepatic responses under controlled exposure conditions . Still, even at this level, translation to human dietary exposure remains indirect.

Exposure detection in humans is more established than intervention evidence. Micro- and nanoplastics have been detected in stool, blood, tissues, and food matrices (Cox et al., 2019), but there is still no corresponding evidence showing that nutritional interventions directly modify these internal exposure markers in a controlled and validated way.

It is also worth noting that there is a substantial epidemiological literature supporting antioxidant-rich dietary patterns in general health contexts (Hemila & Chalker, 2013; Herweyers et al., 2020). However, this evidence is largely indirect with respect to microplastic exposure specifically, and should not be over-interpreted as direct protection against MNP-related effects.

From this perspective, Table 3 should not be read as a ranking of efficacy in a clinical sense. Rather, it reflects a structured grouping of compounds by the strength and type of available evidence, while intentionally avoiding an assumption of equivalence across all interventions. Some compounds—such as vitamin C, vitamin E, curcumin, probiotics, astaxanthin, and sulforaphane—span different levels of evidence, from moderate to experimental.

Public health interpretation therefore needs to remain conservative. The most defensible recommendation is still dietary diversity based on whole foods, with emphasis on gut health-supportive dietary patterns. More targeted or isolated compounds remain within the experimental or emerging evidence category and should be considered primarily for future research rather than immediate policy translation (Koistinen et al., 2018).

At the system level, nutritional strategies should be understood as supportive, not substitutive. They may modulate downstream biological responses, but they do not address exposure sources. For this reason, nutritional policy must remain embedded within broader frameworks of source control, HACCP implementation, and environmental governance (WHO, 2020; Akyildiz et al., 2024). As shown in Table 3, the evidence grading matrix distinguishes between different levels of scientific support for key bioactive compounds.

Table 3. Evidence Grading Matrix for Bioactive Compounds in Relation to Microplastic-Associated Health Protection

Compound	Evidence Level	Key Mechanisms	Human Data Quality
Vitamin C/E	Moderate	ROS scavenging	Clinical / epidemiological
Curcumin	Moderate	Nrf2 activation	Animal + mechanistic
Probiotics	Moderate	Barrier / microbiota modulation	Clinical (non-MNP specific)
Astaxanthin	Experimental	Mitochondrial protection	Animal only
Sulforaphane	Promising	Detoxification pathways	Mechanistic

This table presents an evidence grading matrix for selected dietary bioactive compounds, summarizing their proposed mechanisms of action, evidence levels, and human data quality in the context of microplastic-associated biological effects. The matrix distinguishes between clinical, epidemiological, mechanistic, and experimental evidence categories to support interpretation of current knowledge and research gaps. (Source: WHO, 2020; Akyildiz et al., 2024)

Priority in intervention logic remains upstream control. In primary production systems, reducing entry of pollutants into the agricultural environment can be achieved through improved wastewater treatment, membrane bioreactors, selection of appropriate irrigation water sources, and limiting plastic mulch application (Nizzetto et al., 2016). These measures target contamination before it enters the food chain, which is still the most effective point of intervention.

In processing environments, mitigation relies on engineering controls such as HEPA filtration, ultrafiltration systems, low-shedding surfaces, and preventive maintenance protocols to reduce airborne and waterborne particle release (Hartline et al., 2016). At the packaging level, the use of recyclable or reusable materials, lower-shedding polymers, and emerging smart sensing technologies can contribute to reducing particle release (Geueke et al., 2018; Zhou et al., 2023).

At the consumer level, point-of-use filtration systems and improved food-handling practices may reduce exposure in specific contexts (Schwabl et al., 2019; Zhou et al., 2023), although their impact remains context-dependent rather than universal.

Overall, this represents a layered risk-reduction model rather than a single intervention pathway. Nutritional mitigation is positioned as a downstream supportive layer, not a replacement for exposure reduction. The central logic remains that primary control of exposure sources must precede biological mitigation strategies (Jambeck et al., 2015).

Additional control can also occur during distribution, particularly through packaging integrity and cold-chain stability, which may reduce secondary particle release during storage and transport (Lusher et al., 2015). When combined, these layers form a multi-level defense system aimed at improving food safety, reducing environmental load, and supporting broader sustainability targets, including SDG 2 and SDG 12 ([UNEP], 2023; Völkl, 2023).

14. GAPS IN KNOWLEDGE AND FUTURE DIRECTIONS

Three main gaps remain central to future research.

First, there is a need for standardized, polymer-confirmed detection methods that are scalable and applicable across diverse food matrices. Current approaches remain heterogeneous in cost, sensitivity, and reproducibility.

Second, stronger linkage is needed between exposure measurements and biomarker-based health outcomes. Without this connection, dose–response relationships cannot be reliably established.

Third, intervention-based studies are still largely missing. Well-designed trials are required to evaluate whether dietary patterns or specific bioactive compounds can measurably influence markers such as oxidative stress, gut barrier integrity, or inflammatory responses linked to MNP exposure (McAllister, 2018).

Future work should also integrate feasibility dimensions, including cost, scalability, and implementation burden. This includes comparing source control strategies, packaging redesign, filtration systems, and consumer-level interventions in a unified framework. Without this comparative perspective, the field risks remaining descriptive rather than decision-oriented.

A more integrated research structure is therefore needed. Analytical chemistry should focus on improving detection limits and polymer identification; toxicology should refine mechanistic understanding; nutrition science should test realistic dietary patterns; public health research should identify vulnerable populations; and policy research should evaluate regulatory feasibility.

This type of interdisciplinary alignment is not optional in this field—it is a direct requirement for moving from environmental observation to actionable governance.

15 CONCLUSION

Micro- and nanoplastic contamination of food systems is not a purely descriptive environmental issue. It is a multi-pathway, system-level exposure problem that spans soil, water, air, and food chains, ultimately leading to human ingestion and inhalation exposure.

The most consistently proposed biological mechanisms include oxidative stress, intestinal barrier disruption, inflammatory responses, metabolic alterations, and potential effects on multiple organs. However, the current evidence base remains uneven, and strong causal claims in humans are still not justified.

Nutritional strategies may contribute as supportive mechanisms, mainly through modulation of oxidative and inflammatory pathways. However, they should not be interpreted as preventive substitutes. Their role is adjunctive and mechanistic, not primary.

Accordingly, the chapter emphasizes upstream source control, standardized analytical methods, HACCP-based food safety systems, One Health coordination, and policy-level interventions. Emerging technologies, including AI-assisted spectroscopy, predictive monitoring systems, traceability tools, and selected biotechnological approaches, may strengthen implementation, but only when embedded within validated governance structures.

16. RECOMMENDATIONS

Future work should move toward testable and policy-relevant evidence generation. This includes harmonized analytical protocols for detection, biomonitoring studies linking exposure to biological markers, and controlled intervention trials evaluating realistic dietary strategies.

Food producers should be required to define and monitor critical control points related to micro- and nanoplastic contamination within HACCP systems, including documentation of corrective actions. Regulatory frameworks should also address water quality, packaging design, biosolid management, and plastic waste control more explicitly.

At the communication level, public guidance should avoid over-framing dietary advice as protective against contamination. Instead, it should be presented within a broader food safety and health resilience context.

The One Health framework remains a suitable integrative approach for aligning consumer protection, environmental management, and sustainable food systems.

A practical implementation sequence can be summarized as: measure, prioritize, control, communicate, and evaluate. Measurement ensures evidence quality. Prioritization identifies critical foods, processes, and populations. Control targets key contamination pathways. Communication provides balanced consumer guidance. Evaluation assesses whether interventions reduce exposure over time.

This sequence does not replace existing food safety systems but adds a structured operational layer connecting microplastic research to governance and practice.

While the preceding discussion emphasizes risk characterization, control strategies, and One Health-based mitigation frameworks for micro- and nanoplastic contamination in food systems, an additional dimension becomes essential for real-world implementation. This is the economic feasibility and sustainability of these interventions. Translating scientific and regulatory recommendations into operational policies requires not only technical validation but also assessment of cost-effectiveness, resource allocation, and long-term public health benefits. In this context, the evaluation of mitigation strategies through economic and health-system perspectives provides a critical bridge between environmental risk management and practical policy adoption.

REFERENCES

. Akyildiz, S. H., Fiore, S., Bruno, M., Sezgin, H., Yalcin-Enis, I., Yalcin, B., & Bellopede, R. (2024). Release of microplastic fibers from synthetic textiles during household washing. *Environmental pollution (Barking, Essex: 1987), 357*, 124455. https://doi.org/DOI: 10.1016/j.envpol.2024.124455

Andrady, A. L. (2011). Microplastics in the marine environment. *Marine Pollution Bulletin, 62*(8), 1596–1605. DOI: 10.1016/j.marpolbul.2011.05.030 PMID: 21742351

Anigbogu, C. B., Obiora, C. U., Kolawole, O. O., & Okorafor-Nwosu, A. (2025). Nutritional assessment and sensory properties of dried pap fortified with soybean and date palm fruit flours. *E-Proceedings of the Faculty of Agriculture International Conference*, 164–169. https://doi.org/DOI: 10.5281/da8s6652

Au, S. Y., Lee, C. M., Weinstein, J. E., van den Hurk, P., & Klaine, S. J. (2017). Trophic transfer of microplastics in aquatic ecosystems: Identifying critical research needs. *Integrated Environmental Assessment and Management, 13*(3), 505–509. DOI: 10.1002/ieam.1907 PMID: 28440939

Belzagui Elder, F., & Gutiérrez Bouzán, M. C. (2022). Review of alternatives for reducing textile microfiber emissions to water. *Journal of Environmental Management, 317*, 115347. Advance online publication. DOI: 10.1016/j.jenvman.2022.115347

Bessa, F., Barria, P., Neto, J. M., Frias, J. P., Otero, V., Sobral, P., & Marques, J. C. (2018). Occurrence of microplastics in commercial fish from a wide range of waters. *Environmental Pollution, 243*, 1098–1107. DOI: 10.1016/j.envpol.2018.09.01

Brennecke, D., Duarte, B., Paiva, F., Caçador, I., & Canning-Clode, J. (2016). Microplastics as vectors for heavy metal contamination from the marine environment. *Estuarine, Coastal and Shelf Science, 178*, 189–195. DOI: 10.1016/j.ecss.2015.12.003

Cheng, J., Li, J., Xiong, R. G., Zhou, D. D., Huang, S. Y., Wu, S. X., Shang, A., Tang, G. Y., Li, H. B., & Gan, R. Y. (2025). Potentially harmful effects of micro-/nanoplastics on humans as well as protective actions of dietary natural products. *Trends in Food Science & Technology, 156*, 104841. DOI: 10.1016/j.tifs.2024.104841

Cheung, L. T. O., Lui, C. Y., & Fok, L. (2018). Microplastic Contamination of Wild and Captive Flathead Grey Mullet (Mugil cephalus). *International Journal of Environmental Research and Public Health, 15*(4), 597. DOI: 10.3390/ijerph15040597 PMID: 29587444

Chuang, W.-Y., Young, T.-H., Chiu, W.-Y., & Lin, C.-Y. (2000). Effect of polymeric additives on the structure and permeability of poly (vinyl alcohol) asymmetric membranes. *Additives for Polymers*, *2000*(10–11), 17–18. DOI: 10.1016/S0306-3747(00)90314-2

Cole, M., Lindeque, P., Halsband, C., & Galloway, T. S. (2011). Microplastics as contaminants in the marine environment: A review. *Marine Pollution Bulletin*, *62*(12), 2588–2597. DOI: 10.1016/j.marpolbul.2011.09.025 PMID: 22001295

Coman, L. I., Balaban, D. V., Dumbravă, B. F., Păunescu, H., Marin, R. C., Costescu, M., Dima, L., Jinga, M., & Coman, O. A. (2025). Targeting Oxidative Stress in Acute Pancreatitis: A Critical Review of Antioxidant Strategies. *Nutrients*, *17*(15), 2390. DOI: 10.3390/nu17152390 PMID: 40805976

Cox, K. D., Covernton, G. A., Davies, H. L., Dower, J. F., Juanes, F., & Dudas, S. E. (2019). Human Consumption of Microplastics. *Environmental Science & Technology*, *53*(12), 7068–7074. DOI: 10.1021/acs.est.9b01517 PMID: 31184127

Crosta, A., Parolini, M., & De Felice, B. (2023). Microplastics contamination in nonalcoholic beverages from the Italian market. *International Journal of Environmental Research and Public Health*, *20*(5), 4122. DOI: 10.3390/ijerph20054122 PMID: 36901131

De Corte, W., Hermabessiere, L., & Dekeyser, S. (2019). Microplastics in drinking water: Recommended solutions and open questions. *Current Opinion in Environmental Science & Health*, *12*, 72–82. DOI: 10.1016/j.coesh.2019.10.008

de Oliveira, V. M., Zarzur, H. R., & de Souza, A. C. (2022). Artificial intelligence applications in food safety systems: A review of enabled monitoring technologies. *Trends in Food Science & Technology*, *128*, 245–256. DOI: 10.1016/j.tifs.2022.08.001

Demestichas, K., Peppes, N., Alexakis, T., & Adamopoulou, E. (2020). Blockchain in agriculture traceability systems: A review. *Applied Sciences (Basel, Switzerland)*, *10*(12), 4113. DOI: 10.3390/app10124113

Deng, Y., Zhang, Y., Lemos, B., & Ren, H. (2017). Tissue accumulation of microplastics in mice and biomarker responses suggest widespread health effects of exposure to microplastics. *Journal of Hazardous Materials*, *344*, 1026–1033. DOI: 10.1016/j.jhazmat.2017.11.047

Dris, R., Gasperi, J., Saad, M., Mirande, C., & Tassin, B. (2016). Synthetic fibers in atmospheres and fallouts. *Water, Air, and Soil Pollution*, *228*(8), 311. DOI: 10.1007/s11270-017-3467-5

. Galloway, T. S., Cole, M., & Lewis, C. (2017). Interactions of microplastics throughout the marine ecosystem. Nature Ecology & Evolution, 1(5), 0116. https://doi.org/DOI: 10.1038/s41559-017-0116

Geueke, B., Groh, K. J., & Muncke, J. (2018). Food packaging in the circular economy: Overview of chemical safety aspects for commonly used materials. *Journal of Cleaner Production, 193*, 491–505. DOI: 10.1016/j.jclepro.2018.05.005

Gewert, B., Plassmann, M. M., & MacLeod, M. (2015). Pathways for degradation of plastic polymers floating in the marine environment. *Environmental Science. Processes & Impacts, 17*(9), 1513–1521. DOI: 10.1039/C5EM00207A PMID: 26216708

Geyer, R., Jambeck, J. R., & Law, K. L. (2017). Production, use, and fate of all plastics ever made. *Science Advances, 3*(7), e1700782. DOI: 10.1126/sciadv.1700782 PMID: 28776036

Glüge, J., Scheringer, M., Cousins, I. T., DeWitt, J. C., Goldenman, G., Herzke, D., Lohmann, R., Ng, C. A., Trier, X., & Wang, Z. (2020). An overview of the uses of per- and polyfluoroalkyl substances (PFAS). *Environmental Science. Processes & Impacts, 22*(12), 2345–2373. DOI: 10.1039/D0EM00291G PMID: 33125022

Gundogdu, S., Çevik, C., Guzel, E., & Kilercioğlu, S. (2018). Sources and fate of microplastics in marine environment: A review of recent studies. *Marine Pollution Bulletin, 137*, 393–404. DOI: 10.1016/j.marpolbul.2018.10.045

Guo, X., & Wang, N. (2019). Neurotoxicity of microplastic particles. *Progress in Molecular Biology and Translational Science, 168*, 333–349. DOI: 10.1016/bs.pmbts.2019.11.011

Hartline, N. L., Bruce, N. J., Karba, S. N., Ruff, E. O., Sonar, S. U., & Holden, P. A. (2016). Microfiber Masses Recovered from Conventional Machine Washing of New or Aged Garments. *Environmental Science & Technology, 50*(21), 11532–11538. DOI: 10.1021/acs.est.6b03045 PMID: 27689236

Hemila, H., & Chalker, E. (2013). Vitamin C for preventing and treating the common cold. *Cochrane Database of Systematic Reviews, 2013*(1), CD000980. Advance online publication. DOI: 10.1002/14651858.CD000980.pub4 PMID: 23440782

Hermabessiere, L., Dehaut, A., Paul-Pont, I., Lacroix, C., Jezequel, R., Soudant, P., & Duflos, G. (2017). Occurrence and effects of plastic additives on marine environments and organisms: A review. *Chemosphere, 182*, 781–793. DOI: 10.1016/j.chemosphere.2017.05.096 PMID: 28545000

. Hernandez, L. M., Xu, E. G., Larsson, H. C., Tahara, R., Maisuria, V. B., & Tufenkji, N. (2019). Plastic teabags release billions of microparticles and nanoparticles into tea. *Environmental science & technology, 53*(21), 12300-12310.d

Herweyers, L., Catarci Carteny, C., Scheelen, L., Watts, R., & Du Bois, E. (2020). Consumers' Perceptions and Attitudes toward Products Preventing Microfiber Pollution in Aquatic Environments as a Result of the Domestic Washing of Synthetic Clothes. *Sustainability (Basel), 12*(6), 2244. DOI: 10.3390/su12062244

Ho, K. T. H., Kwok, P. W. H., Chang, S. S. Y., & Chu, A. M. Y. (2023). Gaps between attitudes and behavior in the use of disposable plastic tableware (DPT) and factors influencing sustainable DPT consumption: A study of Hong Kong undergraduates. *Sustainability (Basel), 15*(11), 8958. DOI: 10.3390/su15118958

Huang, M., Gong, Y., Grondolsky, J., & Hoover-Plow, J. (2014). Lp(a)/apo(a) modulate MMP-9 activation and neutrophil cytokines in vivo in inflammation to regulate leukocyte recruitment. *The American Journal of Pathology, 184*(5), 1503–1517. DOI: 10.1016/j.ajpath.2014.01.010 PMID: 24650562

. Ivar do Sul, J. A., & Costa, M. F. (2014). The present and future of microplastic pollution in the marine environment. *Environmental pollution (Barking, Essex: 1987), 185*, 352–364. https://doi.org/DOI: 10.1016/j.envpol.2013.10.036

Jambeck, J. R., Geyer, R., Wilcox, C., Siegler, T. R., Perryman, M., Andrady, A., Narayan, R., & Law, K. L. (2015). Marine pollution. Plastic waste inputs from land into the ocean. *Science, 347*(6223), 768–771. DOI: 10.1126/science.1260352 PMID: 25678662

Jee, D., Kang, S., Yuan, C., Cho, E., & Arroyo, J. G. (2016). Serum 25-Hydroxyvitamin D Levels and Dry Eye Syndrome: Differential Effects of Vitamin D on Ocular Diseases. *PLoS One, 11*(2), e0149294. DOI: 10.1371/journal.pone.0149294 PMID: 26894581

Karami, A., Golieskardi, A., Keong Choo, C., Larat, V., Galloway, T. S., & Salamatinia, B. (2017). The presence of microplastics in commercial salts from different countries. *Scientific Reports, 7*(1), 46173. DOI: 10.1038/srep46173 PMID: 28383020

Kateris, D., & Bochtis, D. (2023). Applied Agri-Technologies for Agriculture 4.0—Part I. *Applied Sciences (Basel, Switzerland), 13*(7), 4180. DOI: 10.3390/app13074180

Kim, J. S., Lee, H. J., Kim, S. K., & Kim, H. J. (2018). Global Pattern of Microplastics (MPs) in Commercial Food-Grade Salts: Sea Salt as an Indicator of Seawater MP Pollution. *Environmental Science & Technology*, *52*(21), 12819–12828. DOI: 10.1021/acs.est.8b04180 PMID: 30285421

. Koelmans, A. A., Besseling, E., & Foekema, E. M. (2014). Leaching of plastic additives to marine organisms. *Environmental pollution (Barking, Essex: 1987)*, *187*, 49–54. https://doi.org/DOI: 10.1016/j.envpol.2013.12.013

Lambert, S., & Wagner, M. (2016). Characterisation of nanoplastics during the degradation of polystyrene. *Chemosphere*, *145*, 265–268. DOI: 10.1016/j.chemosphere.2015.11.078 PMID: 26688263

Leslie, H. A., van Velzen, M. J. M., Brandsma, S. H., Vethaak, A. D., Garcia-Vallejo, J. J., & Lamoree, M. H. (2022). Discovery and quantification of plastic particle pollution in human blood. *Environment International*, *163*, 107199. DOI: 10.1016/j.envint.2022.107199 PMID: 35367073

Liebezeit, G., & Liebezeit, E. (2013). Non-pollen particulates in honey and sugar. *Food Additives & Contaminants. Part A, Chemistry, Analysis, Control, Exposure & Risk Assessment*, *30*(12), 2136–2140. DOI: 10.1080/19440049.2013.843025 PMID: 24160778

Liu, Y., Chen, S., Xiong, X., Wen, Z., Zhao, L., Xu, B., Guo, Q., Xia, J., & Pei, J. (2025). Artificial intelligence guided Raman spectroscopy in biomedicine: Applications and prospects. *Journal of Pharmaceutical Analysis*, *15*(11), 101271. DOI: 10.1016/j.jpha.2025.101271 PMID: 41377140

Lusher, A. L., Tirelli, V., O'Connor, I., & Officer, R. (2015). Microplastics in Arctic polar waters: The first reported values of particles in surface and sub-surface samples. *Scientific Reports*, *5*(1), 14947. Advance online publication. DOI: 10.1038/srep14947 PMID: 26446348

Marfella, R., Prattichizzo, F., Sardu, C., Fulgenzi, G., Graciotti, L., Spadoni, T., D'Onofrio, N., Scisciola, L., La Grotta, R., Frigé, C., Pellegrini, V., Municinò, M., Siniscalchi, M., Spinetti, F., Vigliotti, G., Vecchione, C., Carrizzo, A., Accarino, G., Squillante, A., & Paolisso, G. (2024). Microplastics and Nanoplastics in Atheromas and Cardiovascular Events. *The New England Journal of Medicine*, *390*(10), 900–910. DOI: 10.1056/NEJMoa2309822 PMID: 38446676

Mason, S. A., Welch, V. G., & Neratko, J. (2018). Synthetic Polymer Contamination in Bottled Water. *Frontiers in Chemistry*, *6*, 407. DOI: 10.3389/fchem.2018.00407 PMID: 30255015

McAllister, S. R. (2018). *Implementation of food safety regulations in food service establishments* (Publication No. 5902) [Doctoral dissertation, Walden University]. ScholarWorks. https://scholarworks.waldenu.edu/dissertations/5902

Mintenig, S. M., Int-Veen, I., Löder, M. G. J., Primpke, S., & Gerdts, G. (2017). Identification of Microplastic in Effluents of Waste Water Treatment Plants Using Focal Plane Array-Based Micro-Fourier-Transform Infrared Imaging. *Water Research, 108*, 365–372. DOI: 10.1016/j.watres.2016.11.015 PMID: 27838027

Monahan, K., & Hoch, J. M. (2021). Comparison of phosphate uptake by epiphyton, metaphyton, and epipelon in the Everglades. *Florida Scientist, 84*(1), 35–44. https://nsuworks.nova.edu/occ_facarticles/1184

Neves, D., Sobral, P., Ferreira, J. L., & Pereira, T. (2015). Ingestion of microplastics by commercial fish off the Portuguese coast. *Marine Pollution Bulletin, 101*(1), 119–126. DOI: 10.1016/j.marpolbul.2015.11.008 PMID: 26608506

Nishida, C., Uauy, R., Kumanyika, S., & Shetty, P. (2004). The joint WHO/FAO expert consultation on diet, nutrition and the prevention of chronic diseases: Process, product and policy implications. *Public Health Nutrition, 7*(1a, 1A), 245–250. DOI: 10.1079/PHN2003592 PMID: 14972063

Nizzetto, L., Futter, M., & Langaas, S. (2016). Are Agricultural Soils Dumps for Microplastics of Urban Origin? *Environmental Science & Technology, 50*(20), 10777–10779. DOI: 10.1021/acs.est.6b04140 PMID: 27682621

Oßmann, B. E., Sarau, G., Holtmannspötter, H., Pischetsrieder, M., Christiansen, S. H., & Dicke, W. (2018). Small-sized microplastics and pigmented particles in bottled mineral water. *Water Research, 141*, 307–316. DOI: 10.1016/j.watres.2018.05.027 PMID: 29803096

Prata, J. C., da Costa, J. P., Lopes, I., Duarte, A. C., & Rocha-Santos, T. (2020). Environmental exposure to microplastics: An overview on possible human health effects. *The Science of the Total Environment, 702*, 134455. DOI: 10.1016/j.scitotenv.2019.134455 PMID: 31733547

Primpke, S., Christiansen, S. H., Cowger, W., De Frond, H., Deshpande, A., Fischer, M., Holland, E. B., Meyns, M., O'Donnell, B. A., Ossmann, B. E., Pittroff, M., Sarau, G., Scholz-Böttcher, B. M., & Wiggin, K. J. (2020). Critical Assessment of Analytical Methods for the Harmonized and Cost-Efficient Analysis of Microplastics. *Applied Spectroscopy, 74*(9), 1012–1047. DOI: 10.1177/0003702820921465 PMID: 32249594

Ragusa, A., Svelato, A., Santacroce, C., Catalano, P., Notarstefano, V., Carnevali, O., Papa, F., Rongioletti, M. C. A., Baiocco, F., Draghi, S., D'Amore, E., Rinaldo, D., Matta, M., & Giorgini, E. (2021). Plasticenta: First evidence of microplastics in human placenta. *Environment International*, *146*, 106274. DOI: 10.1016/j.envint.2020.106274 PMID: 33395930

Rinninella, E., Raoul, P., Cintoni, M., Palombaro, M., Pulcini, G., Gasbarrini, A., & Mele, M. C. (2021). Nutritional Interventions Targeting Gut Microbiota during Cancer Therapies. *Microorganisms*, *9*(7), 1469. DOI: 10.3390/microorganisms9071469 PMID: 34361904

Rubiños, S. L., Grados, J. H., Medina, J. T., Chávez, E. N., Huarcaya, E., Rocha, V. E., Grados, H. J., & Neyra, M. R. (2023). Microplastics in the ecosystem: A systematic review of the methods for their detection and removal. *International Journal of Ecology*, *2023*(1), 8826716. DOI: 10.1155/2023/8826716

Schirinzi, G. F., Pérez-Pomeda, I., Sanchís, J., Rossini, C., Farré, M., & Barceló, D. (2017). Cytotoxic effects of commonly used nanomaterials and microplastics on cerebral and epithelial human cells. *Environmental Research*, *159*, 579–587. DOI: 10.1016/j.envres.2017.08.043 PMID: 28898803

Schwabl, P., Köppel, S., Königshofer, P., Bucsics, T., Trauner, M., Reiberger, T., & Liebmann, B. (2019). Detection of Various Microplastics in Human Stool: A Prospective Case Series. *Annals of Internal Medicine*, *171*(7), 453–457. DOI: 10.7326/M19-0618 PMID: 31476765

Scoditti, E., Tumolo, M. R., & Garbarino, S. (2022). Mediterranean Diet on Sleep: A Health Alliance. *Nutrients*, *14*(14), 2998. DOI: 10.3390/nu14142998 PMID: 35889954

Shan, Q., Li, H., Chen, N., Qu, F., & Guo, J. (2020). Understanding the Multiple Effects of PCBs on Lipid Metabolism. *Diabetes, Metabolic Syndrome and Obesity*, *13*, 3691–3702. DOI: 10.2147/DMSO.S264851 PMID: 33116719

Shim, W. J., Hong, S. H., & Eo, S. E. (2017). Identification Methods in Microplastic Analysis: A Review. *Analytical Methods : Advancing Methods and Applications*, *9*(9), 1384–1391. DOI: 10.1039/C6AY02558G

Shruti, V. C., Kutralam-Muniasamy, G., Pérez-Guevara, F., Roy, P. D., & Elizalde Martínez, I. (2022). Occurrence and Characteristics of Atmospheric Microplastics in Mexico City. *The Science of the Total Environment*, *847*, 157601. Advance online publication. DOI: 10.1016/j.scitotenv.2022.157601 PMID: 35882345

Smith, M., Love, D. C., Rochman, C. M., & Neff, R. A. (2018). Microplastics in Seafood and the Implications for Human Health. *Current Environmental Health Reports*, *5*(3), 375–386. DOI: 10.1007/s40572-018-0206-z PMID: 30116998

Tao, Y., Cui, W., Song, W., Xu, H., Zeng, J., & Li, H. (2025). Major adverse cardiovascular events in angiogenesis inhibitor-based cancer therapy: A systematic review and meta-analysis. *Journal of the American Heart Association*, *14*(1), e042896. Advance online publication. DOI: 10.1161/JAHA.125.042896 PMID: 41181998

Tchounwou, P. B., & Zhou, J. (2014). *International Journal of Environmental Research and Public Health* Best Paper Award 2014. *International Journal of Environmental Research and Public Health*, *11*(1), 1192–1194. DOI: 10.3390/ijerph110101192 PMID: 24434595

Thompson, R. C., Olsen, Y., Mitchell, R. P., Davis, A., Rowland, S. J., John, A. W., McGonigle, D., & Russell, A. E. (2004). Lost at sea: Where is all the plastic? *Science*, *304*(5672), 838. DOI: 10.1126/science.1094559 PMID: 15131299

Tran-Dinh, A., Diallo, D., Delbosc, S., Varela-Perez, L. M., Dang, Q. B., Lapergue, B., Burillo, E., Michel, J. B., Levoye, A., Martin-Ventura, J. L., & Meilhac, O. (2013). HDL and endothelial protection. *British Journal of Pharmacology*, *169*(3), 493–511. DOI: 10.1111/bph.12174 PMID: 23488589

Tse, Y. T., Chan, S. M., & Sze, E. T. (2022). Quantitative Assessment of Full Size Microplastics in Bottled and Tap Water Samples in Hong Kong. *International Journal of Environmental Research and Public Health*, *19*(20), 13432. DOI: 10.3390/ijerph192013432 PMID: 36294013

United Nations Environment Programme. (2023, May). *Chemicals in plastics: A technical report.* United Nations Environment Programme; Secretariat of the Basel, Rotterdam and Stockholm Conventions. https://wedocs.unep.org/handle/20.500.11822/42366

. Van Cauwenberghe, L., & Janssen, C. R. (2014). Microplastics in bivalves cultured for human consumption. *Environmental pollution (Barking, Essex: 1987)*, *193*, 65–70. https://doi.org/DOI: 10.1016/j.envpol.2014.06.010

Visalli, G., Facciolà, A., Pruiti Ciarello, M., De Marco, G., Maisano, M., & Di Pietro, A. (2021). Acute and Sub-Chronic Effects of Microplastics (3 and 10 µm) on the Human Intestinal Cells HT-29. *International Journal of Environmental Research and Public Health*, *18*(11), 5833. DOI: 10.3390/ijerph18115833 PMID: 34071645

Völkl, M. (2023). *From single cells to microtissues: Using advanced culture techniques to create a better understanding of microplastic toxicity* (Publication No. 7182) [Doctoral dissertation, University of Bayreuth]. EPub Bayreuth. https://doi.org/DOI: 10.15495/EPub_UBT_00007182

Wagner, S., Hüffer, T., Klöckner, P., Wehrhahn, M., Hofmann, T., & Reemtsma, T. (2018). Tire wear particles in the aquatic environment - A review on generation, analysis, occurrence, fate and effects. *Water Research*, *139*, 83–100. DOI: 10.1016/j.watres.2018.03.051 PMID: 29631188

Wang, Z., Wang, Y., Zhang, J., Feng, G., Miao, S., Lu, R., Tian, X., & Ye, Y. (2025). Antioxidant Intervention Against Microplastic Hazards. *Antioxidants*, *14*(7), 797. DOI: 10.3390/antiox14070797 PMID: 40722901

Yang, J., Guo, X., Li, T., Xie, Y., Wang, D., & Long, Y. (2023). Sulforaphane Inhibits Exhaustive Exercise-Induced Liver Injury and Transcriptome-Based Mechanism Analysis. *Nutrients*, *15*(14), 3220. DOI: 10.3390/nu15143220 PMID: 37513640

Yang, X., Zhou, Y., Li, Y., & Wang, S. (2024). International Law and Regulation of Marine Microplastics: Current Situation, Problems, and Development. *Sustainability (Basel)*, *16*(21), 9337. DOI: 10.3390/su16219337

Yee, M. S., Hii, L. W., Looi, C. K., Lim, W. M., Wong, S. F., Kok, Y. Y., Tan, B. K., Wong, C. Y., & Leong, C. O. (2021). Impact of Microplastics and Nanoplastics on Human Health. *Nanomaterials (Basel, Switzerland)*, *11*(2), 496. DOI: 10.3390/nano11020496 PMID: 33669327

Zhang, G. S., & Liu, Y. F. (2018). The distribution of microplastics in soil aggregate fractions in southwestern China. *The Science of the Total Environment*, *642*, 12–20. DOI: 10.1016/j.scitotenv.2018.06.004 PMID: 29894871

Zhang, W., Li, H., Zheng, H., Lan, H., & Zhuang, Y. (2022). Arsenic speciation and concentration in the urine of infants in Xiamen, China: A cross-sectional survey [Preprint]. *Research Square*. https://doi.org/DOI: 10.21203/rs.3.rs-1522535/v1

Zhou, A., Zhao, Y., Liu, M., Suyamud, B., Yuan, W., & Yang, Y. (2023). Occurrence and risk assessment of microplastics in the Lhasa River-a remote plateau river on the Qinghai-Tibet Plateau, China. *Environmental Monitoring and Assessment*, *195*(3), 433. DOI: 10.1007/s10661-023-11040-z PMID: 36856933

Zhou, X., Wang, Q., Wang, J., Li, H., Ren, J., & Tang, S. (2023). Quantification of Microplastics in Plastic-Bottled Chinese Baijiu Using Micro-FTIR in Imaging Mode. *Applied Sciences (Basel, Switzerland)*, *13*(20), 11142. DOI: 10.3390/app132011142

Chapter 7
Translational One Health Strategies for Nanoplastic Exposure Mitigation

P. Praveen Kumar
http://orcid.org/0000-0002-0707-039X
SRM Institute of Science and Technology, Vadapalani, India

K. C. Raja Shree
http://orcid.org/0009-0009-1029-910X
Saveetha School of Management, Saveetha Institute of Medical and Technical Sciences, India

ABSTRACT

This chapter per the authors addresses translational strategies for mitigating nanoplastic exposure across environmental, technological, economic, and public health domains within a One Health framework. Nanoplastics are evaluated as mobile particulate contaminants capable of crossing biological barriers, accumulating across environmental compartments, and contributing to chronic ecological and physiological exposure burdens. The chapter analyzes mitigation pathways involving exposure surveillance, biotechnology-assisted reduction systems, computational exposure forecasting, and implementation-oriented environmental management models. Integration of exposure analytics, environmental management, and implementation economics contributes operational perspectives for reducing nanoplastic-associated ecological and health burdens across interconnected environments emphasizing integrated governance and resilience-based environmental action for sustainability goals.

DOI: 10.4018/979-8-3373-4490-4.ch007

INTRODUCTION

The transition from nutritional intervention strategies under nanoplastic exposure to broader systems-level consequences highlights a critical dimension that extends beyond individual or clinical response frameworks. While nutritional approaches focus on mitigating biological and physiological disruptions at the organism level, persistent and widespread exposure scenarios inevitably generate cumulative pressures that propagate into public health systems.

As exposure continues across populations and time scales, the burden shifts from individual-level intervention to systemic healthcare demand. This progression introduces escalating public health challenges, particularly in terms of resource allocation, long-term care requirements, and preventive strategy sustainability.

Within this context, the economic implications of exposure mitigation become central to understanding the feasibility and scalability of intervention strategies. Costs associated with monitoring, treatment, and long-term management begin to shape decision-making processes in both clinical and policy domains.

This transition naturally positions the discussion toward a prevention-versus-remediation framework, where the relative efficiency of early intervention strategies is weighed against the increasing economic burden of delayed or reactive responses.

Plastics proliferate at different levels, including micro and nano particles. This has highlighted the importance of addressing public health challenges alongside environmental degradation. Compared with macro-level plastic waste, nanoplastics are smaller in nature, invisible to the naked eye, and capable of passing into human membranes (Bhat et al., 2023; Kandaswamy et al., 2024). Their minute size and tendency to accumulate in tissue traces impose serious concerns for long-term human health. Therefore, nanoplastic pollution should be addressed from a global perspective, and a One Health approach is crucial to protect human, animal, and environmental health. Here, economic evaluation is critical for measuring costs related to environmental exposure and for prioritising resource allocation for mitigation processes.

The contamination of nanoplastics can be reduced through technological implementation, and stronger collaboration is required among different stakeholders involved in the process. Technological innovation alone will not be sufficient to address the problem; economic evaluation is required to improve health resource efficiency and strengthen long-term return on investment. Economic evaluation tools, such as cost-effectiveness analysis, cost-benefit analysis, and regulatory impact analysis, can be applied in this regard. However, investment in AI-based monitoring may appear expensive. Still, when considering long-term healthcare savings, the disease burden suggests that it is highly cost-efficient.

Additionally, remediation methods based on biotechnology must be applied with attention to affordability and scalability across different socioeconomic contexts. Evaluating affordability potential ensures that interventions remain scalable and accessible to wider populations. These approaches should also align with environmental challenges and Sustainable Development Goals to achieve stronger outcomes.

Accordingly, this chapter develops an interdisciplinary framework bridging environmental exposure, public health systems, technological innovation, and economic evaluation within a One Health structure. AI and biotechnology are not supportive add-ons but core mechanisms capable of reducing health burden, operational costs, and long-term societal losses associated with nanoplastic exposure.

LITERATURE REVIEW

The health burden associated with nanoplastics has become a serious issue for environmental and public health systems. Nanoplastics refer to plastic particles less than one micrometre in size and are ubiquitous contaminants produced through the degradation of larger plastics and industrial processes (Dube & Okuthe, 2023). Unlike larger microplastics, nanoplastics penetrate cellular barriers and translocate through the bloodstream, accumulating in multiple organs (Vignesh et al., 2024). Evidence is now found in blood, placenta, breast milk, and faeces. These findings clearly indicate both biological impacts and associated economic costs.

Evidence also shows that nanoplastics may induce oxidative stress, inflammatory responses, metabolic modifications, and neurotoxicity (Velumani et al., 2025). These materials affect food chains and water systems, creating serious risks for wildlife and ecosystems. Consequently, these interconnections reinforce the importance of a One Health perspective, where human, animal, and environmental health are treated as integrated systems (Waring et al., 2018).

EXPOSURE PATHWAYS & BIODISTRIBUTION

Humans are increasingly exposed to nanoplastics through multiple pathways. The primary route is ingestion through contaminated seafood, agricultural products, and drinking water, as plastics are absorbed through soil and irrigation systems. Filter-feeding organisms such as mussels, fish, and oysters transfer plastic particles to humans through dietary consumption (Kibria et al., 2022).

Inhalation is another pathway, where airborne plastics penetrate deeper into the respiratory system. This occurs through household dust, synthetic textiles, tyre abrasion, industrial emissions, and related sources. These particles can become lodged in pulmonary tissues and affect respiratory function.

Dermal absorption is less significant, but when total exposure is considered, personal care products, synthetic fabrics, and cosmetics contribute to cumulative body burden (Alnuqaydan, 2024).

Nanoplastics may accumulate in vital organs such as the reproductive organs, kidneys, liver, and most importantly the lungs. Their nanoscale properties allow interaction with sub-cellular structures such as mitochondria and nuclei, and in severe cases they may cross the blood-brain barrier, potentially affecting neurological function (Rhaman et al., 2022). Collectively, these findings show that human exposure to nanoplastics is not only biologically invasive but also capable of affecting multiple organ systems with possible intergenerational implications.

Biological Effects and their Disease Associations

The epidemiological data explains that nanoplastic produces harms in multiple physiological systems affecting human health (Bastyans et al., 2022). Nanoplastics provoke oxidative stress and cause inflammation as well as cellular damage, altering signalling pathways at various exposure levels. This is plausible with human contact and recurs across endpoints starting with endocrine disruption leading to immune dysregulation (Bossio et al., 2025). Endocrine disorders also arise as chemical additives are carried, disrupting endocrine system functioning. At the cellular level, steroidogenic enzymes are altered, creating modifications in the hypothalamic-pituitary-thyroid axes as well as the hypothalamic-pituitary-gonadal axes (Egalini et al., 2022). Clinical syndromes such as polycystic ovarian disorder and hypothyroidism also arise. Metabolic consequences include obesity, fatty liver, and insulin resistance, while inflammatory processes occur due to nanoplastics, resulting in increased metabolic disease burdens (Zheng et al., 2024). Another major problem is reproductive impairment, including reduced ovarian reserve, follicular damage, disrupted menstrual cycling in females, and in males reduced sperm count, testicular histopathology, and reduced motility. These nanoplastics also affect neurological systems, leading to modifications in synaptic function and affecting behavioural endpoints in humans. Moreover, they affect the immune system and gastrointestinal tract, resulting in loss of epithelial integrity, microbial imbalance, irritable bowel syndrome, and immune dysregulation (Du et al., 2024).

Synergistic Effects

Nanoplastics absorb toxic components such as heavy metals, organic pollutants, bisphenols, flammable substances, and per- and polyfluoroalkyl substances, and transport them to humans (Santhanam et al., 2024). These particulates possess toxicological profiles that disrupt the biological system. These are combined with nanoplastics, and the risks are amplified in multiple ways. Plastics act as delivery vehicles, enhancing the bioavailability of toxicants and facilitating their entry into cells through crossing biological barriers (Lamparelli et al., 2023). Studies (Santhanam et al., 2024; Le et al., 2023) using organoid models demonstrate that microplastics bound with bisphenol A (BPA) produce stronger hepatotoxic and endocrine-disrupting effects. Additionally, formation of a biocorona produces a layer of proteins, lipids, and secretions that alters immune responses, influences biodistribution, and results in toxic effects. These nanoplastics act as risk multipliers, combining direct toxic effects and amplifying disruption within the human system (Le et al., 2023). Despite emerging concerns, significant challenges remain with respect to human epidemiological data and causal links to chronic diseases. Nanoplastics therefore act through multi-layered effects, creating constraints on human health (Jahedi et al., 2025).

Detection and Monitoring of Nanoplastics

One significant challenge is addressing exposure risks of nanoplastics and their detection and monitoring within environmental matrices and biological systems (Peng et al., 2022). Larger plastic debris such as microplastics and nanoplastics are less than one micrometre in size, making them invisible under standard microscopy and difficult to differentiate from naturally formed nanoparticles such as clay and organic matter. Their nanoscale size, irregular shapes, and chemical heterogeneity further complicate identification and decision-making. Nanoplastic detection is classified into spectroscopic, microscopic, and particle-tracking techniques, each with strengths and limitations (Nene et al., 2025). Fourier-transform infrared spectroscopy (FTIR) and Raman spectroscopy are used to identify chemical signatures of polymers, but their resolution often struggles at the nanoscale (Xiao & Schultz, 2017). Further innovations such as surface-enhanced Raman scattering (SERS) improve sensitivity for detecting nanoplastics at very low concentrations and small sizes. Although electron microscopy provides high-resolution visualisation, it is resource-intensive and requires specialised sample preparation, making it less feasible for large-scale monitoring (Peddie et al., 2022). Subsequently, nanoparticle tracking analysis (NTA) and dynamic light scattering (DLS) provide particle size distribution in aqueous samples, but these instruments struggle to differentiate plastics from non-plastic nanoparticles (Zhang et al., 2019). Another technique is thermal analysis using pyrolysis–gas chromatography mass spectrometry (Pyr-GC-MS), which breaks nanoplastics down into identifiable monomers (Regasa, 2024). However, these methods are destructive, preventing the possibility of tracking the same particles in biological systems. To overcome these constraints, researchers have developed labelling and tracer techniques, including fluorescent tagging, allowing visualisation of particle uptake inside the human body.

CONCEPTUAL FRAMEWORK

This chapter employs a cohesive conceptual framework demonstrating a techno-economic model for evaluating nanoplastic exposure mitigation within artificial intelligence and biotechnological interventions in a One Health perspective. This framework deliberately relates environmental contamination processes with methodological responses, health outcomes, and economic implications, thereby supporting evidence-based strategies and investment decisions.

Nanoplastic contamination is the primary variable in this framework, representing an environmental stressor originating from plastic degradation across multiple systems. This can be detected through AI-based detection systems such as smart sensors and machine learning platforms. This leads to biotechnological remediation strategies where interventions are implemented to achieve long-term environmental benefits. These interventions result in health outcomes measured through metrics such as disability-adjusted life years (DALYs) and quality-adjusted life years (QALYs). Thus, the framework, illustrated in Figure 1, translates health gains into economic outcomes operationalised through return on investment (ROI) and cost-effectiveness analysis (CEA), among others. By linking technological implementation directly to economic assessment, the structure clarifies how AI and biotechnology function as economic enablers.

Figure 1. Conceptual framework linking nanoplastic contamination with AI-based detection systems, biotechnological remediation strategies, health outcomes, and economic–environmental benefits within a One Health perspective.

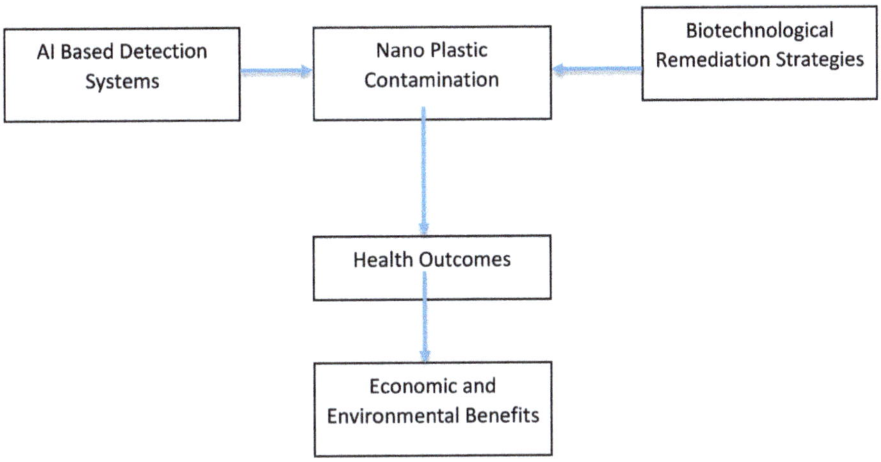

The framework illustrates how nanoplastic contamination acts as the central environmental stressor driving the need for AI-enabled monitoring and biotechnological remediation approaches. These interventions are connected with downstream health outcomes and broader economic and environmental benefits, demonstrating the interdependence between technological implementation, public health protection, and sustainability-oriented mitigation strategies.

Economic Evaluation Tools and Frameworks

Economic evaluation is important in determining intervention value and how it offers value for money. It assesses the association between costs and consequences and helps policymakers make informed decisions. These costs are considered inputs, and the associated consequences are outcomes. In order to mitigate nanoplastic exposure, the use of certain tools is required. Some of the tools explained here are cost of illness, return on investment, and budget impact analysis.

Cost of illness refers to an economic evaluation method estimating the cumulative burden of diseases attributed to a specific exposure, and in this scenario, nanoplastics sum all monetary values, including resources used and losses incurred during illness (Panizzolo, 2025). The main purpose of this is to explore the magnitude of the economic problem and also create mitigation strategies for it. Cost of illness typically includes components such as direct and indirect medical as well as non-medical costs. The direct medical costs comprise hospitalisation, diagnostics, outpatient visits, procedures, long-term care, and medicines, and these aspects can be measured through data obtained from hospital billing records. The direct non-medical costs include transportation charges of patients to the hospital, accommodation, and home modifications, and can be measured using household surveys. The indirect costs demonstrate productivity losses such as time off work, premature death, and reduced work capabilities, and are measured using the human capital method with lost wages and the friction cost method, which counts the losses until a replacement worker is found. The intangible costs are usually difficult to quantify but can be indirectly studied by quality-adjusted life years (QALYs), willingness to pay, and value of a statistical life (VSL) (Krause et al., 2024). The basic formula for COI is the total sum of direct, indirect, and intangible costs for every case, multiplied by the total number of attributable cases. The calculation for attributable case number is usually taken from the population attributable fraction (PAF) by epidemiologists. The advantage of COI is its communicability, as it translates abstract risks into larger monetary values that highlight the urgency of the situation, although there are certain limitations, including uncertainty in epidemiological estimates and constraints in quantifying intangibles.

Return on Investment (ROI) accompanies COI by evaluating whether interventions yield more financial savings than their costs (Turner et al., 2023). ROI is basically calculated using net savings divided by intervention costs, expressed as a ratio or percentage. In cases where benefits accumulate over years, savings are discounted to net present value (NPV) even before ROI is estimated. When an advanced filtration system is implemented against nanoplastics, the costs of installation and discounted savings are considered in terms of ROI. Therefore, ROI helps policymakers and hospital administrators decide whether to adopt interventions that are worthwhile. The advantage of this calculation is mainly in understanding whether interventions pay back in monetary terms, although it undervalues non-monetary benefits such as environmental and equity benefits.

Budget Impact Analysis (BIA) demonstrates the affordability of the entire situation. When an intervention possesses high ROI, insurers and government should decide whether financing can be done within short-term budgets (Smith & Levy, 2024). This BIA addresses the overall cumulative cash flow for a budget holder annually or up to five years under different scenarios. It also multiplies the overall population by uptake rate and the incremental cost for each patient to estimate the overall annual impact. Consequently, BIA results become crucial for finance ministries to show cash requirements every year, though they do not account for broader social value or long-term cost unless this is explicitly included.

Cost-effectiveness and utility analysis are used as critical tools for strengthening the economic evaluation of nanoplastic mitigation (Rasheed et al., 2025). These analyses compare the relative costs of various interventions with measurable healthcare outcomes and are expressed as cost per life year or as cost per case prevented. Therefore, policymakers can rank different strategies not only by affordability but also by efficiency in achieving better health outcomes. The values are expressed as Quality-Adjusted Life Years (QALYs) estimates or Disability-Adjusted Life Years (DALYs) estimates. These calculations provide a better understanding of sustainable health and also assist in mitigating the economic constraints of nanoplastics.

Together, these tools work in sequence: COI computes the economic burden of nanoplastics to demonstrate the immediate need for action, ROI explains interventions through financial return to identify efficient strategies, and BIA demonstrates whether interventions are affordable within actual budgets and timelines. One of the best ways is to integrate these analyses with cost-effectiveness or utility analysis to evaluate value per health gain. These values assist in drawing useful insights from national disease registries, hospital accounts, environmental monitoring, labour surveys, and insurance claims.

ARTIFICIAL INTELLIGENCE APPLICATIONS

Artificial Intelligence (AI) has evolved as one of the transformative forces in environmental health and supports efforts to combat nanoplastic pollution, gaining increasing global attention (Ratre et al., 2024). Nanoplastics are usually smaller than 1 micrometre and pose serious threats to ecosystems, human health, and food chains. To address this, proper detection methods such as spectroscopy, microscopy, and chromatography are required, which are effective in controlled laboratory environments and for real-time monitoring. AI creates a bridge by enabling automation, decision-support systems, and operation across diversified environments ranging from water treatment plants to national public health networks. Artificial intelligence is implemented in various dimensions, which are explained below and summarized in Figure 2.

Smart Sensor Networks: AI-powered smart sensors can assist in continuous monitoring of water quality in various facilities such as rivers and lakes for detecting nanoplastic contamination. Periodic manual sampling is complemented by real-time data, assisting in enabling rapid intervention. AI algorithms identify abnormal patterns and detect sudden spikes in particle contamination, indicating upstream dumping, industrial leakage, as well as stormwater overflow (Rathour & Pathania, 2025). In European countries, pilot smart buoys equipped with AI-enhanced optical sensors have been deployed in rivers. These buoys help detect fluctuating loads of microplastics during seasonal rainfall and assist regulators in coordinating with industries to prevent contamination.

Machine Learning Algorithms: Nanoplastic contamination is not uniform in distribution. Some regions, such as industrial zones, are more prone to contamination, and machine learning models are used to predict future hotspots. Supervised machine learning algorithms such as random forests and gradient boosting combine multi-source data, including satellite imagery, population density, and hydrological flow patterns (Xie et al., 2025). These predictions help policymakers prioritise inspections and remediation strategies by estimating the probability of nanoplastic contamination.

Image Recognition Tools: Major challenges in nanoplastic detection include distinguishing tiny plastic particles from organic debris under microscopes and spectrometers. AI-powered recognition tools provide improved image analysis across different particle sizes (Bhagat & Chandrakar, 2025). At the University of Toronto, a deep neural network analyses scanning electron microscopy images and helps distinguish nanoplastics with about 95% accuracy. Conventional neural networks are also being used to automatically distinguish polymers from natural particles in complex environments. This reduces human error and analysis time. These tools assist in high-throughput screening by enabling researchers to process many images within minutes and identify irregular-shaped particles that are difficult to detect using traditional methods. Integration of these machine learning techniques with spectroscopy and microscopy enables real-time detection and monitoring of nanoplastics.

Early Warning Systems: Integration of environmental monitoring with health records can enhance early warning systems for predicting potential health risks due to nanoplastic exposure (Panizzolo, 2025). These systems integrate environmental data from sensors and satellites, epidemiological data from public health surveillance and hospital admissions, and socio-demographic data to issue alert messages. In countries such as Singapore, AI-based early warning systems are linked to water management systems to track microplastic surges and support preventive measures, with similar frameworks in Japan and Europe. Predictive analytics are used to assess climate variables and wastewater sampling for identifying nanoplastic hotspots. These systems enable governments to act promptly, empower communities through localised alerts, and support long-term sustainable health policies.

Robotics and Drones: AI-enabled robots and drones are increasingly deployed for environmental monitoring and cleanup operations. Autonomous underwater drones equipped with AI-based cameras and sensors are capable of scanning rivers, lakes, and coastal areas for nanoplastic and microplastic hotspots. On land, robotic systems equipped with computer vision identify soil patches with plastic contamination and assist in remediation. Japan has experimented with AI-guided drones used underwater to trace plastic flows from urban areas into oceans due to industrial activities, enabling more accurate mapping of contamination pathways (Bansal et al., 2025).

Digital Twins: Digital twins, which are virtual replicas of environmental systems, enable the simulation of nanoplastic movement through water, soil, and air under various conditions. These models assist researchers and policymakers in simulating long-term contamination pathways and testing "what-if" scenarios such as changes in industrial discharge, rainfall intensity, or urban waste management policies. European countries have implemented frameworks for simulating urban river basins and predicting nanoplastic accumulation under different climatic conditions (Xi et al., 2025).

Natural Language Processing: With increasing concern about nanoplastic exposure, natural language processing (NLP) tools are being used by researchers to scan scientific literature, policy reports, and environmental datasets to identify emerging trends, knowledge gaps, and best practices. These systems summarise information and help policymakers and scientists stay updated on global evidence. In addition to summarisation, NLP can analyse correlations across disciplines. Integrating toxicology findings in food chains with environmental studies helps align reports with local pollution data and global climate datasets. Multilingual NLP models also help bridge regional barriers through translation and harmonisation of publications. Some governments and international bodies are implementing NLP-driven policy dashboards for continuous tracking of regulations and exposure trends, creating a dynamic ecosystem for decision-making (Kumar et al., 2023).

Blockchain-Integrated Systems: AI is increasingly combined with blockchain to track the entire lifecycle of plastics from production to consumption, disposal, and recycling. These systems provide tamper-proof and auditable records of plastic movement and contamination, helping industries demonstrate compliance with environmental regulations and enabling governments to enforce accountability. Beyond tracking, these networks analyse real-time data and detect unusual patterns, such as sudden spikes in plastic waste in specific regions, indicating illegal dumping or mismanagement. Several projects in South Korea and the European Union have been launched to verify waste disposal and recycling rates to reduce environmental leakage and nanoplastic pollution. These integrated platforms can automatically trigger notifications, and regulatory penalties may be issued by relevant authorities to promote proactive compliance. Linking blockchain data with environmental monitoring systems enables stakeholders to track contamination pathways from source to destination, forming a more comprehensive framework for sustainable plastic management (Verma et al., 2022).

Smart Grid Monitoring: South Korea is a notable example of implementing advanced technology for monitoring nanoplastics and protecting public health. AI-enabled sensors have been integrated into water facilities, creating a smart grid for maintaining water quality. This system allows real-time detection and prevention of micro- and nanoplastics entering the environment. It has reportedly reduced contamination in water by around 30% within three years, and the return on investment was estimated at approximately 270%, as healthcare savings and productivity gains exceeded implementation costs.

Figure 2. Major artificial intelligence applications associated with nanoplastic monitoring and environmental health systems.

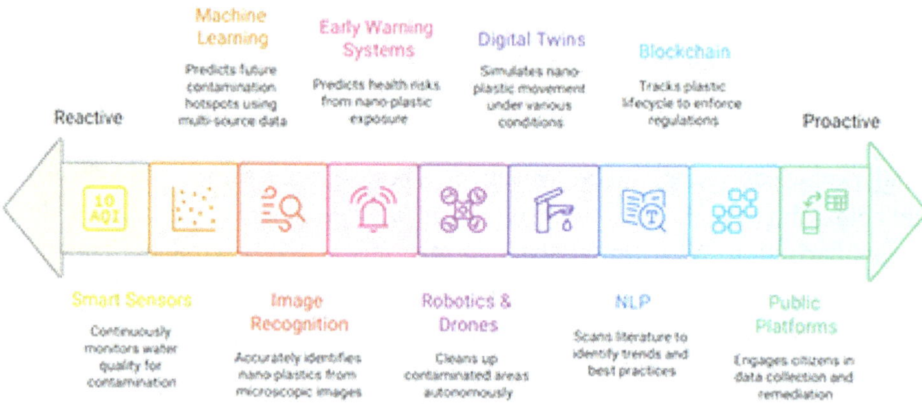

The figure illustrates the progression from reactive monitoring approaches towards more proactive AI-driven systems, including smart sensors, machine learning, image recognition, robotics, digital twins, blockchain systems, and public participation platforms. Together, these approaches support real-time detection, prediction, environmental surveillance, and remediation planning for nanoplastic contamination.

BIOTECHNOLOGICAL REMEDIATION STRATEGIES

Biotechnological remediation employs living organisms and their enzymes to transform and mineralise nanoplastics. Compared with purely physical and chemical methods, biotechnological approaches generally require lower energy input and have greater potential for industrial application, with the possibility of achieving more complete mineralisation under appropriate conditions.

Microbial and Enzymatic Remediation

This is one of the most significant biotechnological strategies for addressing nanoplastic pollution. Some microorganisms and enzymes possess a natural ability to degrade polymers into smaller and less harmful compounds, and this field has been strengthened by recent advances in synthetic biology and expanded catalytic capacity. The bacterium *Ideonella sakaiensis* was found to metabolise polyethylene terephthalate (PET), a widely used plastic, through the production of specialised enzymes such as PETase and MHETase. PETase hydrolyses PET into smaller fragments, whereas MHETase further breaks these down into terephthalic acid and ethylene glycol, which can then be metabolised as carbon sources by the bacteria. This dual-enzyme mechanism has been optimised in laboratory settings to increase efficiency and perform under diverse strains and temperature conditions. Beyond bacteria, fungi such as *Pleurotus ostreatus* have been used for mycoremediation, with fungal enzymes including peroxidases and laccases that trap and oxidise nanoplastics in soils and sediments. Experimental trials conducted in Europe demonstrated that fungal inoculation in contaminated composting areas accelerates the breakdown of fragmented polymers and reduces plastic persistence. Furthermore, microbial consortia, which are groups of organisms that oxidise plastics and metabolise by-products, have been effective when used in bioreactors for wastewater treatment plants. Through these approaches, technology can help transform nanoplastics from persistent pollutants into biodegradable compounds in specific environments (Qiu et al., 2025).

Life Cycle Costing (LCC)

Life cycle costing provides a structured framework for evaluating the financial feasibility of biological remediation, accounting for all costs associated with a technology over its full lifespan. These costs typically include capital expenditure (CAPEX) for infrastructure and equipment, operational expenditure (OPEX) for labour, energy, and enzyme replacement, microbial maintenance, and end-of-life costs for decommissioning and safe disposal. When deploying immobilised PETase enzyme cartridges in water treatment units, high upfront CAPEX is required for reactor installation and modification, but lower OPEX can reduce energy demands compared with chemical degradation methods. In contrast, bioreactors that utilise engineered bacteria entail higher OPEX because of continuous nutrient input, monitoring, and compliance with biosafety regulations. One example comes from Japan, where wastewater plants were used for enzyme-based systems. The study examined enzyme systems with higher installation costs but operational savings and reduced waste, showing cost-effectiveness over a ten-year period. Likewise, mycoremediation in contaminated soil involves comparatively low CAPEX because fungi can be deployed directly, but slower degradation rates may extend the time required to achieve meaningful results. By applying life cycle costing, policymakers can compare these options more effectively and ensure that investments remain economically sustainable over the long term. Developing countries can implement low-cost fungal and enzyme-based remediation to improve affordability and overcome resource constraints, especially when combined with environmental impact assessments. LCC can therefore serve as a powerful decision-making tool for balancing economics with environmental sustainability (Manoharan et al., 2023; Zibonas, 2024).

Policy Integration

The success of remediation strategies depends not only on scientific innovation but also on the establishment of stronger and more supportive policy frameworks. Traditional waste management methods are becoming obsolete, and biotechnological solutions should be implemented for better results, which requires government intervention to overcome financial and regulatory barriers. Policy integration enables technology to move from the laboratory to real-world implementation for safeguarding environmental and public health. One such policy is financial incentivisation, in which governments can offer subsidies and tax benefits to lower the costs of adopting technologies and encourage private institutions to invest in projects that would otherwise be too costly. The Horizon 2020 programme in Europe was funded by several agencies and supported research on enzymatic degradation of plastics, contributing to reduced plastic persistence. Similarly, in countries such as South Korea, green procurement policies have been implemented to support eco-friendly technologies. Tax incentives can reward industries that invest in zero-waste facilities for wastewater treatment, including plastic residues. Companies that use enzyme-based filtration and microbial remediation units would benefit from tax reductions, while also reducing nanoplastic release from effluents.

Performance-based grants are also effective because they help ensure that public funds are used appropriately. Beyond financial aspects, regulatory frameworks must address biosafety risks through containment standards and ecological assessment. Before approving the use of engineered bacteria, governments should require independent trials to demonstrate potential adverse effects on microbial systems. Clear regulatory pathways are essential for reducing uncertainty and unintended consequences in natural environments. Furthermore, establishing certification and monitoring standards will improve the performance of implemented technologies and build trust among the general population. Another important element is international collaboration, since nanoplastic pollution has no boundaries and harmonised policies are essential to prevent loopholes. International platforms such as the UN

Environment Programme (UNEP) can facilitate guidelines for global benchmarks, ensuring safety and sustainability. Joint funding programmes are also essential for large-scale testing of remediation technologies (Sorensen et al., 2023). Globally, various regulatory frameworks address microplastics and nanoplastics, although this area is still emerging. This is reflected in European Union REACH regulations, UNEP plastic pollution resolutions, and developing WHO guidelines on environmental health risks, which indicate early-stage governance. In India, provisions under the Plastic Waste Management Rules (2016, amended 2022) point towards reducing plastic outflow into ecosystems. However, specific standards for nanoplastics remain limited, representing a serious governance gap. Strengthening regulatory alignment will help nations ensure proper monitoring and mitigation of this exposure.

Enzymatic Bioremediation

The European Union has implemented a project in the Baltic Sea using PETase enzymes to break down plastic pollutants, including both micro- and nanoplastics, more effectively than traditional clean-up methods. These approaches improve sustainability at a larger scale. Economic evaluation revealed that this project achieved life-cycle cost savings through reduced environmental degradation and a longer-term cleaning process. These approaches can also be applied in coastal zones to reduce high levels of plastic contamination in those areas.

Figure 3. Biotechnology-Driven Solutions

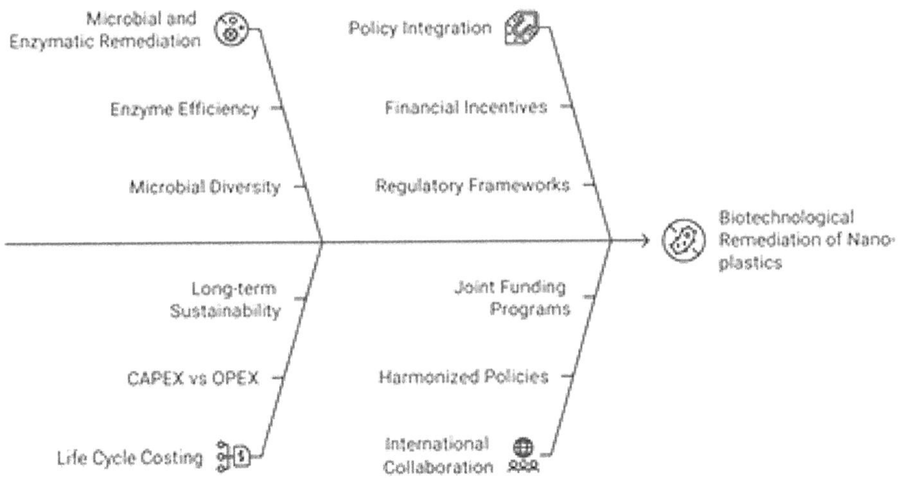

Source: *(Authors' own elaboration)*

Techno-Economic Analysis

Although artificial intelligence and biotechnological interventions provide numerous benefits for mitigating nanoplastic contamination, these should be evaluated using techno-economic analysis. This method helps to clarify the relationship between technological implementation and health and economic outcomes.

Table 1. Detection and Monitoring

S.No	Detection Method	Accuracy of Results	Cost Effectiveness
1	Manual Sampling, Microscopy	Low due to limited resolution for nanoparticles	Low
2	Raman Spectroscopy, μ-FTIR	Moderate, up to ~10 μm particle size	Moderate
3	AI-based real-time monitoring	High initial investment but lower marginal costs	High due to continuous detection

The table above summarizes the principal differences between AI-based detection systems and conventional methods. Within marine monitoring contexts, conventional approaches to microplastic detection are generally associated with higher annual operational costs due to labour demands and equipment requirements. In contrast, AI-based systems may enhance throughput and facilitate more continuous monitoring by improving detection sensitivity and analytical efficiency (Meyers et al., 2024).

Table 2. Remediation Approaches

S. No	Remediation Method	Costs	Environmental Risk
1	Physical & Chemical flocculation	High	Moderate to High
2	Nanotechnology-based adsorption	Moderate	Moderate
3	Microbial & Enzymatic bioremediation	Low	Low

The table demonstrates both costs and associated environmental risks across remediation methods. Strategies vary depending on the approach adopted, with some requiring higher initial investment but offering lower long-term environmental risk. These trade-offs should be assessed at an early stage to support effective nanoplastic mitigation planning (Pandit et al., 2025).

NUTRITION AS A SUPPLEMENTARY RISK-MITIGATION STRATEGY

Nutrition-based interventions do not substitute artificial intelligence or biotechnological mitigation approaches. Instead, they function as a complementary and low-cost measure to reduce health impacts within a One Health framework, where nutrition helps to reduce disease burden alongside technological interventions.

Nanoplastic exposure has become a growing global concern, contributing to oxidative stress, chronic inflammation, disruption of epithelial layers, and metabolic dysfunction. These biological pathways are activated through increased reactive oxygen species production, mitochondrial damage, and microbiome alterations. Nutritional modulation may help in reducing some of these effects. Diets rich in antioxidants can counteract oxidative stress, including fruits, green tea, berries, grapes, vegetables, and micronutrients such as vitamins C and E, which enhance endogenous antioxidant enzyme activity. Vitamin C also regenerates oxidised vitamin E, contributing to a synergistic antioxidant system. Vitamin E, in turn, protects cell membranes as a lipid-soluble antioxidant (Rai, 2022).

The benefits of nutrition-based interventions include mitigation of specific nanoplastic-associated harms while improving overall health outcomes by reducing risks of cardiovascular diseases, diabetes, and infections. This makes nutritional programmes relevant for policymakers due to their broader public health benefits. Japan's healthy meal initiative promotes antioxidant-rich school lunches, not specifically for nanoplastic exposure but for broader environmental health resilience. Coastal fishing communities and industrial workers with higher exposure risk may benefit from targeted nutritional interventions (Panizzolo, 2025).

In India, interventions have focused on urban slums with higher exposure to environmental toxins. Nutrition campaigns are used to address metabolic and inflammatory disorders in these populations. These programmes are low-cost and utilise locally available antioxidant-rich foods, helping reduce oxidative stress associated with nanoplastic ingestion. Reported outcomes indicate a reduction in metabolic syndrome within one year. From a cost perspective, these interventions are economically efficient and cost less per household annually. Therefore, nanoplastic mitigation strategies do not necessarily need to be technology-based only; community-based interventions can also contribute significantly to health outcomes and environmental sustainability.

Health Economic Metrics

Health economic outcomes are translated at the population level using disability-adjusted life years (DALY) and quality-adjusted life years (QALY). This intervention would prevent approximately 0.005 DALY per person per year, equivalent to about 1.8 days of healthy life. In a population of one lakh, this corresponds to approximately 500 DALYs averted annually.

If the programme is considered cost-saving, the cost per DALY becomes negative, indicating that the intervention not only improves health outcomes but also generates net economic savings. This positions it as a dominant strategy in economic evaluation. This is particularly relevant for policy decision-making, where cost-effectiveness thresholds are used to guide resource allocation. Nutritional interventions against nanoplastic exposure can therefore qualify as cost-saving strategies.

The choice of delivery model significantly affects both effectiveness and cost. Direct supplementation provides precise dosing and easier monitoring with reliable supply chains. Food fortification, such as adding micronutrients to staple foods, spreads costs and ensures wide population coverage. Behavioural change programmes promote dietary diversity and home-based cooking practices, which may be culturally sustainable and support long-term outcomes.

In India, the Mid-Day Meal Scheme is an established example providing micronutrient-rich meals to millions of children, reducing anaemia and malnutrition. Similar models can be applied in coastal regions to improve omega-3 intake and address nutritional deficiencies. In Europe, community-based dietary programmes promoting Mediterranean diets show evidence of healthcare cost savings and cardiovascular disease prevention.

Policy interventions should be supported by pilot studies and biomarker assessment of oxidative stress to maximise health and economic returns. This approach strengthens integration of dietary interventions into national health strategies. Policymakers can support multi-benefit programmes that not only address nanoplastic exposure but also improve overall population health resilience.

Collaboration between health and agricultural ministries can enhance production and distribution of nutrient-dense crops, supported by subsidies and tax incentives to improve affordability. Government support for research on interactions between nanoplastics and diet is also necessary. Public engagement through education campaigns can further increase awareness of nutritional strategies for improving general health and resilience to nanoplastic exposure (Beresniak et al., 2025).

POLICY INSTRUMENTS AND ECONOMIC REGULATION

The major challenge in addressing nanoplastic pollution lies in the fact that environmental and health costs are largely borne by society, while the benefits are concentrated among producers of plastics.

Internalising Externalities

The costs of nanoplastic pollution are externalised into society in multiple forms, including environmental degradation, healthcare expenses, and reduced quality of life. This creates a classic market failure scenario. To address this, policy instruments such as plastic taxes can be implemented to reduce excessive production and consumption. When plastics become relatively more expensive compared to eco-friendly substitutes, countries such as Ireland and the UK have effectively used plastic taxes and achieved reductions in single-use plastic consumption.

In addition, regulatory registration mechanisms can ensure that industries remain responsible for plastic waste management, including cleanup and remediation costs. This approach aligns with the extended producer responsibility (EPR) framework, where manufacturers take full responsibility for end-of-life product management. EPR encourages circular economy principles and eco-design strategies, shifting firms from a cost-externalisation model to producer accountability, thereby reducing environmental and economic burden. Together, these instruments can also generate public revenue. Plastic tax revenues can be reinvested into wastewater treatment infrastructure, contributing to improved environmental outcomes (Bithas, 2011).

Financing Mechanisms

Addressing nanoplastic pollution requires substantial financial investment in research, infrastructure, monitoring, and remediation. Public funding alone is often insufficient, as innovation and large-scale implementation require additional financial mobilisation.

Therefore, contemporary financing approaches are increasingly relevant. Green bonds can serve as debt instruments to fund environmentally beneficial projects. Governments can issue green bonds to support the development and deployment of biodegradable plastic alternatives and related mitigation technologies. For example, the World Bank has implemented green bond initiatives aimed at environmental resilience, which can be adapted for nanoplastic mitigation contexts.

Blended finance models can also be applied, combining public funds with private capital to reduce risk for private investors. In parallel, public–private partnerships (PPPs) supported by government subsidies can facilitate the development of AI-based monitoring systems for nanoplastic detection. Investment structures should also encourage socially responsible investors, where returns are linked not only to financial performance but also to measurable environmental and health outcomes. These models can contribute to reduced disease burden, including gastrointestinal illnesses associated with contaminated water, while simultaneously improving water quality. Overall, such financing mechanisms support sustained investment in innovation and remediation systems (Kumar et al., 2021).

Regulatory Impact Analysis (RIA)

When implementing regulations, governments should assess economic costs alongside expected social and environmental outcomes. Regulatory impact analysis (RIA) provides a structured framework for quantifying both direct and indirect effects of proposed interventions.

For example, AI-based water quality monitoring systems may increase municipal budgets due to high initial equipment costs. However, over time, these systems can reduce disease incidence associated with nanoplastic-contaminated water. In one study, interventions resulted in approximately a 40% reduction in healthcare claims, translating into direct cost savings for both individuals and governments.

RIA also captures secondary impacts such as improved labour productivity in healthier populations and reduced environmental degradation. This broader evaluation ensures that policymakers consider societal value when selecting regulatory options and supports more sustainable decision-making (Kangalawe, 2012).

Integration into Sustainable Development Goals (SDGs)

Nanoplastic pollution intersects directly with global sustainability challenges and the United Nations Sustainable Development Goals (SDGs). Aligning economic evaluations with SDG metrics ensures that national policies address local environmental health issues while contributing to global accountability.

Reducing nanoplastic exposure supports SDG 3 (Good Health and Well-being) by targeting reductions in waterborne and metabolic diseases. Improvements in waste management and pollution control align with SDG 6 (Clean Water and Sanitation) and SDG 14 (Life Below Water), which focus on protecting aquatic ecosystems from plastic toxicity.

Figure 4 illustrates how SDG integration may support regulatory agencies in monitoring environmental health using standardized indicators, such as nanoplastic concentrations in water systems, while also supporting healthcare cost reductions. Overall, SDG integration provides a strategic framework for linking nanoplastic mitigation to broader global sustainability and policy agendas (Abudurexiti et al., 2025).

Figure 4. **Nanoplastic pollution and its implications for Global Sustainability.**

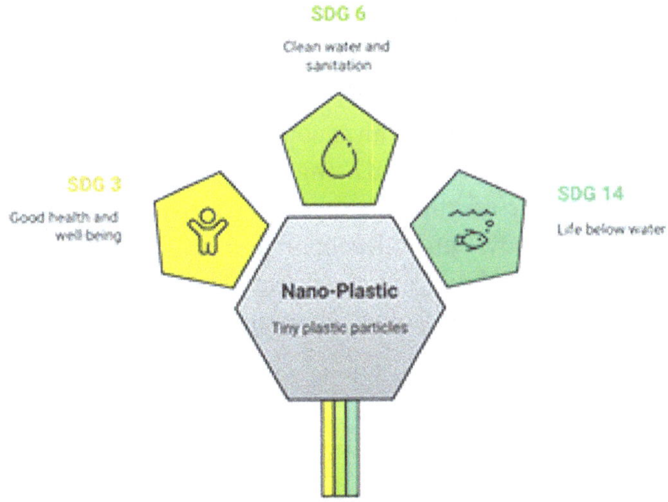

Source: Adapted from the United Nations Sustainable Development Goals framework.

ETHICAL AND GOVERNANCE CONSIDERATIONS

Various initiatives are being strengthened through AI mobile applications, allowing individuals to upload water and soil sample data along with photos and location information. AI algorithms help in validating, cleaning, and analysing submissions and converting them into more reliable, high-quality datasets through data validation steps. The system provides immediate feedback to participants, including estimated contamination levels and suggested remediation actions, which increases awareness and engagement. Smartphone-based spectrometry techniques are also available and integrated with AI to support detection of micro and nanoplastics in household water samples, extending data coverage beyond official monitoring networks. Additionally, gamification elements combined with AI-driven dashboards can motivate community participation by showing collective progress and identifying local hotspots. By integrating citizen-generated data with municipal monitoring systems, real-time inputs can be achieved, empowering communities to actively contribute to environmental stewardship and improving transparency and public trust (Zhao et al., 2024). While these measures are implemented, government authorities should remain cautious regarding data privacy and informed consent.

Future Scope

Nanoplastic pollution should be approached as an interdisciplinary issue, addressed from technological, biotechnological, and sustainability perspectives, considering all environmental dimensions. Future studies can focus on quantifying long-term health outcomes associated with nanoplastic exposure. The establishment of standard metrics can also be developed alongside economic evaluation methods, with applications explored across different income and socioeconomic settings. By combining technological innovation with economic evidence, global policy frameworks can be strengthened to build more resilient systems for protecting human health, economies, and ecosystems from nanoplastic risks.

CONCLUSION

Nanoplastics demonstrate a systemic threat across human and environmental systems. Their extremely small size enables infiltration into biological systems and disruption of ecological balance. Therefore, this chapter proposes a One Health approach that recognises the interconnectedness of environmental, human, and animal health. The integration of technological innovation with economic valuation methodologies may support more affordable and scalable strategies for mitigating nanoplastic pollution. Alignment with the Sustainable Development Goals further reinforces global accountability frameworks. From a policy perspective, prevention strategies may benefit from closer integration with dietary interventions and emerging technological solutions to reduce both public health impacts and healthcare-related economic burdens. Overall, nanoplastic pollution represents not only an environmental concern but also a broader socio-economic challenge, requiring evidence-informed mechanisms for effective mitigation.

Within this context, a key methodological question emerges regarding how techno-economic models and AI-based detection tools can be more effectively integrated into comprehensive life cycle assessment frameworks that capture the interconnected dimensions of environmental systems, animal health, and human health under a unified One Health perspective.

REFERENCES

Abudurexiti, A. S. J., Abulikemu, Z., & Bavumiragira, J. P. (2025). Circular Economy and Sustainable Development Goals in the Reduction of Global Nano (Micro) Plastic Pollution. *Polish Journal of Environmental Studies*, *34*(4), 4489–4498. DOI: 10.15244/pjoes/190280

Alnuqaydan, A. M. (2024). The dark side of beauty: An in-depth analysis of the health hazards and toxicological impact of synthetic cosmetics and personal care products. *Frontiers in Public Health*, *12*, 1439027. DOI: 10.3389/fpubh.2024.1439027 PMID: 39253281

Bansal, D., Bhattacharya, N., & Shandilya, P. (2025). Exploring Applications: Intelligent Drones and Robots in Industrial Settings. In *Building Embodied AI Systems: The Agents, the Architecture Principles, Challenges, and Application Domains* (pp. 159-180). Cham: Springer Nature Switzerland. https://doi.org/DOI: 10.1007/978-3-031-68256-8_7

Bastyans, S., Jackson, S., & Fejer, G. (2022). Micro and nano-plastics, a threat to human health? *Emerging Topics in Life Sciences*, *6*(4), 411–422. DOI: 10.1042/ETLS20220024 PMID: 36453919

Beresniak, A., Bremond-Gignac, D., Dupont, D., & Duru, G. (2025). Reevaluating health metrics: Unraveling the limitations of disability-adjusted life years as an indicator in disease burden assessment. *World Journal of Methodology*, *15*(1), 95796. DOI: 10.5662/wjm.v15.i1.95796 PMID: 40115408

Bhagat, N., & Chandrakar, P. (2025). AI-Powered Image Recognition. *International Journal of Innovative Science and Research Technology*, *10*(5), 4521–4526. DOI: 10.38124/ijisrt/25may2243

Bhat, M. A., Gedik, K., & Gaga, E. O. (2023). Atmospheric micro (nano) plastics: Future growing concerns for human health. *Air Quality, Atmosphere & Health*, *16*(2), 233–262. DOI: 10.1007/s11869-022-01272-2 PMID: 36276170

Bithas, K. (2011). Sustainability and externalities: Is the internalization of externalities a sufficient condition for sustainability? *Ecological Economics*, *70*(10), 1703–1706. DOI: 10.1016/j.ecolecon.2011.05.014

Bossio, S., Ruffolo, S. A., Lofaro, D., Perri, A., & La Russa, M. F. (2025). Endocrine Toxicity of Micro-and Nanoplastics, and Advances in Detection Techniques for Human Tissues: A Comprehensive Review. *Endocrines*, *6*(2), 23. DOI: 10.3390/endocrines6020023

Du, L., Liu, H., Song, X., Feng, X., Xu, H., Tang, W., & Yang, J. (2024). Developments in the field of intestinal toxicity and signaling pathways associated with rodent exposure to micro (nano) plastics. *Toxicology*, *507*, 153883. DOI: 10.1016/j.tox.2024.153883 PMID: 38996996

Dube, E., & Okuthe, G. E. (2023). Plastics and micro/nano-plastics (MNPs) in the environment: Occurrence, impact, and toxicity. *International Journal of Environmental Research and Public Health*, *20*(17), 6667. DOI: 10.3390/ijerph20176667 PMID: 37681807

Egalini, F., Marinelli, L., Rossi, M., Motta, G., Prencipe, N., Rossetto Giaccherino, R., Pagano, L., Grottoli, S., & Giordano, R. (2022). Endocrine disrupting chemicals: Effects on pituitary, thyroid and adrenal glands. *Endocrine*, *78*(3), 395–405. DOI: 10.1007/s12020-022-03076-x PMID: 35604630

Jahedi, F., Frad, N. J. H., Khaksar, M. A., Rashidi, P., Safdari, F., & Mansouri, Z. (2025). Nano and microplastics: Unveiling their profound impact on endocrine health. *Toxicology Mechanisms and Methods*, *35*(8), 1–29. DOI: 10.1080/15376516.2025.2509745 PMID: 40432394

Kandaswamy, K., Ajay, G., Panda, S. P., Anahas, A. M. P., Kari, Z. A., Giri, J. P., Almutairi, B. O., Arokiyaraj, S., Malafaia, G., & Jesu Arockiaraj, J. (2024). Polystyrene nanoplastics synergistically exacerbate diclofenac toxicity in embryonic development and the health of adult zebrafish. *Comparative Biochemistry and Physiology. Toxicology & Pharmacology : CBP*, *281*, 109926. Advance online publication. DOI: 10.1016/j.cbpc.2024.109926 PMID: 38641085

Kangalawe, R. Y. (2012). Food security and health in the southern highlands of Tanzania: A multidisciplinary approach to evaluate the impact of climate change and other stress factors. *African Journal of Environmental Science and Technology*, *6*(1), 50–66. DOI: 10.5897/AJEST11.003

Kibria, G., Nugegoda, D., & Haroon, A. Y. (2022). Microplastic pollution and contamination of seafood (including fish, sharks, mussels, oysters, shrimps and seaweeds): a global overview. *Microplastic pollution: environmental occurrence and treatment technologies*, 277-322. https://doi.org/DOI: 10.1007/978-3-030-89220-3_14

Krause, S., Ouellet, V., Allen, D., Allen, S., Moss, K., Nel, H. A., Manaseki-Holland, S., & Lynch, I. (2024). The potential of micro-and nanoplastics to exacerbate the health impacts and global burden of non-communicable diseases. *Cell Reports Medicine*, *5*(6), 101581. Advance online publication. DOI: 10.1016/j.xcrm.2024.101581 PMID: 38781963

Kumar, A., Bakshi, B. R., Ramteke, M., & Kodamana, H. (2023). Recycle-BERT: Extracting knowledge about plastic waste recycling by natural language processing. *ACS Sustainable Chemistry & Engineering*, *11*(32), 12123–12134. https://pubs.acs.org/doi/10.1021/acssuschemeng.3c03162?goto=supporting-info. DOI: 10.1021/acssuschemeng.3c03162

Kumar, R., Verma, A., Shome, A., Sinha, R., Sinha, S., Jha, P. K., Kumar, R., Kumar, P., Shubham, , Das, S., Sharma, P., & Vara Prasad, P. V. (2021). Impacts of plastic pollution on ecosystem services, sustainable development goals, and need to focus on circular economy and policy interventions. *Sustainability (Basel)*, *13*(17), 9963. DOI: 10.3390/su13179963

Lamparelli, E. P., Marino, M., Szychlinska, M. A., Della Rocca, N., Ciardulli, M. C., Scala, P., D'Auria, R., Testa, A., Viggiano, A., Cappello, F., Meccariello, R., Della Porta, G., & Santoro, A. (2023). The other side of plastics: Bioplastic-based nanoparticles for drug delivery systems in the brain. *Pharmaceutics*, *15*(11), 2549. DOI: 10.3390/pharmaceutics15112549 PMID: 38004530

Le, V.-G., Nguyen, M.-K., Nguyen, H.-L., Lin, C., Hadi, M., Hung, N. T. Q., Hoang, H.-G., Nguyen, K. N., Tran, H.-T., Hou, D., Zhang, T., & Bolan, N. S. (2023). A comprehensive review of micro-and nano-plastics in the atmosphere: Occurrence, fate, toxicity, and strategies for risk reduction. *The Science of the Total Environment*, *904*, 166649. DOI: 10.1016/j.scitotenv.2023.166649 PMID: 37660815

Manoharan, K., Pandey, M., & Bhattacharya, S. (2023). Life cycle assessment of nanoscale polymer-based coatings. In *Polymer-Based Nanoscale Materials for Surface Coatings* (pp. 613–631). Elsevier., DOI: 10.1016/B978-0-32-390778-1.00033-5

Meyers, N., Kopke, K., Buhhalko, N., Mattsson, K., Janssen, C. R., Everaert, G., & De Witte, B. (2024). Value for money: A cost-effectiveness analysis of microplastic analytics in seawater. *Microplastics and Nanoplastics*, *4*(1), 4. DOI: 10.1186/s43591-024-00081-x

Nene, A., Sadeghzade, S., Viaroli, S., Yang, W., Uchenna, U. P., Kandwal, A., Xinghui, L., Somani, P., & Galluzzi, M. (2025). Recent advances and future technologies in nano-microplastics detection. *Environmental Sciences Europe*, *37*(1), 7. DOI: 10.1186/s12302-024-01044-y

Pandit, S., Yadav, N., Sharma, P., Prakash, A., & Kuila, A. (2025). Life cycle assessment and techno-economic analysis of nanotechnology-based wastewater treatment: Status, challenges and future prospectives. *Journal of the Taiwan Institute of Chemical Engineers*, *166*, 105567. DOI: 10.1016/j.jtice.2024.105567

Panizzolo, M. (2025). Epidemiological applications of inflammation and oxidative stress biomarkers to assess the early effects of human exposure to micro and nano-plastics. DOI: 10.1016/j.ecoenv.2023.115645

Peddie, C. J., Genoud, C., Kreshuk, A., Meechan, K., Micheva, K. D., Narayan, K., Pape, C., Parton, R. G., Schieber, N. L., Schwab, Y., Titze, B., Verkade, P., Aubrey, A., & Collinson, L. M. (2022). Volume electron microscopy. *Nature Reviews. Methods Primers*, *2*(1), 51. DOI: 10.1038/s43586-022-00131-9 PMID: 37409324

Peng, L., Mehmood, T., Bao, R., Wang, Z., & Fu, D. (2022). An overview of micro (nano) plastics in the environment: Sampling, identification, risk assessment and control. *Sustainability (Basel)*, *14*(21), 14338. DOI: 10.3390/su142114338

Qiu, Y., Mintenig, S., Barchiesi, M., & Koelmans, A. A. (2025). Using artificial intelligence tools for data quality evaluation in the context of microplastic human health risk assessments. *Environment International*, *197*, 109341. DOI: 10.1016/j.envint.2025.109341 PMID: 39987688

Rai, A. (2022). Nanoplastics, Gut Microbiota, and Neurodegeneration. In *Gut Microbiome in Neurological Health and Disorders* (pp. 211–234). Springer Nature Singapore., DOI: 10.1007/978-981-19-4530-4_13

Rasheed, A., Bashir, Y., & Das, S. (2025). Are (bio) electrochemical techniques sustainable solutions to combat micro-and nano-plastic pollution? *Current Opinion in Chemical Engineering*, *47*, 101091. DOI: 10.1016/j.coche.2025.101091

Rathour, A., & Pathania, S. (2025). Sensing of Microplastics Using Advanced Materials: A Comprehensive Review. *Sensing and Imaging*, *26*(1), 80. DOI: 10.1007/s11220-025-00613-6

Ratre, P., Nazeer, N., Soni, N., Kaur, P., Tiwari, R., & Mishra, P. K. (2024). Smart carbon-based sensors for the detection of non-coding RNAs associated with exposure to micro (nano) plastics: An artificial intelligence perspective. *Environmental Science and Pollution Research International*, *31*(6), 8429–8452. DOI: 10.1007/s11356-023-31779-9 PMID: 38182954

Regasa, M. B. (2024). New Approaches for Micro (Nano) Plastics Analysis. Toxic Effects of Micro-and Nanoplastics: Environment, Food and Human Health, 471-492. https://doi.org/DOI: 10.1002/9781394238163.ch20

Rhaman, M. M., Islam, M. R., Akash, S., Mim, M., Noor Alam, M., Nepovimova, E., Valis, M., Kuca, K., & Sharma, R. (2022). Exploring the role of nanomedicines for the therapeutic approach of central nervous system dysfunction: At a glance. *Frontiers in Cell and Developmental Biology*, *10*, 989471. DOI: 10.3389/fcell.2022.989471 PMID: 36120565

Santhanam, S. D., Ramamurthy, K., Priya, P. S., Sudhakaran, G., Guru, A., & Arockiaraj, J. (2024). A combinational threat of micro-and nano-plastics (MNPs) as potential emerging vectors for per-and polyfluoroalkyl substances (PFAS) to human health. *Environmental Monitoring and Assessment*, *196*(12), 1182. DOI: 10.1007/s10661-024-13292-9 PMID: 39514026

Smith, N. R., & Levy, D. E. (2024). Budget impact analysis for implementation decision making, planning, and financing. *Translational Behavioral Medicine*, *14*(1), 54–59. DOI: 10.1093/tbm/ibad059 PMID: 37776567

Sorensen, R. M., Kanwar, R. S., & Jovanovi, B. (2023). Past, present, and possible future policies on plastic use in the United States, particularly microplastics and nanoplastics: A review. *Integrated Environmental Assessment and Management*, *19*(2), 474–488. DOI: 10.1002/ieam.4678 PMID: 36036190

Turner, H. C., Hori, Y., Revill, P., Rattanavipapong, W., Arai, K., Nonvignon, J., Jit, M., & Teerawattananon, Y. (2023). Analyses of the return on investment of public health interventions: A scoping review and recommendations for future studies. *BMJ Global Health*, *8*(8), e012798. Advance online publication. DOI: 10.1136/bmjgh-2023-012798 PMID: 37648275

Velumani, K., Thomas, A. G., Shaik, M. R., Hussain, S. A., Shaik, B., Guru, A., & Issac, P. K. (2025). Functionalized Zinc Oxide Nanoparticles Conjugated With Artemisinin Against Luperox-Induced Oxidative Stress and Their Impact on Antioxidant Defense Mechanism. *Journal of Biochemical and Molecular Toxicology*, *39*(7), e70380. DOI: 10.1002/jbt.70380 PMID: 40586296

Verma, D., Okhawilai, M., Dalapati, G. K., Ramakrishna, S., Sharma, A., Sonar, P., Krishnamurthy, S., Biring, S., & Sharma, M. (2022). Blockchain technology and AI-facilitated polymers recycling: Utilization, realities, and sustainability. *Polymer Composites*, *43*(12), 8587–8601. DOI: 10.1002/pc.27054

Vignesh, K. S., Prapanchan, V. N., Selvan, V. I., Karmegam, N., Kim, W., Barcelo, D., & Govarthanan, M. (2024). Microplastics, their abundance, and distribution in water and sediments in North Chennai, India: An assessment of pollution risk and human health impacts. *Journal of Contaminant Hydrology*, *263*, 104339. DOI: 10.1016/j.jconhyd.2024.104339 PMID: 38564944

Waring, R. H., Harris, R. M., & Mitchell, S. C. (2018). Plastic contamination of the food chain: A threat to human health? *Maturitas*, *115*, 64–68. DOI: 10.1016/j.maturitas.2018.06.010 PMID: 30049349

Xi, Z., Nicolas, R., & Wei, J. (2025). An Edge-Deployable Multi-Modal Nano-Sensor Array Coupled with Deep Learning for Real-Time, Multi-Pollutant Water Quality Monitoring. *Water (Basel)*, *17*(14), 2065. DOI: 10.3390/w17142065

Xiao, L., & Schultz, Z. D. (2017). Spectroscopic imaging at the nanoscale: Technologies and recent applications. *Analytical Chemistry*, *90*(1), 440–458. DOI: 10.1021/acs.analchem.7b04151 PMID: 29028297

Xie, L., Ma, M., Ge, Q., Liu, Y., & Zhang, L. (2025). Machine learning advancements and strategies in microplastic and nanoplastic detection. *Environmental Science & Technology*, *59*(18), 8885–8899. DOI: 10.1021/acs.est.4c11888 PMID: 40293506

Zhang, M., Yang, J., Cai, Z., Feng, Y., Wang, Y., Zhang, D., & Pan, X. (2019). Detection of engineered nanoparticles in aquatic environments: Current status and challenges in enrichment, separation, and analysis. *Environmental Science. Nano*, *6*(3), 709–735. DOI: 10.1039/C8EN01086B

Zhao, B., Richardson, R. E., & You, F. (2024). Advancing microplastic analysis in the era of artificial intelligence: From current applications to the promise of generative AI. *Nexus*, *1*(4), 100043. Advance online publication. DOI: 10.1016/j.ynexs.2024.100043

Zheng, P. C., Li, R., Lai, K. P., & Zhang, X. X. (2024). Biological exposure to microplastics and nanoplastics and plastic additives: Impairment of glycolipid metabolism and adverse effects on metabolic diseases. *Environmental Science and Pollution Research International*, *31*(51), 60778–60791. DOI: 10.1007/s11356-024-35220-7 PMID: 39412714

ADDITIONAL READING

Babaniyi, B. R., Olamide, I. G., Fagbamigbe, D. E., Adebomi, J. I., & Areo, I. F. (2024). Environmental pollution and the entrance of toxic elements into the food chain. In *Phytoremediation in food safety* (pp. 109–124). CRC Press., DOI: 10.1201/9781032683751-11

Galloway, T. S. (2015). Micro-and nano-plastics and human health. In *Marine anthropogenic litter* (pp. 343–366). Springer International Publishing., DOI: 10.1007/978-3-319-16510-3_13

Ghosh, A., & Gorain, B. (2025). Mechanistic insight of neurodegeneration due to micro/nano-plastic-induced gut dysbiosis. *Archives of Toxicology*, *99*(1), 83–101. DOI: 10.1007/s00204-024-03875-3 PMID: 39370473

Roy, T. (2024). Aquatic ecosystem toxicity and food chain. In *Toxicity of aquatic system and remediation* (pp. 25–36). CRC Press., DOI: 10.1201/9781003297901-3

Yadav, M., Pandey, A., & Nghiem, X. H. (2025). Policy and regulatory approaches to mitigating micro-and nano plastic pollution. In *Global Impacts of Micro-and Nano-Plastic Pollution* (pp. 309–330). IGI Global., DOI: 10.4018/979-8-3693-3447-8.ch013

KEY TERMS AND DEFINITIONS

Budget Impact Analysis (BIA): BIA examines the short- to medium-term financial consequences of implementing a new intervention or a technology within given budget.

Cost of Illness (COI): COI is a metric that demonstrates the cumulative financial burden of one disease, including the direct costs (medical expenses, hospitalization) and also indirect costs (productivity loss, premature mortality), along with intangible costs (reduced quality of life).

Digital Twins: This is a type of virtual model usually used for continuously updating the sensor data from the source.

Health Economics Metrics: These instruments are used to evaluate the health impact and also economic efficiency of the implemented interventions among a population.

Image Recognition Tools: These devices use the computer vision and also AI to analyze the microscopic or spectroscopic images to quantify nanoplastics.

Machine Learning Algorithms: They are computational models used for identifying patterns and also make predictions from the larger datasets

Microbial and Enzymatic Remediation: This demonstrates the usage of microbes or enzymes for breaking down plastics into harmless compounds.

Nanoplastics: These are plastic particles lesser than 1 μm. They are pollutants and cross the biological barriers, gets accumulated in organs, and disrupt the ecosystems.

Return on Investment (ROI): ROI estimates economic benefit obtained to the cost of any intervention.

Smart Sensor Networks: These consists of interconnected, real-time monitoring devices detecting and transmitting the environmental data.

Chapter 8
Economic and Life Cycle Assessment of Nanoplastics Remediation Within a One Health Framework:
Biotechnological and AI-Driven Solutions

Odangowei Inetiminebi Ogidi
http://orcid.org/0000-0002-9961-893X
Bayelsa Medical University, Yenagoa, Nigeria

Ejeomo Christopher
Michael and Cecilia Ibru University, Agbarha-Otor, Nigeria

Obienu Anayochukwu Chukwunonso
http://orcid.org/0000-0001-5503-5911
Bayelsa Medical University, Yenagoa, Nigeria

Emmanuel Aziagba
Sheda Science and Technology, Abuja, Nigeria

ABSTRACT

This chapter per the authors analyzes economic and life cycle dimensions of nanoplastic remediation through analytical frameworks linking environmental burden remediation efficiency and resource allocation dynamics within One Health systems.

DOI: 10.4018/979-8-3373-4490-4.ch008

Nanoplastic generation pathways environmental transport behavior and cross matrix exposure routes across aquatic terrestrial and food associated environments are examined using life cycle assessment methodologies to identify environmental and economic hotspots associated with remediation deployment. Biotechnological mitigation approaches including microbial degradation pathways and enzyme mediated transformation systems are analyzed alongside computational optimization tools supporting predictive remediation modeling cost efficiency analysis and contaminant forecasting. Integration of ecological risk metrics public health considerations and socioeconomic performance indicators contributes a techno economic perspective for sustainable nanoplastic management environmentally responsive remediation planning and long term exposure mitigation.

INTRODUCTION AND RATIONALE

Nanoplastics, typically defined as plastic particles smaller than 100 nm, have emerged as a critical environmental and public health concern due to their high mobility, persistence, and biological reactivity (Gil-Díaz et al., 2016). Unlike larger microplastics, nanoplastics can penetrate biological membranes, circulate within organisms, and accumulate across trophic levels, making them a significant risk within the One Health framework. Nanoplastics originate from both primary and secondary sources. Primary sources include intentional industrial production for applications such as cosmetics, coatings, and nanocomposites. Secondary sources arise from the fragmentation of macro- and microplastics through processes such as ultraviolet radiation, mechanical abrasion, and biodegradation. (Thome et al., 2015; EPA, 2010).

Once released, nanoplastics are transported across interconnected environmental systems. In aquatic environments, they are present in rivers, oceans, groundwater, and even drinking water supplies. In soil systems, they accumulate in agricultural lands through pathways such as sewage sludge application and degradation of plastic mulches. Additionally, nanoplastics can become airborne, facilitating long-range transport through atmospheric pathways. Exposure routes within a One Health context are multifaceted. Humans are exposed primarily through ingestion of contaminated food and water, as well as inhalation of airborne particles. Animals are exposed via contaminated water, feed, and habitats, while ecosystems experience impacts through microbial disruption and trophic transfer across food webs (Cecchin et al., 2017; Zhao et al., 2016).

These interconnected pathways demonstrate that nanoplastics contamination is not an isolated environmental issue but a systemic challenge affecting environmental integrity, animal health, and human well-being. Consequently, addressing this issue requires integrated and multidisciplinary solutions. While nanoparticle-based remediation technologies have been extensively studied, their relevance in this chapter is reframed toward nanoplastics-specific applications, particularly in evaluating economic feasibility, environmental impacts, and sustainability within a One Health paradigm (Cecchin et al., 2017; Francy & Shanthakumar, 2015; Thome et al., 2015).

Existing literature on nanoplastics spans three dominant domains: (i) environmental distribution and remediation technologies, (ii) human and ecological health implications, and (iii) emerging socioeconomic and governance perspectives. However, current studies remain fragmented, with limited integration across these domains, resulting in a lack of holistic understanding of system-wide impacts and feedback mechanisms.

Socioeconomic Context and Public Perception of Nanoplastics

Even though nanotechnology has been explored comprehensively regarding technical and environmental aspects, the socioeconomic impact and perception of the public regarding nanoplastics have been given inadequate representation in the current state of literature. Public perception of the danger and mitigating potential of nanoplastics directly impacts the acceptance of technological change and policy as practices of the society. According to studies, the particular concern among the public regarding nanoplastics is the potential toxicity and long-term effects of nanotoxicity (Bhattacharya & Kim, 2024). However, socioeconomic disparity may widen the sensitivity of the underprivileged classes, which are generally found near industrial outlets.

The integration of the socioeconomic analysis within the life cycle assessment of nanoplastics offers broader insight into risk management and sustainability. As asserted by Ma & Lin (2024), the successful pursuit of sustainable remediation is not possible without dealing with distributional equity, participation, and the welfare of the affected communities. Thus, aside from the measurements of the environmental and technical issues, the broader framework shall incorporate the indirect socioeconomic costs on healthcare, life, and jobs. Perception on nanoplastics by key stakeholders such as industry, government, and local community members determines the viability for technology implementation. Engaging with stakeholders, risk communication, and decision-making schemes have improved clarity and acceptability within environmental management. They also harmonize well with schemes underpinned by One Health by ensuring different stakeholders' well-being, human, environmental, and economic well-being, are assessed together.

Various methods exist for synthesizing nanoscale iron with no valentine. The technologies used, properties of physical or chemical processes, and the quality of the end product are the bases on which these methods are classified into those producing nZVI (Thome et al., 2015). While top-down approaches start with larger particles, such as bulk materials, to generate nZVI, bottom-up approaches begin with smaller entities, such as atoms and molecules, and extend into larger structures, often tens to hundreds of nanometres in size (Crane &Scott, 2012). Milling iron particles is a physical process that does not rely on chemical interactions to synthesize nZVI (Li et al., 2009). However, there are numerous chemical stages involved in the synthesis of nZVI (Crane & Scott, 2012). Green synthesis utilizing polyphenol-rich plant extracts, electrolysis of ferrous salts, microemulsion techniques, carbothermal reduction, ultrasonic waves, sol-gel processes, hydrogen gas, and traditional bottom-up methods of creation of nZVI all belong to this category. Therein, nanomaterials, such as nZVI, are physicochemically produced, which exposes workers to high levels of chemicals, produces hazardous waste, and requires huge amounts of water and energy.

American Polyflon, Canadian MKnano, and American Sky Spring Nano Materials, Inc. are prominent commercializers of nZVI, alongside Japan's Toda Kogyo Ltd. and the Czech Republic's NanoIron. Zhao et al. (2016); Stefaniuk et al. (2016) Golder Associates Inc. of the United States oversees the large-scale manufacture of nZVI, but not its commercialisation. In conjunction with other modifications in the management of polluted sites, there has been a growing concern in recent years regarding the effects of cleanup activities (Rizzo et al., 2016). The propagation of sustainable remediation practices is increasingly significant. This methodology encompasses the complete restoration process, from inception to conclusion, with its social, economic, and environmental ramifications (Eizenberg & Jabareen, 2017). The evaluation of remediation methods and tactics can be enhanced by the application of technologies like the nZVI case study. Producing and disposing of any product or process incurs environmental and financial costs. LCA and LCC can quantify and study these effects, accordingly. Many fields use these tools to enable process and product development, feasibility studies, as well as decision-making processes, by Pranjic et al. 2018; Heijungs and Wiloso 2014; Singh et al., 2008.

Soil remediation uses life cycle assessment (LCA) and life cycle cost (LCC) to compare different technologies and choose the one that has the least negative effect on the environment and the cheapest price tag (Baalousha et al., 2023; Song et al., 2018; Martins et al., 2017). Researchers have vast knowledge about soil and groundwater remediation; however, the impact of its production methods has received considerably less attention. Few studies have investigated the economic and ecological impact of nZVI synthesis methods, and these are just by Joshi et al. (2018) and Martins et al. (2017). In the work by Martins et al. (2017), conventional methods, which applied sodium boron hydride for the chemical reduction, were compared to green, environmentally friendly approaches using plant extracts. Based on this work, Joshi et al. (2018) studied the cost and impacts of a green synthesis process by reductive microbes using naturally occurring Fe (III). Several current methods of producing nZVI were assessed in these studies, but their appropriateness was only limited to a controlled laboratory environment (Ogidi & Richard-Nwachukwu, 2024). This chapter has sought to investigate the environmental impact and the related costs of some of the methods of producing nZVI regarding remediation applications. The investigated production techniques have been based on a wide range of production and application technologies available in commercial and academic environments.

Nanoparticle–Nanoplastic Interrelationship

Nanoplastics share physicochemical properties with engineered nanoparticles, including high surface area, reactivity, and mobility. These similarities allow existing nanotechnology-based remediation strategies to be adapted for nanoplastics detection, capture, and degradation. Key mechanisms relevant to nanoplastics remediation include: Adsorption using carbon-based nanomaterials and biochar-supported systems. Photocatalytic degradation using TiO_2 and ZnO nanoparticles. Membrane filtration using graphene oxide and nanocomposite membranes. Bioremediation, including microbial colonization and enzymatic polymer breakdown.

However, unlike conventional contaminants, nanoplastics present unique challenges due to their heterogeneous composition, persistence, and ability to adsorb toxic co-contaminants. Therefore, remediation strategies must be evaluated not only for efficiency but also for their environmental and health implications within a life cycle and One Health framework. This reframing shifts the focus from general nanoparticle remediation toward nanoplastics-specific management, aligning with the broader objectives of sustainable environmental intervention.

Overview of Nanoparticle Remediation Technologies

Nanoparticle remediation technologies (Figure 1) are emerging as a revolutionary innovation for environmental engineering, exploiting the unique properties of nanoscale materials toward contamination in water, soil, and air systems. Because of their size, usually less than 100 nm, NPs have high surface area-to-volume ratios and tunable physicochemical properties that enable them to interact with pollutants much more effectively compared to their bulk counterparts (Wang & Chen, 2016). It is this that makes them particularly useful in the effective immobilization, transformation, or degradation of a wide range of contaminants such as heavy metals, chlorinated solvents, pesticides, pharmaceuticals, and VOCs (Karn et al., 2009). nZVI, among intensively studied and applied nanomaterials for in situ remediation, especially in the case of groundwater contaminated with chlorinated hydrocarbons and heavy metals, nanoscale zero-valent iron has huge reduction potential, thus allowing electron transfer reactions to occur in the cleavage of complex pollutants into less toxic or inert byproducts (Mondal et al., 2023; Zhang, 2003). Challenges such as rapid oxidation, particle aggregation, and limited mobility in porous media have spurred the development of surface-modified and stabilized nZVI formulations. Coatings, such as polyacrylic acid, starch, or biochar, have proved quite effective in enhancing particle dispersion, longevity, and remediation performance under real conditions (Tratnyek & Johnson, 2006; Ali et al., 2021; Ogidi et al., 2024b, c; Wang & Chen, 2016; Karn et al., 2009).

Hybrid nanocomposites combining nanoparticles with substrates like zeolites, clays, or polymers find increasing applications owing to their synergistic properties. Most of them combine adsorption and catalysis with regeneration in one system, thus providing efficiency and sustainability in the treatment technology. For example, iron-modified zeolites have given excellent performance in the removal of dyes and chromium from industrial effluents. Their high ion-exchange capacity and structural stability make them attractive in continuous flow applications. Magnetic nanoparticles such as Fe_3O_4 and $CoFe_2O_4$ allow for quick separation of the treated pollutant with the aid of an external magnetic field. This simply enables recovery and reutilization to minimize operational cost and potential risks of environmental release. Functionalization of MNPs using ligands or chelating agents, such as EDTA, enhances their selectivity to certain metal ions and improves sorption kinetics. Such systems are of particular importance in the case of industrial wastewater, where fast treatment and recyclability of the sorbing material are important.

Other recent research is also focused on integrative systems that couple nanomaterials with either biological or electrochemical processes (Ogidi & Akpan, 2024). This includes plant-based nanophytoremediation, microbe-nanoparticle consortia, and nano-Fe_3O_4-catalyzed electro-Fenton reactions. These hybrid approaches often improve contaminant specificity and degradation rates, while enabling self-regenerating treatment cycles (Mondal et al., 2023). Such designs epitomize the trend in next-generation, multifunctional, adaptive remediation platforms. Despite these promising developments, a number of challenges have to be surmounted before field deployment could become widespread. Key challenges involve particle stability, transport limitations in heterogeneous subsurface environments, long-term ecotoxicity, and cost of production and deployment (Ogidi et al., 2025a; Zhang, 2003; Ogidi et al., 2024a). Furthermore, regulatory lacunas, data gaps with regard to nanoparticle fate after remediation, and lack of harmonized protocols are looming barriers to scalability and public confidence. Therefore, comprehensive life cycle assessments, field validation, and interdisciplinary collaboration will be required to ensure that nanoparticle remediation technologies transition safely and sustainably from the laboratory into large-scale environmental restoration projects (Karn et al.,

2009; Wang & Chen, 2016). These technological foundations provide a framework for adapting nanoparticle systems toward emerging contaminants such as nanoplastics.

Figure 1. Types of nanoparticle remediation technologies

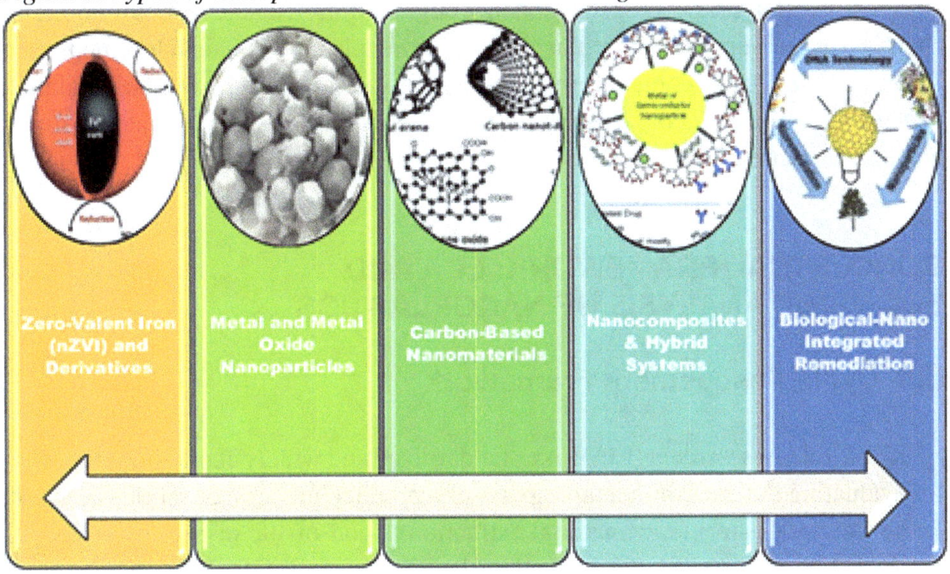

BIOTECHNOLOGICAL APPROACHES TO NANOPLASTICS REMEDIATION

Biotechnological strategies offer environmentally sustainable alternatives for the degradation and transformation of nanoplastics. These approaches leverage biological systems to break down polymer structures into less harmful compounds. Biotechnological approaches complement nanomaterial-based remediation by offering biologically driven pathways that operate under milder and more sustainable environmental conditions.

Microbial Degradation: Certain microorganisms, including bacteria and fungi, can colonize plastic surfaces and degrade polymer chains through metabolic processes. Species such as *Pseudomonas* and *Ideonella sakaiensis* have demonstrated the ability to degrade polyethylene terephthalate (PET) and related polymers.

Enzymatic Breakdown: Plastic-degrading enzymes such as PETase, MHETase, laccases, and peroxidases enable targeted cleavage of polymer bonds. These enzymes provide controlled and selective degradation pathways, reducing environmental persistence.

Bioengineered and Hybrid Systems: Emerging innovations include: Microbe–nanoparticle consortia, Enzyme-functionalized nanomaterials, Nano-bio catalytic systems. These hybrid approaches enhance degradation efficiency while minimizing chemical usage, supporting circular economy principles and sustainable remediation. Within a One Health context, biotechnological solutions are particularly valuable because they reduce secondary pollution risks and align with ecological and human health protection goals.

CONCEPTUAL FRAMEWORK: LCA AND ECONOMIC ANALYSIS OF NANOPLASTICS

Life Cycle Assessment Framework

Life Cycle Assessment (LCA) is a standardized and widely applied methodology for evaluating the environmental impacts of a product, process, or service across its entire life cycle—from raw material extraction to end-of-life management, including recycling or disposal. As defined by Jacquemin et al. (2012), LCA provides a comprehensive framework for assessing environmental burdens associated with all stages of a system, thereby supporting sustainable design, production, and decision-making. By systematically examining each phase, LCA enables the identification of critical impact points and uncovers opportunities for environmental performance improvement that may otherwise remain unrecognized.

Figure 2 presents the conceptual framework integrating economic analysis with life cycle assessment for nanoplastics. The framework adopted in this study follows the ISO 14040/44 guidelines. The system boundary encompasses raw material extraction, nanomaterial synthesis, operation of remediation systems, and waste management. A functional unit of 1 m^3 of treated water is applied to ensure consistency in comparative evaluation. Environmental indicators, including energy consumption (MJ/m^3), carbon emissions (kg CO_2-eq/m^3), and solid waste generation (g/m^3), were quantified using OpenLCA software in conjunction with the Ecoinvent 3.9 database.

LCA typically considers a broad range of environmental impact categories, including emissions to air, water, and soil, as well as resource depletion (Zhang et al., 2013; Vince et al., 2008; Garrigues et al., 2012; Klinglmair et al., 2014; Hossain & Poon, 2018). It is extensively applied in policy development, industrial process optimization, and product design to evaluate trade-offs and minimize environmental burdens.

According to ISO 14040:2006, LCA consists of three core phases. The first phase, Life Cycle Inventory (LCI), involves the systematic collection and quantification of inputs and outputs associated with the system. This forms the analytical foundation of the assessment. The second phase, Life Cycle Impact Assessment (LCIA), evaluates the potential environmental impacts based on inventory data, including categories such as global warming potential, acidification, eutrophication, and ozone depletion. The final phase, interpretation, integrates the findings from both inventory and impact assessment stages to support informed decision-making and identify priority areas for environmental improvement.

These phases collectively correspond to key stages within the product life cycle, including raw material acquisition, manufacturing, utilization, and end-of-life management. Each stage contributes differently to overall environmental impact, allowing targeted interventions such as process optimization, improved energy efficiency, and the adoption of renewable resources where the highest burdens are identified (Okereke et al., 2016; Angaye et al., 2018, 2019).

Within the context of nanoplastics, LCA provides a critical framework for quantifying the environmental implications of nanomaterial production and application in remediation systems. However, increasing data complexity and uncertainty associated with nanoplastics behavior and fate present significant challenges. These limitations highlight the need for integrating advanced computational tools, including artificial intelligence and data-driven modeling, to enhance the accuracy, robustness, and predictive capacity of LCA-based evaluations.

Figure 2. Conceptual framework of economic and life cycle assessment for nanoplastics

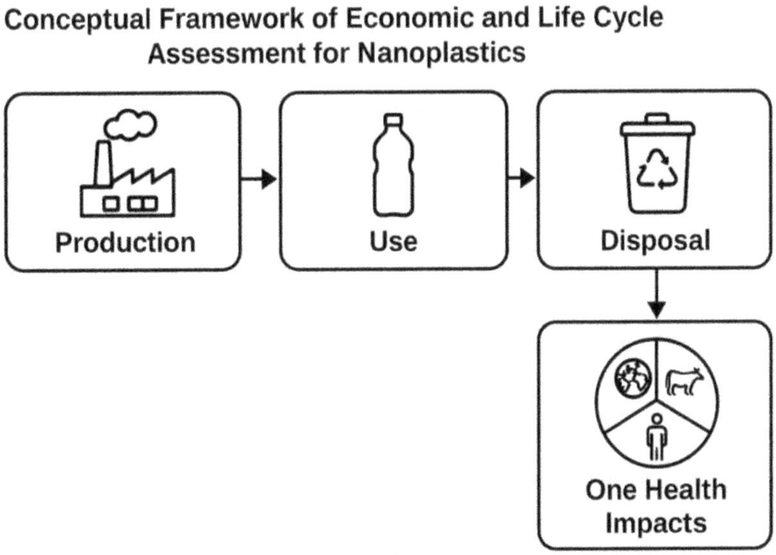

ARTIFICIAL INTELLIGENCE APPLICATIONS IN NANOPLASTICS MANAGEMENT

Artificial intelligence (AI) plays a critical role in advancing nanoplastics research beyond traditional life cycle assessment (LCA) models by enabling more precise detection, monitoring, and predictive analysis. AI-enabled tools, including machine learning algorithms and sensor-based systems, significantly improve the identification and quantification of nanoplastics in complex environmental matrices. In addition, AI supports predictive modeling of environmental fate and transport, bioaccumulation across food chains, and long-term ecological and health impacts. These capabilities allow for a deeper understanding of nanoplastics behavior within interconnected environmental and biological systems. Furthermore, AI-driven decision-support platforms facilitate the optimization of remediation strategies, as well as cost–benefit and scenario analyses, thereby enhancing policy evaluation and planning. By integrating large datasets across environmental, biological, and economic domains, AI enables real-time, data-driven decision-making, strengthening the application of the One Health framework in nanoplastics management (Aßen et al., 2014).

Beyond monitoring and prediction, AI is increasingly being integrated into economic and life cycle assessment frameworks to improve the evaluation of nanoplastics remediation strategies. However, the effectiveness of AI applications remains dependent on data quality, availability, and model transparency. Tools such as neural networks and regression-based simulators enhance LCA by improving data accuracy, reducing computational uncertainty, and enabling multi-scenario simulations of environmental and economic impacts. Within a One Health context, AI-assisted LCA models can identify intervention strategies that simultaneously minimize ecological damage and human health risks. However, the growing reliance on predictive modeling also highlights the need for robust and standardized sustainability indicators to quantitatively assess outcomes. Establishing such metrics is essential for enabling meaningful comparisons of nanoremediation systems across their life cycles and ensuring that AI-driven insights translate into practical, sustainable solutions (Peters & Weil 2016).

Environmental Metrics

Environmental indicators are essential tools for evaluating the environmental impacts associated with each stage of a product life cycle, including raw material extraction, manufacturing, use, and end-of-life management. These indicators support the identification and quantification of large-scale environmental burdens, thereby enabling more informed decision-making aimed at reducing overall impacts

(García-García et al., 2021). Within the context of life cycle assessment (LCA), key environmental performance indicators include carbon footprint, energy consumption, and resource depletion.

The carbon footprint is one of the most widely recognized environmental metrics, representing the total greenhouse gas (GHG) emissions associated with a product, process, or system, typically expressed in terms of CO_2-equivalents. LCA-based carbon footprint analysis accounts for emissions across all life cycle stages, including raw material extraction, production, transportation, usage, and final disposal (Aßen et al., 2014).

Energy consumption refers to the total amount of energy required throughout the life cycle of a product, encompassing mechanical, electrical, and thermal energy inputs. This metric is critical for assessing resource efficiency, identifying opportunities for energy conservation, and supporting the transition toward renewable energy systems (Klinglmair et al., 2014).

Resource depletion captures the consumption of natural resources, including water, minerals, and fossil fuels, across the entire life cycle. This indicator is particularly important for evaluating the long-term sustainability of resource use and the environmental consequences associated with the extraction and utilization of non-renewable resources (Peters & Weil, 2016).

Collectively, these environmental indicators enable the identification of life cycle stages that contribute most significantly to environmental degradation. Such insights support targeted interventions, including process optimization, improved energy efficiency, and sustainable material selection. In addition, environmental indicators play a critical role in policy development, industrial decision-making, and sustainability assessment by facilitating comparisons across technologies and guiding strategic planning.

Furthermore, increasing regulatory requirements have emphasized the importance of environmental reporting and compliance. LCA provides a standardized framework for quantifying environmental performance, thereby enhancing transparency in corporate reporting and strengthening stakeholder confidence, including that of policymakers and investors. This comprehensive perspective allows organizations to assess and improve their environmental performance while aligning with sustainability goals.

However, environmental indicators alone may not fully capture the practical feasibility of remediation strategies without integrating economic considerations. An integrated assessment that combines environmental and economic perspectives is therefore essential for evaluating the overall sustainability of nanotechnology-based solutions, particularly in the context of nanoplastics management.

ECONOMIC ASSESSMENT METHODOLOGIES FOR NANOPARTICLE REMEDIATION TECHNOLOGIES

The economic evaluation of nanoparticle-based remediation technologies typically begins with Life Cycle Costing (LCC), which quantifies costs across the entire system life cycle—from initial capital investment to end-of-life management. These assessments commonly employ detailed, bottom-up approaches to disaggregate expenditures into capital, operational, and external cost components. For nanomaterials such as magnetite (Fe_3O_4), LCC models incorporate raw material sourcing, utility consumption, labor, and depreciation to determine cost-efficient production and deployment strategies.

Capital expenditure (CAPEX) constitutes a major component of LCC, encompassing infrastructure requirements such as reactor systems, land, and processing equipment. In green nanoparticle production, high initial investments in bioreactors or specialized microbial systems can significantly influence overall economic feasibility. These upfront costs must therefore be evaluated in relation to potential efficiency gains, including reduced energy consumption and smaller reactor footprints enabled by advanced nanomaterials.

Operational expenditure (OPEX) includes costs associated with raw materials, energy consumption, labor, and waste management. The cost of TiO_2 nanoparticles, for example, varies considerably, ranging from approximately US$176 to $198 per kilogram (US$0.176–0.198 per gram) based on techno-economic data (Demir et al., 2025). In contrast, bulk titanium dioxide is substantially less expensive, at approximately US$2.21 per kilogram (US$0.00221 per gram) (Demir et al., 2025), highlighting the significant cost differential associated with nanoscale production.

Similarly, the cost of nanoscale zero-valent iron (nZVI) has decreased over time due to technological advancements. Historical estimates indicate vendor prices ranging from US$20 to $77 per pound (approximately US$0.044–0.17 per gram) (EPA, 2012; Crane, 2012). Despite these reductions, bulk zero-valent iron remains considerably cheaper, although up-to-date comparative pricing data remain limited.

Total Cost of Ownership (TCO) and whole-life costing extend traditional LCC frameworks by incorporating indirect and long-term costs, including maintenance, disposal, and post-remediation environmental management. These approaches encourage decision-makers to consider long-term economic implications rather than focusing solely on initial investment costs.

Cost–Benefit Analysis (CBA), traditionally centered on direct financial returns, has evolved to incorporate environmental and public health externalities. Hybrid approaches, such as triple bottom line CBA, integrate economic, environmental, and social dimensions to support more comprehensive sustainability assessments. In parallel, Cost-Effectiveness Analysis (CEA) provides a comparative framework based on non-monetized outcomes, such as cost per unit of pollutant removed, thereby enabling benchmarking across remediation strategies without relying exclusively on financial valuation.

The integration of Life Cycle Costing with Life Cycle Assessment (LCA) offers a more holistic evaluation of both economic and environmental performance. Monetized LCA, for instance, assigns economic value to environmental impacts, enhancing traditional appraisal methods such as social CBA. Techno-economic analyses of specific nanomaterials, including biogenic magnetite, further contribute to this integrated perspective. The production of magnetite via magnetotactic bacteria has been estimated at US$10–11k per kilogram, corresponding to market prices ranging from approximately $21 to $120 per gram. These estimates may be competitive with certain synthetic alternatives, particularly when accounting for environmental and sustainability benefits. Such analyses also incorporate factors such as economies of scale, feedstock variability, and process optimization in projecting viable production pathways.

Sensitivity and scenario analyses play a critical role in addressing uncertainty within economic evaluations. Techniques such as Monte Carlo simulation enable the assessment of parameter variability and robustness of model outcomes. In life cycle sustainability assessments (LCSA) of nZVI production, such analyses have demonstrated that green synthesis approaches maintain sustainability advantages across a range of economic scenarios, with relatively low sensitivity to weighting assumptions (Silva Trentin et al., 2021).

However, despite the increasing sophistication of economic assessment tools, significant challenges remain in capturing the full economic implications of nanoparticle-based remediation technologies. Variability in material costs, scale-up uncertainties, and limited long-term performance data introduce considerable uncertainty into cost projections. Furthermore, discrepancies between laboratory-scale efficiency and field-scale performance may complicate the translation of economic models into practical applications. These limitations highlight the importance of incorporating flexible, scenario-based modeling approaches to support more robust and context-sensitive decision-making.

For example, Zhang et al. (2022) reported that bioremediation using immobilized enzymes reduced overall treatment costs by approximately 25% compared to chemical oxidation, without compromising removal efficiency. In addition, the integration of predictive tools, including AI-assisted life cycle costing models implemented in platforms such as OpenLCA, offers further opportunities to improve decision-making and reduce uncertainty in economic evaluations (Pini & Rossi, 2024).

Overall, these frameworks collectively highlight the importance of balancing economic feasibility with environmental sustainability in the development of nanoremediation technologies. The following section outlines the methodological approach adopted for the analysis and discussion presented in this chapter.

METHODOLOGY AND ANALYTICAL DETAIL

Methodological Approach

The approach adopted in this chapter is based on a qualitative Life Cycle Assessment (LCA) framework supported by an economic evaluation of various nanoplastic management strategies. Data were collected from peer-reviewed literature, governmental databases, and published life cycle inventories covering the period from 2018 to 2025.

The system boundary includes nanoplastic production, environmental release, exposure pathways, and end-of-life management. The functional unit is defined as "one kilogram of nanoplastic material processed or remediated." The assessment considers key impact categories, including greenhouse gas (GHG) emissions, energy demand, aquatic ecotoxicity, and potential impacts on human health.

The economic evaluation incorporates Cost–Benefit Analysis (CBA) alongside the valuation of ecosystem service losses. Due to data limitations, quantitative modeling was applied selectively; however, comparative insights between microplastic and nanoplastic life cycle assessments were developed where feasible.

Overall, the integration of economic and life cycle assessment frameworks provides a structured basis for evaluating the environmental and economic implications of nanoplastic management strategies. To further contextualize these findings and enhance practical relevance, the following section presents selected case studies.

FINDINGS AND DISCUSSION

Notable Case Studies and Real-World Applications of Nanoparticle Remediation Technologies

Environmental applications of nanoscale zero-valent iron (nZVI) have demonstrated significant success in the remediation of chlorinated solvents in groundwater systems. For example, the injection of carboxymethyl cellulose-coated nZVI into DNAPL-contaminated aquifers has shown sustained contaminant degradation over extended periods, supported by integrated simulation frameworks that model plume dynamics and remediation kinetics. Field-scale demonstrations across North America and Europe have confirmed the effectiveness of nZVI in large-scale remediation.

More than 70 sites—including 44 documented by 2009—have implemented in situ nZVI injections tailored to site-specific hydrogeological conditions, achieving cost-effective treatment with minimal excavation requirements.

Green-synthesized iron oxide nanoparticles represent another important example of sustainable remediation. Comparative life cycle assessments have shown that biogenic iron oxides produced via green synthesis exhibit lower environmental impacts than conventional coprecipitation methods, primarily due to reduced chemical usage and lower energy demand (Bafana et al., 2018).

Silver nanoparticles (AgNPs) have also demonstrated strong antimicrobial and pollutant degradation capabilities. For instance, glucose- and starch-assisted microwave synthesis routes have been associated with reduced impacts in categories such as acidification, human toxicity, and particulate matter formation. Life cycle analyses indicate that these impacts are largely driven by electricity consumption and silver precursor inputs, rather than the synthesis reagents themselves (Bafana et al., 2018).

Comparative assessments of AgNP synthesis methods reveal that more than 90% of environmental burdens originate from bulk silver production. Among synthesis techniques, flame spray pyrolysis exhibits the highest environmental impacts, whereas bio-based chemical reduction offers comparatively lower energy consumption and emissions. However, when normalized based on antimicrobial efficacy, the relative performance rankings shift, highlighting the importance of functional unit selection in life cycle assessment (Pourzahedi & Eckelman, 2015).

In the textile sector, life cycle assessments of nanosilver-treated garments produced via flame spray pyrolysis and plasma sputtering indicate that, although production stages are energy-intensive, overall environmental benefits may be achieved under conditions of reduced washing frequency and extended product use (Walser et al., 2011). Similarly, cradle-to-grave analyses of nanosilver medical bandages show that upstream electricity consumption and silver production dominate environmental impacts, while disposal-related burdens remain comparatively minor (Pourzahedi et al., 2014).

AgNPs have also been applied in industrial wastewater treatment, demonstrating rapid removal of phenolic compounds, with reported reductions exceeding 90% within short reaction times. Biogenic AgNPs synthesized using *Conocarpus erectus* leaf extract or *Pseudomonas* cultures have achieved comparable removal efficiencies for dyes such as Reactive Black 5 and Reactive Red 120 in textile effluents. In addition to high remediation efficiency, these green synthesis approaches offer advantages in terms of cost-effectiveness and reduced toxicity relative to conventional chemical synthesis methods (Mahajan et al., 2026).

Modeling studies further indicate that AgNPs released through wastewater treatment systems or accidental discharges may reach surface water bodies at nanogram-per-liter concentrations, with potential implications for aquatic ecosystems. These findings underscore the need for risk-informed deployment strategies and the development of standardized protocols for nanoparticle application in environmental remediation (Arundhathi et al., 2022). Key case studies and applications are summarized in Table 1.

Comparative case analyses also highlight trade-offs between remediation technologies. For instance, photocatalytic degradation using TiO_2 nanoparticles has been reported to incur higher initial treatment costs (approximately US$150/m^3) compared to bioremediation (approximately US$95/m^3), yet it produces 30% less sludge and achieves an 18% reduction in greenhouse gas emissions. These findings demonstrate that economic and environmental benefits can be aligned, supporting One Health objectives through reduced ecological toxicity (Li & Liu, 2023; OECD, 2013).

While the majority of existing case studies focus on nanoparticle-based remediation, their findings provide valuable insights for emerging nanoplastics management strategies. Mechanisms such as adsorption, catalytic degradation, and nano-bio hybrid systems can be adapted for nanoplastics treatment. However, future research should prioritize case studies specifically addressing nanoplastics in environmental and food systems to ensure stronger thematic alignment and practical applicability.

Table 1. Notable case studies and real-world applications of nanoparticle remediation

Technology / System	Target Contaminant(s)	Location / Study	Outcome / Notes	Reference
Emulsified Zero-Valent Iron (EZVI)	Chlorinated VOCs (CVOCs)	Field site, USA	2.5-year monitoring showed ~86% CVOC mass reduction	(Su et al., 2012)
Carbo-Iron (activated carbon + nZVI)	PCE	Germany	Initial PCE drop, but rebound after ~1000 days	(Mackenzie et al., 2016)
Bimetallic Fe/Pd Nanoparticles	TCE, PCE	U.S. Navy sites	Rapid dechlorination, improved over nZVI	(He et al., 2010)
EZVI (Parrick Island, SC)	PCE, TCE	Field trial	Both injection methods reduced solvent levels	(Quinn et al., 2005)
nZVI in Sediment Columns	Zn, Cd, Pb, Cu, As	Minas Gerais, Brazil	Removal rates: Zn 95–99%, As 90–99%	(Gomes et al., 2022)
Nano-Bio Hybrid (Magnetite NPs + microbes)	Hydrocarbons	Egypt	Accelerated petroleum degradation in water	(El-Sheikh et al., 2025)
Biogenic FeO NPs (Halomonas sp.)	Pb, Cd	Lab study	Nearly 100% Pb removal (24h), Cd (48h)	(Salem et al., 2022)
Fungus + NPs (TiO_2, Ag, Fe_2O_3, Se)	Phenols, 2,4-DCP, Cd^{2+}	Lab-scale	Enhanced degradation of organic & metal pollutants	(Rizwan et al., 2022)

continued on following page

Table 1. Continued

Technology / System	Target Contaminant(s)	Location / Study	Outcome / Notes	Reference
Nano-Bioremediation (various)	PCBs, PAHs, crude oil	Soil & water studies	Combined systems >2× removal vs. single	(Sharma et al., 2023)
PRBs (Granular Fe)	PCE, TCE, Uranium	USA (Sunnyvale, Moffett, Fry Canyon)	Reduced chlorinated solvents to non-detect; >99.9% U removal	(Blowes et al., 2000)
Nanosponges	PAHs (phenanthrene)	Column tests (Cornell)	Sequestered PAHs effectively in soil	(Blowes et al., 2000)

SOCIOECONOMIC DIMENSIONS AND PUBLIC PERCEPTION OF NANOPLASTICS

The socioeconomic implications of nanoplastics contamination extend beyond environmental risks, influencing public health, economic stability, and social equity. Public perception is primarily shaped by concerns related to toxicity, long-term exposure, and uncertainty in emerging scientific evidence. These perceptions significantly affect policy acceptance, regulatory support, and the adoption of remediation technologies. In some cases, limited awareness or distrust of nanotechnology can hinder the implementation of innovative solutions, even when they offer clear environmental benefits.

Exposure to nanoplastics is also unevenly distributed, raising important environmental justice concerns. Populations located near industrial zones, waste disposal sites, or polluted water systems are often disproportionately affected, reinforcing existing socioeconomic inequalities. Comparative evidence suggests that high-income regions are more likely to implement advanced technological interventions and robust regulatory frameworks, while low- and middle-income regions face constraints related to infrastructure, enforcement capacity, and vulnerability to contamination. In this context, participatory approaches such as citizen science and community-based monitoring have demonstrated effectiveness in improving environmental outcomes and strengthening public trust. These considerations underscore the importance of integrating equity, risk communication, and stakeholder engagement into nanoplastics management within a One Health framework.

Comparison of Technologies Based on Cost and Environmental Impact

Comparative analyses of remediation technologies show clear trade-offs between cost, efficiency, and environmental performance. For instance, in situ chemical oxidation (ISCO) and pump-and-treat methods, when compared with enhanced zero-valent iron (EZVI) injections, demonstrate that although EZVI has a slightly higher upfront cost (approximately \$1.28 million for a 2,718 m^3 plume) than ISCO (\$1.15 million), it remains significantly more cost-effective than pump-and-treat systems (\$2.61 million). This advantage is primarily attributed to reduced waste generation and shorter treatment durations (Malik et al., 2022). However, injection costs and long-term monitoring remain major cost drivers (Malik et al., 2022).

Similarly, Fe/Cu nanoparticle-based coagulation and adsorption in wastewater treatment is estimated at approximately \$6.35/m^3, with raw material and adsorbent costs accounting for nearly \$6.1/m^3 of operational expenses (Malik et al., 2022). When combined with electrocoagulation and ozonation processes, operational costs decrease slightly to \$5.80/m^3, although nanoparticle losses remain a significant economic limitation (Malik et al., 2022).

Scale effects play a critical role in cost reduction. For example, the production cost of magnetic-core nanoparticles decreases substantially from approximately \$70/kg at laboratory scale to around \$4/kg at industrial scale. Regeneration and reuse further improve lifecycle cost efficiency. However, composite nanoparticles such as SiO_2–TiO_2–nZVI remain relatively expensive (approximately \$13.46 per batch of reagents), although their high adsorption capacity and reusability may compensate for higher initial investment through extended operational life (RSC Publishing, 2024).

Green synthesis approaches for TiO_2 nanoparticles have been shown to significantly reduce greenhouse gas emissions and toxicity, with life cycle assessment results indicating lower global warming potential even under comparable material cost conditions (Rodríguez Rojas et al., 2024). Likewise, techno-economic and LCA studies of engineered nanoparticles used in ocean fertilization estimate average production costs of approximately US\$72/kg and a carbon footprint of about 38 kg CO_2e/kg ENP. Green synthesis methods reduce both indicators by roughly 20%, although costs remain 4–10 times higher than conventional iron sulfate-based alternatives.

Green-synthesized nanoscale zero-valent iron (nZVI), such as that derived from palm-petiole extract, demonstrates reduced reagent and energy costs while maintaining high Cr(VI) removal efficiency. However, comprehensive economic assessments incorporating process efficiency and supply chain considerations remain limited. Carbon-based nanoparticles such as carbon nanotubes (CNTs) and graphene oxide present additional trade-offs: despite high synthesis costs and elevated embodied energy, their strong adsorption capacity can reduce overall cost per unit pollutant removed when regeneration is feasible.

Coating and composite systems, including iron-modified zeolites and TiO_2-functionalized float glass, generally exhibit higher production-phase environmental burdens but offer extended operational lifetimes and sustained reactivity. For instance, life cycle assessment of TiO_2-coated float glass indicates that although production dominates environmental impact, long-term operational benefits can offset these initial burdens.

Overall, nanoparticle remediation technologies involve inherent trade-offs between cost, environmental burden, and performance. While green synthesis and composite materials often require higher initial investment and exhibit greater embodied impacts, they can deliver superior contaminant removal efficiency and reduced pollutant rebound over extended use. Consequently, economic and environmental performance can only be optimized through appropriate scaling strategies, regeneration practices, and carefully defined functional units.

Despite promising advancements, large-scale implementation of nanoremediation technologies remains constrained by uncertainties in methodology, environmental variability, and data limitations. These challenges are further discussed in the following section, which identifies key gaps necessary for improving reliability and policy integration. Table 2 presents a comparative overview of nanoparticle remediation technologies in terms of cost and environmental impact.

Table 2. Comparison of nanoparticle remediation technologies by cost and environmental impact

Technology	Relative Cost	Environmental Impact	Notes / Considerations
nZVI (Nanoscale Zero-Valent Iron)	**Medium** – production and stabilization moderately costly	**Moderate impact** – iron oxides formed are benign; possible aquifer clogging; transport limited	Widely used in groundwater; requires stabilization to improve mobility.
Bimetallic nZVI (Fe–Pd, Fe–Ni, Fe–Cu)	**High** – cost of noble metals (Pd, Ni, Cu) increases expenses	**Low–Moderate** – highly reactive, smaller doses required; but metal leaching risk (esp. Ni, Cu)	Effective for chlorinated solvents; not always economically feasible at scale.
Stabilized / Polymer-coated nZVI	**Medium–High** – cost increases due to polymeric stabilizers	**Low–Moderate** – enhanced dispersion reduces required doses; possible persistence of polymers	Better transport in subsurface; reduced aggregation.
Magnetite Nanoparticles (Fe_3O_4, γ-Fe_2O_3)	**Low–Medium** – relatively cheaper, scalable	**Low impact** – environmentally benign, easily recovered magnetically	Often used in wastewater for heavy metal removal.

continued on following page

Table 2. Continued

Technology	Relative Cost	Environmental Impact	Notes / Considerations
TiO_2 Nanoparticles (Photocatalysis)	**Medium** – requires UV/visible light sources	**Low impact** – inert, stable; but potential ecotoxicity of free NPs if not recovered	Widely used in wastewater; sunlight-driven versions cheaper & greener.
Carbon Nanotubes (CNTs)	**High** – expensive synthesis and scaling	**Potentially High** – persistence in environment; health/ecotoxicity concerns	Excellent adsorption capacity but recovery is a major challenge.
Graphene Oxide (GO, rGO)	**Medium–High** – depends on synthesis route	**Moderate** – possible toxicity to microbes & aquatic organisms	Enhances sorption of metals and organics; often used in composites.
Biochar-supported Nanoparticles	**Low–Medium** – biochar cheap, NP synthesis adds cost	**Low impact** – biochar stabilizes particles, reduces toxicity and mobility	Sustainable option; often used for As, Pb remediation.
Nano-Bioremediation (NPs + microbes/ enzymes)	**Medium** – requires both microbial cultures and nanoparticles	**Low–Moderate** – synergistic, reduces NP dosage; less ecotoxicity	Enhances degradation of hydrocarbons and chlorinated solvents.
Nanosponges (Cyclodextrin-based)	**Medium** – higher production cost than raw NPs	**Low impact** – biodegradable, non-toxic carriers	Good for PAHs and hydrophobic pollutants.

CHALLENGES AND LIMITATIONS IN THE ASSESSMENT OF NANOPARTICLE REMEDIATION TECHNOLOGIES

The end-of-life (EOL) stage introduces substantial uncertainty in the assessment of nanoparticle remediation technologies. Most life cycle assessment (LCA) studies assume that nanoparticles follow the same disposal pathways as conventional materials. However, in real environmental systems, nanoparticles may aggregate, dissolve, or undergo complex transformations in landfills and soil environments that are rarely captured in existing models. As a result, end-of-life emissions and secondary transformation processes are often underrepresented, contributing to systematic gaps in current assessments.

Additional methodological limitations arise from modeling choices and decision subjectivity. Variations in system boundaries, allocation methods, and functional units across studies reduce comparability and may introduce bias if methodological transparency is insufficient. Furthermore, reliance on global or averaged datasets introduces temporal and spatial uncertainty, potentially masking regional differences in energy systems, emission intensities, and ecological sensitivities, which may lead to inaccurate environmental characterizations of nanoparticle systems.

Conventional LCA frameworks also struggle to capture dynamic system effects, including feedback loops and rebound effects. For instance, improvements in remediation efficiency may indirectly influence policy decisions or land-use changes that offset environmental gains, representing system dynamics beyond linear modeling approaches. In addition, a significant gap exists in toxicity and ecotoxicity characterization. Current life cycle impact assessment (LCIA) methods do not fully incorporate the complex biological responses associated with engineered nanoparticles, limiting their relevance for comprehensive risk–benefit evaluations.

Further uncertainty arises from methodological constraints in sampling and analytical detection. Environmental matrices such as water, soil, and biota contain numerous background particles, making it difficult to isolate and characterize engineered nanoparticles accurately. This measurement uncertainty directly affects data quality and model reliability.

Variability in green synthesis approaches further complicates assessment outcomes. Differences in plant extracts, microbial cultures, and synthesis conditions introduce inconsistencies in nanoparticle properties, affecting reproducibility and associated environmental and economic impacts. The absence of standardized protocols for comparing green and conventional synthesis routes remains a major limitation in current literature.

Finally, social and economic dimensions remain insufficiently integrated into most LCA studies. Issues such as labor conditions, community-level impacts, and broader cost–benefit trade-offs are often excluded, limiting the capacity of assessments to support fully holistic decision-making.

REGULATORY FRAMEWORKS AND POLICY CHALLENGES

Existing regulatory frameworks for nanomaterials, including REACH and OECD guidelines, provide a foundational structure for risk assessment; however, they remain insufficient for addressing the specific challenges posed by nanoplastics. Key limitations include the absence of standardized definitions and detection protocols, limited incorporation of life cycle perspectives, and inadequate integration of combined human and ecological risk assessments. Consequently, current regulatory systems often fail to fully capture the environmental behavior and systemic impacts of nanoplastics (OECD, 2013).

These limitations are further reinforced by inconsistencies in regulatory implementation across regions, insufficient environmental monitoring capacity, and persistent uncertainties in toxicity data. Addressing these gaps requires the development of nanoplastics-specific regulatory frameworks supported by advanced analytical tools, including AI-based monitoring systems and predictive modeling approaches. In addition, international harmonization of standards and the integration of One Health principles into regulatory design are essential for achieving coherent and adaptive governance structures. Such developments are critical for ensuring that regulatory systems remain responsive to the evolving scientific and technological landscape of nanoplastics (Zhao & Peters, 2024; Omondi & Chen, 2024).

INTEGRATION WITH ONE HEALTH AND AI-BASED TOOLS

Integrating life cycle assessment with economic analysis under the One Health framework enables a comprehensive evaluation of impacts across environmental, animal, and human systems associated with nanoplastics. The economic consequences of pollution—ranging from degraded fisheries to increased healthcare expenditures—reflect broader ecological and health burdens. Table 3 presents a comparative overview of the environmental and economic impacts of microplastics and nanoplastics.

Exposure-related health effects contribute to increased healthcare costs, representing indirect economic externalities of environmental contamination. In this context, artificial intelligence (AI) enhances One Health-based LCA through predictive modeling, sensor data integration, and large-scale data analytics. AI-driven models can identify emission hotspots, optimize remediation costs, and support cross-sectoral benefit analyses. Machine learning approaches further enable simulation of resource use patterns and prediction of environmental and health outcomes under different policy scenarios.

Overall, the integration of One Health principles and AI-based modeling represents an important advancement toward more adaptive and interdisciplinary nanoplastics management. This integrated approach not only enhances optimization of remediation strategies but also expands the analytical scope of current methodologies. The application of these concepts is further illustrated in Figure 3, which presents the integration of One Health and AI-based modeling for sustainable nanoplastics management.

Figure 3. Integration of one health and AI-based modeling for sustainable nanoplastic management

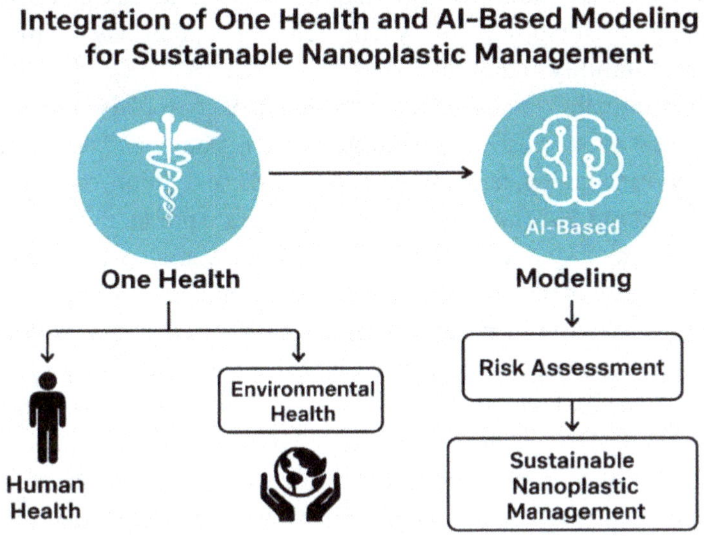

Table 3. Comparison of microplastics and nanoplastics by environmental and economic impact indicators.

Indicator	Microplastics (Approx.)	Nanoplastics (Estimated)	Environmental/Economic Impact
GHG emissions (kg CO_2-eq/kg)	1.8–3.2	4.5–6.1	Higher due to production and detection energy demand
Energy consumption (MJ/kg)	70–85	80–90	High fossil fuel input
Toxicity factor (USEtox model)	Moderate	High	Greater bioavailability
Cost of removal (USD/kg)	1.5–3.0	3.5–5.0	Higher detection and filtration cost

FUTURE DIRECTIONS AND RECOMMENDATIONS

Future nanoparticle remediation research is expected to shift from laboratory-scale validation toward full-scale field implementation. While numerous studies have demonstrated the effectiveness of engineered nanoparticles (ENPs) in removing heavy metals, organic pollutants, and pathogens under controlled conditions, field-scale evidence remains limited. Future research should therefore prioritize real-world applications characterized by heterogeneous subsurface environments, variable contaminant mixtures, and dynamic climatic conditions in order to better assess long-term stability and performance.

A second key direction involves the advancement of prospective life cycle assessment (LCA). Most existing LCAs are retrospective in nature and do not adequately incorporate future technological improvements, market evolution, or realistic end-of-life scenarios. Future LCA frameworks should integrate production scaling effects, advancements in green synthesis, and circular economy strategies to better support sustainable innovation pathways and investment planning.

A persistent limitation in current research is the uncertainty associated with nanoparticle fate and transport, particularly in soil and groundwater systems. Future studies should focus on predictive, data-driven models that integrate nanoparticle behavior with site-specific hydrogeological and biogeochemical conditions. These models should incorporate time-dependent processes such as aggregation, dissolution, and interactions with natural colloids to ensure both effective performance and long-term environmental safety.

Standardization of testing protocols represents another critical area for development. Variations in synthesis methods, experimental conditions, and target contaminants currently limit reproducibility and comparability across studies. The establishment of harmonized international testing frameworks—particularly for green-synthesized nanoparticles—would improve benchmarking, facilitate collaboration, and accelerate regulatory acceptance.

Economic assessment methodologies also require further advancement beyond simple cost-per-unit removal metrics. Future frameworks should incorporate uncertainty analysis, sensitivity testing, and scenario-based forecasting to account for fluctuations in raw material availability, market dynamics, and evolving regulatory environments. These models should be aligned with sustainability indicators and life cycle trade-offs, particularly in the context of net-zero transition strategies.

In the near future, the integration of nanoparticle remediation technologies into hybrid and multifunctional systems is expected to expand significantly. Emerging designs may combine engineered nanoparticles with biological agents such as microbes or plants, or integrate electrochemical and membrane-based systems. Such hybrid approaches can enhance treatment efficiency, increase system resilience, and reduce operational costs while enabling adaptive responses to complex contamination scenarios.

Safety-by-design principles will play an increasingly central role in nanoparticle development. Future engineering strategies should prioritize environmental degradability, minimal bioaccumulation potential, and reduced ecotoxicity. Early-stage toxicity screening during research and development phases will be essential to ensure that remediation materials do not introduce new environmental risks, particularly in sensitive ecosystems and agricultural areas.

Policy frameworks will also need to evolve to support responsible innovation. Governments and environmental agencies should develop regulatory systems that balance technological advancement with precautionary principles, including risk-based performance thresholds, post-deployment monitoring requirements, and enhanced transparency mechanisms. Public investment in pilot-scale demonstrations and public–private research partnerships will be essential for scaling promising technologies.

Public engagement and environmental education represent additional priorities. Given the potential for public concern regarding nanotechnology, transparent communication of risks, benefits, and safety measures is essential. Participatory governance approaches involving local communities, NGOs, and municipal authorities can strengthen trust and improve the acceptance of nanoremediation technologies.

Finally, long-term progress will depend on interdisciplinary collaboration. The development of integrated assessment frameworks that combine environmental, economic, and social dimensions requires coordinated input from environmental scientists, engineers, toxicologists, economists, and policy experts. Such a systems-level approach is essential for ensuring that nanoparticle remediation technologies contribute effectively to global sustainability goals and the protection of both ecological and human health.

CONCLUSION

Nanoplastics contamination represents a complex and interconnected challenge affecting environmental, animal, and human health systems. This chapter demonstrates that its effective management requires an integrated framework combining life cycle assessment, economic evaluation, and One Health principles. Biotechnological innovations and artificial intelligence–driven tools offer promising pathways for improved detection, efficient degradation, and data-informed decision-making in nanoplastics remediation.

However, important gaps remain in regulatory frameworks, data availability, and socioeconomic integration. Future progress will depend on the development of nanoplastics-specific policies, stronger interdisciplinary collaboration, and the adoption of data-driven analytical approaches. Ultimately, aligning economic feasibility with ecological sustainability and health resilience is essential for advancing responsible and effective nanoplastics management.

Authorship and AI Disclosure Statement

The authors affirm that all research design, data interpretation, and analytical writing in this chapter were conducted by the human author. AI tools were used only for minor language editing and formatting purposes. All intellectual content, conclusions, and interpretations are solely the responsibility of the author, in compliance with IGI Global's Publication Ethics and Integrity Policy (2025).

REFERENCES

Ali, I., Alharbi, O. M. L., Alothman, Z. A., Badjah, A. Y., & Alwarthan, A. (2021). Nanomaterials for water remediation: Adsorption and photocatalysis. *Desalination and Water Treatment*, *215*, 85–106. DOI: 10.5004/dwt.2021.26974

Angaye, T. C. N., & Ogidi, O. I. (2018). The environmental impacts of electronic wastes in Bayelsa State, Nigeria. *MedCrave Online Journal of Toxicology*, *4*(5), 331–332.

Angaye, T. C. N., Oyinke, G. N., & Ogidi, O. I. (2019). Environmental and health risk assessment of particulate matter associated with sporting activities in Yenagoa metropolis, Nigeria. *General Internal medicine and clinical. Innovations*, *4*(3), 1–3. DOI: 10.15761/GIMCI.1000180

Arundhathi, V., Nagalakshmi Manasa, S., Kalyani Lakshmi, S., Varsha, V., & Sundaram, B. (2022). Life-cycle assessment of various synthesis routes of silver nanoparticles. In *Life-Cycle Assessment of Nanomaterials* (pp. 267–277). https://doi.org/DOI: 10.1007/978-981-15-3662-5_22

Aßen, N. V. D., Voll, P., Peters, M., & Bardow, A. (2014). Life Cycle Assessment of CO2capture and Utilization: A Tutorial Review. *Chemical Society Reviews*, *43*(23), 7982–7994. DOI: 10.1039/C3CS60373C PMID: 24441866

Baalousha, M., Escobar, M., Unz, R. F., Shipley, H. J., Kothalawala, N., Lo, P., & Perreault, F. (2023). LCA and LCC analysis of engineered nanoparticles for ocean fertilization. *Environmental Science & Technology*. PubMed Central. https://pmc.ncbi.nlm.nih.gov/articles/PMC9747614

Bafana, A., Kumar, S. V., Temizel-Sekeryan, S., Dahoumane, S. A., Haselbach, L., & Jeffryes, C. S. (2018). Evaluating microwave-synthesized silver nanoparticles from silver nitrate with life cycle assessment techniques. *The Science of the Total Environment*, *636*, 936–943. DOI: 10.1016/j.scitotenv.2018.04.345 PMID: 29729511

Bhattacharya, R., & Kim, S. (2024). Socioeconomic dimensions of nanoplastic remediation in global coastal communities. In Gupta, A., & Kim, H. (Eds.), *Nanotechnology for Sustainable Development* (pp. 255–276). Springer., DOI: 10.1007/978-981-97-83632_12

Blowes, D. W., Ptacek, C. J., Benner, S. G., McRae, C., Bennett, T. A., & Puls, R. W. (2000). Treatment of inorganic contaminants using permeable reactive barriers. *Journal of Contaminant Hydrology*, *45*(1–2), 123–137. DOI: 10.1016/S0169-7722(00)00122-4

Cecchin, I., Reddy, K. R., Thome, A., Tessaro, E. F., & Schnaid, F. (2017). Nano bioremediation: Integration of nanoparticles and bioremediation for sustainable remediation of chlorinated organic contaminants in soils. *International Biodeterioration & Biodegradation*, *119*, 419–428. DOI: 10.1016/j.ibiod.2016.09.027

Crane, R. A. (2012). *Nanoscale zero-valent iron: Future prospects for an emerging water treatment technology*. U.S. EPA.

Crane, R. A., & Scott, T. B. (2012). Nanoscale zero-valent iron: Future prospects for an emerging water treatment technology. *Journal of Hazardous Materials*, *211–212*, 112–125. DOI: 10.1016/j.jhazmat.2011.11.073 PMID: 22305041

Demir, U., Gülcan, H. E., & Özer, S. (2025). The effect of TiO_2 nanoadditive on emissions, exergetic performance, and environmental/social/economic indicators in a small UAV jet engine fueled with kerosene. *Fuel*, *390*, 134725. DOI: 10.1016/j.fuel.2025.134725

Eizenberg, E., & Jabareen, Y. (2017). Social sustainability: A new conceptual framework. *Sustainability (Basel)*, *9*(1), 68. DOI: 10.3390/su9010068

El-Sheikh, R., El-Sayed, H., Ibrahim, A., & Salem, W. (2025). Biosynthesized magnetite nanoparticles combined with hydrocarbon-degrading bacteria for enhanced petroleum pollutant degradation. *Frontiers in Microbiology*, *16*, 1590636. DOI: 10.3389/fmicb.2025.1590636

Environmental Protection Agency. (2010). *Nanotechnology database fact sheet*. U.S. Environmental Protection Agency. https://clu-in.org/download/misc/NanoTechnology_FactSheet.pdf

Francy, N., & Shanthakumar, S. (2015). Application of nanoscale zero-valent iron (nZVI) for contaminated soil remediation: A review. In *Proceedings of the 50th Indian Geotechnical Conference* (Vol. 19, pp. 550–558). Springer. https://doi.org/ DOI: 10.1007/s11356-011-0576-3

García-García, G., Fernandez, M. C., Armstrong, K., Woolass, S., & Styring, P. (2021). Analytical Review of Life-Cycle Environmental Impacts of Carbon Capture and Utilization Technologies. *ChemSusChem*, *14*(4), 995–1015. Advance online publication. DOI: 10.1002/cssc.202002126 PMID: 33314601

Garrigues, E., Corson, M. S., Angers, D. A., Van Der Werf, H. M. G., & Walter, C. (2012). Soil quality in Life Cycle Assessment: Towards development of an indicator. *Ecological Indicators*, *18*, 434–442. DOI: 10.1016/j.ecolind.2011.12.014

Gil-Díaz, M., Diez-Pascual, S., Gonzalez, A., Alonso, J., Rodríguez-Valdes, E., Gallego, J. R., & Lobo, M. C. (2016). A nanoremediation strategy for the recovery of an As-polluted soil. *Chemosphere, 149*, 137–145. DOI: 10.1016/j.chemosphere.2016.01.106 PMID: 26855217

Gomes, H. I., Dias-Ferreira, C., & Ribeiro, A. B. (2022). Nanoscale zero-valent iron (nZVI) for remediation of contaminated soils and sediments: Field column study in Brazil. *Geosciences, 12*(8), 287. DOI: 10.3390/geosciences12080287

He, F., Zhao, D., Liu, J., & Roberts, C. (2010). Stabilization of Fe–Pd nanoparticles with sodium carboxymethyl cellulose for enhanced dechlorination of chlorinated hydrocarbons in soil and groundwater. *Industrial & Engineering Chemistry Research, 49*(22), 11284–11292. DOI: 10.1021/ie101258z

Heijungs, R., & Wiloso, E. I. (2014). Life cycle assessment of bioenergy systems. In Wang, L. (Ed.), *Sustainable bioenergy production* (pp. 234–259). CRC Press.

Hossain, M., & Poon, C. S. (2018). Comparative LCA of wood waste management strategies generated from building construction activities. *Journal of Cleaner Production, 177*, 387–397. DOI: 10.1016/j.jclepro.2017.12.233

Jacquemin, L., Pontalier, P.-Y., & Sablayrolles, C. (2012). Life cycle assessment (LCA) applied to the process industry: A review. *The International Journal of Life Cycle Assessment, 17*(8), 1028–1041. DOI: 10.1007/s11367-012-0432-9

Joshi, N., Filip, J., Coker, V., Sadhukhan, J., Safarik, I., Bagshaw, H., & Lloyd, J. R. (2018). Microbial reduction of natural Fe(III) minerals: Towards the sustainable production of functional magnetic nanoparticles. *Frontiers in Environmental Science, 6*, 127. DOI: 10.3389/fenvs.2018.00127

Karn, B., Kuiken, T., & Otto, M. (2009). Nanotechnology and in situ remediation: A review of the benefits and potential risks. *Environmental Health Perspectives, 117*(12), 1813–1831. DOI: 10.1289/ehp.0900793 PMID: 20049198

Klinglmair, M., Sala, S., & Brandão, M. (2014). Assessing resource depletion in LCA: A review of methods and methodological issues. *The International Journal of Life Cycle Assessment, 19*(3), 580–592. DOI: 10.1007/s11367-013-0650-9

Li, C., & Liu, T. (2023). Circular economy strategies for nanomaterial sustainability. *Environmental Science & Policy, 147*, 52–63.

Li, S., Yan, W., & Zhang, W. (2009). Solvent-free production of nanoscale zero-valent iron (nFeZ) with precision milling. *Green Chemistry, 11*(10), 1628–1636. DOI: 10.1039/b913056j

Ma, J., & Lin, Y. (2024). Regulatory frameworks for nanomaterial governance: Comparative policy perspectives. *ChemistrySelect*, *9*(12), e202400959. DOI: 10.1002/slct.202400959

Mackenzie, K., Bleyl, S., Georgi, A., & Kopinke, F. D. (2016). Carbo-Iron—An Fe/AC composite for in situ treatment of groundwater contaminated with chlorinated hydrocarbons. *Environmental Science & Technology*, *50*(6), 3159–3168. DOI: 10.1021/acs.est.5b05031

Mahajan, P., Gusain, M., & Singh, P. (2026). Analytical strategies for nanoplastics detection and remediation: Advances, challenges, and perspectives. *Microchemical Journal : Devoted to the Application of Microtechniques in All Branches of Science*, *222*, 117228. DOI: 10.1016/j.microc.2026.117228

Malik, M., Anwar, M. A., Ullah, A., Hafeez, A., Shakoor, M. B., & Arif, M. (2022). A review of the techno-economic feasibility of nanoparticle application for wastewater treatment. *Water (Basel)*, *14*(10), 1550. DOI: 10.3390/w14101550

Martins, F., Machado, S., Albergaria, T., & Delerue-Matos, C. (2017). LCA applied to nanoscale zero valent iron synthesis. *The International Journal of Life Cycle Assessment*, *22*(5), 707–714. DOI: 10.1007/s11367-016-1258-7

Mondal, A., Moradi, S., Rahman, A., & Su, C. (2023). Recent advances of nanoremediation technologies for soil and groundwater remediation: A review. *Water (Basel)*, *13*(16), 2186. DOI: 10.3390/w13162186

OECD. (2013). *"Regulatory Frameworks for Nanotechnology in Foods and Medical Products: Summary Results of a Survey Activity", OECD Science, Technology and Industry Policy Papers, No. 4*. OECD Publishing., DOI: 10.1787/5k47w4vsb4s4-

Ogidi, O. I., Ajoko, I. T., & Tawariowei, A. M. (2024a). Sustainable Application of Genetic Ecotoxicological Techniques. In *Biomonitoring For Environmental Sustainability. In: Biomonitoring of Pollutants in the Global South. Publisher*. Springer Nature Singapore Pte Ltd. DOI: 10.1007/978-981-97-1658-6_19

Ogidi, O. I., & Akpan, U. M. (2024). Microbial and Phyto-Remediation of Crude Oil Contaminated Soil. In: Eco-Restoration of the Environment. Published by *CRC Press, Routledge Taylor and Francis Group.*

Ogidi, O. I., Angaye, T. C. N., & Carbom, H. E. (2024c). Microorganisms in the monitoring of terrestrial and aquatic ecosystems: applications, challenges and future prospects. In *Biomonitoring of Pollutants in the Global South. Publisher*. Springer Nature Singapore Pte Ltd. DOI: 10.1007/978-981-97-1658-6_18

Ogidi, O. I., Ijere, N. D., & Felagha, I. (2025a). Ecotoxicity and Human Health Risks of Improper E-Waste Handling. In Güngör, A., & Polat, O. (Eds.), *Integrated Approaches for Sustainable E-Waste Management* (pp. 243–282). IGI Global Scientific Publishing., DOI: 10.4018/979-8-3693-7383-5.ch009

Ogidi, O. I., Onwuagba, C. G., & Richard-Nwachukwu, N. (2024b). Biomonitoring Tools, Techniques and Approaches For Environmental Assessments. In *Biomonitoring of Pollutants in the Global South. Publisher.* Springer Nature Singapore Pte Ltd. DOI: 10.1007/978-981-97-1658-6_7

Ogidi, O. I., & Richard-Nwachukwu, N. (2024). Microbe Assisted Remediation of Pesticide Residues from Soil and Water. In: Eco-Restoration of the Environment. Published by *CRC Press, Routledge Taylor and Francis Group.* DOI: 10.1201/9781003423393-9

Okereke, J. N., Ogidi, O. I., & Obasi, K. O. (2016). Environmental and Health Impact of Industrial Wastewater Effluents in Nigeria- A Review. *International Journal of Advanced Research in Biological Science*, *3*(6), 55–67.

Omondi, P., & Chen, Q. (2024). Public perception and risk communication in nanotechnology applications. *Zenodo*. https://doi.org/DOI: 10.5281/zenodo.10845914

Peters, J., & Weil, M. (2016). A Critical Assessment of the Resource Depletion Potential of Current and Future Lithium Ion Batteries. *Resources*, *5*(4), 46. DOI: 10.3390/resources5040046

Pini, M., & Rossi, G. (2024). AI-driven life cycle costing in environmental remediation. *Sustainability*, *16*(2), 945.

Pourzahedi, L., & Eckelman, M. J. (2015). Comparative life cycle assessment of silver nanoparticle synthesis routes. *Environmental Science. Nano*, *2*(4), 361–369. DOI: 10.1039/C5EN00075K

Pourzahedi, L., Vance, M., & Eckelman, M. J. (2014). Life cycle assessment of environmental impacts of nanosilver-enabled bandages. *Environmental Science & Technology*. Advance online publication. DOI: 10.1021/es504655y PMID: 25489789

Pranjic, A. M., Oprckal, P., Mladenovic, A., Zapusek, P., Urleb, M., & Turk, J. (2018). Comparative life cycle assessment of possible methods for the treatment of contaminated soil at an environmentally degraded site. *Journal of Environmental Management*, *218*, 497–508. DOI: 10.1016/j.jenvman.2018.04.051 PMID: 29709818

Quinn, J., Geiger, C., Clausen, C., Brooks, K., Coon, C., O'Hara, S., Krug, T., Major, D., Yoon, W. S., Gavaskar, A., & Holdsworth, T. (2005). Field demonstration of DNAPL dehalogenation using emulsified zero-valent iron. *Environmental Science & Technology*, *39*(5), 1309–1318. DOI: 10.1021/es0490018 PMID: 15787371

Rizwan, M., Ali, S., Rehman, M. Z. U., Qayyum, M. F., Ok, Y. S., & Ahmad, N. (2022). Nanoparticle-fungus integrated remediation of organic and inorganic contaminants. *Frontiers in Microbiology*, *13*, 982611. DOI: 10.3389/fmicb.2022.982611

Rizzo, E., Bardos, P., Pizzol, L., Critto, A., Giubilato, E., Marcomini, A., Albano, D., Darmendrail, D., Döberl, G., Harclerode, M., Harries, N., Nathanail, P., Pachon, C., Rodriguez, A., Slenders, H., & Smith, G. (2016). Comparison of international approaches to sustainable remediation. *Journal of Environmental Management*, *184*, 4–17. DOI: 10.1016/j.jenvman.2016.07.062 PMID: 27520125

Rodríguez-Rojas, M. D. P., Bustos-Terrones, V., Díaz-Cárdenas, M. Y., Vázquez-Vélez, E., & Martínez, H. (2024). Life cycle assessment of green synthesis of TiO_2 nanoparticles vs. chemical synthesis. *Sustainability (Basel)*, *16*(17), 7751. DOI: 10.3390/su16177751

Salem, W. M., Haridy, M., & El-Azab, I. H. (2022). Biosynthesized iron oxide nanoparticles by Halomonas sp. for the bioremediation of Pb and Cd in contaminated soils. *Frontiers in Sustainable Food Systems*, *6*, 932424. DOI: 10.3389/fsufs.2022.932424

Sharma, R., Patel, M., & Singh, A. (2023). Nanobioremediation: Synergistic effects of nanoparticles and microbes in pollutant removal. *Frontiers in Environmental Science*, *11*, 10204040. DOI: 10.3389/fenvs.2023.10204040

Silva Trentin, A. W. de S., Visentin, C., Braun, A. B., & Thomé, A. (2021). Sustainability assessment of nanoscale zero-valent iron production methods. *Environmental Engineering Science*, *38*(2), 123–137.

Singh, A. K., Lou, H. H., Pike, R. W., Agboola, A., Li, X., Hopper, J. R., & Yaws, C. L. (2008). Environmental impact assessment for potential continuous process for the production of carbon nanotubes. *American Journal of Environmental Sciences*, *4*(5), 522–534. DOI: 10.3844/ajessp.2008.522.534

Song, Y., Hou, D., Zhang, J., O'Connor, D., Li, G., Gu, Q., Li, S., & Liu, P. (2018). Environmental and socio-economic sustainability appraisal of contaminated land remediation strategies: A case study at a mega-site in China. *The Science of the Total Environment*, *610–611*, 391–401. DOI: 10.1016/j.scitotenv.2017.08.016 PMID: 28806555

Stefaniuk, M., Oleszczuk, P., & Ok, Y. S. (2016). Review on nano zerovalent iron (nZVI): From synthesis to environmental applications. *Chemical Engineering Journal*, *287*, 618–632. DOI: 10.1016/j.cej.2015.11.046

Su, C., Puls, R. W., & Krug, T. (2012). A two-and-a-half-year performance evaluation of field application of emulsified zero-valent iron for remediation of chlorinated volatile organic compounds in groundwater. *Environmental Science & Technology*, *46*(24), 12962–12970. DOI: 10.1021/es302725t

Thome, A., Reddy, K. R., Reginatto, C., & Cecchin, I. (2015). Review of nanotechnology for soil and groundwater remediation: Brazilian perspectives. *Water, Air, and Soil Pollution*, *226*(4), 1–20. DOI: 10.1007/s11270-014-2243-z

Tratnyek, P. G., & Johnson, R. L. (2006). Nanotechnologies for environmental cleanup. *Nano Today*, *1*(2), 44–48. DOI: 10.1016/S1748-0132(06)70048-2

Vince, F., Aoustin, E., Bréant, P., & Marechal, F. (2008). LCA tool for the environmental evaluation of potable water production. *Desalination*, *220*(1–3), 37–56. DOI: 10.1016/j.desal.2007.01.021

Walser, T., Demou, E., Lang, D. J., & Hellweg, S. (2011). Prospective environmental life cycle assessment of nanosilver T-shirts. *Environmental Science & Technology*, *45*(10), 4570–4578. DOI: 10.1021/es2001248 PMID: 21506582

Wang, J., & Chen, C. (2016). Nanomaterials for environmental remediation. *Environmental Science. Nano*, *3*(2), 240–266. DOI: 10.1039/C5EN00230C

Zhang, W. X. (2003). Nanoscale iron particles for environmental remediation: An overview. *Journal of Nanoparticle Research*, *5*(3–4), 323–332. DOI: 10.1023/A:1025520116015

Zhang, X., Shen, L., & Zhang, L. (2013). Life cycle assessment of the air emissions during building construction process: A case study in Hong Kong. *Renewable & Sustainable Energy Reviews*, *17*, 160–169. DOI: 10.1016/j.rser.2012.09.024

Zhang, Y.. (2022). Economic optimization of nano-enabled remediation technologies. *Journal of Cleaner Production*, *377*, 134512. DOI: 10.1016/j.jclepro.2022.134512

Zhao, L., & Peters, D. (2024). Innovations in inorganic nanomaterial synthesis and environmental management. *Inorganic Chemistry Communications*, *162*, 112611. DOI: 10.1016/j.inoche.2024.112611

Zhao, X., Liu, W., Cai, Z., Han, B., Qian, T., & Zhao, D. (2016). An overview of preparation and applications of stabilized zero-valent iron nanoparticles for soil and groundwater remediation. *Water Research*, *100*, 245–266. DOI: 10.1016/j.watres.2016.05.019 PMID: 27206054

Conclusion

SYSTEMS-LEVEL STRATEGIC OUTLOOK TOWARD 2040

1. From Contaminant Toxicology to Environmental Signal Intelligence

By the end of the present decade, the scientific interpretation of nanoplastic exposure will likely undergo a substantial epistemic transition. The field is progressively moving beyond the historical confines of classical contaminant toxicology, where plastic particulates were primarily evaluated as isolated material pollutants measured through static concentration metrics and compartment-specific exposure endpoints. Such reductionist frameworks, although operationally useful during earlier stages of environmental toxicology, appear increasingly insufficient for interpreting the dynamic, transboundary, and signal-mediated behavior of nanoscale polymeric systems across interconnected biological and ecological infrastructures. Nanoplastics are gradually emerging not merely as passive contaminants, but as environmentally distributed signal-generating entities capable of modulating physicochemical interactions, microbiome stability, cellular communication pathways, and adaptive biological responses across multiple scales simultaneously.

2. Structural Limitations and the Progressive Collapse of Classical Exposure Assessment Models

This transition carries significant implications for the future architecture of exposure science itself. Conventional compartment-based risk assessment models remain structurally dependent upon fragmented monitoring systems, delayed computational synthesis, and linear toxicological assumptions that often fail to capture temporal variability, chronic low-dose perturbation dynamics, and cumulative multi-interface exposure propagation. The persistent separation between environmental chemistry, computational decoding, biological signaling analysis, and predictive ecological modeling has created a methodological lag that is becoming increasingly difficult to reconcile with the complexity of real-world nanoplastic dissemination patterns. Under such conditions, exposure assessment may progressively shift from retrospective contaminant quantification toward adaptive environmental signal intelligence capable of integrating continuous biosurveillance frameworks, ecological variability mapping, and predictive analytical integration within unified analytical environments.

3. Emergence of Cyber-Biotechnological Ecosystems and Environmental Digital Twins by 2040

Toward 2040, emerging cyber-biotechnological ecosystems may substantially redefine how environmental perturbations are monitored, interpreted, and operationally governed. Environmental digital twins, AI-assisted ecological interpretation systems, autonomous biosensor grids, and multiomics-linked biosurveillance infrastructures are expected to become increasingly integrated into environmental health analytics. These systems will likely function not as isolated technological instruments, but as continuously adaptive computational interfaces capable of correlating atmospheric transport patterns, aquatic contaminant dispersion, microbiome perturbation signatures, food-chain transmission dynamics, and physiological exposure indicators in near real time. Importantly, the future scientific challenge will not simply involve generating larger environmental datasets. The more consequential challenge will involve constructing computationally interpretable ecological intelligence systems capable of distinguishing transient environmental noise from biologically meaningful exposure trajectories.

4. Transition from Reactive Environmental Regulation to Predictive One Health Governance

Simultaneously, environmental governance itself may experience a gradual but decisive transformation. Current reactive environmental regulation systems remain predominantly defensive, frequently responding to contamination only after measurable ecological or physiological damage has already occurred. Such regulatory architectures are poorly aligned with the distributed and persistent nature of nanoscale particulate exposure. By contrast, future One Health governance systems may increasingly adopt anticipatory operational logic based upon predictive environmental computation, exposure forecasting, adaptive intervention modeling, and cross-sector ecological coordination frameworks. In this evolving framework, regulatory action may no longer depend exclusively upon delayed toxicological confirmation. Instead, intervention thresholds could emerge from continuously updated environmental probability models integrating computational toxicology, ecological trend forecasting, population exposure analytics, and systems-level biosurveillance indicators.

5. Nanoplastic Intelligence Infrastructures and Ecological Security Networks

The development of nanoplastic intelligence infrastructures may therefore become strategically linked to broader issues extending beyond toxicology alone. Ecological security networks, agro-ecological resilience systems, food-system stability, microbiome preservation, and long-term environmental health forecasting are likely to converge within interconnected exposure intelligence platforms capable of monitoring perturbation propagation across planetary biological networks. This convergence may fundamentally alter the role of environmental monitoring agencies, transforming them from passive observational institutions into computationally active ecological interpretation systems supporting dynamic public health and environmental resilience strategies. Within such frameworks, the distinction between environmental monitoring, computational epidemiology, ecological forecasting, and biosecurity surveillance may gradually become increasingly blurred.

6. Ethical, Computational, and Geopolitical Constraints of Environmental Intelligence Systems

Nevertheless, the emergence of large-scale environmental intelligence systems will introduce substantial ethical, computational, and geopolitical constraints. The expansion of continuous environmental biosurveillance infrastructures raises unresolved concerns regarding environmental data ownership, algorithmic bias, unequal technological accessibility, ecological surveillance asymmetry between high-income and resource-limited regions, and the political governance of predictive environmental analytics. Future computational exposure models may inadvertently reproduce existing global inequities if environmental intelligence infrastructures remain concentrated within technologically dominant economies. Furthermore, the increasing reliance upon algorithmically mediated environmental interpretation introduces additional concerns surrounding transparency, model reproducibility, ecological uncertainty propagation, and automated decision-making authority within public health governance systems. These issues cannot remain peripheral considerations. They will likely become central structural determinants shaping the credibility and legitimacy of future environmental intelligence architectures.

7. Toward Planetary Bio-Computational Governance: Strategic Environmental Futures Beyond 2040

By 2040 and moving into the strategic environmental futures beyond, environmental monitoring may no longer operate as a fragmented process centered solely upon pollutant detection and isolated toxicological reporting. Instead, it may progressively evolve into a core component of a broader planetary bio-computational governance architecture designed to continuously interpret environmental perturbation signals across interconnected ecological, biological, technological, and societal infrastructures.

Figure 1. presents the proposed governance-centric systems architecture underlying the transition from conventional contaminant monitoring toward adaptive planetary bio-digital exposure intelligence. The framework conceptualizes the Planetary Bio-Digital Dashboard (PBDD) as an interoperable coordination core integrating multi-scale environmental sensing, computational analytics, predictive risk orchestration, and adaptive One Health governance across interconnected ecological and societal infrastructures.

Figure 1.

Within this emerging scientific landscape, nanoplastic exposure will likely represent not only a toxicological challenge, but also a systems-level indicator of environmental instability, adaptive biological stress, and ecological governance capacity.

To operationalize this macro-governance framework, future iterations of exposure science could benefit from a conceptual blueprint centered on a Planetary Bio-Digital Dashboard (PBDD). Designed as a non-centralized, cross-calibrated cyber-infrastructure, this proposed architecture aims to complement standard mass concentration registries by exploring the integration of continuous biochemical feedback loops and real-time kinetic modeling. The technical goal of such a system is to aggregate heterogeneous data streams—such as cellular stress markers, localized epigenetic shifts, and multi-trophic bio-accumulation metrics—into a harmonized data protocol. Rather than offering deterministic predictions, the PBDD is envisioned as an iterative computational environment to help distinguish stochastic environmental variability from biologically interpretable signaling trajectories, thereby serving as an integrative analytical coordination layer for global health risk assessments.

Consequently, this systematic transition reframes the future direction of the field, shifting the primary scientific objective away from the passive accumulation of fragmented exposure datasets and toward the rigorous deployment of computationally disciplined, biologically interpretable, and ethically governable environmental intelligence systems capable of supporting adaptive One Health decision-making under conditions of increasing planetary complexity. As a future-oriented extension of the systems-level perspectives developed throughout this volume, the following visual scientific roadmap presents a strategic conceptual outlook on emerging interdisciplinary trajectories expected to influence environmental health, computational biotechnology, microbiome science, sustainable nutrition, and translational One Health systems toward 2040.

One Health 2040:
Biological Systems Medicine, Sustainable Nutrition, Polybiome Dynamics, and Computational Biotechnology in the Nanoplastic Era

The visual scientific roadmap presented in this volume was developed to provide an integrated conceptual synthesis of the interconnected biological, environmental, technological, nutritional, and translational dimensions associated with nanoplastic contamination within a One Health framework. Rather than functioning as conventional illustrative figures, these conceptual architectures are intended to support systems-level interpretation across the thematic structure of the book and to facilitate interdisciplinary integration across the eight chapters.

The three figures collectively explore the convergence among nanoplastic exposure ecology, computational biotechnology, microbiome dynamics, sustainable nutrition, antimicrobial resistance, environmental intelligence, and precision medicine. Each framework reflects a distinct systems perspective, ranging from exposure-system medicine and biological resilience to food-system ecology and environmental health integration. Together, they illustrate how chronic environmental exposure may increasingly interact with microbial adaptation, metabolic regulation, ecological instability, and emerging computational approaches within future One Health research landscapes.

These visual frameworks are presented as strategic scientific foresight models rather than linear mechanistic pathways. Their objective is to strengthen conceptual continuity across the volume while highlighting emerging interdisciplinary research directions relevant to biological systems medicine, sustainable health governance, exposure biology, and computational biotechnology toward 2040.

CONCEPTUAL DESIGN PRINCIPLES OF THE VISUAL SCIENTIFIC ROADMAP

Figure 1. The One Health 2040 Convergence Architecture.

Figure 1 is a conceptual systems-level framework illustrating the convergence among nanoplastic exposure ecology, polybiome dynamics, sustainable nutrition, computational biotechnology, antimicrobial resistance, environmental intelligence, precision medicine, and systemic disease pathways within an integrated One Health perspective.

Figure 2. Exposure-System Medicine in the Nanoplastic Era.

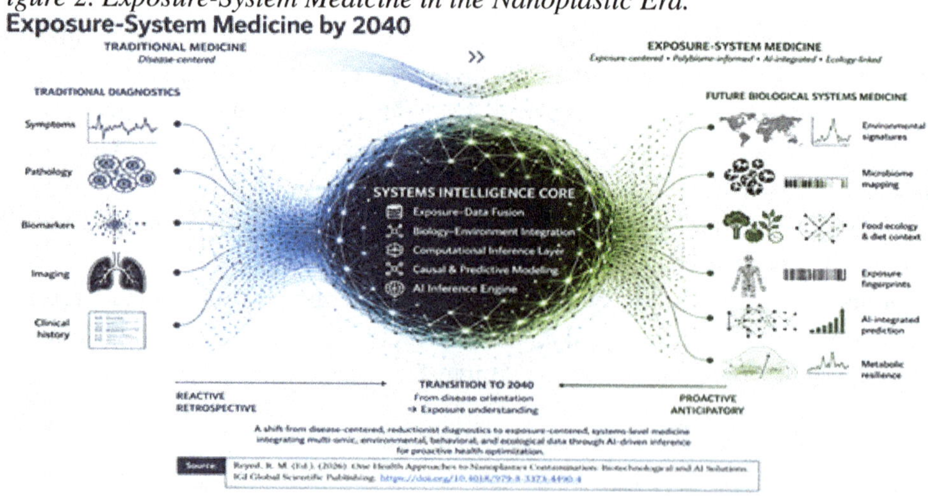

Figure 2 is a strategic conceptual model illustrating the transition from conventional disease-centered medicine toward integrated exposure-system medicine incorporating environmental exposure biology, microbiome signatures, metabolic fingerprints, nutritional ecology, and computational biotechnology within future precision One Health systems.

Figure 3. The Polybiome–Nutrition–Nanoplastic Axis.

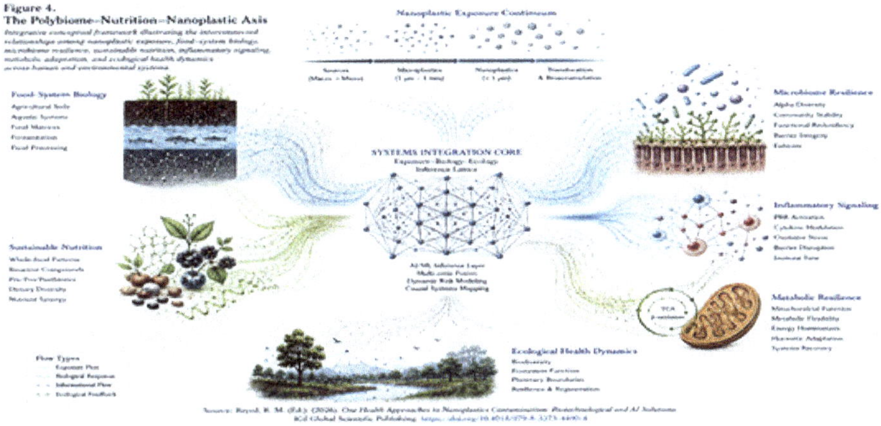

Figure 3 is an integrative conceptual framework illustrating the interconnected relationships between nanoplastic exposure, food-system biology, microbiome resilience, sustainable nutrition, inflammatory signaling, metabolic adaptation, and ecological health dynamics across human and environmental systems.

TRANSITIONAL INTEGRATIVE NOTE

The systems-oriented progression developed throughout this volume ultimately necessitates a unified operational vocabulary capable of stabilizing communication across environmental science, computational intelligence, biotechnology, toxicology, systems biology, and translational One Health governance.

Because many of the concepts examined throughout the chapters emerge at the intersection of multiple analytical domains, the following glossary is provided to support terminological consistency, cross-disciplinary interpretive continuity, and production-stage reference standardization across the broader cyber-biotechnological framework developed within this monograph.

Integrated One Health Systems Taxonomy:
A Cross-Disciplinary Conceptual and Operational Knowledge Framework

This section functions as an integrated cross-disciplinary systems taxonomy consolidating the biological, environmental, computational, and translational terminologies developed throughout this volume. It provides unified conceptual definitions, operational interpretations, and functional systems roles designed to support terminological consistency, interdisciplinary navigation, and analytical continuity across the volume's multi-layered One Health framework.

PART A: Core Biological, Ecological, and One Health Concepts

1. Foundational System Concepts

One Health Architecture

- *Systems Role:* Adaptive Governance
- Functional Definition: Transdisciplinary governance monitoring the functional interdependence of human, animal, and ecosystem health networks.
- Reference: [Ch. 1, 5, 7]

Environmental Interface

- *Systems Role:* Flux Regulation
- Functional Definition: Border zones regulating physicochemical transport and chemical partitioning between distinct ecosystem compartments.
- Reference: [Ch. 1, 3]

Exposure Continuum

- *Systems Role:* Pathway Tracking
- Functional Definition: Structural framework modeling a pollutant's complete trajectory from environmental emission points to internal biological receptors.
- Reference: [Ch. 1, 4]

Ecosystem Health

- *Systems Role:* Baseline Evaluation
- Functional Definition: Structural state assessing ecological resilience, biodiversity maintenance, and active self-regulatory buffering capacities under anthropogenic pressure.
- Reference: [Ch. 1, 3]

Nanoplastic Exposure

- *Systems Role:* Target Interaction
- Functional Definition: Chronic cellular and tissue contact with submicron plastic fragments via inhalation, ingestion, or dermal translocation routes.
- Reference: [Ch. 1, 4, 6]

Biological Barrier Integrity

- *Systems Role:* Systemic Defense
- Functional Definition: Selective endothelial or mucosal cell sheets regulating and restricting particulate translocation into vital internal target organs.
- Reference: [Ch. 1, 4]

2. Applied Targeted Mechanisms

Human Biomonitoring

- *Systems Role:* Burden Quantification
- Functional Definition: Systemic assessment quantifying internal chemical or polymeric bodies within biological specimens to map target organ concentration loads.
- Reference: [Ch. 1, 4]

Bioaccumulation Kinetics

- *Systems Role:* Cellular Retention
- Functional Definition: Cellular retention of non-biodegradable synthetic polymers occurring when pollutant uptake rates exceed biological clearance velocities.
- Reference: [Ch. 1, 6]

Toxicodynamic Responses

- *Systems Role:* Phenotypic Onset
- Functional Definition: Biochemical interactions of internalized pollutants with target cell receptors, governing downstream phenotypic changes and cellular injury severity.
- Reference: [Ch. 4]

Oxidative Stress Cascade

- *Systems Role:* Homeostatic Failure
- Functional Definition: Intracellular molecular imbalance caused by excessive accumulation of reactive oxygen species, triggering lipid peroxidation and enzymatic denaturation.
- Reference: [Ch. 1, 4]

Endocrine Disruption

- *Systems Role:* Signaling Interference
- Functional Definition: Hormonal pathway interference driven by submicron polymeric binding, destabilizing cellular communication and reproductive kinetics.
- Reference: [Ch. 1, 4]

Genotoxic Aberrations

- *Systems Role:* Macromolecular Damage
- Functional Definition: DNA alterations inducing uncompensated strand breaks, chromosomal anomalies, or permanent mutagenic lesions under inflammatory stress.
- Reference: [Ch. 1, 4]

Gut Microbiome Perturbation

- *Systems Role:* Microbial Dysbiosis
- Functional Definition: Taxonomic and functional shifts in gastrointestinal microbial ecologies that induce internal dysbiosis and alter metabolic equilibrium.
- Reference: [Ch. 5]

Nutrient Assimilation

- *Systems Role:* Metabolic Uptake
- Functional Definition: Epithelial transport and uptake mechanisms directly impaired by submicron plastic surface attachment and localized tissue stress.
- Reference: [Ch. 5, 6]

Xenobiotic Processing

- *Systems Role:* Metabolic Clearance
- Functional Definition: Microbial and hepatic enzymatic sequences transforming and detoxifying foreign chemical compounds to dictate host retention kinetics.
- Reference: [Ch. 5]

Oral Exposure Burdens

- *Systems Role:* Dietary Ingestion
- Functional Definition: Cumulative quantities of internalized contaminants ingested through food chains, triggering localized metabolic and barrier stress.
- Reference: [Ch. 6]

PART B: Computational Modeling and Environmental Systems Analytics

B1. Computational Intelligence Infrastructure

AI Biosurveillance Networks

- *Systems Role:* Threat Detection

- Functional Definition: Predictive profiling platforms processing multidimensional environmental and epidemiological data for early anomaly detection.
- Reference: [Ch. 1, 2, 7]

Data Fusion Architectures

- *Systems Role:* Signal Integration
- Functional Definition: Integration layers combining disparate spectroscopic, molecular, and spatial datasets into unified computational matrices.
- Reference: [Ch. 2, 6]

AI-Driven Food Surveillance

- *Systems Role:* Supply Traceability
- Functional Definition: Automated computer vision screening systems executing tracking protocols across commercial food supply networks.
- Reference: [Ch. 6]

Targeted Biosensing Systems

- *Systems Role:* Real-time Transduction
- Functional Definition: Analytical devices pairing selective biological recognition elements with physical transducers to output high-resolution electrical signals.
- Reference: [Ch. 1, 3]

Biological Signatures

- *Systems Role:* Biomarker Isolation
- Functional Definition: Biologically interpretable physiological patterns and molecular fingerprints distinguishing authentic biological responses from environmental background noise.
- Reference: [Ch. 1, 5]

B2. Predictive Modeling and Exposure Analytics

Geospatial Risk Mapping

- *Systems Role:* Hotspot Identification

- Functional Definition: Visual risk models fusing geospatial telemetry with toxicological probability arrays to project spatial-temporal hazard densities.
- Reference: [Ch. 1, 7]

Computational Modeling Hierarchy

Predictive Fate Simulation

- *Systems Role:* Flux Forecasting
- Functional Definition: Long-term contaminant transport models integrating degradation and environmental partitioning variables.
- Reference: [Ch. 1, 3]

Exposure Reconstruction

- *Systems Role:* Dose Interpolation
- Functional Definition: Retrospective toxicological equations interpolating historic internal dose kinetics based on contemporary bioanalytical signatures.
- Reference: [Ch. 2, 4]

Source Attribution Models

- *Systems Role:* Provenance Tracking
- Functional Definition: Inverse probabilistic networks processing weathering profiles and transport variables to calculate original manufacturing sources.
- Reference: [Ch. 2]

Removal Optimization Protocols

- *Systems Role:* Process Refinement
- Functional Definition: Iterative engineering simulations refining processing efficiency, fluid dynamics, and material clearance velocities.
- Reference: [Ch. 3, 8]

Risk-Stratification Processing

- *Systems Role:* Cohort Segmentation
- Functional Definition: Statistical screening tools sorting ecological or human populations into risk groups based on cumulative exposure loads.
- Reference: [Ch. 7]

Cost-Efficiency Forecasting

- *Systems Role:* Resource Allocation
- Functional Definition: Techno-economic models evaluating remediation expenditures against contaminant clearance efficiency under variable environmental conditions.
- Reference: [Ch. 8]

B3. Environmental Transport Dynamics

Environmental Compartment Flux

- *Systems Role:* Environmental Exchange
- Functional Definition: Intercompartmental mass transfer rates across connected atmospheric, aquatic, and terrestrial environmental compartments.
- Reference: [Ch. 4, 7]

Atmospheric Deposition

- *Systems Role:* Aerial Dispersion
- Functional Definition: Gravitational fallout and settling mechanisms transporting airborne submicron synthetic particles onto aquatic and terrestrial landscapes.
- Reference: [Ch. 1, 4]

Cross-Matrix Propagation

- *Systems Role:* Transboundary Mobility
- Functional Definition: Migration mechanisms driving pollutants across successive biological, hydrological, and agricultural boundary layers.
- Reference: [Ch. 2, 4]

Trophic Transfer

- *Systems Role:* Trophic Amplification
- Functional Definition: Dietary transmission pathways delivering bioavailable contaminants across successive consumer tiers, escalating upper-level predator burdens.
- Reference: [Ch. 1, 6]

Physicochemical Determinants

- *Systems Role:* Colloidal Regulation
- Functional Definition: Structural and chemical parameters—including particle size, surface charge, and density—regulating colloidal aggregation velocities.
- Reference: [Ch. 3]

Heterogeneous Colloidal Dynamics

- *Systems Role:* Partitioning Constraints
- Functional Definition: Multi-component suspensions with non-uniform geometries creating significant constraints for thermodynamic partitioning analysis.
- Reference: [Ch. 3]

Plastic Additive Leachates

- *Systems Role:* Chemical Migration
- Functional Definition: Aqueous chemical amalgams and plasticizers migrating outward from degraded polymers into natural hydrological networks.
- Reference: [Ch. 1, 3]

Emerging Chemical Stressors

- *Systems Role:* Hazard Classification
- Functional Definition: Anthropogenic chemical and particulate threats recently identified as persistent risk vectors lacking comprehensive regulatory baselines.
- Reference: [Ch. 1, 3, 4]

Stochastic Environmental Variation

- *Systems Role:* Analytical Noise
- Functional Definition: Random environmental fluctuations and variable natural background responses that confound predictive transport models.
- Reference: [Ch. 1, 3]

PART C: Biotechnological Mitigation and Translational Implementation Economics

1. Foundational System Concepts

Bioremediation Frameworks

- *Systems Role:* Biological Remediation
- Functional Definition: Ecological mitigation schemes introducing natural or modified microbial consortia to metabolize, degrade, or neutralize persistent polymer materials.
- Reference: [Ch. 1, 8]

Hybrid Mitigation Technologies

- *Systems Role:* Integrated Extraction
- Functional Definition: Combined engineering architectures coupling physical separation membranes with specialized biological degradation cascades.
- Reference: [Ch. 3]

Translational Intervention

- *Systems Role:* Policy Implementation
- Functional Definition: Translational pathways converting laboratory toxicological findings into field-ready environmental management policies.
- Reference: [Ch. 7]

Health-Economic Evaluation

- *Systems Role:* Asset Allocation
- Functional Definition: Analytical matrices assessing long-term public health mitigation measures by balancing infrastructure costs against diagnostic savings.
- Reference: [Ch. 7, 8]

Sustainability-Oriented Measures

- *Systems Role:* Circular Governance

- Functional Definition: Regulatory and technological interventions engineered to reduce long-term pollutant accumulation without inducing secondary eco-toxicological trade-offs.
- Reference: [Ch. 7, 8]

Life Cycle Assessment Methodologies

- *Systems Role:* Footprint Auditing
- Functional Definition: ISO-standardized frameworks measuring life cycle inputs, energy outputs, and chemical footprints from cradle to grave.
- Reference: [Ch. 3, 8]

2. Applied Targeted Mechanisms

PETase-Based Catalysis

- *Systems Role:* Biocatalytic Cleavage
- Functional Definition: Bacterial hydrolase enzymes optimized to catalytically cleave ester bonds within crystalline polymers under ambient temperatures.
- Reference: [Ch. 1, 8]

Microbial Degradation

- *Systems Role:* Polymer Degradation
- Functional Definition: Biological depolymerization processes mediated through bacterial or fungal enzyme cascades to break down high-molecular-weight matrices.
- Reference: [Ch. 1, 5, 8]

Enzymatic Recycling Systems

- *Systems Role:* Resource Recovery
- Functional Definition: Closed-loop industrial strategies deploying specialized enzyme cocktails to reduce post-consumer plastics back to pristine constituent monomers.
- Reference: [Ch. 1, 8]

Synthetic Biology Chassis

- *Systems Role:* Metabolic Optimization

- Functional Definition: Genetically altered microbial hosts tailored to maximize target enzyme secretion and accelerate synthetic polymer consumption.
- Reference: [Ch. 1, 8]

Enzyme-Mediated Transformation

- *Systems Role:* Protein Engineering
- Functional Definition: Engineered protein systems accelerating polymer breakdown under ambient environmental conditions.
- Reference: [Ch. 8]

Bioanalytical Discrimination

- *Systems Role:* Affinity Profiling
- Functional Definition: Molecular assays isolating target materials from complex organic backgrounds through specific affinity interactions and surface coatings.
- Reference: [Ch. 2]

Microbiome-Targeted Strategies

- *Systems Role:* Ecosystem Restoration
- Functional Definition: Therapeutic deployments of probiotics or prebiotics to restore disrupted microbial ecosystems and reinforce barrier protection.
- Reference: [Ch. 5, 6]

Nutritional Mitigation Strategies

- *Systems Role:* Cellular Defense
- Functional Definition: Dietary interventions deploying bioactive compounds or polyphenols to counteract particle-induced oxidative injury and inflammatory pathways.
- Reference: [Ch. 6]

Socioeconomic Indicators

- *Systems Role:* Equity Auditing
- Functional Definition: Public monitoring metrics assessing implementation equity, compliance cost distribution, and labor safety factors of mitigation deployments.

- Reference: [Ch. 8]

Process-Specific Hotspots

- *Systems Role:* Process Optimization
- Functional Definition: Specific engineering sequence steps that generate disproportionately high toxic emissions, secondary pollutants, or excessive capital costs.
- Reference: [Ch. 8]

Contaminant Provenance Analysis

- *Systems Role:* Forensic Attribution
- Functional Definition: Forensic determination of an industrial material's origin through structural degradation modeling and chemical fingerprinting.
- Reference: [Ch. 2]

Compilation of References

Abbott, L. C., & Maynard, A. D. (2010). Exposure Assessment Approaches for Engineered Nanomaterials. *Risk Analysis : An Official Publication of the Society for Risk Analysis, 30*(11), 1634–1644. DOI: 10.1111/j.1539-6924.2010.01446.x PMID: 20626687

Abeltino, A., Hatem, D., Serantoni, C., Riente, A., De Giulio, M. M., De Spirito, M., De Maio, F., & Maulucci, G. (2024). Unraveling the gut microbiota: Implications for precision nutrition and personalized medicine. *Nutrients, 16*(22), 3806. DOI: 10.3390/nu16223806 PMID: 39599593

Abudurexiti, A. S. J., Abulikemu, Z., & Bavumiragira, J. P. (2025). Circular Economy and Sustainable Development Goals in the Reduction of Global Nano (Micro) Plastic Pollution. *Polish Journal of Environmental Studies, 34*(4), 4489–4498. DOI: 10.15244/pjoes/190280

Agbasi, J. C., Egbueri, J. C., Pande, C. B., Khan, M. Y. A., Ighalo, J. O., Uwajingba, H. C., & Abba, S. I. (2025). Review of the Potential Effects and Remediation Strategies of Microplastic Pollutants in Drinking Water Sources. Analytical Letters, 58(5), 799 839. DOI: 10.1080/00032719.2024.2343366

Ali, I., Alharbi, O. M. L., Alothman, Z. A., Badjah, A. Y., & Alwarthan, A. (2021). Nanomaterials for water remediation: Adsorption and photocatalysis. *Desalination and Water Treatment, 215*, 85–106. DOI: 10.5004/dwt.2021.26974

Ali, I., Tan, X., Li, J., Peng, C., Wan, P., Naz, I., Duan, Z., & Ruan, Y. (2023). Innovations in the development of promising adsorbents for the remediation of microplastics and nanoplastics – A critical review. *Water Research, 230*, 119526. DOI: 10.1016/j.watres.2022.119526 PMID: 36577257

Alimi, O. S., Farner Budarz, J., Hernandez, L. M., & Tufenkji, N. (2018). Microplastics and nanoplastics in aquatic environments: Aggregation, deposition, and enhanced contaminant transport. Environmental Science & Technology, 52(4), 1704 1724. https://doi.org/DOI: 10.1021/acs.est.7b05559

Allan, J., Belz, S., Hoeveler, A., Hugas, M., Okuda, H., Patri, A., Rauscher, H., Silva, P., Slikker, W., Sokull-Kluettgen, B., Tong, W., & Anklam, E. (2021). Regulatory landscape of nanotechnology and nanoplastics from a global perspective. *Regulatory Toxicology and Pharmacology : RTP*, *122*, 104885. DOI: 10.1016/j.yrtph.2021.104885 PMID: 33617940

Alnuqaydan, A. M. (2024). The dark side of beauty: An in-depth analysis of the health hazards and toxicological impact of synthetic cosmetics and personal care products. *Frontiers in Public Health*, *12*, 1439027. DOI: 10.3389/fpubh.2024.1439027 PMID: 39253281

Al-Sid-Cheikh, M., Rowland, S. J., Stevenson, K., Rouleau, C., Henry, T. B., & Thompson, R. C. (2018). Uptake, whole-body distribution, and depuration of nanoplastics by the scallop *Pecten maximus* at environmentally realistic concentrations. *Environmental Science & Technology*, *52*(24), 14480–14486. DOI: 10.1021/acs.est.8b05266 PMID: 30457844

Alum, E. U., Aloh, H. E., Obasi, D. C., Okoroh, P. N., Aniokete, U. C., & Emeruwa, A. P. (2025). Maternal Nutrition, Toxicants, and Epigenetic Programming of Obesity Across Generations. *Diabetes, Metabolic Syndrome and Obesity*, *18*, 4873–4911. DOI: 10.2147/DMSO.S579409 PMID: 41497887

Amaral-Zettler, L. A., Zettler, E. R., & Mincer, T. J. (2020). Ecology of the plastosphere. *Nature Reviews. Microbiology*, *18*(3), 139–151. DOI: 10.1038/s41579-019-0308-0 PMID: 31937947

Amer, Y., Soufali, A., & Zaghwan, A. (2026). A digital twin-based framework for predictive quality assurance and supply chain resilience in the automotive industry. *Advanced Engineering Informatics*, *69*(Part C), 103969. DOI: 10.1016/j.aei.2025.103969

Andrady, A. L. (2011). Microplastics in the marine environment. *Marine Pollution Bulletin*, *62*(8), 1596–1605. DOI: 10.1016/j.marpolbul.2011.05.030 PMID: 21742351

Angaye, T. C. N., & Ogidi, O. I. (2018). The environmental impacts of electronic wastes in Bayelsa State, Nigeria. *MedCrave Online Journal of Toxicology*, *4*(5), 331–332.

Angaye, T. C. N., Oyinke, G. N., & Ogidi, O. I. (2019). Environmental and health risk assessment of particulate matter associated with sporting activities in Yenagoa metropolis, Nigeria. *General Internal medicine and clinical. Innovations*, *4*(3), 1–3. DOI: 10.15761/GIMCI.1000180

Anigbogu, C. B., Obiora, C. U., Kolawole, O. O., & Okorafor-Nwosu, A. (2025). Nutritional assessment and sensory properties of dried pap fortified with soybean and date palm fruit flours. *E-Proceedings of the Faculty of Agriculture International Conference*, 164–169. https://doi.org/DOI: 10.5281/da8s6652

Ankley, G. T., & Edwards, S. W. (2018). The Adverse Outcome Pathway: A Multifaceted Framework Supporting 21st Century Toxicology. *Current Opinion in Toxicology*, *9*, 1–7. DOI: 10.1016/j.cotox.2018.03.004 PMID: 29682628

Araújo, A. M., Mota, C., Ramos, H., Faria, M. A., Carvalho, M., & Ferreira, I. M. P. L. V. O. (2025). The neurotoxic threat of micro- and nanoplastics: Evidence from In Vitro and In Vivo models. *Archives of Toxicology*, *99*(9), 3505–3525. DOI: 10.1007/s00204-025-04091-3 PMID: 40459748

Arundhathi, V., Nagalakshmi Manasa, S., Kalyani Lakshmi, S., Varsha, V., & Sundaram, B. (2022). Life-cycle assessment of various synthesis routes of silver nanoparticles. In *Life-Cycle Assessment of Nanomaterials* (pp. 267–277). https://doi.org/DOI: 10.1007/978-981-15-3662-5_22

Arya, N., & Kumar, A. (2025). Micro/nano-plastics pollution: Challenges to agriculture productivity. In *Global impacts of micro- and nano-plastic pollution* (pp. 206–225). IGI Global., DOI: 10.4018/979-8-3693-3447-8.ch009

Aßen, N. V. D., Voll, P., Peters, M., & Bardow, A. (2014). Life Cycle Assessment of CO2capture and Utilization: A Tutorial Review. *Chemical Society Reviews*, *43*(23), 7982–7994. DOI: 10.1039/C3CS60373C PMID: 24441866

Au, S. Y., Lee, C. M., Weinstein, J. E., van den Hurk, P., & Klaine, S. J. (2017). Trophic transfer of microplastics in aquatic ecosystems: Identifying critical research needs. *Integrated Environmental Assessment and Management*, *13*(3), 505–509. DOI: 10.1002/ieam.1907 PMID: 28440939

Azeem, I., Shakoor, N., Chaudhary, S., Adeel, M., Zain, M., Ahmad, M. A., Li, Y., Zhu, G., Shah, S. A. A., Khan, K., Khan, A. A., Xu, M., & Rui, Y. (2023). Analytical challenges in detecting microplastics and nanoplastics in soil-plant systems. *Plant Physiology and Biochemistry : PPB*, *204*, 108132. DOI: 10.1016/j.plaphy.2023.108132 PMID: 37918078

Baalousha, M., Escobar, M., Unz, R. F., Shipley, H. J., Kothalawala, N., Lo, P., & Perreault, F. (2023). LCA and LCC analysis of engineered nanoparticles for ocean fertilization. *Environmental Science & Technology*. PubMed Central. https://pmc.ncbi.nlm.nih.gov/articles/PMC9747614

Bafana, A., Kumar, S. V., Temizel-Sekeryan, S., Dahoumane, S. A., Haselbach, L., & Jeffryes, C. S. (2018). Evaluating microwave-synthesized silver nanoparticles from silver nitrate with life cycle assessment techniques. *The Science of the Total Environment, 636*, 936–943. DOI: 10.1016/j.scitotenv.2018.04.345 PMID: 29729511

Bain, A. (2022). Buoyancy and Brownian motion of plastics in aqueous media: Predictions and implications for density separation and aerosol internal mixing state. *Environmental Science. Nano, 9*(11), 4249–4254. DOI: 10.1039/D2EN00525E

Ballesteros, S., Domenech, J., Barguilla, I., Cortes, C., Marcos, R., & Hernandez, A. (2020). Genotoxic and immunomodulatory effects in human white blood cells after ex vivo exposure to polystyrene nanoplastics. *Environmental Science. Nano, 7*(11), 3431–3446. DOI: 10.1039/D0EN00748J

Bansal, D., Bhattacharya, N., & Shandilya, P. (2025). Exploring Applications: Intelligent Drones and Robots in Industrial Settings. In *Building Embodied AI Systems: The Agents, the Architecture Principles, Challenges, and Application Domains* (pp. 159-180). Cham: Springer Nature Switzerland. https://doi.org/DOI: 10.1007/978-3-031-68256-8_7

Barguilla, I., Domenech, J., Ballesteros, S., Rubio, L., Marcos, R., & Hernández, A. (2022). Long-term exposure to nanoplastics alters molecular and functional traits related to the carcinogenic process. *Journal of Hazardous Materials, 438*, 129470. DOI: 10.1016/j.jhazmat.2022.129470 PMID: 35785738

Bassan, S., & Harel, O. (2018). The ethics in synthetics: Statistics in the service of ethics and law in health-related research in big data from multiple sources. *Journal of Law and Health, 31*(1), 87–117. PMID: 30889335

Bastyans, S., Jackson, S., & Fejer, G. (2022). Micro and nanoplastics, a threat to human health? *Emerging Topics in Life Sciences, 6*(4), 411–422. DOI: 10.1042/ETLS20220024 PMID: 36453919

Basumatary, T., Biswas, D., Boro, S., Nava, A. R., Narayan, M., & Sarma, H. (2025). Dynamics and Impacts of Microplastics (MPs) and Nanoplastics (NPs) on Ecosystems and Biogeochemical Processes: The Need for Robust Regulatory Frameworks. *ACS Omega, 10*(17), 17051–17069. DOI: 10.1021/acsomega.5c01175 PMID: 40352536

Baysal, A., Soyocak, A., Saygin, H., & Saridag, A. M. (2024). Exposure to phagolysosomal simulated fluid altered the cytotoxicity of PET micro(nano)plastics to human lung epithelial cells. *Toxicology Mechanisms and Methods*, *34*(1), 72–97. DOI: 10.1080/15376516.2023.2256847 PMID: 37697451

Bel Hassen, M., Bellaaj Zouari, A., Abdennadher, M., Assaf, J.-C., Nakad, M., Abboud, R., Khammeri, Y., Banni, M., Panzeri, A., Gomes, L., & Hamd, W. (2025). Plastics pollution: Pathways, impacts, and regulatory challenges in marine environments. *Frontiers in Environmental Science*, *13*, 1635230. DOI: 10.3389/fenvs.2025.1635230

Belle, J., Velásquez, C., Oxley, M., & Getiashvili, K. (2024). Understanding and integrating systemic risk (SR) into disaster risk reduction (DRR) and risk informed development (RID) for long-term resilient development. In *Sustainable Development*. IntechOpen. https://doi.org/DOI: 10.5772/intechopen.1004915

Bell, I. R., Ives, J. A., & Jonas, W. B. (2013). Nonlinear effects of nanoparticles: Biological variability from hormetic doses, small particle sizes, and dynamic adaptive interactions. *Dose-Response*, *12*(2), 202–232. DOI: 10.2203/dose-response.13-025. Bell PMID: 24910581

Belzagui Elder, F., & Gutiérrez Bouzán, M. C. (2022). Review of alternatives for reducing textile microfiber emissions to water. *Journal of Environmental Management*, *317*, 115347. Advance online publication. DOI: 10.1016/j.jenvman.2022.115347

Benitez, A. J., Tanes, C., Friedman, E. S., Zackular, J. P., Ford, E., Gerber, J. S., DeRusso, P. A., Kelly, A., Li, H., Elovitz, M. A., Wu, G. D., Zemel, B., & Bittinger, K. (2025). Antibiotic exposure is associated with minimal gut microbiome perturbations in healthy term infants. *Microbiome*, *13*(1), 21. DOI: 10.1186/s40168-024-01999-3 PMID: 39856742

Beresniak, A., Bremond-Gignac, D., Dupont, D., & Duru, G. (2025). Reevaluating health metrics: Unraveling the limitations of disability-adjusted life years as an indicator in disease burden assessment. *World Journal of Methodology*, *15*(1), 95796. DOI: 10.5662/wjm.v15.i1.95796 PMID: 40115408

Bergamaschi, E., Craig, P., Guseva Canu, I., & Prina-Mello, A. (2015). The role of biological monitoring in nanosafety. *Nano Today*, *10*(3), 274–277. DOI: 10.1016/j.nantod.2015.02.001

Berg, G., Rybakova, D., Fischer, D., Cernava, T., Vergès, M. C., Charles, T., Chen, X., Cocolin, L., Eversole, K., Corral, G. H., Kazou, M., Kinkel, L., Lange, L., Lima, N., Loy, A., Macklin, J. A., Maguin, E., Mauchline, T., McClure, R., & Schloter, M. (2020). Microbiome definition re-visited: Old concepts and new challenges. *Microbiome*, *8*(1), 103. DOI: 10.1186/s40168-020-00875-0 PMID: 32605663

Bergkamp, L., Michaux, G., & Herbatschek, N. (2014). Nanotechnology regulation in Europe: From REACH and nano-registries to cosmetics, biocides, and medical devices. *Nanotechnology Law & Business*, *11*(2), 136–159.

Bessa, F., Barria, P., Neto, J. M., Frias, J. P., Otero, V., Sobral, P., & Marques, J. C. (2018). Occurrence of microplastics in commercial fish from a wide range of waters. *Environmental Pollution*, *243*, 1098–1107. DOI: 10.1016/j.envpol.2018.09.01

Besseling, E., Quik, J. T. K., Sun, M., & Koelmans, A. A. (2017). Fate of nano- and microplastics in freshwater systems: A modeling study. Environmental Pollution, 220, 540 548.DOI: 10.1016/j.envpol.2016.10.001

Bhagat, N., & Chandrakar, P. (2025). AI-Powered Image Recognition. *International Journal of Innovative Science and Research Technology*, *10*(5), 4521–4526. DOI: 10.38124/ijisrt/25may2243

Bhat, M. A., Gedik, K., & Gaga, E. O. (2022). Environmental toxicity of emerging micro and nanoplastics: A lesson learned from nanomaterials. In *Nanotechnology interventions in food packaging and shelf life* (pp. 311–327). CRC Press., DOI: 10.1201/9781003207641-18

Bhat, M. A., Gedik, K., & Gaga, E. O. (2023). Atmospheric micro (nano) plastics: Future growing concerns for human health. *Air Quality, Atmosphere & Health*, *16*(2), 233–262. DOI: 10.1007/s11869-022-01272-2 PMID: 36276170

Bhattacharya, R., & Kim, S. (2024). Socioeconomic dimensions of nanoplastic remediation in global coastal communities. In Gupta, A., & Kim, H. (Eds.), *Nanotechnology for Sustainable Development* (pp. 255–276). Springer., DOI: 10.1007/978-981-97-83632_12

Bhattacharyya, S., Greer, M. L., & Salehi, M. (2025). Impact of micro- and nanoplastics exposure on human health: Focus on neurological effects from ingestion. *Frontiers in Public Health*, *13*, 1681776. DOI: 10.3389/fpubh.2025.1681776 PMID: 41246084

Bhuller, Y., Bancroft, X., Deonandan, R., Grudniewicz, A., Wiles, A., & Krewski, D. (2025). Key attributes of health and environmental risk decision-making: A scoping review. *Risk Analysis : An Official Publication of the Society for Risk Analysis, 45*(7), 1926–1939. DOI: 10.1111/risa.17715 PMID: 39894676

Bithas, K. (2011). Sustainability and externalities: Is the internalization of externalities a sufficient condition for sustainability? *Ecological Economics, 70*(10), 1703–1706. DOI: 10.1016/j.ecolecon.2011.05.014

Bläsing, M., & Amelung, W. (2018). Plastics in soil: Analytical methods and possible sources. *The Science of the Total Environment, 612*, 422–435. DOI: 10.1016/j.scitotenv.2017.08.086 PMID: 28863373

Blowes, D. W., Ptacek, C. J., Benner, S. G., McRae, C., Bennett, T. A., & Puls, R. W. (2000). Treatment of inorganic contaminants using permeable reactive barriers. *Journal of Contaminant Hydrology, 45*(1–2), 123–137. DOI: 10.1016/S0169-7722(00)00122-4

Boakye, G. A., Trotta, E., Ambagahawatta, N., Venkataraman, A., Cheeptham, N., & Papadopoulos, C. (2025). Nanoscale plastic pollution: Sources, identification and potential mitigation. *Nanotechnology, 36*(42), 422001. DOI: 10.1088/1361-6528/ae0c1b PMID: 41005319

Boctor, J., Hoyle, F. C., Farag, M. A., Ebaid, M., Walsh, T., Whiteley, A. S., & Murphy, D. V. (2025). Microplastics and nanoplastics: Fate, transport, and governance from agricultural soil to food webs and humans. *Environmental Sciences Europe, 37*(1), 68. DOI: 10.1186/s12302-025-01104-x

Bodzek, M. (2023). Remediation of nano- and microplastics in water environment using nanomaterials. *Desalination and Water Treatment, 316*, 557–573. DOI: 10.5004/dwt.2023.30172

Bodzek, M., & Bodzek, P. (2025). Remediation of micro- and nanoplastics by membrane technologies. *Membranes (Basel), 15*(3), 82. DOI: 10.3390/membranes15030082 PMID: 40137034

Bodzek, M., & Pohl, A. (2023a). Possibilities of removing microplastics from the aquatic environment using membrane processes. *Desalination and Water Treatment, 288*, 104–120. DOI: 10.5004/dwt.2023.29154

Bodzek, M., & Pohl, A. (2023b). Removal of microplastics in unit processes used in water and wastewater treatment: A review. *Archives of Environmental Protection, 48*(2), 102–128. DOI: 10.24425/aep.2022.143713

Bondarenko, O. M., Heinlaan, M., Sihtmäe, M., Ivask, A., Kurvet, I., Joonas, E., Jemec, A., Mannerström, M., Heinonen, T., Rekulapelly, R., Singh, S., Zou, J., Pyykkö, I., Drobne, D., & Kahru, A. (2016). Multilaboratory evaluation of 15 bioassays for (eco)toxicity screening and hazard ranking of engineered nanomaterials: FP7 project NANOVALID. *Nanotoxicology*, *10*(9), 1229–1242. DOI: 10.1080/17435390.2016.1196251 PMID: 27259032

Bossio, S., Ruffolo, S. A., Lofaro, D., Perri, A., & La Russa, M. F. (2025). Endocrine Toxicity of Micro-and Nanoplastics, and Advances in Detection Techniques for Human Tissues: A Comprehensive Review. *Endocrines*, *6*(2), 23. DOI: 10.3390/endocrines6020023

Brachner, A., Fragouli, D., Duarte, I. F., Farias, P. M. A., Dembski, S., Ghosh, M., Barisic, I., Zdzieblo, D., Vanoirbeek, J., Schwabl, P., & Neuhaus, W. (2020). Assessment of Human Health Risks Posed by Nano-and Microplastics Is Currently Not Feasible. *International Journal of Environmental Research and Public Health*, *17*(23), 8832. DOI: 10.3390/ijerph17238832 PMID: 33261100

Brennecke, D., Duarte, B., Paiva, F., Caçador, I., & Canning-Clode, J. (2016). Microplastics as vectors for heavy metal contamination from the marine environment. *Estuarine, Coastal and Shelf Science*, *178*, 189–195. DOI: 10.1016/j.ecss.2015.12.003

Brito, A. L. C. P., Mattioni, J. V., Ramos, G. R., Nakamura, M., & Toma, H. E. (2024). Direct monitoring of the enzymatically sequestering and degrading of PET microplastics using hyperspectral Raman microscopy. *Micron : The International Research and Review Journal for Microscopy*, *187*, 103722. DOI: 10.1016/j.micron.2024.103722 PMID: 39306918

Bruno, A., Dovizio, M., Milillo, C., Aruffo, E., Pesce, M., Gatta, M., Chiacchiaretta, P., Di Carlo, P., & Ballerini, P. (2024). Orally ingested micro- and nano-plastics: A hidden driver of inflammatory bowel disease and colorectal cancer. *Cancers (Basel)*, *16*(17), 3079. DOI: 10.3390/cancers16173079 PMID: 39272937

Bubnov, R. V., Babenko, L. P., Lazarenko, L. M., Mokrozub, V. V., Demchenko, O. A., Nechypurenko, O. V., & Spivak, M. Y. (2017). Comparative study of probiotic effects of *Lactobacillus* and *Bifidobacteria* strains on cholesterol levels, liver morphology and the gut microbiota in obese mice. *The EPMA Journal*, *8*(4), 357–376. DOI: 10.1007/s13167-017-0117-3 PMID: 29209439

Callewaert, E., Louisse, J., Kramer, N., Sanz-Serrano, J., & Vinken, M. (2025). Adverse Outcome Pathways Mechanistically Describing Hepatotoxicity. *Methods in Molecular Biology (Clifton, N.J.)*, *2834*, 249–273. DOI: 10.1007/978-1-0716-4003-6_12 PMID: 39312169

Campbell, J. M., Balhoff, J. B., Landwehr, G. M., Rahman, S. M., Vaithiyanathan, M., & Melvin, A. T. (2018). Microfluidic and Paper-Based Devices for Disease Detection and Diagnostic Research. *International Journal of Molecular Sciences*, *19*(9), 2731. DOI: 10.3390/ijms19092731 PMID: 30213089

Capodaglio, A. G. (2025). Micro- and Nano-Plastics in Drinking Water: Threat or Hype? Critical State-of-the-Art Analysis of Risks and Approaches. *Journal of Xenobiotics*, *15*(3), 85. DOI: 10.3390/jox15030085 PMID: 40558868

Caporalini, S., Azimi, B., Zergat, S., Ansari Chaharsoughi, M., Maleki, H., Batoni, G., & Danti, S. (2025). Electrospinning Enables Opportunity for Green and Effective Antibacterial Coatings of Medical Devices. *Journal of Functional Biomaterials*, *16*(7), 249. DOI: 10.3390/jfb16070249 PMID: 40710463

Caputo, F., Vogel, R., Savage, J., Vella, G., Law, A., Della Camera, G., Hannon, G., Peacock, B., Mehn, D., Ponti, J., Geiss, O., Aubert, D., Prina-Mello, A., & Calzolai, L. (2021). Measuring particle size distribution and mass concentration of nanoplastics and microplastics: Addressing some analytical challenges in the sub-micron size range. *Journal of Colloid and Interface Science*, *588*, 401–417. DOI: 10.1016/j.jcis.2020.12.039 PMID: 33422789

Cardoso, A. M. (2024). Microbial influence on blood pressure: Unraveling the complex relationship for health insights. *Microbiome Research Reports*, *3*(2), 22. DOI: 10.20517/mrr.2023.73 PMID: 38841410

Catarino, A. I., Frutos, A., & Henry, T. B. (2019). Use of fluorescent-labelled nanoplastics (NPs) to demonstrate NP absorption is inconclusive without adequate controls. *The Science of the Total Environment*, *670*, 915–920. DOI: 10.1016/j.scitotenv.2019.03.194 PMID: 30921723

Cavazzoli, S., Ferrentino, R., Scopetani, C., Monperrus, M., & Andreottola, G. (2023). Analysis of micro- and nanoplastics in wastewater treatment plants: Key steps and environmental risk considerations. *Environmental Monitoring and Assessment*, *195*(12), 1483. DOI: 10.1007/s10661-023-12030-x PMID: 37971551

Cecchin, I., Reddy, K. R., Thome, A., Tessaro, E. F., & Schnaid, F. (2017). Nano bioremediation: Integration of nanoparticles and bioremediation for sustainable remediation of chlorinated organic contaminants in soils. *International Biodeterioration & Biodegradation*, *119*, 419–428. DOI: 10.1016/j.ibiod.2016.09.027

Chabi, K., Ye, C., Gad, M., Rizk, M., & Yu, X. (2026). Toxicity of microplastics and nanoplastics in aquatic systems in interaction with biological–chemical contaminants: Advances in remediations. In *Environmental Sciences*. IntechOpen., DOI: 10.5772/intechopen.1013428

Chandran, P. J. I., Ranjani, M., & Veerasingam, S. (2025). Challenges of microplastic characterization and mitigation. In *Handbook of Microplastic Pollution in the Environment* (pp. 1–22). CRC Press., DOI: 10.1201/9781003487593-2

Chang, W.-H., Herianto, S., Lee, C.-C., Hung, H., & Chen, H.-L. (2021). The effects of phthalate ester exposure on human health: A review. *The Science of the Total Environment, 786*, 147371. DOI: 10.1016/j.scitotenv.2021.147371 PMID: 33965815

Cheng, J., Li, J., Xiong, R. G., Zhou, D. D., Huang, S. Y., Wu, S. X., Shang, A., Tang, G. Y., Li, H. B., & Gan, R. Y. (2025). Potentially harmful effects of micro-/nanoplastics on humans as well as protective actions of dietary natural products. *Trends in Food Science & Technology, 156*, 104841. DOI: 10.1016/j.tifs.2024.104841

Cheng, K., Chang, Y., Lee, X., Ji, D., Qiao, L., Zou, Z., Duan, Y., & Huang, R. J. (2024). Life-Course Health Risk Assessment of $PM_{2.5}$ Elements in China: Exposure Disparities by Species, Source, Age, Gender, and Location. *Environmental Science & Technology, 58*(8), 3629–3640. DOI: 10.1021/acs.est.3c05404 PMID: 38354315

Chen, M., Nan, J., & Breider, F. (2024). A comparative study on the stability and coagulation removal of aged vs. nonaged nanoplastics in surface water. *Journal of Hazardous Materials, 480*, 136490. DOI: 10.1016/j.jhazmat.2024.136490 PMID: 39547043

Chen, X., Zhuang, J., Chen, Q., Xu, L., Yue, X., & Qiao, D. (2022). Polyvinyl chloride microplastics induced gut barrier dysfunction, microbiota dysbiosis and metabolism disorder in adult mice. *Ecotoxicology and Environmental Safety, 241*, 113809. DOI: 10.1016/j.ecoenv.2022.113809 PMID: 36068740

Chepuri, K., Tulasi, C. D. S. L. N., & Manikantha, D. (2023). Cellular and animal toxicities of micro- and nanoplastics. In *Micro and nanoplastics in soil* (pp. 261–292). Springer Nature., DOI: 10.1007/978-3-031-21195-9_12

Cheung, L. T. O., Lui, C. Y., & Fok, L. (2018). Microplastic Contamination of Wild and Captive Flathead Grey Mullet (Mugil cephalus). *International Journal of Environmental Research and Public Health, 15*(4), 597. DOI: 10.3390/ijerph15040597 PMID: 29587444

Choi, D., Hwang, J., Bang, J., Han, S., Kim, T., Oh, Y., Hwang, Y., Choi, J., & Hong, J. (2021). In vitro toxicity from a physical perspective of polyethylene microplastics based on statistical curvature change analysis. *The Science of the Total Environment, 752*, 142242. DOI: 10.1016/j.scitotenv.2020.142242 PMID: 33207500

Choi, S., Lee, S., Kim, M.-K., Yu, E.-S., & Ryu, Y.-S. (2024). Challenges and recent analytical advances in micro(nano)plastics analysis. *Analytical Chemistry*, *96*(22), 8846–8854. DOI: 10.1021/acs.analchem.3c05948 PMID: 38758170

Christopher, E. A., Christopher-de Vries, Y., Devadoss, A., Mandemaker, L. D. B., van Boxel, J., Copsey, H. M., Dusza, H. M., Legler, J., Meirer, F., Muncke, J., Nawrot, T. S., Saenen, N. D., Scholz-Böttcher, B. M., Tran, L., Weckhuysen, B. M., Zou, R., Zimmermann, L., Galea, K. S., Vermeulen, R., & Boyles, M. S. P. (2024). Impacts of micro- and nanoplastics on early-life health: A roadmap toward risk assessment. *Microplastics and Nanoplastics*, *4*(1), 13. DOI: 10.1186/s43591-024-00089-3

Chuang, W.-Y., Young, T.-H., Chiu, W.-Y., & Lin, C.-Y. (2000). Effect of polymeric additives on the structure and permeability of poly (vinyl alcohol) asymmetric membranes. *Additives for Polymers*, *2000*(10–11), 17–18. DOI: 10.1016/S0306-3747(00)90314-2

Chulkov, V., Gasanov, M., Isakov, V., Denisenko, A., Nwosu, C., & Rodkin, S. (2025). Molecular and Cellular Effects of Microplastics and Nanoplastics in the Pathogenesis of Cardiovascular, Nervous, Urinary, Digestive, and Reproductive System Diseases: A Global Systematic Review. *International Journal of Molecular Sciences*, *26*(22), 11194. DOI: 10.3390/ijms262211194 PMID: 41303677

Clarke, R., Shajahan, A. N., Wang, Y., Tyson, J. J., Riggins, R. B., Weiner, L. M., Bauman, W. T., Xuan, J., Zhang, B., Facey, C., Aiyer, H., Cook, K., Hickman, F. E., Tavassoly, I., Verdugo, A., Chen, C., Zwart, A., Wärri, A., & Hilakivi-Clarke, L. A. (2011). Endoplasmic reticulum stress, the unfolded protein response, and gene network modeling in antiestrogen resistant breast cancer. *Hormone Molecular Biology and Clinical Investigation*, *5*(1), 35–44. DOI: 10.1515/HMBCI.2010.073 PMID: 23930139

Cole, M., Lindeque, P., Halsband, C., & Galloway, T. S. (2011). Microplastics as contaminants in the marine environment: A review. *Marine Pollution Bulletin*, *62*(12), 2588–2597. DOI: 10.1016/j.marpolbul.2011.09.025 PMID: 22001295

Coman, L. I., Balaban, D. V., Dumbravă, B. F., Păunescu, H., Marin, R. C., Costescu, M., Dima, L., Jinga, M., & Coman, O. A. (2025). Targeting Oxidative Stress in Acute Pancreatitis: A Critical Review of Antioxidant Strategies. *Nutrients*, *17*(15), 2390. DOI: 10.3390/nu17152390 PMID: 40805976

Corbett, L. N., & Amaral-Zettler, L. A. (2022). Nanoplastics and the marine environment: The unseen majority? In Andrady, A. L. (Ed.), *Plastics and the Ocean: Origin, Characterization, Fate, and Impacts* (pp. 389–428). Wiley., DOI: 10.1002/9781119768432.ch14

Correia, M., & Loeschner, K. (2018). Detection of nanoplastics in food by asymmetric flow field-flow fractionation coupled to multi-angle light scattering: Possibilities, challenges and analytical limitations. *Analytical and Bioanalytical Chemistry, 410*(22), 5603–5615. DOI: 10.1007/s00216-018-0919-8 PMID: 29411085

Corte Pause, F., Urli, S., Crociati, M., Stradaioli, G., & Baufeld, A. (2024). Connecting the Dots: Livestock Animals as Missing Links in the Chain of Microplastic Contamination and Human Health. *Animals (Basel), 14*(2), 350. DOI: 10.3390/ani14020350 PMID: 38275809

Cox, K. D., Covernton, G. A., Davies, H. L., Dower, J. F., Juanes, F., & Dudas, S. E. (2019). Human consumption of microplastics. Environmental Science & Technology, 53(12), 7068 7074. https://doi.org/DOI: 10.1021/acs.est.9b01517

Crane, R. A. (2012). *Nanoscale zero-valent iron: Future prospects for an emerging water treatment technology*. U.S. EPA.

Crane, R. A., & Scott, T. B. (2012). Nanoscale zero-valent iron: Future prospects for an emerging water treatment technology. *Journal of Hazardous Materials, 211–212*, 112–125. DOI: 10.1016/j.jhazmat.2011.11.073 PMID: 22305041

Croock, D., Swart, Y., Schurz, H., Petersen, D. C., Möller, M., & Uren, C. (2024). Data harmonization guidelines to combine multiplatform genomic data from admixed populations and boost power in genome-wide association studies. *Current Protocols, 4*(6), e1055. DOI: 10.1002/cpz1.1055 PMID: 38837690

Crosta, A., Parolini, M., & De Felice, B. (2023). Microplastics contamination in nonalcoholic beverages from the Italian market. *International Journal of Environmental Research and Public Health, 20*(5), 4122. DOI: 10.3390/ijerph20054122 PMID: 36901131

Cuciniello, R., Di Meo, F., Filosa, S., Crispi, S., & Bergamo, P. (2023). The antioxidant effect of dietary bioactives arises from the interplay between the physiology of the host and the gut microbiota: Involvement of short-chain fatty acids. *Antioxidants, 12*(5), 1073. DOI: 10.3390/antiox12051073 PMID: 37237938

Cui, Y., Eccles, K. M., Kwok, R. K., Joubert, B. R., Messier, K. P., & Balshaw, D. M. (2022). Integrating multiscale geospatial environmental data into large population health studies: Challenges and opportunities. *Toxics, 10*(7), 403. DOI: 10.3390/toxics10070403 PMID: 35878308

Cunningham, B. E., Sharpe, E. E., Brander, S. M., Landis, W. G., & Harper, S. L. (2023). Critical gaps in nanoplastics research and their connection to risk assessment. *Frontiers in Toxicology*, *5*, 1154538. DOI: 10.3389/ftox.2023.1154538 PMID: 37168661

Cunningham, E. M., Cuthbert, R. N., Coughlan, N. E., Kregting, L., Cairnduff, V., & Dick, J. T. A. (2021). Microplastics do not affect the feeding rates of a marine predator. *The Science of the Total Environment*, *779*, 146487. DOI: 10.1016/j.scitotenv.2021.146487 PMID: 34030230

Cverenkárová, K., Valachovičová, M., Mackuľak, T., Žemlička, L., & Bírošová, L. (2021). Microplastics in the Food Chain. *Life (Basel, Switzerland)*, *11*(12), 1349. DOI: 10.3390/life11121349 PMID: 34947879

D'Angelo, C., Sudakaran, S., Asimakopoulos, F., Hematti, P., El-Gamal, D., Safdar, N., & Callander, N. (2023). Perturbation of the gut microbiome and association with outcomes following autologous stem cell transplantation in patients with multiple myeloma. *Leukemia & Lymphoma*, *64*(1), 87–97. DOI: 10.1080/10428194.2022.2131410 PMID: 36218226

Dai, Y., Shen, Z., Khachatryan, L. G., Vadiyan, D. E., Karampoor, S., & Mirzaei, R. (2023). Unraveling mechanistic insights into the role of microbiome in neurogenic hypertension: A comprehensive review. *Pathology, Research and Practice*, *249*, 154740. DOI: 10.1016/j.prp.2023.154740 PMID: 37567034

Danasekaran, R. (2024). One Health: A Holistic Approach to Tackling Global Health Issues. *Indian Journal of Community Medicine : Official Publication of Indian Association of Preventive & Social Medicine*, *49*(2), 260–263. DOI: 10.4103/ijcm.ijcm_521_23 PMID: 38665439

Dang, F., Wang, Q., Huang, Y., Wang, Y., & Xing, B. (2022). Key knowledge gaps for One Health approach to mitigate nanoplastic risks. *Eco-Environment & Health*, *1*(1), 11–22. DOI: 10.1016/j.eehl.2022.02.001 PMID: 38078201

Dang, M., Yang, C., Jin, G., Deng, Q., Wu, L., & Teh, G. B. (2026). From biomolecular condensates to functional nanomaterials: LLPS-inspired frameworks for nanoscale hydrogels and adaptive materials. *Journal of Nanobiotechnology*, *24*(1), 280. Advance online publication. DOI: 10.1186/s12951-026-04151-5 PMID: 41721403

Dar, A. I., Randhawa, S., Verma, M., Saini, T. C., & Acharya, A. (2025). Debugging the dynamics of protein corona: Formation, composition, challenges, and applications at the nano-bio interface. *Advances in Colloid and Interface Science*, *342*, 103535. DOI: 10.1016/j.cis.2025.103535 PMID: 40319752

Davaasuren, N., Marino, A., Boardman, C., Alparone, M., Nunziata, F., Ackermann, N., & Hajnsek, I. (2018, July). Detecting microplastics pollution in world oceans using SAR remote sensing. In *IGARSS 2018-2018 IEEE International Geoscience and Remote Sensing Symposium* (pp. 938-941). IEEE. DOI: 10.1109/IGARSS.2018.8517281

Day, M., Cooney, J. D., Klein, C., & Fox, J. (1996). *Thermal degradation of automotive plastics: A possible recycling opportunity*. American Chemical Society.

De Corte, W., Hermabessiere, L., & Dekeyser, S. (2019). Microplastics in drinking water: Recommended solutions and open questions. *Current Opinion in Environmental Science & Health*, *12*, 72–82. DOI: 10.1016/j.coesh.2019.10.008

de Oliveira, V. M., Zarzur, H. R., & de Souza, A. C. (2022). Artificial intelligence applications in food safety systems: A review of enabled monitoring technologies. *Trends in Food Science & Technology*, *128*, 245–256. DOI: 10.1016/j.tifs.2022.08.001

Demestichas, K., Peppes, N., Alexakis, T., & Adamopoulou, E. (2020). Blockchain in agriculture traceability systems: A review. *Applied Sciences (Basel, Switzerland)*, *10*(12), 4113. DOI: 10.3390/app10124113

Demir, U., Gülcan, H. E., & Özer, S. (2025). The effect of TiO_2 nanoadditive on emissions, exergetic performance, and environmental/social/economic indicators in a small UAV jet engine fueled with kerosene. *Fuel*, *390*, 134725. DOI: 10.1016/j.fuel.2025.134725

Deng, D., Wang, L., Zeng, Z., Lu, L., Yang, M., Wang, Y., Shi, X., Lovell, J. F., Zhang, J., & Jin, H. (2026). Visualizing extracellular vesicles in cancer: From biogenesis to theranostic applications. *Journal of Nanobiotechnology*, *24*(1), 334. DOI: 10.1186/s12951-026-04223-6 PMID: 41776618

Deng, Y., Zhang, Y., Lemos, B., & Ren, H. (2017). Tissue accumulation of microplastics in mice and biomarker responses suggest widespread health effects of exposure to microplastics. *Journal of Hazardous Materials*, *344*, 1026–1033. DOI: 10.1016/j.jhazmat.2017.11.047

Deng, Y., Zhang, Y., Lemos, B., & Ren, H. (2017). Tissue accumulation of microplastics in mice and biomarker responses suggest widespread health risks of exposure. *Scientific Reports*, *7*(1), 46687. DOI: 10.1038/srep46687 PMID: 28436478

Dennis, J., & Kumar Mistri, T. (2026). Nanoplastics in the Biological System: A Microscopic Menace with an Impact. In Environmental Sciences. IntechOpen. https://doi.org/DOI: 10.5772/intechopen.1013429

Der, O., Khaksar, H., & Gnecco, E. (2025). On the formation and characterization of nanoplastics during surface wear processes. *Surfaces*, *8*(2), 27. DOI: 10.3390/surfaces8020027

Destoumieux-Garzón, D., Mavingui, P., Boetsch, G., Boissier, J., Darriet, F., Duboz, P., Fritsch, C., Giraudoux, P., Le Roux, F., Morand, S., Paillard, C., Pontier, D., Sueur, C., & Voituron, Y. (2018). The one health concept: 10 years old and a long road ahead. *Frontiers in Veterinary Science*, *5*, 14. DOI: 10.3389/fvets.2018.00014 PMID: 29484301

Devi, M. K., Karmegam, N., Manikandan, S., Subbaiya, R., Song, H., Kwon, E. E., Sarkar, B., Bolan, N., Kim, W., Rinklebe, J., & Govarthanan, M. (2022). Removal of nanoplastics in water treatment processes: A review. *The Science of the Total Environment*, *845*, 157168. DOI: 10.1016/j.scitotenv.2022.157168 PMID: 35817120

Dewika, M., Markandan, K., Nor Ruwaida, J., Sara, Y. Y., Deb, A., Ahmad Irfan, N., & Khalid, M. (2024). Integrating the quintuple helix approach into atmospheric microplastics management policies for planetary health preservation. *The Science of the Total Environment*, *954*, 176063. Advance online publication. DOI: 10.1016/j.scitotenv.2024.176063 PMID: 39245389

Díaz-Villanueva, J. F., Díaz-Molina, R., & García-González, V. (2015). Protein Folding and Mechanisms of Proteostasis. *International Journal of Molecular Sciences*, *16*(8), 17193–17230. DOI: 10.3390/ijms160817193 PMID: 26225966

Ding, Q., Liu, K., Xu, K., Sun, R., Zhang, J., Yin, L., & Pu, Y. (2018). Further understanding of degradation pathways of microcystin-LR by an indigenous *Sphingopyxis* sp. in environmentally relevant pollution concentrations. *Toxins*, *10*(12), 536. DOI: 10.3390/toxins10120536 PMID: 30558170

Ding, X., Zhou, C., Zhong, W., & Tang, P. (2019). Addressing uncertainty of environmental governance in environmentally sensitive areas in developing countries: A precise-strike and spatial-targeting adaptive governance framework. *Sustainability (Basel)*, *11*(16), 4510. DOI: 10.3390/su11164510

Dong, X., Liu, X., Hou, Q., & Wang, Z. (2023). From natural environment to animal tissues: A review of microplastics (nanoplastics) translocation and hazards studies. *The Science of the Total Environment*, *855*, 158686. DOI: 10.1016/j.scitotenv.2022.158686 PMID: 36099943

Dris, R., Gasperi, J., Mirande, C., Mandin, C., & Guerrouache, M. (2018). A first overview of textile fibers, including microplastics, in indoor and outdoor environments. *Environmental Pollution*, *234*, 645–655. DOI: 10.1016/j.envpol.2017.12.013 PMID: 27989388

Dris, R., Gasperi, J., Saad, M., Mirande, C., & Tassin, B. (2016). Synthetic fibers in atmospheres and fallouts. *Water, Air, and Soil Pollution*, *228*(8), 311. DOI: 10.1007/s11270-017-3467-5

Dube, E., & Okuthe, G. E. (2023). Plastics and micro/nano-plastics (MNPs) in the environment: Occurrence, impact, and toxicity. *International Journal of Environmental Research and Public Health*, *20*(17), 6667. DOI: 10.3390/ijerph20176667 PMID: 37681807

Dugershaw, B. B., Aengenheister, L., Hansen, S. S. K., Hougaard, K. S., & Buerki-Thurnherr, T. (2020). Recent insights on indirect mechanisms in developmental toxicity of nanomaterials. *Particle and Fibre Toxicology*, *17*(1), 31. Advance online publication. DOI: 10.1186/s12989-020-00359-x PMID: 32653006

Du, L., Liu, H., Song, X., Feng, X., Xu, H., Tang, W., & Yang, J. (2024). Developments in the field of intestinal toxicity and signaling pathways associated with rodent exposure to micro (nano) plastics. *Toxicology*, *507*, 153883. DOI: 10.1016/j.tox.2024.153883 PMID: 38996996

Durden, J. M. (2025). Environmental management using a digital twin. *Environmental Science & Policy*, *164*, 104018. DOI: 10.1016/j.envsci.2025.104018

Du, T., Shao, S., Qian, L., Meng, R., Li, T., Wu, L., & Li, Y. (2023). Effects of photochlorination on the physicochemical transformation of polystyrene nanoplastics: Mechanism and environmental fate. *Water Research*, *243*, 120367. Advance online publication. DOI: 10.1016/j.watres.2023.120367 PMID: 37499544

Egalini, F., Marinelli, L., Rossi, M., Motta, G., Prencipe, N., Rossetto Giaccherino, R., Pagano, L., Grottoli, S., & Giordano, R. (2022). Endocrine disrupting chemicals: Effects on pituitary, thyroid and adrenal glands. *Endocrine*, *78*(3), 395–405. DOI: 10.1007/s12020-022-03076-x PMID: 35604630

Eizenberg, E., & Jabareen, Y. (2017). Social sustainability: A new conceptual framework. *Sustainability (Basel)*, *9*(1), 68. DOI: 10.3390/su9010068

Elawady, A., Mohamed, J. A. H., & Abid, M. B. (2026). Advancements in microplastics detection techniques and their multidimensional impacts on aquatic ecosystems and human health. *Journal of Hazardous Materials: Plastics*, *2*, 100025. DOI: 10.1016/j.hazmp.2025.100025

Elgharbawy, A. A. M., Noor, N. M., Azmi, N. A. N., & Dewanti, B. S. D. (2026). Beyond the Solvent: Engineering Ionic Liquids for Biomedical Applications-Advances, Challenges, and Future Directions. *Molecules (Basel, Switzerland)*, *31*(2), 305. DOI: 10.3390/molecules31020305 PMID: 41599355

El-Sheikh, R., El-Sayed, H., Ibrahim, A., & Salem, W. (2025). Biosynthesized magnetite nanoparticles combined with hydrocarbon-degrading bacteria for enhanced petroleum pollutant degradation. *Frontiers in Microbiology*, *16*, 1590636. DOI: 10.3389/fmicb.2025.1590636

Elzayat, H., Mesto, G., & Al-Marzooq, F. (2023). Unraveling the impact of gut and oral microbiome on gut health in inflammatory bowel diseases. *Nutrients*, *15*(15), 3377. DOI: 10.3390/nu15153377 PMID: 37571313

Environmental Protection Agency. (2010). *Nanotechnology database fact sheet*. U.S. Environmental Protection Agency. https://clu-in.org/download/misc/NanoTechnology_FactSheet.pdf

Enyoh, C. E., Devi, A., Maduka, T. O., Tyagi, L., Rana, S., Akuwudike, I. S., & Wang, Q. (2025). A review of materials for the removal of micro- and nanoplastics from different environments. *Micro*, *5*(2), 17. DOI: 10.3390/micro5020017

Enyoh, C. E., Wang, Q., Chowdhury, T., Wang, W., Lu, S., Xiao, K., & Chowdhury, M. A. H. (2021). New analytical approaches for effective quantification and identification of nanoplastics in environmental samples. *Processes (Basel, Switzerland)*, *9*(11), 2086. Advance online publication. DOI: 10.3390/pr9112086

Erhirhie, E. O., Ihekwereme, C. P., & Ilodigwe, E. E. (2018). Advances in acute toxicity testing: Strengths, weaknesses and regulatory acceptance. *Interdisciplinary Toxicology*, *11*(1), 5–12. DOI: 10.2478/intox-2018-0001 PMID: 30181707

Esawy, M. A., & Reyed, R. M. (2026). Biointelligent reframing of colorectal cancer: From somatic mutation to systemic coherence. In Reyed, R., Prabhakar, P., & Haghi, A. (Eds.), *Advancing personalized medicine with AI-driven microbiome and nutrigenomics: Innovations and future perspectives* (pp. 173–218). IGI Global Scientific Publishing., DOI: 10.4018/979-8-3373-3121-8.ch003

Everaert, G., Van Cauwenberghe, L., De Rijcke, M., Koelmans, A. A., Mees, J., Vandegehuchte, M., & Janssen, C. R. (2018). Risk assessment of microplastics in the ocean. Environmental Pollution, 242, 1930 1938. https://doi.org/DOI: 10.1016/j.envpol.2018.07.069

Eyles, J. E., Bramwell, V. W., Williamson, E. D., & Alpar, H. O. (2001). Microsphere translocation and immunopotentiation in systemic tissues following intranasal administration. *Vaccine*, *19*(32), 4732–4742. DOI: 10.1016/S0264-410X(01)00220-1 PMID: 11535324

Facchin, S., Bertin, L., Bonazzi, E., Lorenzon, G., De Barba, C., Barberio, B., Zingone, F., Maniero, D., Scarpa, M., Ruffolo, C., Angriman, I., & Savarino, E. V. (2024). Short-Chain Fatty Acids and Human Health: From Metabolic Pathways to Current Therapeutic Implications. *Life (Basel, Switzerland)*, *14*(5), 559. DOI: 10.3390/life14050559 PMID: 38792581

Falsafi, A., Togiani, A., Colley, A., Varis, J., & Horttanainen, M. (2025). Life cycle assessment in circular design process: A systematic literature review. *Journal of Cleaner Production*, *521*, 146188. DOI: 10.1016/j.jclepro.2025.146188

Fang, C., Luo, Y., & Naidu, R. (2023). Microplastics and nanoplastics analysis: Options, imaging, advancements and challenges. *Trends in Analytical Chemistry*, *166*, 117158. DOI: 10.1016/j.trac.2023.117158

Fanghella, F., Pesce, M., Franceschelli, S., Panella, V., Elsallabi, O., Lupi, T., Rizza, B., Di Battista, M. G., Bruno, A., Ballerini, P., Patruno, A., & Speranza, L. (2025). Biological modulation of autophagy by nanoplastics: A current overview. *International Journal of Molecular Sciences*, *26*(15), 7035. DOI: 10.3390/ijms26157035 PMID: 40806168

Fan, W., Salmond, J. A., Dirks, K. N., Cabedo Sanz, P., Miskelly, G. M., & Rindelaub, J. D. (2022). Evidence and mass quantification of atmospheric microplastics in a coastal New Zealand city. *Environmental Science & Technology*, *56*(29), 17556–17568. DOI: 10.1021/acs.est.2c05850 PMID: 36459143

Ferrari, A. (2021). Regulatory aspects of nanotechnology in the EU. In S. S. Mudliar & K. K. Pant (Eds.), *Environmental nanotechnology volume 4: Applications of nanotechnology for energy and the environment* (pp. 185–210). Springer. https://doi.org/DOI: 10.1007/978-3-030-79808-6_8

Fishbein, S. R., Mahmud, B., & Dantas, G. (2023). Antibiotic perturbations to the gut microbiome. *Nature Reviews. Microbiology*, *21*(12), 772–788. DOI: 10.1038/s41579-023-00933-y PMID: 37491458

Francy, N., & Shanthakumar, S. (2015). Application of nanoscale zero-valent iron (nZVI) for contaminated soil remediation: A review. In *Proceedings of the 50th Indian Geotechnical Conference* (Vol. 19, pp. 550–558). Springer. https://doi.org/DOI: 10.1007/s11356-011-0576-3

Free, C. M., Jensen, O. P., Mason, S. A., Eriksen, M., Williamson, N. J., & Boldgiv, B. (2014). High levels of microplastic pollution in a large, remote, mountain lake. *Marine Pollution Bulletin*, *85*(1), 156–163. DOI: 10.1016/j.marpolbul.2014.06.001 PMID: 24973278

Frias, J. P. G. L., & Nash, R. (2019). Microplastics: Finding a consensus on the definition. Marine Pollution Bulletin, 138, 145 147 https://doi.org/DOI: 10.1016/j.marpolbul.2018.11.022

Furxhi, I., & Murphy, F. (2020). Predicting In Vitro Neurotoxicity Induced by Nanoparticles Using Machine Learning. *International Journal of Molecular Sciences*, *21*(15), 5280. DOI: 10.3390/ijms21155280 PMID: 32722414

Gagné, F. (2019). Detection of polystyrene nanoplastics in biological tissues with a fluorescent molecular rotor probe. *Journal of Xenobiotics*, *9*(2), 8147. DOI: 10.4081/xeno.2019.8147 PMID: 31178985

Galloway, T. S., Cole, M., & Lewis, C. (2017). Interactions of microplastic debris throughout the marine ecosystem. Nature Ecology & Evolution, 1, 0116. https://doi.org/DOI: 10.1038/s41559-017-0116

Galloway, T. (2015). *Micro- and nano-plastics and human health*. Springer., DOI: 10.1007/978-3-319-16510-3_13

Gambhir, A., Albert, M. J., Doe, S. S. P., Donges, J. F., Farajalla, N., Giatti, L. L., Gundimeda, H., Hendel-Blackford, S., Homer-Dixon, T., Hoyer, D., Adan, S., Jacome-Polit, D., Kemp, L., Korowicz, D., Kovacic, Z., Kwakkel, J., Laybourn, L., Lempert, R., Mahamoud, A., & Richardson, R. (2025). A systemic risk assessment methodological framework for the global polycrisis. *Nature Communications*, *16*(1), 7382. DOI: 10.1038/s41467-025-62029-w PMID: 40813573

Garcia Rodriguez, A., Rocabert, A., Hernández, A., & Marcos, R. (2026). Micronanoplastics induce dysbiosis of intestinal gut microbiota in living beings. In Chandrasekaran, N., Marcos, R., & Manzo, S. (Eds.), *Micronanoplastics: From Source to Sink* (pp. 87–118). Elsevier., DOI: 10.1016/B978-0-443-24782-8.00014-5

García-García, G., Fernandez, M. C., Armstrong, K., Woolass, S., & Styring, P. (2021). Analytical Review of Life-Cycle Environmental Impacts of Carbon Capture and Utilization Technologies. *ChemSusChem*, *14*(4), 995–1015. Advance online publication. DOI: 10.1002/cssc.202002126 PMID: 33314601

Garrigues, E., Corson, M. S., Angers, D. A., Van Der Werf, H. M. G., & Walter, C. (2012). Soil quality in Life Cycle Assessment: Towards development of an indicator. *Ecological Indicators*, *18*, 434–442. DOI: 10.1016/j.ecolind.2011.12.014

Garske, B., Holz, W., & Ekardt, F. (2024). Digital twins in sustainable transition: Exploring the role of EU data governance. *Frontiers in Research Metrics and Analytics*, *9*, 1303024. DOI: 10.3389/frma.2024.1303024 PMID: 38515644

Gasperi, J., Wright, S. L., Dris, R., Collard, F., Mandin, C., Guerrouache, M., Langlois, V., Kelly, F. J., & Tassin, B. (2018). Microplastics in air: Are we breathing it in? *Current Opinion in Environmental Science & Health*, *1*, 1–5. DOI: 10.1016/j.coesh.2017.10.002

Gawey, B. J., Mars, R. A., & Kashyap, P. C. (2025). The role of the gut microbiome in disorders of gut-brain interaction. *The FEBS Journal*, *292*(6), 1357–1377. DOI: 10.1111/febs.17200 PMID: 38922780

Geiser, M., Rothen-Rutishauser, B., Kapp, N., Schürch, S., Kreyling, W., Schulz, H., Semmler, M., Im Hof, V., Heyder, J., & Gehr, P. (2005). Ultrafine particles cross cellular membranes by nonphagocytic mechanisms in lungs and cultured cells. *Environmental Health Perspectives*, *113*(11), 1555–1560. DOI: 10.1289/ehp.8006 PMID: 16263511

Generalova, A., Davidova, S., & Satchanska, G. (2025). The mechanisms of lead toxicity in living organisms. *Journal of Xenobiotics*, *15*(5), 146. DOI: 10.3390/jox15050146 PMID: 40981357

Geueke, B., Groh, K. J., & Muncke, J. (2018). Food packaging in the circular economy: Overview of chemical safety aspects for commonly used materials. *Journal of Cleaner Production*, *193*, 491–505. DOI: 10.1016/j.jclepro.2018.05.005

Gewert, B., Plassmann, M. M., & MacLeod, M. (2015). Pathways for degradation of plastic polymers floating in the marine environment. *Environmental Science. Processes & Impacts*, *17*(9), 1513–1521. DOI: 10.1039/C5EM00207A PMID: 26216708

Geyer, R., Jambeck, J. R., & Law, K. L. (2017). Production, use, and fate of all plastics ever made. *Science Advances*, *3*(7), e1700782. DOI: 10.1126/sciadv.1700782 PMID: 28776036

Gigault, J., Halle, A. T., Baudrimont, M., Pascal, P. Y., Gauffre, F., Phi, T. L., El Hadri, H., Grassl, B., & Reynaud, S. (2018). Current opinion: What is a nanoplastic? Environmental Pollution, 235, 1030 1034. DOI: 10.1016/j.envpol.2018.01.024

Gigault, J., & Davranche, M. (2025). Nanoplastics in focus: Exploring interdisciplinary approaches and future directions. *NanoImpact*, *37*, 100544. DOI: 10.1016/j.impact.2025.100544 PMID: 39848489

Gil-Díaz, M., Diez-Pascual, S., Gonzalez, A., Alonso, J., Rodríguez-Valdes, E., Gallego, J. R., & Lobo, M. C. (2016). A nanoremediation strategy for the recovery of an As-polluted soil. *Chemosphere*, *149*, 137–145. DOI: 10.1016/j.chemosphere.2016.01.106 PMID: 26855217

Gillibert, R., Balakrishnan, G., Deshoules, Q., Tardivel, M., Magazzù, A., Donato, M. G., Maragò, O. M., Lamy de La Chapelle, M., Colas, F., Lagarde, F., & Gucciardi, P. G. (2019). Raman tweezers for small microplastics and nanoplastics identification in seawater. *Environmental Science & Technology*, *53*(16), 9003–9013. DOI: 10.1021/acs.est.9b03105 PMID: 31259538

Glüge, J., Scheringer, M., Cousins, I. T., DeWitt, J. C., Goldenman, G., Herzke, D., Lohmann, R., Ng, C. A., Trier, X., & Wang, Z. (2020). An overview of the uses of per- and polyfluoroalkyl substances (PFAS). *Environmental Science. Processes & Impacts*, *22*(12), 2345–2373. DOI: 10.1039/D0EM00291G PMID: 33125022

Gomes, H. I., Dias-Ferreira, C., & Ribeiro, A. B. (2022). Nanoscale zero-valent iron (nZVI) for remediation of contaminated soils and sediments: Field column study in Brazil. *Geosciences*, *12*(8), 287. DOI: 10.3390/geosciences12080287

Gonçalves, J., Pequeno, J., Krzisnik, D., Sobral, P., & Antunes, J. (2026). Nanoplastics: An Emerging Threat to Human Health—A Perspective Review. *Pollutants*, *6*(2), 21. DOI: 10.3390/pollutants6020021

Gouin, T., Ellis-Hutchings, R., Thornton Hampton, L. M., Lemieux, C. L., & Wright, S. L. (2022). Screening and prioritization of nano- and microplastic particle toxicity studies for evaluating human health risks - development and application of a toxicity study assessment tool. *Microplastics and Nanoplastics*, *2*(1), 2. DOI: 10.1186/s43591-021-00023-x PMID: 35098152

Gou, Z., Wu, H., Li, S., Liu, Z., & Zhang, Y. (2024). Airborne micro- and nanoplastics: Emerging causes of respiratory diseases. [REMOVED HYPERLINK FIELD]. *Particle and Fibre Toxicology*, *21*(1), 50. DOI: 10.1186/s12989-024-00613-6 PMID: 39633457

Grafmueller, S., Manser, P., Diener, L., Diener, P. A., Maeder-Althaus, X., Maurizi, L., Jochum, W., Krug, H. F., Buerki-Thurnherr, T., von Mandach, U., & Wick, P. (2015). Bidirectional transfer study of polystyrene nanoparticles across the placental barrier in an ex vivo human placental perfusion model. *Environmental Health Perspectives*, *123*(12), 1280–1286. DOI: 10.1289/ehp.1409271 PMID: 25956008

Gul, A., Gallus, I., Tegginamath, A., Maryska, J., & Yalcinkaya, F. (2021). Electrospun Antibacterial Nanomaterials for Wound Dressings Applications. *Membranes (Basel)*, *11*(12), 908. DOI: 10.3390/membranes11120908 PMID: 34940410

Gundogdu, S., Çevik, C., Guzel, E., & Kilercioğlu, S. (2018). Sources and fate of microplastics in marine environment: A review of recent studies. *Marine Pollution Bulletin*, *137*, 393–404. DOI: 10.1016/j.marpolbul.2018.10.045

Guo, X., & Wang, N. (2019). Neurotoxicity of microplastic particles. *Progress in Molecular Biology and Translational Science, 168*, 333–349. DOI: 10.1016/bs.pmbts.2019.11.011

Guo, Y., Tang, N., Lu, L., Li, N., Hu, T., Guo, J., Zhang, J., Zeng, Z., & Liang, J. (2024). Aggregation behavior of polystyrene nanoplastics: Role of surface functional groups and protein and electrolyte variation. *Chemosphere, 350*, 140998. DOI: 10.1016/j.chemosphere.2023.140998 PMID: 38142881

Gutow, L., Ehlers, S. M., Pittroff, M., Bartl, K., Wurm, Y., Galgani, F., & Saborowski, R. (2016). Microplastics in the marine environment: A review of the methods used for identification and quantification. *Environmental Science & Technology, 50*(24), 12327–12340. DOI: 10.1021/acs.est.6b04499

Habumugisha, T., Zhang, Z., Uwizewe, C., Yan, C., Ndayishimiye, J. C., Rehman, A., Zhang, X., & Zhang, X. (2024). Toxicological review of micro- and nanoplastics in aquatic environments: Risks to ecosystems, food web dynamics and human health. *Ecotoxicology and Environmental Safety, 278*, 116426. Advance online publication. DOI: 10.1016/j.ecoenv.2024.116426 PMID: 38718727

Halfon, N., & Hochstein, M. (2002). Life course health development: An integrated framework for developing health, policy, and research. *The Milbank Quarterly, 80*(3), 433–iii. DOI: 10.1111/1468-0009.00019 PMID: 12233246

Hammond, C. B., Qadikolaei, A. F., Aghaaminiha, M., Sharma, S., & Gigault, J. (2024). New insights into the formation of aggregates of bidisperse microplastics in water based on the analysis of in situ microscopy and molecular simulation. *Langmuir : The ACS Journal of Surfaces and Colloids, 40*(28), 14455–14466. DOI: 10.1021/acs.langmuir.4c01216 PMID: 38967440

Han, F., Wang, Y., An, S. Y., Liu, L., Ma, L. Q., Yang, L., & Liu, S. Y. (2024). Poly(dimethyldiallylammonium chloride)-modified magnetic bentonite for removal of polystyrene nanoplastics. *Journal of Environmental Chemical Engineering, 12*(5), 114013. Advance online publication. DOI: 10.1016/j.jece.2024.114013

Hanif, M. A., Ibrahim, N., Dahalan, F. A., Md Ali, U. F., Hasan, M., & Abdul Jalil, A. (2022). Microplastics and nanoplastics: Recent literature studies and patents on their removal from aqueous environment. *The Science of the Total Environment, 810*, 152115. DOI: 10.1016/j.scitotenv.2021.152115 PMID: 34896138

Hanif, M. A., Ibrahim, N., Hayazi, N. A., Dahalan, F. A., Ali, U. F. M., Jalil, A. A., & Syafiuddin, A. (2024). Enhancement of microplastics and nanoplastics removal via filtration method using surface-engineered palm kernel shell biochar. *Separation and Purification Technology, 360*, 130596. DOI: 10.1016/j.seppur.2024.130596

Hartline, N. L., Bruce, N. J., Karba, S. N., Ruff, E. O., Sonar, S. U., & Holden, P. A. (2016). Microfiber Masses Recovered from Conventional Machine Washing of New or Aged Garments. *Environmental Science & Technology*, *50*(21), 11532–11538. DOI: 10.1021/acs.est.6b03045 PMID: 27689236

Hartmann, N. B., Hüffer, T., Thompson, R. C., Hassellöv, M., Verschoor, A., Daugaard, A. E., Rist, S., Karlsson, T., Brennholt, N., Cole, M., Herrling, M. P., Hess, M. C., Ivleva, N. P., Lusher, A. L., & Wagner, M. (2019). Are we speaking the same language? Recommendations for a definition and categorization framework for plastic debris. *Environmental Science & Technology*, *53*(3), 1039–1047. DOI: 10.1021/acs.est.8b05297 PMID: 30608663

Hayat, R., Khan, A., Muhammad Awais, , Muhammad Farhan, , Ahmad, U., Abdul Manan, , Abdul Basit, , Yasin Asad, M. T., Shah Nawaz, , & Hafeez, U. (2026). The Silent Sink: A Comprehensive Review of Microplastic Accumulation in Agricultural Soils and its Impact on Crop Physiology. *Indus Journal of Bioscience Research*, *4*(1), 90–102. DOI: 10.70749/ijbr.v4i1.2838

Hays, K. E., Pfaffinger, J. M., & Ryznar, R. (2024). The interplay between gut microbiota, short-chain fatty acids, and implications for host health and disease. *Gut Microbes*, *16*(1), 2393270. DOI: 10.1080/19490976.2024.2393270 PMID: 39284033

He, F., Zhao, D., Liu, J., & Roberts, C. (2010). Stabilization of Fe–Pd nanoparticles with sodium carboxymethyl cellulose for enhanced dechlorination of chlorinated hydrocarbons in soil and groundwater. *Industrial & Engineering Chemistry Research*, *49*(22), 11284–11292. DOI: 10.1021/ie101258z

Heijungs, R., & Wiloso, E. I. (2014). Life cycle assessment of bioenergy systems. In Wang, L. (Ed.), *Sustainable bioenergy production* (pp. 234–259). CRC Press.

Helal, M., Hartmann, N. B., Khan, F. R., & Xu, E. G. (2023). Time to integrate "One Health Approach" into nanoplastic research. *Eco-Environment & Health*, *2*(1), 18–20. DOI: 10.1016/j.eehl.2023.01.001 PMID: 38074454

Hemila, H., & Chalker, E. (2013). Vitamin C for preventing and treating the common cold. *Cochrane Database of Systematic Reviews*, *2013*(1), CD000980. Advance online publication. DOI: 10.1002/14651858.CD000980.pub4 PMID: 23440782

Hermabessiere, L., Dehaut, A., Paul-Pont, I., Lacroix, C., Jezequel, R., Soudant, P., & Duflos, G. (2017). Occurrence and effects of plastic additives on marine environments and organisms: A review. *Chemosphere*, *182*, 781–793. DOI: 10.1016/j.chemosphere.2017.05.096 PMID: 28545000

Hernandez, L. M., Xu, E. G., Larsson, H. C. E., Tahara, R., Maisuria, V. B., & Tufenkji, N. (2019). Plastic teabags release billions of microparticles and nanoparticles into tea. Environmental Science & Technology, 53(21), 12300 12310. DOI: 10.1021/acs.est.9b02540

Hernandez, E., Nowack, B., & Mitrano, D. M. (2017). Polyester textiles as a source of microplastics from households: A mechanistic study to understand microfiber release during washing. *Environmental Science & Technology*, *51*(12), 7036–7046. DOI: 10.1021/acs.est.7b01750 PMID: 28537711

Herweyers, L., Catarci Carteny, C., Scheelen, L., Watts, R., & Du Bois, E. (2020). Consumers' Perceptions and Attitudes toward Products Preventing Microfiber Pollution in Aquatic Environments as a Result of the Domestic Washing of Synthetic Clothes. *Sustainability (Basel)*, *12*(6), 2244. DOI: 10.3390/su12062244

He, Y., Li, Q., Sun, Q., Li, H., Yu, T., Chen, M., Zhang, G., Zhang, B., Wang, W., & Ju, S. (2025). Gut microbiota contributes to polystyrene nanoplastics-induced fetal growth restriction by disturbing placental nicotinamide metabolism. *Journal of Nanobiotechnology*, *23*(1), 561. DOI: 10.1186/s12951-025-03650-1 PMID: 40796882

Higashisaka, K., Yoshioka, Y., Yamashita, K., Morishita, Y., Pan, H., Ogura, T., Nagano, T., Kunieda, A., Nagano, K., Abe, Y., Kamada, H., Tsunoda, S., Nabeshi, H., Yoshikawa, T., & Tsutsumi, Y. (2012). Hemopexin as biomarkers for analyzing the biological responses associated with exposure to silica nanoparticles. *Nanoscale Research Letters*, *7*(1), 555. DOI: 10.1186/1556-276X-7-555 PMID: 23039107

Ho, K. L., Yeap, S. P., & Lee, K. M. (2025). A critical review of microplastics and nanoplastics in wastewater: Insights into adsorbent-based remediation strategies. *Environmental Pollution*, *382*, 126658. DOI: 10.1016/j.envpol.2025.126658 PMID: 40513794

Ho, K. T. H., Kwok, P. W. H., Chang, S. S. Y., & Chu, A. M. Y. (2023). Gaps between attitudes and behavior in the use of disposable plastic tableware (DPT) and factors influencing sustainable DPT consumption: A study of Hong Kong undergraduates. *Sustainability (Basel)*, *15*(11), 8958. DOI: 10.3390/su15118958

Hossain, M., & Poon, C. S. (2018). Comparative LCA of wood waste management strategies generated from building construction activities. *Journal of Cleaner Production*, *177*, 387–397. DOI: 10.1016/j.jclepro.2017.12.233

Hossain, R., & Sahajwalla, V. (2026). A comprehensive toolkit for micro- to nanoplastic analysis. *Environmental Science. Nano*, *13*(1), 122–149. Advance online publication. DOI: 10.1039/D5EN00856E

Hosseinnejad, M., & Jafari, S. M. (2016). Evaluation of different factors affecting antimicrobial properties of chitosan. *International Journal of Biological Macromolecules*, *85*, 467–475. DOI: 10.1016/j.ijbiomac.2016.01.022 PMID: 26780706

Hou, Z., Meng, R., Chen, G., Lai, T., Qing, R., Hao, S., Deng, J., & Wang, B. (2022). Distinct accumulation of nanoplastics in human intestinal organoids. *The Science of the Total Environment*, *838*, 155811. DOI: 10.1016/j.scitotenv.2022.155811 PMID: 35597345

Huang, J., Tan, X., Ali, I., Ok, Y. S., Duan, Z., Liang, J., & Zhu, R. (2024). Efficient removal of nanoplastics by iron-modified biochar: Understanding the removal mechanisms. *Environmental Pollution*, *363*, 125121. DOI: 10.1016/j.envpol.2024.125121 PMID: 39426478

Huang, M., Gong, Y., Grondolsky, J., & Hoover-Plow, J. (2014). Lp(a)/apo(a) modulate MMP-9 activation and neutrophil cytokines in vivo in inflammation to regulate leukocyte recruitment. *The American Journal of Pathology*, *184*(5), 1503–1517. DOI: 10.1016/j.ajpath.2014.01.010 PMID: 24650562

HuBMAP Consortium. (2023). The Human BioMolecular Atlas Program. *Nature*, *622*(7985), 567–578.

Hüffer, T., Hofmann, T., & Asbach, C. (2017). The great unknown—Nanoplastics in personal care products. *Environmental Science. Nano*, *4*(4), 730–742. DOI: 10.1039/C7EN00471A

Huppertsberg, S., & Knepper, T. P. (2018). Instrumental analysis of microplastics Benefits and challenges. Analytical and Bioanalytical Chemistry, 410(25), 6343 6352. https://doi.org/DOI: 10.1007/s00216-018-1210-8

Iqbal, B., Khan, A. A., Hu, J., Liu, Q., Wang, C., Li, G., & Ye, M. (2026). Soil microplastics hidden web: Interaction of microbes and viruses as a frontier for sustainable ecosystem recovery. *Agricultural Ecology and Environment*, *2*(1), e006. DOI: 10.48130/aee-0026-0003

Isibor, P. O., Kayode-Edwards, I. I., & Taiwo, O. S. (2024). Emerging technology and future directions in environmental nanotoxicology Enuneku (Eds.), Environmental nanotoxicology. Springer. https://doi.org/DOI: 10.1007/978-3-031-54154-4_16

Isigonis, P., Hristozov, D., Benighaus, C., Giubilato, E., Grieger, K., Pizzol, L., Semenzin, E., Linkov, I., Zabeo, A., & Marcomini, A. (2019). Risk Governance of Nanomaterials: Review of Criteria and Tools for Risk Communication, Evaluation, and Mitigation. [PDF]

Jaafarzadeh, N., & Talepour, N. (2023). Microplastics as carriers of antibiotic resistance genes and pathogens in municipal solid waste (MSW) landfill leachate and soil: A review. *Journal of Environmental Health Science & Engineering*, *22*(1), 1–12. DOI: 10.1007/s40201-023-00879-6 PMID: 38887766

Jacquemin, L., Pontalier, P.-Y., & Sablayrolles, C. (2012). Life cycle assessment (LCA) applied to the process industry: A review. *The International Journal of Life Cycle Assessment*, *17*(8), 1028–1041. DOI: 10.1007/s11367-012-0432-9

Jahedi, F., Frad, N. J. H., Khaksar, M. A., Rashidi, P., Safdari, F., & Mansouri, Z. (2025). Nano and microplastics: Unveiling their profound impact on endocrine health. *Toxicology Mechanisms and Methods*, *35*(8), 1–29. DOI: 10.1080/15376516.2025.2509745 PMID: 40432394

Jahedi, F., & Jaafarzadeh Haghighi Fard, N. (2025). Micro- and nanoplastic toxicity in humans: Exposure pathways, cellular effects, and mitigation strategies. *Toxicology Reports*, *14*, 102043. DOI: 10.1016/j.toxrep.2025.102043 PMID: 40487378

Jalón-Rojas, I., Figueroa, D., & Rivas, J. (2022). Effects of biofilms and particle physical properties on the vertical dynamics of microplastic sheets. *Environmental Science & Technology*, *56*(17), 12195–12204. DOI: 10.1021/acs.est.2c01302 PMID: 35593651

Jambeck, J. R., Geyer, R., Wilcox, C., Siegler, T. R., Perryman, M., Andrady, A., Narayan, R., & Law, K. L. (2015). Marine pollution. Plastic waste inputs from land into the ocean. *Science*, *347*(6223), 768–771. DOI: 10.1126/science.1260352 PMID: 25678662

Jani, V., & Wu, S. (2025). Nanoplastics (NPs): Environmental presence, ecological implications, and mitigation approaches *Microplastics, 4*(3), 48. https://doi.org/DOI: 10.3390/microplastics4030048

Jansen, M. A. K., Andrady, A. L., Bornman, J. F., Aucamp, P. J., Bais, A. F., Banaszak, A. T., Barnes, P. W., Bernhard, G. H., Bruckman, L. S., Busquets, R., Häder, D. P., Hanson, M. L., Heikkilä, A. M., Hylander, S., Lucas, R. M., Mackenzie, R., Madronich, S., Neale, P. J., Neale, R. E., & Zhu, L. (2024). Plastics in the environment in the context of UV radiation, climate change and the Montreal Protocol: UNEP Environmental Effects Assessment Panel, Update 2023. *Photochemical & Photobiological Sciences : Official Journal of the European Photochemistry Association and the European Society for Photobiology*, *23*(4), 629–650. DOI: 10.1007/s43630-024-00552-3 PMID: 38512633

Jarmakiewicz-Czaja, S., Ferenc, K., Sokal-Dembowska, A., Wojdyła, D., & Filip, R. (2026). Effects of microplastics on the intestines. *BMC Gastroenterology*, *26*(1), 289. DOI: 10.1186/s12876-026-04749-x PMID: 41928110

Jee, D., Kang, S., Yuan, C., Cho, E., & Arroyo, J. G. (2016). Serum 25-Hydroxyvitamin D Levels and Dry Eye Syndrome: Differential Effects of Vitamin D on Ocular Diseases. *PLoS One*, *11*(2), e0149294. DOI: 10.1371/journal.pone.0149294 PMID: 26894581

Jeong, J., Im, J., & Choi, J. (2024). Integrating aggregate exposure pathway and adverse outcome pathway for micro/nanoplastics: A review on exposure, toxicokinetics, and toxicity studies. *Ecotoxicology and Environmental Safety*, *272*, 116022. DOI: 10.1016/j.ecoenv.2024.116022 PMID: 38309230

Jiang, F., Gao, C., Chan, A. W. H., Topping, D. O., Zhang, H., Li, W., Coe, H., Liu, X., Ching, J., & Zheng, Z. (2025). A review of atmospheric micro/nanoplastics: Insights into source and fate for modeling studies. *Current Pollution Reports*, *11*(53), 53. Advance online publication. DOI: 10.1007/s40726-025-00375-5

Jian, K., Li, Z., Xiaoguang, D., Hongqi, S., & Zhimin, A. (2019). Degradation of cosmetic microplastics via functionalized carbon nanosprings. *Matter*, *1*(3), 745–758. DOI: 10.1016/j.matt.2019.06.004

Jiao, H., Ali, S. S., Alsharbaty, M. H. M., Elsamahy, T., Abdelkarim, E., Schagerl, M., Al-Tohamy, R., & Sun, J. (2024). A critical review on plastic waste life cycle assessment and management: Challenges, research gaps, and future perspectives. *Ecotoxicology and Environmental Safety*, *271*, 115942. DOI: 10.1016/j.ecoenv.2024.115942 PMID: 38218104

Jiménez-Arroyo, C., Tamargo, A., Molinero, N., & Moreno-Arribas, M. V. (2023). The gut microbiota, a key to understanding the health implications of micro (nano) plastics and their biodegradation. *Microbial Biotechnology*, *16*(1), 34–53. DOI: 10.1111/1751-7915.14182 PMID: 36415969

Jin, Y., Lu, L., Tu, W., Luo, T., & Fu, Z. (2019). Impacts of polystyrene microplastic on the gut barrier, microbiota and metabolism of mice. *The Science of the Total Environment*, *649*, 308–317. DOI: 10.1016/j.scitotenv.2018.08.353 PMID: 30176444

Joshi, N., Filip, J., Coker, V., Sadhukhan, J., Safarik, I., Bagshaw, H., & Lloyd, J. R. (2018). Microbial reduction of natural Fe(III) minerals: Towards the sustainable production of functional magnetic nanoparticles. *Frontiers in Environmental Science*, *6*, 127. DOI: 10.3389/fenvs.2018.00127

Junaid, M., & Wang, J. (2021). Interaction of nanoplastics with extracellular polymeric substances (EPS) in the aquatic environment: A special reference to eco-corona formation and associated impacts. *Water Research*, *201*, 117319. DOI: 10.1016/j.watres.2021.117319 PMID: 34130084

Kamath, S., Sokolenko, E., Collins, K., Chan, N. S. L., Mills, N., Clark, S. R., Marques, F. Z., & Joyce, P. (2025). IUPHAR themed review: The gut microbiome in schizophrenia. *Pharmacological Research*, *211*, 107561. DOI: 10.1016/j.phrs.2024.107561 PMID: 39732352

Kandaswamy, K., Ajay, G., Panda, S. P., Anahas, A. M. P., Kari, Z. A., Giri, J. P., Almutairi, B. O., Arokiyaraj, S., Malafaia, G., & Jesu Arockiaraj, J. (2024). Polystyrene nanoplastics synergistically exacerbate diclofenac toxicity in embryonic development and the health of adult zebrafish. *Comparative Biochemistry and Physiology. Toxicology & Pharmacology : CBP*, *281*, 109926. Advance online publication. DOI: 10.1016/j.cbpc.2024.109926 PMID: 38641085

Kangalawe, R. Y. (2012). Food security and health in the southern highlands of Tanzania: A multidisciplinary approach to evaluate the impact of climate change and other stress factors. *African Journal of Environmental Science and Technology*, *6*(1), 50–66. DOI: 10.5897/AJEST11.003

Kang, Y., Gao, S. H., Pan, Y., Gao, R., Li, T., Fan, L., Su, Y., Zhang, W., Yu, Z., Liang, B., Su, J. Q., Luo, Y., Wang, Y., Guo, J., & Wang, A. (2025). Roles of micro/nanoplastics in the spread of antimicrobial resistance through conjugative gene transfer. *Nature Communications*, *17*(1), 1118. DOI: 10.1038/s41467-025-67879-y PMID: 41444368

Kara Murdoch, F., Sun, Y., Fuller, M. E., Mullins, L., Hill, A., Lilly, J., Wilson, J., Löffler, F. E., & Kucharzyk, K. H. (2025). Presence of Micro- and Nanoplastics Affects Degradation of Chlorinated Solvents. *Toxics*, *13*(8), 656. DOI: 10.3390/toxics13080656 PMID: 40863933

Karami, A., Golieskardi, A., Keong Choo, C., Larat, V., Galloway, T. S., & Salamatinia, B. (2017). The presence of microplastics in commercial salts from different countries. *Scientific Reports*, *7*(1), 46173. DOI: 10.1038/srep46173 PMID: 28383020

Karim, A., Yadav, A., Sweety, U. H., Kumar, J., Delgado, S. A., Hernandez, J. A., White, J. C., Vukovic, L., & Narayan, M. (2024). Interfacial interactions between nanoplastics and biological systems: Toward an atomic and molecular understanding of plastics-driven biological dyshomeostasis. *ACS Applied Materials & Interfaces*, *16*(20), 25740–25756. DOI: 10.1021/acsami.4c03008 PMID: 38722759

Karn, B., Kuiken, T., & Otto, M. (2009). Nanotechnology and in situ remediation: A review of the benefits and potential risks. *Environmental Health Perspectives*, *117*(12), 1813–1831. DOI: 10.1289/ehp.0900793 PMID: 20049198

Kateris, D., & Bochtis, D. (2023). Applied Agri-Technologies for Agriculture 4.0—Part I. *Applied Sciences (Basel, Switzerland)*, *13*(7), 4180. DOI: 10.3390/app13074180

Katsoulakis, E., Wang, Q., Wu, H., Shahriyari, L., Fletcher, R., Liu, J., Achenie, L., Liu, H., Jackson, P., Xiao, Y., Syeda-Mahmood, T., Tuli, R., & Deng, J. (2024). Digital twins for health: A scoping review. *NPJ Digital Medicine*, *7*(1), 77. DOI: 10.1038/s41746-024-01073-0 PMID: 38519626

Kazmi, S. S. U. H., Batool, S. M., Pastorino, P., Barcelò, D., Grossart, H.-P., Yaseen, Z. M., Khan, Z. H., Azeem, M., & Li, G. (2026). The plastisphere as a nexus for antimicrobial resistance: Micro(nano)plastics in pathogen colonization, gene transfer, and global health risks. *Biological Reviews of the Cambridge Philosophical Society*, brv.70163. Advance online publication. DOI: 10.1002/brv.70163 PMID: 41873837

Khanam, M. M., Uddin, M. K., & Kazi, J. U. (2025). Advances in machine learning for the detection and characterization of microplastics in the environment. *Frontiers in Environmental Science*, *13*, 1573579. DOI: 10.3389/fenvs.2025.1573579

Khan, D., & Ali, S. A. (2023). On the Novel Process of Pristine Microplastic Biofragmentation by Zebrafish (*Danio rerio*). *Archives of Environmental Contamination and Toxicology*, *84*(3), 299–306. DOI: 10.1007/s00244-023-00987-2 PMID: 36929014

Kibria, G., Nugegoda, D., & Haroon, A. Y. (2022). Microplastic pollution and contamination of seafood (including fish, sharks, mussels, oysters, shrimps and seaweeds): a global overview. *Microplastic pollution: environmental occurrence and treatment technologies*, 277-322. https://doi.org/DOI: 10.1007/978-3-030-89220-3_14

Kibria, M. G., Masuk, N. I., Safayet, R., Nguyen, H. Q., & Mourshed, M. (2023). Plastic waste: Challenges and opportunities to mitigate pollution and effective management. *International Journal of Environmental Research*, *17*(1), 20. DOI: 10.1007/s41742-023-00507-z PMID: 36711426

Kim, J. S., Lee, H. J., Kim, S. K., & Kim, H. J. (2018). Global Pattern of Microplastics (MPs) in Commercial Food-Grade Salts: Sea Salt as an Indicator of Seawater MP Pollution. *Environmental Science & Technology*, *52*(21), 12819–12828. DOI: 10.1021/acs.est.8b04180 PMID: 30285421

Kitamura, S. (2005). Comparative study of the endocrine-disrupting activity of bisphenol A and 19 related compounds. *Toxicological Sciences : An Official Journal of the Society of Toxicology, 84*(2), 249–259. DOI: 10.1093/toxsci/kfi074 PMID: 15635150

Klein, M., Fischer, E. K., & Neophytou, C. (2021). Nanoplastics: A growing threat to marine organisms. *Frontiers in Marine Science, 8*, 627888. DOI: 10.3389/fmars.2021.627888

Klinglmair, M., Sala, S., & Brandão, M. (2014). Assessing resource depletion in LCA: A review of methods and methodological issues. *The International Journal of Life Cycle Assessment, 19*(3), 580–592. DOI: 10.1007/s11367-013-0650-9

Kochanek, A., Grąz, K., Kochanek, A., Grąz, K., Potok, H., Gronba-Chyła, A., Kwaśny, J., Wiewiórska, I., Ciuła, J., Basta, E., & Łapiński, J. (2025). Micro- and Nanoplastics in the Environment: Current State of Research, Sources of Origin, Health Risks, and Regulations—A Comprehensive Review. *Toxics, 13*(7), 564. DOI: 10.3390/toxics13070564 PMID: 40711009

Koelmans, A. A., Mohamed Nor, N. H., Hermsen, E., Kooi, M., Mintenig, S. M., & De France, J. (2019). Microplastics in freshwaters and drinking water: Critical review and assessment of data quality. *Water Research, 155*, 410–422. DOI: 10.1016/j.watres.2019.02.054 PMID: 30861380

Koelmans, A. A., Redondo-Hasselerharm, P. E., Nor, N. H. M., de Ruijter, V. N., Mintenig, S. M., & Kooi, M. (2022). Risk assessment of microplastic particles. *Nature Reviews. Materials, 7*(2), 138–152. DOI: 10.1038/s41578-021-00411-y

Kögel, T., Bjorøy, Ø., Toto, B., Bienfait, A. M., & Sanden, M. (2020). Micro- and nanoplastic toxicity on aquatic life: Determining factors. *The Science of the Total Environment, 709*, 136050. DOI: 10.1016/j.scitotenv.2019.136050 PMID: 31887526

Kole, P. J., Lahr, J., Van Belleghem, F. G. A. J., & Ragas, A. M. J. (2019, May 26–30). Wear and tear of tyres in the global environment: Size distribution, emission, pathways and health effects. In *Proceedings of the SETAC Europe 29th Annual Meeting*. Helsinki, Finland.

Kopczyńska, J., & Kowalczyk, M. (2024). The potential of short-chain fatty acid epigenetic regulation in chronic low-grade inflammation and obesity. *Frontiers in Immunology, 15*, 1380476. DOI: 10.3389/fimmu.2024.1380476 PMID: 38605957

Korenhof, P., Metcalfe, B., & Henschke, A. (2025). Cyber-governance of the natural world: The implications of digital twins in environmental governance. *Earth System Governance, 25*, 100276. DOI: 10.1016/j.esg.2025.100276

Kosuth, M., Mason, S. A., & Wattenberg, E. V. (2018). Anthropogenic contamination of tap water, beer, and sea salt. *PLoS One*, *13*(4), e0194970. DOI: 10.1371/journal.pone.0194970 PMID: 29641556

Krause, S., Ouellet, V., Allen, D., Allen, S., Moss, K., Nel, H. A., Manaseki-Holland, S., & Lynch, I. (2024). The potential of micro-and nanoplastics to exacerbate the health impacts and global burden of non-communicable diseases. *Cell Reports Medicine*, *5*(6), 101581. Advance online publication. DOI: 10.1016/j.xcrm.2024.101581 PMID: 38781963

Krewski, D., Acosta, D.Jr, Andersen, M., Anderson, H., Bailar, J. C.III, Boekelheide, K., Brent, R., Charnley, G., Cheung, V. G., Green, S.Jr, Kelsey, K. T., Kerkvliet, N. I., Li, A. A., McCray, L., Meyer, O., Patterson, R. D., Pennie, W., Scala, R. A., Solomon, G. M., & Zeise, L.Staff of Committee on Toxicity Testing and Assessment of Env. (2010). Toxicity testing in the 21st century: A vision and a strategy. *Journal of Toxicology and Environmental Health. Part B, Critical Reviews*, *13*(2–4), 51–138. DOI: 10.1080/10937404.2010.483176 PMID: 20574894

Kumar, A., Bakshi, B. R., Ramteke, M., & Kodamana, H. (2023). Recycle-BERT: Extracting knowledge about plastic waste recycling by natural language processing. *ACS Sustainable Chemistry & Engineering*, *11*(32), 12123–12134. https://pubs.acs.org/doi/10.1021/acssuschemeng.3c03162?goto=supporting-info. DOI: 10.1021/acssuschemeng.3c03162

Kumar, A., Nehra, M., Singhal, N. K., & Kumar, S. (2026). Recent progress and technological advancements for detection of micro/nanoplastics in the environment. *Advances in Colloid and Interface Science*, *351*, 103817. DOI: 10.1016/j.cis.2026.103817 PMID: 41678884

Kumar, C., Singh, H., Kaur, M., Arya, S. K., & Khatri, M. (2025). Remediation strategies for micro/nanoplastic pollution using magnetic nanomaterials. *Environmental Science and Pollution Research International*, *32*(26), 15574–15603. DOI: 10.1007/s11356-025-36663-2 PMID: 40562997

Kumar, R., Verma, A., Shome, A., Sinha, R., Sinha, S., Jha, P. K., Kumar, R., Kumar, P., Shubham, , Das, S., Sharma, P., & Vara Prasad, P. V. (2021). Impacts of plastic pollution on ecosystem services, sustainable development goals, and need to focus on circular economy and policy interventions. *Sustainability (Basel)*, *13*(17), 9963. DOI: 10.3390/su13179963

Lackner, M., & Branka, M. (2024). Microplastics in farmed animals—A review. *Microplastics*, *3*(4), 35. Advance online publication. DOI: 10.3390/microplastics3040035

Lambert, S., & Wagner, M. (2016). Characterisation of nanoplastics during the degradation of polystyrene. *Chemosphere, 145*, 265–268. DOI: 10.1016/j.chemosphere.2015.11.078 PMID: 26688263

Lamichhane, G., Acharya, A., Marahatha, R., Modi, B., Paudel, R., Adhikari, A., Raut, B. K., Aryal, S., & Parajuli, N. (2022). Microplastics in environment: Global concern, challenges, and controlling measures. *International Journal of Environmental Science and Technology*. Advance online publication. DOI: 10.1007/s13762-022-04261-1 PMID: 35638092

Lamparelli, E. P., Marino, M., Szychlinska, M. A., Della Rocca, N., Ciardulli, M. C., Scala, P., D'Auria, R., Testa, A., Viggiano, A., Cappello, F., Meccariello, R., Della Porta, G., & Santoro, A. (2023). The other side of plastics: Bioplastic-based nanoparticles for drug delivery systems in the brain. *Pharmaceutics, 15*(11), 2549. DOI: 10.3390/pharmaceutics15112549 PMID: 38004530

Landrigan, P. J., Raps, H., Cropper, M., Bald, C., Brunner, M., Canonizado, E. M., Charles, D., Chiles, T. C., Donohue, M. J., Enck, J., Fenichel, P., Fleming, L. E., Ferrier-Pages, C., Fordham, R., Gozt, A., Griffin, C., Hahn, M. E., Haryanto, B., Hixson, R., & Dunlop, S. (2023). The Minderoo-Monaco Commission on Plastics and Human Health. *Annals of Global Health, 89*(1), 23. DOI: 10.5334/aogh.4056 PMID: 36969097

Landrigan, P. J., Stegeman, J. J., Fleming, L. E., Allemand, D., Anderson, D. M., Backer, L. C., Brucker-Davis, F., Chevalier, N., Corra, L., Czerucka, D., Bottein, M.-Y. D., Demeneix, B., Depledge, M., Deheyn, D. D., Dorman, C. J., Fénichel, P., Fisher, S., Gaill, F., Galgani, F., & Rampal, P. (2020). Human health and ocean pollution. *Annals of Global Health, 86*(1), 151. DOI: 10.5334/aogh.2831 PMID: 33354517

Leads, R. R., & Weinstein, J. E. (2019). Occurrence of tire wear particles and other microplastics within the tributaries of the Charleston Harbor Estuary, South Carolina, USA. *Marine Pollution Bulletin, 145*, 569–582. DOI: 10.1016/j.marpolbul.2019.06.061 PMID: 31590826

Lee, J., Jang, H., Kim, B., & Jung, J. (2025). Comparative Metabolomic Approaches to Nanoplastic Toxicity in Mammalian and Aquatic Systems. *International Journal of Molecular Sciences, 27*(1), 50. DOI: 10.3390/ijms27010050 PMID: 41515930

Lee, S. H., & Kim, Y. (2026). Amine-functionalized copper mesh for the efficient removal and colorimetric quantification of polypropylene nanoplastics. *Journal of Environmental Chemical Engineering, 14*(1), 120610. DOI: 10.1016/j.jece.2025.120610

Lee, W. S., Cho, H. J., Kim, E., Huh, Y. H., Kim, H. J., Kim, B., Kang, T., Lee, J. S., & Jeong, J. (2019). Bioaccumulation of polystyrene nanoplastics and their effect on the toxicity of Au ions in zebrafish embryos. *Nanoscale, 11*(7), 3173–3185. DOI: 10.1039/C8NR09321K PMID: 30534785

Lee, Y. H., Zheng, C. M., Wang, Y. J., Wang, Y. L., & Chiu, H. W. (2025). Effects of microplastics and nanoplastics on the kidney and cardiovascular system. *Nature Reviews. Nephrology, 21*(9), 585–596. DOI: 10.1038/s41581-025-00971-0 PMID: 40537559

Lehner, R., Petri-Fink, A., & Rothen-Rutishauser, B. (2017). Nanoplastic impact on human health—A 3D intestinal model to study the interaction with nanoplastic particles. In Deidun, A. C. (Ed.), *Proceedings of the International Conference on Microplastic Pollution in the Mediterranean Sea* (pp. 219–221). Springer. https://doi.org/DOI: 10.1007/978-3-319-71279-6_33

Lehner, R., Weder, C., Petri-Fink, A., & Rothen-Rutishauser, B. (2019). Emergence of nanoplastic in the environment and possible impact on human health. *Environmental Science & Technology, 53*(4), 1748–1765. DOI: 10.1021/acs.est.8b05512 PMID: 30629421

Lehner, R., Wohlleben, W., Septiadi, D., Landsiedel, R., Petri-Fink, A., & Rothen-Rutishauser, B. (2020). A novel 3D intestine barrier model to study the immune response upon exposure to microplastics. *Archives of Toxicology, 94*(7), 2463–2479. DOI: 10.1007/s00204-020-02750-1 PMID: 32307674

Lenz, R., Enders, K., & Nielsen, T. G. (2016). Microplastic exposure studies should be environmentally realistic. *Proceedings of the National Academy of Sciences of the United States of America, 113*(29), E4121–E4122. DOI: 10.1073/pnas.1606615113 PMID: 27407153

Lerchner, T., Jedlička, J., Kripnerová, M., Dejmek, J., & Kuncová, J. (2024). Influence of micro- and nanoplastics on mitochondrial function in the cardiovascular system: A review of the current literature. *Physiological Research, 73*(S3), S685–S695. DOI: 10.33549/physiolres.935500 PMID: 39808171

Leslie, H. A., van Velzen, M. J. M., Brandsma, S. H., Vethaak, A. D., Garcia-Vallejo, J. J., & Lamoree, M. H. (2022). Discovery and quantification of plastic particle pollution in human blood. *Environment International, 163*, 107199. DOI: 10.1016/j.envint.2022.107199 PMID: 35367073

Leso, V., Battistini, B., Vetrani, I., Reppuccia, L., Fedele, M., Ruggieri, F., Bocca, B., & Iavicoli, I. (2023). The endocrine disrupting effects of nanoplastic exposure: A systematic review. *Toxicology and Industrial Health*, *39*(11), 613–629. DOI: 10.1177/07482337231203053 PMID: 37753827

Leso, V., Nowack, B., Karakoltzidis, A., Nikiforou, F., Karakitsios, S., Sarigiannis, D., & Iavicoli, I. (2025). Next generation risk assessment and new approach methodologies for safe and sustainable by design chemicals and materials: Perspectives and challenges for occupational health. *Toxicology*, *517*, 154211. DOI: 10.1016/j.tox.2025.154211 PMID: 40490209

Le, V.-G., Nguyen, M.-K., Nguyen, H.-L., Lin, C., Hadi, M., Hung, N. T. Q., Hoang, H.-G., Nguyen, K. N., Tran, H.-T., Hou, D., Zhang, T., & Bolan, N. S. (2023). A comprehensive review of micro-and nano-plastics in the atmosphere: Occurrence, fate, toxicity, and strategies for risk reduction. *The Science of the Total Environment*, *904*, 166649. DOI: 10.1016/j.scitotenv.2023.166649 PMID: 37660815

Liang, X., Andrikopoulos, N., Tang, H., Wang, Y., Ding, F., & Ke, P. C. (2024). Nanoplastic stimulates the amyloidogenesis of Parkinson's alpha-synuclein NACore. *Small*, *20*(14), 2308753. DOI: 10.1002/smll.202308753 PMID: 37988678

Li, C., & Liu, T. (2023). Circular economy strategies for nanomaterial sustainability. *Environmental Science & Policy*, *147*, 52–63.

Liebezeit, G., & Liebezeit, E. (2013). Non-pollen particulates in honey and sugar. *Food Additives & Contaminants. Part A, Chemistry, Analysis, Control, Exposure & Risk Assessment*, *30*(12), 2136–2140. DOI: 10.1080/19440049.2013.843025 PMID: 24160778

Li, H., Wang, J., Hao, L., & Huang, G. (2024). Exploring the interconnection between metabolic dysfunction and gut microbiome dysbiosis in osteoarthritis: A narrative review. *Biomedicines*, *12*(10), 2182. DOI: 10.3390/biomedicines12102182 PMID: 39457494

Li, J., Zhang, S., & Zhang, Y. (2023). Removal of pristine and aged microplastics from water by magnetic biochar: Adsorption and magnetization. *The Science of the Total Environment*, *875*, 162647. DOI: 10.1016/j.scitotenv.2023.162647 PMID: 36889392

Likpalimor, O. N., Zhang, S., Adagba, F. I., Fatoye, S., Adega, K. J., & Njue, J. (2025). Mechanisms of microplastic sorption of antibiotics and impacts on aquatic ecosystems for sustainable development goals. *Discover Environment*, *3*(1), 215. DOI: 10.1007/s44274-025-00431-3

Lin, R. H., Chen, H. T., Lee, I. T., Vo, T. T., & Wang, Y. L. (2025). Microplastics and Nanoplastics in Health Concerning Cellular Toxicity Mechanisms, Exposure Pathways, and Global Mitigation Strategies. *Life (Basel, Switzerland)*, *15*(9), 1449. DOI: 10.3390/life15091449 PMID: 41010391

Li, S., Yan, W., & Zhang, W. (2009). Solvent-free production of nanoscale zero-valent iron (nFeZ) with precision milling. *Green Chemistry*, *11*(10), 1628–1636. DOI: 10.1039/b913056j

Lisuzzo, L., Cavallaro, G., Milioto, S., & Lazzara, G. (2020). Effects of halloysite content on the thermo-mechanical performances of composite bioplastics. *Applied Clay Science*, *185*, 105416. DOI: 10.1016/j.clay.2019.105416

Liu, B., Guo, K., Feng, Q., Zhang, Y., Zhang, X., & Zhang, Y. (2025). Revealing the removal behavior of polystyrene nanoplastics and natural organic matter by AlTi-based coagulant from the perspective of functional groups. *Journal of Hazardous Materials*, *487*, 137258. DOI: 10.1016/j.jhazmat.2025.137258 PMID: 39837030

Liu, L., Sun, Y., Du, S., Li, Y., & Wang, J. (2023). Nanoplastics promote the dissemination of antibiotic resistance genes and diversify their bacterial hosts in soil. *Eco-Environment & Health*, *3*(1), 1–10. DOI: 10.1016/j.eehl.2023.09.005 PMID: 38187015

Liu, Q., & Khor, S. M. (2024). Emerging absorption-based techniques for removing microplastics and nanoplastics from actual water bodies. *Trends in Analytical Chemistry*, *170*, 117465. DOI: 10.1016/j.trac.2023.117465

Liu, S., Chen, Q., Ding, H., Song, Y., Pan, Q., Deng, H., & Zeng, E. Y. (2024). Differences of microplastics and nanoplastics in urban waters: Environmental behaviors, hazards, and removal. *Water Research*, *260*, 121895. DOI: 10.1016/j.watres.2024.121895 PMID: 38875856

Liu, Y., Chen, S., Xiong, X., Wen, Z., Zhao, L., Xu, B., Guo, Q., Xia, J., & Pei, J. (2025). Artificial intelligence guided Raman spectroscopy in biomedicine: Applications and prospects. *Journal of Pharmaceutical Analysis*, *15*(11), 101271. DOI: 10.1016/j.jpha.2025.101271 PMID: 41377140

Liu, Y., Liu, L., Wang, X., Shao, M., Wei, Z., Wang, L., Li, B., Li, C., Luo, X., Li, F., & Zheng, H. (2025). Microplastics enhance the prevalence of antibiotic resistance genes in mariculture sediments by enriching host bacteria and promoting horizontal gene transfer. *Eco-Environment & Health*, *4*(1), 100136. DOI: 10.1016/j.eehl.2025.100136 PMID: 40052062

Liu, Y., Luo, R., Sun, Z., Zhang, Y., Guo, Y., Chen, Y., Li, L., & Yue, Z. (2025). Synergistic toxicity of combined exposure to acrylamide and polystyrene nanoplastics on the gut–liver axis in mice. *Biology (Basel)*, *14*(5), 523. DOI: 10.3390/biology14050523 PMID: 40427712

Liu, Z. (2025). A maritime community with a shared future: Bridging fragmented legal regulations for marine plastic pollution. *Frontiers in Marine Science*, *12*, 1680110. DOI: 10.3389/fmars.2025.1680110

Li, Y., Han, F., & Zheng, Y. (2026, January). Artificial Intelligence in Surface Water Quality Research and Management: Recent Progress and Future Directions. *Ecosystem Health and Sustainability*, *12*, 0474. Advance online publication. DOI: 10.34133/ehs.0474

Li, Y., Ling, W., Yang, J., & Xing, Y. (2025). Risk assessment of microplastics in humans: Distribution, exposure, and toxicological effects. *Polymers*, *17*(12), 1699. DOI: 10.3390/polym17121699 PMID: 40574225

Li, Y., Zhang, C., Tian, Z., Cai, X., & Guan, B. (2024). Identification and quantification of nanoplastics (20–1000 nm) in a drinking water treatment plant using AFM-IR and Pyr-GC/MS. *Journal of Hazardous Materials*, *463*, 132933. DOI: 10.1016/j.jhazmat.2023.132933 PMID: 37951177

Li, Z., Zhu, S., Liu, Q., Wei, J., Jin, Y., Wang, X., & Zhang, L. (2020). Polystyrene microplastics cause cardiac fibrosis by activating Wnt/β-catenin signaling pathway and promoting cardiomyocyte apoptosis in rats. *Environmental Pollution*, *265*, 115025. DOI: 10.1016/j.envpol.2020.115025 PMID: 32806417

Long, Y., Wang, P., Lei, J., Su, B., Wei, Q., & Liu, X. (2025). Mechanical Signaling: Molecular Mechanisms, Biological Functions, Diseases, and Therapeutic Targets. *MedComm*, *6*(12), e70523. DOI: 10.1002/mco2.70523 PMID: 41377763

Lovric, S., Zee, J., Woroniecki, R., & Susztak, K. (2022). The Human Kidney Atlas: Integrative omics for precision nephrology. *Nature Reviews. Nephrology*, *18*(9), 583–601.

Luo, H., Xiang, Y., He, D., Li, Y., Zhao, Y., Wang, S., & Pan, X. (2019). Leaching behavior of fluorescent additives from microplastics and the toxicity of leachate to *Chlorella vulgaris*. *The Science of the Total Environment*, *678*, 1–9. DOI: 10.1016/j.scitotenv.2019.04.401 PMID: 31075575

Luo, H., Zhao, X., Zhang, Y., & Cai, B. (2024). A framework for measuring the implementation of the Sustainable Development Goal 4.2 (Quality Preprimary Education) at the local scale: An empirical study in Nanjing, China. *Sustainable Development (Bradford)*, *32*(4), 3979–4000. DOI: 10.1002/sd.2876

Lusher, A. L., Tirelli, V., O'Connor, I., & Officer, R. (2015). Microplastics in Arctic polar waters: The first reported values of particles in surface and sub-surface samples. *Scientific Reports*, *5*(1), 14947. Advance online publication. DOI: 10.1038/srep14947 PMID: 26446348

Mackenzie, K., Bleyl, S., Georgi, A., & Kopinke, F. D. (2016). Carbo-Iron—An Fe/AC composite for in situ treatment of groundwater contaminated with chlorinated hydrocarbons. *Environmental Science & Technology*, *50*(6), 3159–3168. DOI: 10.1021/acs.est.5b05031

MacLeod, M., Arp, H. P. H., Tekman, M. B., & Jahnke, A. (2021). The global threat from plastic pollution. *Science*, *373*(6550), 61–65. DOI: 10.1126/science.abg5433 PMID: 34210878

Macrì, S., & Di-Poï, N. (2024). The SmARTR pipeline: A modular workflow for the cinematic rendering of 3D scientific imaging data. *iScience*, *27*(12), 111475. DOI: 10.1016/j.isci.2024.111475 PMID: 39720527

Mahajan, P., Gusain, M., & Singh, P. (2026). Analytical strategies for nanoplastics detection and remediation: Advances, challenges, and perspectives. *Microchemical Journal : Devoted to the Application of Microtechniques in All Branches of Science*, *222*, 117228. DOI: 10.1016/j.microc.2026.117228

Mahalingaiah, S., Nadeau, K. C., & Christiani, D. C. (2025). Microplastics and Human Health. *Journal of the American Medical Association*, *334*(21), 1941–1942. DOI: 10.1001/jama.2025.14718 PMID: 41091491

Mahapatra, S., Maity, J. P., Singha, S., Mishra, T., Dey, G., Samal, A. C., Banerjee, P., Biswas, C., Chattopadhyay, S., Patra, R. R., Patnaik, S., & Bhattacharya, P. (2024). Microplastics and nanoplastics in environment: Sampling, characterization and analytical methods. *Groundwater for Sustainable Development*, *20*, 101267. DOI: 10.1016/j.gsd.2024.101267

Mahendran, R., & Ramaswamy, S. N. (2024). Nanoplastics as Trojan horses: Deciphering complex connections and environmental ramifications: A review. *Chemistry Africa*, *7*(1), 2265–2282. DOI: 10.1007/s42250-023-00865-3

Ma, J., & Lin, Y. (2024). Regulatory frameworks for nanomaterial governance: Comparative policy perspectives. *ChemistrySelect*, *9*(12), e202400959. DOI: 10.1002/slct.202400959

Majid, S., Reus, A., Hoondert, R., Blokker, M., Dash, A., Houtman, C., Schriks, M., & Dingemans, M. M. L. (2025). A framework for evaluating less-than-lifetime exposures: Advancing toxicological risk assessment for drinking water quality. *Archives of Toxicology*, *99*(8), 3147–3168. DOI: 10.1007/s00204-025-04061-9 PMID: 40415114

Malik, F. A., Kushwah, S. V., Aslam, M. S., Reyed, R. M., & Haghi, A. K. (Eds.). (2026). *Aligning ESG Goals with the UNs Sustainable Development Goals Planning: Strategies for Equitable Financing and World Sustainability (World Sustainability Series)*. Springer Cham., DOI: 10.1007/978-3-032-01715-4

Malik, M., Anwar, M. A., Ullah, A., Hafeez, A., Shakoor, M. B., & Arif, M. (2022). A review of the techno-economic feasibility of nanoparticle application for wastewater treatment. *Water (Basel)*, *14*(10), 1550. DOI: 10.3390/w14101550

Manoharan, K., Pandey, M., & Bhattacharya, S. (2023). Life cycle assessment of nanoscale polymer-based coatings. In *Polymer-Based Nanoscale Materials for Surface Coatings* (pp. 613–631). Elsevier., DOI: 10.1016/B978-0-32-390778-1.00033-5

Marfella, R., Prattichizzo, F., Sardu, C., Fulgenzi, G., Graciotti, L., Spadoni, T., D'Onofrio, N., Scisciola, L., La Grotta, R., Frigé, C., Pellegrini, V., Municinò, M., Siniscalchi, M., Spinetti, F., Vigliotti, G., Vecchione, C., Carrizzo, A., Accarino, G., Squillante, A., & Paolisso, G. (2024). Microplastics and Nanoplastics in Atheromas and Cardiovascular Events. *The New England Journal of Medicine*, *390*(10), 900–910. DOI: 10.1056/NEJMoa2309822 PMID: 38446676

Marik, A., Biswas, S., & Banerjee, E. R. (2024). Exploring the relationship between gut microbial ecology and inflammatory disease: An insight into health and immune function. *World Journal of Immunology*, *14*(1), 96209. DOI: 10.5411/wji.v14.i1.96209

Marras, L., Caputo, M., Bisicchia, S., Soato, M., Bertolino, G., Vaccaro, S., & Inturri, R. (2021). The Role of Bifidobacteria in Predictive and Preventive Medicine: A Focus on Eczema and Hypercholesterolemia. *Microorganisms*, *9*(4), 836. DOI: 10.3390/microorganisms9040836 PMID: 33919907

Martin-Pérez, J., Villacorta, A., Banaei, G., Morataya-Reyes, M., Tavakolpournegari, A., Marcos, R., Hernández, A., & García-Rodríguez, A. (2024). Hazard assessment of nanoplastics is driven by their surface-functionalization: Effects in human-derived primary endothelial cells. *The Science of the Total Environment*, *934*, 173236. DOI: 10.1016/j.scitotenv.2024.173236 PMID: 38761522

Martins, F., Machado, S., Albergaria, T., & Delerue-Matos, C. (2017). LCA applied to nanoscale zero valent iron synthesis. *The International Journal of Life Cycle Assessment, 22*(5), 707–714. DOI: 10.1007/s11367-016-1258-7

Mason, S. A., Welch, V. G., & Neratko, J. (2018). Synthetic polymer contamination in bottled water. *Frontiers in Chemistry, 6*, 407. DOI: 10.3389/fchem.2018.00407 PMID: 30255015

Matus, P., Urquidi, C., Cárcamo, M., & Vidal, V. (2024). Integrating the exposome and one health approach to national health surveillance: An opportunity for Latin American countries in health preventive management. *Frontiers in Public Health, 12*, 1376609. DOI: 10.3389/fpubh.2024.1376609 PMID: 39211902

Ma, X., Tian, Y., Yang, R., Wang, H., Allahou, L. W., Chang, J., Williams, G., Knowles, J. C., & Poma, A. (2024). Nanotechnology in healthcare, and its safety and environmental risks. *Journal of Nanobiotechnology, 22*(1), 715. DOI: 10.1186/s12951-024-02901-x PMID: 39548502

Ma, Y.-B., Xie, Z.-Y., Hamid, N., Tang, Q.-P., Deng, J.-Y., Luo, L., & Pei, D.-S. (2023). Recent advances in micro (nano) plastics in the environment: Distribution, health risks, challenges and future prospects. *Aquatic Toxicology, 261*, 106597. DOI: 10.1016/j.aquatox.2023.106597 PMID: 37311378

Mazzanti, V., Pariante, R., Bonanno, A., Ballesteros, O. D., Mollica, F., & Filippone, G. (2019). Reinforcing mechanisms of natural fibers in green composites: Role of fibers morphology in a PLA/hemp model system. *Composites Science and Technology, 180*, 51–59. DOI: 10.1016/j.compscitech.2019.05.015

McAllister, S. R. (2018). *Implementation of food safety regulations in food service establishments* (Publication No. 5902) [Doctoral dissertation, Walden University]. ScholarWorks. https://scholarworks.waldenu.edu/dissertations/5902

Medley, E. A., Spratlen, J. J., Yan, B., Herbstman, J. B., & Deyssenroth, M. A. (2023). A systematic review of the placental translocation of micro- and nanoplastics. *Current Environmental Health Reports, 10*(2), 99–111. DOI: 10.1007/s40572-023-00391-x PMID: 36848019

Meng, F., Fan, T., Yang, X., Riksen, M., Xu, M., & Geissen, W. (2020). Effect of plastic mulching on the accumulation and distribution of macro- and microplastic particles in soils of two farming systems in North West China. *PeerJ, 8*, e10375. DOI: 10.7717/peerj.10375 PMID: 33344073

Meyers, N., Kopke, K., Buhhalko, N., Mattsson, K., Janssen, C. R., Everaert, G., & De Witte, B. (2024). Value for money: A cost-effectiveness analysis of microplastic analytics in seawater. *Microplastics and Nanoplastics*, *4*(1), 4. DOI: 10.1186/s43591-024-00081-x

Mintenig, S. M., Bauerlein, P. S., Koelmans, A. A., Dekker, S. C., & van Wezel, A. P. (2018). Closing the gap between small and smaller: Towards a framework to analyse nano- and microplastics in aqueous samples. Environmental Science: Nano, 5(7), 1640 1649. https://doi.org/DOI: 10.1039/C8EN00186C

Mintenig, S. M., Int-Veen, I., Löder, M. G. J., Primpke, S., & Gerdts, G. (2017). Identification of Microplastic in Effluents of Waste Water Treatment Plants Using Focal Plane Array-Based Micro-Fourier-Transform Infrared Imaging. *Water Research*, *108*, 365–372. DOI: 10.1016/j.watres.2016.11.015 PMID: 27838027

Mintenig, S. M., Loder, M. G. J., Primpke, S., & Gerdts, G. (2019). Low numbers of microplastics detected in drinking water from groundwater sources. *The Science of the Total Environment*, *648*, 631–635. DOI: 10.1016/j.scitotenv.2018.08.178 PMID: 30121540

Miri, S., Saini, R., Davoodi, S. M., Pulicharla, R., Brar, S. K., & Magdouli, S. (2022). Biodegradation of microplastics: Better late than never. *Chemosphere*, *286*, 131670. DOI: 10.1016/j.chemosphere.2021.131670 PMID: 34351281

Monahan, K., & Hoch, J. M. (2021). Comparison of phosphate uptake by epiphyton, metaphyton, and epipelon in the Everglades. *Florida Scientist*, *84*(1), 35–44. https://nsuworks.nova.edu/occ_facarticles/1184

Mondal, A., Moradi, S., Rahman, A., & Su, C. (2023). Recent advances of nanoremediation technologies for soil and groundwater remediation: A review. *Water (Basel)*, *13*(16), 2186. DOI: 10.3390/w13162186

Monitoring and Enforcement. (2022). Strategies for monitoring nanoplastic pollution. *Environmental Monitoring and Assessment*, *194*(6), 421. DOI: 10.1007/s10661-022-10089-4 PMID: 35543765

Morreale, M., & La Mantia, F. P. (2024). Current concerns about microplastics and nanoplastics: A brief overview. *Polymers*, *16*(11), 1525. Advance online publication. DOI: 10.3390/polym16111525 PMID: 38891471

Moshood, T. D., Nawanir, G., Mahmud, F., Mohamad, F., Ahmad, M. H., & Abdul-Ghani, A. (2022). Sustainability of biodegradable plastics: New problem or solution to solve the global plastic pollution? *Current Research in Green and Sustainable Chemistry*, *5*, 100273. DOI: 10.1016/j.crgsc.2022.100273

Moulton, C., Baroni, A., & Tasciotti, E. (2026). Micro- and Nanoplastics Exposure Across the Lifespan: One Health Implications for Aging and Longevity. *Journal of Xenobiotics*, *16*(2), 52. DOI: 10.3390/jox16020052 PMID: 41874123

Murray, A., & Örmeci, B. (2020). Removal Effectiveness of Nanoplastics (<400 nm) with Separation Processes Used for Water and Wastewater Treatment. *Water (Basel)*, *12*(3), 635. DOI: 10.3390/w12030635

Muxika, A., Etxabide, A., Uranga, J., Guerrero, P., & de la Caba, K. (2017). Chitosan as a bioactive polymer: Processing, properties and applications. *International Journal of Biological Macromolecules*, *105*, 1358–1368. DOI: 10.1016/j.ijbiomac.2017.07.087 PMID: 28735006

Nahiduzzaman, F. N. U., Rahman, M. Z., Akhi, M. A. J., Manik, M., Khatun, M. M., Islam, M. A., Matin, M. N., & Haque, M. A. (2025). Potential biological impacts of microplastics and nanoplastics on farm animals: Global perspectives with insights from Bangladesh. *Animals (Basel)*, *15*(10), 1394. DOI: 10.3390/ani15101394 PMID: 40427269

National Academies of Sciences, Engineering, and Medicine [NASEM] (2016). Appendix E: A brief history of ecological risk assessment. In *Gene drives on the horizon: Advancing science, navigating uncertainty, and aligning research with public values*. The National Academies Press. https://doi.org/DOI: 10.17226/23405

Nawaz, F., Islam, Z. U., Ghori, S. A., Bahadur, A., Ullah, H., Ahmad, M., & Khan, G. U. (2025). Microplastic and nanoplastic pollution: Assessing translocation, impact, and mitigation strategies in marine ecosystems. *Water Environment Research : A Research Publication of the Water Environment Federation*, *97*(3), 70032. Advance online publication. DOI: 10.1002/wer.70032 PMID: 39927485

Nematollahi, M. J., Mobasheri, M., Esmaeili, Z., Mahmoudi, M., Yousefi, N., & Busquets, R. (2025). Distribution and abundance of microplastics in urban and industrial wastewater treatment plants in Tabriz metropolis. *Scientific Reports*, *15*(1), 24577. DOI: 10.1038/s41598-025-10000-6 PMID: 40629021

Nene, A., Sadeghzade, S., Viaroli, S., Yang, W., Uchenna, U. P., Kandwal, A., Liu, X., Somani, P., & Galluzzi, M. (2025). Recent advances and future technologies in nanomicroplastics detection. *Environmental Sciences Europe*, *37*(7), 7. Advance online publication. DOI: 10.1186/s12302-024-01044-y

Neves, D., Sobral, P., Ferreira, J. L., & Pereira, T. (2015). Ingestion of microplastics by commercial fish off the Portuguese coast. *Marine Pollution Bulletin*, *101*(1), 119–126. DOI: 10.1016/j.marpolbul.2015.11.008 PMID: 26608506

Ney, L. M., Wipplinger, M., Grossmann, M., Engert, N., Wegner, V. D., & Mosig, A. S. (2023). Short-chain fatty acids: Key regulators of the local and systemic immune response in inflammatory diseases and infections. *Open Biology*, *13*(3), 230014. DOI: 10.1098/rsob.230014 PMID: 36977462

Ngarize, S., Makuch, K. E., & Pereira, R. (2013). The case for regulating nanotechnologies: International, European and national perspectives. *Review of European, Comparative & International Environmental Law*, *22*(3), 251–264. DOI: 10.1111/reel.12029

Nie, H. Y., Ge, J., Huang, G. X., Liu, K. G., Yue, Y., Li, H., Lin, H. G., Zhang, T., Yan, H. F., Xu, B. X., Sun, H. W., Yang, J. W., Si, S. Y., Zhou, J. L., & Cui, Y. (2024). New insights into the intestinal barrier through "gut-organ" axes and a glimpse of microgravity's effects on intestinal barrier. *Frontiers in Physiology*, *15*, 1465649. DOI: 10.3389/fphys.2024.1465649 PMID: 39450142

Nishida, C., Uauy, R., Kumanyika, S., & Shetty, P. (2004). The joint WHO/FAO expert consultation on diet, nutrition and the prevention of chronic diseases: Process, product and policy implications. *Public Health Nutrition*, *7*(1a, 1A), 245–250. DOI: 10.1079/PHN2003592 PMID: 14972063

Nizzetto, L., Futter, M., & Langaas, S. (2016). Are Agricultural Soils Dumps for Microplastics of Urban Origin? *Environmental Science & Technology*, *50*(20), 10777–10779. DOI: 10.1021/acs.est.6b04140 PMID: 27682621

Nizzetto, L., Futter, M., & Langaas, S. (2016). Investigating the mechanisms of microplastic transport in soil. *Environmental Science & Technology*, *50*(20), 11117–11126. DOI: 10.1021/acs.est.6b03438 PMID: 27682621

Njema, G. G., Agoro, M. A., & Kibet, J. K. (2026). A critical evaluation of nano- and microplastics: Sources, health risks, and mitigation strategies. *Advances in Materials Science and Engineering*, *2026*(1), 1357400. Advance online publication. DOI: 10.1155/amse/1357400

Noornama, N., Zainol Abidin, M. N., Abu Bakar, N. K., & Hashim, N. A. (2024). Innovative solutions for the removal of emerging microplastics from water by utilizing advanced techniques. *Marine Pollution Bulletin*, *206*, 116752. DOI: 10.1016/j.marpolbul.2024.116752 PMID: 39053257

O'Brian, F. P. (2011). The nanotechnology challenge: Creating legal institutions for uncertain risks. In Dana, D. A. (Ed.), *The nanotechnology challenge: Creating legal institutions for uncertain risks* (pp. 121–142). Cambridge University Press., DOI: 10.1017/CBO9780511988554.018

OECD. (2013). *"Regulatory Frameworks for Nanotechnology in Foods and Medical Products: Summary Results of a Survey Activity", OECD Science, Technology and Industry Policy Papers, No. 4.* OECD Publishing., DOI: 10.1787/5k47w4vsb4s4-

OECD. (2025). *"Development co-operation and the elimination of plastic pollution", OECD Policy Briefs, No. 20.* OECD Publishing., DOI: 10.1787/96803ffc-

Ogidi, O. I. (2024). Recent advances in anticancer activity and Bioinformatics Approach from Potential Plants. In: Computational Approaches in Biomaterials, Bioinformatics and Biomedical Engineering Applications. Published *CRC Press, Taylor and Francis Group, Florida, USA*. DOI: 10.1201/9781003354437-3

Ogidi, O. I., & Ajoko, I. (2024). Neuroprotective Potentials of Phytochemicals. In: M.S. Rajput, T. Sarachana, P.D. Sarkar, M. Raval (eds) NeuroPhytomedicine. Published by *CRC Press, Routledge Taylor and Francis Group.*

Ogidi, O. I., & Akpan, U. M. (2024). Microbial and Phyto-Remediation of Crude Oil Contaminated Soil. In: Eco-Restoration of the Environment. Published by *CRC Press, Routledge Taylor and Francis Group.*

Ogidi, O. I., & Emaikwu, N. G. (2024). Plant Phenolic compound isolation and its Bioinformatics Approaches of Molecular Mechanisms in Antimicrobial activities and Resistance. In: Computational Approaches in Biomaterials, Bioinformatics and Biomedical Engineering Applications. Publishered *CRC Press, Taylor and Francis Group, Florida, USA*. DOI: 10.1201/9781003354437-8

Ogidi, O. I., & Richard-Nwachukwu, N. (2024). Microbe Assisted Remediation of Pesticide Residues from Soil and Water. In: Eco-Restoration of the Environment. Published by *CRC Press, Routledge Taylor and Francis Group*. DOI: 10.1201/9781003423393-9

Ogidi, O. I., Orlu, H. A., & Poripo, B. E. (2025g). Evaluation of Hepatoprotective Activities of *Bryophyllum pinnatum* Leaf Extract in Paracetamol Induced Toxicity In Wistar Rats. *Trends in Pharmaceutical Sciences, 11*(1), -. DOI: 10.30476/tips.2025.105055.1272

Ogidi, O. I. (2025). Natural Products Modulating Mitophagy: Implications for Cancer Therapy. In Pandurangan, A. (Ed.), *Roles of Mitophagy in Cancer Regulation* (pp. 325–358). IGI Global Scientific Publishing., DOI: 10.4018/979-8-3693-3836-0.ch012

Ogidi, O. I., Ajoko, I. T., & Tawariowei, A. M. (2024a). Sustainable Application of Genetic Ecotoxicological Techniques. In *Biomonitoring For Environmental Sustainability. In: Biomonitoring of Pollutants in the Global South. Publisher*. Springer Nature Singapore Pte Ltd. DOI: 10.1007/978-981-97-1658-6_19

Ogidi, O. I., Angaye, T. C. N., & Carbom, H. E. (2024c). Microorganisms in the monitoring of terrestrial and aquatic ecosystems: applications, challenges and future prospects. In *Biomonitoring of Pollutants in the Global South. Publisher*. Springer Nature Singapore Pte Ltd. DOI: 10.1007/978-981-97-1658-6_18

Ogidi, O. I., Barwant, M. M., & Okara, E. (2024a). Biomarker Discovery: Protein Phosphorylation State. In *Spatially Variable Genes in Cancer: Development, Progression, and Treatment Response*. IGI Global., DOI: 10.4018/979-8-3693-7728-4.ch005

Ogidi, O. I., Christopher, E., & Orlu, H. A. (2025a). Environmental DNA and RNA Metabarcoding: Advancing Insights in Cancer Research. In Raghavan, R. (Ed.), *Cancer Exposomics and Environmental Influences on Carcinogenesis* (pp. 279–322). IGI Global Scientific Publishing., DOI: 10.4018/979-8-3373-2165-3.ch007

Ogidi, O. I., Erho, S. A., & Omu, O. (2025c). Reactive Oxygen Species in Cancer. In Prabhakar, P. (Ed.), *The Role of Reactive Oxygen Species in Human Health and Disease* (pp. 339–376). IGI Global Scientific Publishing., DOI: 10.4018/979-8-3693-7919-6.ch010

Ogidi, O. I., & Felagha, I. (2024). Gut Microbiome and Diet Interaction. In *Nutrition Controversies and Advances in Autoimmune Disease. Publisher*. IGI Global. DOI: 10.4018/979-8-3693-5528-2.ch012

Ogidi, O. I., Ijere, N. D., & Felagha, I. (2025b). Ecotoxicity and Human Health Risks of Improper E-Waste Handling. In Güngör, A., & Polat, O. (Eds.), *Integrated Approaches for Sustainable E-Waste Management* (pp. 243–282). IGI Global Scientific Publishing., DOI: 10.4018/979-8-3693-7383-5.ch009

Ogidi, O. I., Ijere, N. D., & Sogoi, P. B. (2025f). Neural and Endocrine Regulation of Immunity. In Reddy, K. (Ed.), *Research Methodologies and Practical Applications in Psychoneuroimmunology* (pp. 17–42). IGI Global Scientific Publishing., DOI: 10.4018/979-8-3693-7432-0.ch002

Ogidi, O. I., Ogregade, I., & Joshua, S. (2024e). Investigation of Renal Protective Effects of Bryophyllum pinnatum (Lam) Extract in Acetaminophen Induced Toxicity in Albino Rats. *Medical and Pharmaceutical Journal*, *3*(2), 82–93. DOI: 10.55940/medphar202485

Ogidi, O. I., & Tawariowei, A. M. (2023). Endocrine Disrupting Chemicals and their Role in Cancer-A review. *Molecular Sciences And Applications.*, *3*, 5–14. DOI: 10.37394/232023.2023.3.2

Ogoun, T. R., Ogidi, O. I., & Aye, T. (2022). Toxicity studies of Yoyo Cleanser Bitters Poly-herbal formulation in Albino Rats. *World Journal of Pharmaceutical Research,* 11(1): 1-11. DOI: https://www.doi.DOI: 10.20959/wjpr20221-22534

Okereke, J. N., Ogidi, O. I., & Obasi, K. O. (2016). Environmental and Health Impact of Industrial Wastewater Effluents in Nigeria- A Review. *International Journal of Advanced Research in Biological Science*, *3*(6), 55–67.

Oliveri Conti, G., Ferrante, M., Banni, M., Favara, C., Nicolosi, I., Cristaldi, A., Fiore, M., & Zuccarello, P. (2020). Micro- and nano-plastics in edible fruit and vegetables. *Environmental Research*, *187*, 109677. DOI: 10.1016/j.envres.2020.109677 PMID: 32454310

Omondi, P., & Chen, Q. (2024). Public perception and risk communication in nanotechnology applications. *Zenodo.* https://doi.org/DOI: 10.5281/zenodo.10845914

Oßmann, B. E., Sarau, G., Holtmannspötter, H., Pischetsrieder, M., Christiansen, S. H., & Dicke, W. (2018). Small-sized microplastics and pigmented particles in bottled mineral water. *Water Research*, *141*, 307–316. DOI: 10.1016/j.watres.2018.05.027 PMID: 29803096

Ouda, M., Banat, F., Hasan, S. W., & Karanikolos, G. N. (2023). A 3D CFD study on the effects of feed spacer designs on membrane performance for high-permeance RO membranes. *Journal of Water Process Engineering*, *53*, 103543. DOI: 10.1016/j.jwpe.2023.103543

Paço, A., Duarte, K., da Costa, J. P., Santos, P. S. M., Pereira, R., Pereira, M. E., Freitas, A. C., Duarte, A. C., & Rocha-Santos, T. A. P. (2017). Biodegradation of polyethylene microplastics by the marine fungus *Zalerion maritimum. The Science of the Total Environment*, *586*, 10–15. DOI: 10.1016/j.scitotenv.2017.02.017 PMID: 28199874

Pandit, S., Yadav, N., Sharma, P., Prakash, A., & Kuila, A. (2025). Life cycle assessment and techno-economic analysis of nanotechnology-based wastewater treatment: Status, challenges and future prospectives. *Journal of the Taiwan Institute of Chemical Engineers*, *166*, 105567. DOI: 10.1016/j.jtice.2024.105567

Panizzolo, M. (2025). Epidemiological applications of inflammation and oxidative stress biomarkers to assess the early effects of human exposure to micro and nanoplastics. DOI: 10.1016/j.ecoenv.2023.115645

Panyachanakul, T., Sorachart, B., Lumyong, S., Lorliam, W., Kitpreechavanich, V., & Krajangsang, S. (2019). Development of biodegradation process for poly(DL-lactic acid) degradation by crude enzyme produced by *Actinomadura keratinilytica* strain T16-1. *Electronic Journal of Biotechnology, 40*, 52–57. DOI: 10.1016/j.ejbt.2019.04.005

Park, J. W., Lee, S. J., Hwang, D. Y., & Seo, S. (2021). Removal of microplastics via tannic acid-mediated coagulation and *in vitro* impact assessment. *RSC Advances, 11*(6), 3556–3566. DOI: 10.1039/D0RA09645H PMID: 35424293

Pawak, V. S., Shekhar, C., Loganathan, V. A., & Sabapathy, M. (2023). Separation of nanoplastics from synthetic and industrial wastewater using electrolysis-assisted flotation approach: A green approach for real-time contaminant mitigation. *Chemical Engineering Research & Design, 198*, 184–199. DOI: 10.1016/j.cherd.2023.08.038

Peddie, C. J., Genoud, C., Kreshuk, A., Meechan, K., Micheva, K. D., Narayan, K., Pape, C., Parton, R. G., Schieber, N. L., Schwab, Y., Titze, B., Verkade, P., Aubrey, A., & Collinson, L. M. (2022). Volume electron microscopy. *Nature Reviews. Methods Primers, 2*(1), 51. DOI: 10.1038/s43586-022-00131-9 PMID: 37409324

Pegoraro, A. F., Chen, M., Sakib, S., Jakubek, Z. J., Maan, Z., Prezgot, D., Sallam, A., Corriveau, D., Vandenberg, M., & Zou, S. (2026). Nanoplastic reference materials for biological and methodological assessment. *Microplastics and Nanoplastics, 6*(1), 1. DOI: 10.1186/s43591-025-00157-2

Peng, J., Liu, X., Wang, J., Meng, N., Cai, R., Peng, Y., Han, Y., Liao, J., Li, C., Rubin-Blum, M., Ma, Q., & Dong, X. (2025). Diverse quorum sensing systems regulate microbial communication and biogeochemical processes in deep-sea cold seeps. *Microbiome, 14*(1), 16. DOI: 10.1186/s40168-025-02280-x PMID: 41340071

Peng, L., Mehmood, T., Bao, R., Wang, Z., & Fu, D. (2022). An overview of micro (nano) plastics in the environment: Sampling, identification, risk assessment and control. *Sustainability (Basel), 14*(21), 14338. DOI: 10.3390/su142114338

Perra, N., Giusto, D., & Anedda, M. (2026). A survey of modern data acquisition and analysis systems for environmental risk monitoring in aquatic ecosystems. *Sensors (Basel), 26*(5), 1566. DOI: 10.3390/s26051566 PMID: 41829526

Petersen, E. J., Barrios, A. C., Henry, T. B., Johnson, M. E., Koelmans, A. A., Montoro Bustos, A. R., Matheson, J., Roesslein, M., Zhao, J., & Xing, B. (2022). Potential Artifacts and Control Experiments in Toxicity Tests of Nanoplastic and Microplastic Particles. *Environmental Science & Technology, 56*(22), 15192–15206. DOI: 10.1021/acs.est.2c04929 PMID: 36240263

Peters, J., & Weil, M. (2016). A Critical Assessment of the Resource Depletion Potential of Current and Future Lithium Ion Batteries. *Resources*, *5*(4), 46. DOI: 10.3390/resources5040046

Piergiacomo, F., Brusetti, L., & Pagani, L. (2022). Understanding the Interplay between Antimicrobial Resistance, Microplastics and Xenobiotic Contaminants: A Leap towards One Health? *International Journal of Environmental Research and Public Health*, *20*(1), 42. DOI: 10.3390/ijerph20010042 PMID: 36612363

Pini, M., & Rossi, G. (2024). AI-driven life cycle costing in environmental remediation. *Sustainability*, *16*(2), 945.

Pinto da Costa, J., Reis, V., Paço, A., Costa, M., Duarte, A. C., & Rocha-Santos, T. (2019). Micro(nano)plastics—Analytical challenges towards risk evaluation. *Trends in Analytical Chemistry*, *111*, 173–184. DOI: 10.1016/j.trac.2018.12.013

Pirade, F., Foppen, J. W., van der Hoek, J. P., & Lompe, K. M. (2024). Polystyrene nanoplastics are unlikely to aggregate in freshwater bodies. *Environmental Pollution*, *366*, 125393. DOI: 10.1016/j.envpol.2024.125393 PMID: 39586453

Pitt, J. A., Trevisan, R., Massarsky, A., Kozal, J. S., Levin, E. D., & Di Giulio, R. T. (2018). Maternal transfer of nanoplastics to offspring in zebrafish (*Danio rerio*): A case study with nanopolystyrene. *The Science of the Total Environment*, *643*, 324–334. DOI: 10.1016/j.scitotenv.2018.06.186 PMID: 29940444

Porter, L. L. (2023). Fluid protein fold space and its implications. *BioEssays : News and Reviews in Molecular, Cellular and Developmental Biology*, *45*(9), e2300057. DOI: 10.1002/bies.202300057 PMID: 37431685

Pourzahedi, L., & Eckelman, M. J. (2015). Comparative life cycle assessment of silver nanoparticle synthesis routes. *Environmental Science. Nano*, *2*(4), 361–369. DOI: 10.1039/C5EN00075K

Pourzahedi, L., Vance, M., & Eckelman, M. J. (2014). Life cycle assessment of environmental impacts of nanosilver-enabled bandages. *Environmental Science & Technology*. Advance online publication. DOI: 10.1021/es504655y PMID: 25489789

Pradel, A., Catrouillet, C., & Gigault, J. (2023). The environmental fate of nanoplastics: What we know and what we need to know about aggregation. *NanoImpact*, *29*, 100453. DOI: 10.1016/j.impact.2023.100453 PMID: 36708989

Pradel, A., Ferreres, S., Veclin, C., El Hadri, H., Gautier, M., Grassl, B., & Gigault, J. (2021). Stabilization of fragmental polystyrene nanoplastic by natural organic matter: Insight into mechanisms. *ACS ES&T Water*, *1*(5), 1198–1207. DOI: 10.1021/acsestwater.0c00283

Pranjic, A. M., Oprckal, P., Mladenovic, A., Zapusek, P., Urleb, M., & Turk, J. (2018). Comparative life cycle assessment of possible methods for the treatment of contaminated soil at an environmentally degraded site. *Journal of Environmental Management, 218*, 497–508. DOI: 10.1016/j.jenvman.2018.04.051 PMID: 29709818

Prata, J. C., da Costa, J. P., Lopes, I., Duarte, A. C., & Rocha-Santos, T. (2020). Environmental exposure to microplastics: An overview on possible human health effects. *The Science of the Total Environment, 702*, 134455. DOI: 10.1016/j.scitotenv.2019.134455 PMID: 31733547

Primpke, S., Christiansen, S. H., Cowger, W., De Frond, H., Deshpande, A., Fischer, M., Holland, E. B., Meyns, M., O'Donnell, B. A., Ossmann, B. E., Pittroff, M., Sarau, G., Scholz-Böttcher, B. M., & Wiggin, K. J. (2020). Critical Assessment of Analytical Methods for the Harmonized and Cost-Efficient Analysis of Microplastics. *Applied Spectroscopy, 74*(9), 1012–1047. DOI: 10.1177/0003702820921465 PMID: 32249594

Procopio, A. C., Soggiu, A., Urbani, A., & Roncada, P. (2025). Interactions between microplastics and microbiota in a One Health perspective. One health (Amsterdam, Netherlands), 20, 101002. https://doi.org/DOI: 10.1016/j.onehlt.2025.101002

Qian, N., Gao, X., Lang, X., Deng, H., Bratu, T. M., Chen, Q., Stapleton, P., Yan, B., & Min, W. (2024). Rapid single-particle chemical imaging of nanoplastics by SRS microscopy. *Proceedings of the National Academy of Sciences of the United States of America, 121*(3), e2300582121. DOI: 10.1073/pnas.2300582121 PMID: 38190543

Qin, W., Cui, Y., Wang, H., Zhang, Y., & Cheng, Z. (2026). System architecture control design and clinical translation in closed-loop drug delivery. *Materials Today. Chemistry, 54*, 103601. DOI: 10.1016/j.mtchem.2026.103601

Qiu, Y., Mintenig, S., Barchiesi, M., & Koelmans, A. A. (2025). Using artificial intelligence tools for data quality evaluation in the context of microplastic human health risk assessments. *Environment International, 197*, 109341. DOI: 10.1016/j.envint.2025.109341 PMID: 39987688

Quinn, J., Geiger, C., Clausen, C., Brooks, K., Coon, C., O'Hara, S., Krug, T., Major, D., Yoon, W. S., Gavaskar, A., & Holdsworth, T. (2005). Field demonstration of DNAPL dehalogenation using emulsified zero-valent iron. *Environmental Science & Technology, 39*(5), 1309–1318. DOI: 10.1021/es0490018 PMID: 15787371

Qu, M., Kong, Y., Yuan, Y., & Wang, D. (2019). Neuronal damage induced by nanopolystyrene particles in nematode *Caenorhabditis elegans. Environmental Science. Nano, 6*(9), 2591–2601. DOI: 10.1039/C9EN00473D

Radanliev, P. (2025). AI Ethics: Integrating Transparency, Fairness, and Privacy in AI Development. *Applied Artificial Intelligence*, *39*(1), 2463722. Advance online publication. DOI: 10.1080/08839514.2025.2463722

Ragusa, A., & Fanos, V. (2025). Microplastics and nanoplastics in the brain: A review of the neurodevelopmental risks. [JPNIM]. *Journal of Pediatric and Neonatal Individualized Medicine*, *14*(2), e140206. DOI: 10.7363/140206

Ragusa, A., Svelato, A., Santacroce, C., Catalano, P., Notarstefano, V., Carnevali, O., Papa, F., Rongioletti, M. C. A., Baiocco, F., Draghi, S., D'Amore, E., Rinaldo, D., Matta, M., & Giorgini, E. (2021). Plasticenta: First evidence of microplastics in human placenta. *Environment International*, *146*, 106274. DOI: 10.1016/j.envint.2020.106274 PMID: 33395930

Rahman, A., Sarkar, A., Yadav, O. P., Achari, G., & Slobodnik, J. (2021). Potential human health risks due to environmental exposure to microplastics and knowledge gaps: A scoping review. *The Science of the Total Environment*, *757*, 143872. DOI: 10.1016/j.scitotenv.2020.143872 PMID: 33310568

Rai, A. (2022). Nanoplastics, Gut Microbiota, and Neurodegeneration. In *Gut Microbiome in Neurological Health and Disorders* (pp. 211–234). Springer Nature Singapore., DOI: 10.1007/978-981-19-4530-4_13

Rajtar, N., Starek, M., Vincenti, L., Dąbrowska, M., Romek, M., Rinaldi, R., Lionetto, F., & Kepczynski, M. (2025). Effect of PET micro/nanoplastics on model freshwater zooplankton. *Polymers*, *17*(9), 1256. Advance online publication. DOI: 10.3390/polym17091256 PMID: 40363040

Ramamurthy, M. O., Subramanian, P., Aravindan, S., Periyasamy, L., & Aravindan, N. (2025). Radiation Without Borders: Unraveling Bystander and Non-Targeted Effects in Oncology. *Cells*, *14*(22), 1761. DOI: 10.3390/cells14221761 PMID: 41294814

Ramirez-Arenas, L., Ramseier Gentile, S., Zimmermann, S., & Stoll, S. (2022). Fate and removal efficiency of polystyrene nanoplastics in a pilot drinking water treatment plant. *The Science of the Total Environment*, *813*, 152623. DOI: 10.1016/j.scitotenv.2021.152623 PMID: 34963580

Ramsperger, A. F. R. M., Bergamaschi, E., Panizzolo, M., Fenoglio, I., Barbero, F., Peters, R., Undas, A., Purker, S., Giese, B., Lalyer, C. R., Tamargo, A., Moreno-Arribas, M. V., Grossart, H.-P., Kühnel, D., Dietrich, J., Paulsen, F., Afanou, A. K., Zienolddiny-Narui, S., Eriksen Hammer, S., & Laforsch, C. (2023). Nano- and microplastics: A comprehensive review on their exposure routes, translocation, and fate in humans. *NanoImpact*, *29*, 100441. DOI: 10.1016/j.impact.2022.100441 PMID: 36427812

Rangrazi, A., Bonyadi, Z., & Sarkhosh, M. (2026). Investigation of airborne microplastics emission and characteristics in hospital laundry environments. *Scientific Reports*, *16*(1), 6107. DOI: 10.1038/s41598-026-35421-9 PMID: 41580448

Rasheed, A., Bashir, Y., & Das, S. (2025). Are (bio) electrochemical techniques sustainable solutions to combat micro-and nano-plastic pollution? *Current Opinion in Chemical Engineering*, *47*, 101091. DOI: 10.1016/j.coche.2025.101091

Rasheed, A., Sharma, N., Surampalli, R. Y., & Das, S. (2023). Evaluating treatment solutions: Critical review on technologies employed for microplastic removal from water matrices. *Current Opinion in Environmental Science & Health*, *36*, 100516. DOI: 10.1016/j.coesh.2023.100516

Rasmussen, K., Sokull-Klüttgen, B., Yu, I. J., Kanno, J., & Hirose, A. (2017). Regulation and legislation. In Stone, V., Boxall, A. B. A., & Clift, K. (Eds.), *Adverse effects of engineered nanomaterials: Exposure, toxicology, and impact on human health* (2nd ed., pp. 107–125). Academic Press., DOI: 10.1016/B978-0-12-809199-9.00007-0

Rathour, A., & Pathania, S. (2025). Sensing of Microplastics Using Advanced Materials: A Comprehensive Review. *Sensing and Imaging*, *26*(1), 80. DOI: 10.1007/s11220-025-00613-6

Ratre, P., Nazeer, N., Soni, N., Kaur, P., Tiwari, R., & Mishra, P. K. (2024). Smart carbon-based sensors for the detection of non-coding RNAs associated with exposure to micro (nano) plastics: An artificial intelligence perspective. *Environmental Science and Pollution Research International*, *31*(6), 8429–8452. DOI: 10.1007/s11356-023-31779-9 PMID: 38182954

Ravi Shankar, A. L., Irwan, J., Spaenig, M., Carlier, M., Hansen, T., Zumbülte, N., Islam, B., Lundquist, P., Svensson, R., Gouin, T., Hamers, T., & Escher, S. E. (2026). Extrapolation of in vitro effect concentrations to in vivo bioavailable concentrations using PBK modeling in humans for two classes of persistent and mobile compounds: Triazoles and triazines. *Archives of Toxicology*, *100*(5), 1877–1901. DOI: 10.1007/s00204-025-04268-w PMID: 41699307

Rawat, H., Gaur, A., Singh, N., Selvaraj, M., Karnwal, A., Pant, G., & Malik, T. (2025). Artificial intelligence-driven detection of microplastics in food: A comprehensive review of sources, health risks, detection techniques, and emerging artificial intelligence solutions. *Food Chemistry: X*, *29*, 102687. DOI: 10.1016/j.fochx.2025.102687 PMID: 40677556

Regasa, M. B. (2024). New approaches for micro(nano)plastics analysis. In *Toxic Effects of Micro- and Nanoplastics* (pp. 471–492). https://doi.org/DOI: 10.1002/9781394238163.ch20

Rendeiro, A. F., Ravichandran, H., Bram, Y., Chandar, V., Kim, J., Meydan, C., Park, J., Foox, J., Hether, T., Warren, S., Kim, Y., Reeves, J., Salvatore, S., Mason, C. E., Swanson, E. C., Borczuk, A. C., Elemento, O., & Schwartz, R. E. (2023). Spatial omics and AI in translational pathology. *Nature Medicine*, *29*(3), 497–511. DOI: 10.1038/s41591-023-02117-9

Revel, M., Châtel, A., & Mouneyrac, C. (2018). Micro(nano)plastics: A threat to human health? Current Opinion in Environmental Science & Health, 1, 17 23. https://doi.org/DOI: 10.1016/j.coesh.2017.10.003

Revel-Muroz, A., Akulinin, M., Shilova, P., Tyakht, A., & Klimenko, N. (2023). Stability of human gut microbiome: Comparison of ecological modelling and observational approaches. *Computational and Structural Biotechnology Journal*, *21*, 4456–4468. DOI: 10.1016/j.csbj.2023.08.030 PMID: 37745638

Rex, C. M., Debroy, A., Nirmala, M. J., & Mukherjee, A. (2023). Ecotoxicological significance of bio-corona formation on micro/nanoplastics in aquatic organisms. *RSC Advances*, *13*(33), 22905–22917. DOI: 10.1039/D3RA04054B PMID: 37520083

Reyed, R. M. (2026c). Polybiome systems medicine: Conceptual architecture, methodological foundations, and translational applications — Volume I: Vision and foundational methodology (Version 0.2). Zenodo. https://doi.org/DOI: 10.5281/zenodo.19258533

Reyed, R. M. (2007). The role of Bifidobacteria in health. *Research Journal of Medicine and Medical Sciences*, *2*(1), 14–24.

Reyed, R. M. (2019). Computational biotechnology: An approach in silico based modeling bioprocess. *International Journal of Research Studies in Microbiology and Biotechnology*, *5*(4), 1–5. DOI: 10.22259/2454-9428.0504001

Reyed, R. M. (2023). Simulate the dynamic interactions across nutritional supplements, the polybiome, and the biochemical approach dependent on patient-derived gastrointestinal knowledge. *Journal of Clinical Epidemiology & Toxicology*, *4*(2), 1–13. DOI: 10.47363/JCET/2023(4)133

Reyed, R. M. (2026b). Digital Nano-Plastic Science (DNPS) paradigm: Computational intelligence and proteostasis disruptions. *Zenodo*. https://doi.org/DOI: 10.5281/zenodo.18435353

Reyed, R. M. (2026d). Nanobiotechnology and AI Convergence for Antimicrobial Resistance and Predictive Precision Therapeutics: Precision Therapeutics and One Health. In Reyed, R. (Ed.), *AI, Nanobiotechnology, and the Future of Precision Antibiotics* (pp. 1–74). IGI Global Scientific Publishing., DOI: 10.4018/979-8-3373-5268-8.ch001

Reyed, R. M., Patel, R., Patel, S., & Prajapati, B. G. (2026). Application of synthetic biology in genetic disorder. In *Synthetic biology and its consequences in present healthcare* (pp. 201–213). Academic Press., DOI: 10.1016/B978-0-443-33649-2.00001-0

Reyed, R. M., & Prabhakar, P. K. (2025). Mechanisms, impact, and therapeutic strategies for oxidative stress in metabolic diseases: Understanding and mitigating oxidative stress. In Prabhakar, P. (Ed.), *The role of reactive oxygen species in human health and disease* (pp. 203–256). IGI Global Scientific Publishing., DOI: 10.4018/979-8-3693-7919-6.ch006

Reyed, R. M., & Prabhakar, P. K. (2026a). Modulating neuroinflammation and autophagy in psychiatric disorders through mechanisms and therapies: Therapeutic strategies and pathways. In Mokhtari, T., & Uludag, K. (Eds.), *Autophagy and inflammation in neuropsychological disorders* (pp. 61–124). IGI Global Scientific Publishing., DOI: 10.4018/979-8-3693-5908-2.ch003

Reyed, R. M., & Prabhakar, P. K. (2026b). Panflemmotic triggers to metastatic pathways: Complex signatures of tumorigenesis in the colonic microenvironment. In Kumar, A., Mishra, N., Singh, A. K., Pareek, A., & Prabhakar, P. K. (Eds.), *Inflammation and cancer* (pp. 247–346). Academic Press., DOI: 10.1016/B978-0-443-36500-3.00008-8

Rhaman, M. M., Islam, M. R., Akash, S., Mim, M., Noor Alam, M., Nepovimova, E., Valis, M., Kuca, K., & Sharma, R. (2022). Exploring the role of nanomedicines for the therapeutic approach of central nervous system dysfunction: At a glance. *Frontiers in Cell and Developmental Biology, 10*, 989471. DOI: 10.3389/fcell.2022.989471 PMID: 36120565

Ringeval, M., Etindele Sosso, F. A., Cousineau, M., & Paré, G. (2025). Advancing Health Care With Digital Twins: Meta-Review of Applications and Implementation Challenges. *Journal of Medical Internet Research, 27*, e69544. DOI: 10.2196/69544 PMID: 39969978

Rinninella, E., Raoul, P., Cintoni, M., Palombaro, M., Pulcini, G., Gasbarrini, A., & Mele, M. C. (2021). Nutritional Interventions Targeting Gut Microbiota during Cancer Therapies. *Microorganisms*, *9*(7), 1469. DOI: 10.3390/microorganisms9071469 PMID: 34361904

Rist, S., Almroth, B. C., Hartmann, N. B., Karlsson, T. M., & van Bavel, B. (2018). Evaluating the risk of microplastics in the ocean: The hidden fate of plastic in fish. *Global Change Biology*, *24*(2), 984–992. DOI: 10.1111/gcb.13981

Rizwan, M., Ali, S., Rehman, M. Z. U., Qayyum, M. F., Ok, Y. S., & Ahmad, N. (2022). Nanoparticle-fungus integrated remediation of organic and inorganic contaminants. *Frontiers in Microbiology*, *13*, 982611. DOI: 10.3389/fmicb.2022.982611

Rizzo, E., Bardos, P., Pizzol, L., Critto, A., Giubilato, E., Marcomini, A., Albano, D., Darmendrail, D., Döberl, G., Harclerode, M., Harries, N., Nathanail, P., Pachon, C., Rodriguez, A., Slenders, H., & Smith, G. (2016). Comparison of international approaches to sustainable remediation. *Journal of Environmental Management*, *184*, 4–17. DOI: 10.1016/j.jenvman.2016.07.062 PMID: 27520125

Rodrigues, A. C. B., de Jesus, G. P., Waked, D., Gomes, G. L., Silva, T. M., Yariwake, V. Y., da Silva, M. P., Magaldi, A. J., & Veras, M. M. (2022). Scientific evidence about the risks of micro and nanoplastics (MNPLs) to human health: Their exposure routes through the environment. *Toxics*, *10*(6), 308. DOI: 10.3390/toxics10060308 PMID: 35736916

Rodríguez-Rojas, M. D. P., Bustos-Terrones, V., Díaz-Cárdenas, M. Y., Vázquez-Vélez, E., & Martínez, H. (2024). Life cycle assessment of green synthesis of TiO_2 nanoparticles vs. chemical synthesis. *Sustainability (Basel)*, *16*(17), 7751. DOI: 10.3390/su16177751

Roubert, F., Beuzelin-Ollivier, M.-G., Hofmann-Amtenbrink, M., Hofmann, H., & Hool, A. (2016). "Nanostandardization" in action: Implementing standardization processes in a multidisciplinary nanoparticle-based research and development project. *NanoEthics*, *10*(1), 41–62. DOI: 10.1007/s11569-015-0248-8

Rubiños, S. L., Grados, J. H., Medina, J. T., Chávez, E. N., Huarcaya, E., Rocha, V. E., Grados, H. J., & Neyra, M. R. (2023). Microplastics in the ecosystem: A systematic review of the methods for their detection and removal. *International Journal of Ecology*, *2023*(1), 8826716. DOI: 10.1155/2023/8826716

Saarimäki, L. A., Del Giudice, G., & Greco, D. (2023). Expanding adverse outcome pathways toward one health model for nanosafety. *Frontiers in Toxicology*, *5*, 1176745. DOI: 10.3389/ftox.2023.1176745 PMID: 37692900

Sadek, A. H. S., Abdel-Haleem, F. M., & Ateia, M. A. (2025). Micro/nanoplastics removal from the aquatic environment using micro/nanobubbles-assisted flotation technology: A review. *Separation and Purification Technology*, *377*, 134262. DOI: 10.1016/j.seppur.2025.134262

Saleh, A. M., Selim, R. N., El-Morsy, M. T., Mansour, S. A., Elhady, R., Ismail, S. F. K., Gamal, M., & Kenawy, E. S. (2026). A review of nanofiber-based drug delivery systems: Fabrication, characterization, advances, and future prospects. *Journal of Materials Science: Polymer*, *1*(1), 10. DOI: 10.1007/s44493-026-00009-2

Salem, W. M., Haridy, M., & El-Azab, I. H. (2022). Biosynthesized iron oxide nanoparticles by Halomonas sp. for the bioremediation of Pb and Cd in contaminated soils. *Frontiers in Sustainable Food Systems*, *6*, 932424. DOI: 10.3389/fsufs.2022.932424

Samsudin, M., Azid, A., & Rani, N. L. A. (2023). Nanoplastics in environment. In *Nanofillers for sustainable applications* (pp. 188–197). CRC Press., DOI: 10.1201/9781003400998-9

Santhanam, S. D., Ramamurthy, K., Priya, P. S., Sudhakaran, G., Guru, A., & Arockiaraj, J. (2024). A combinational threat of micro-and nano-plastics (MNPs) as potential emerging vectors for per-and polyfluoroalkyl substances (PFAS) to human health. *Environmental Monitoring and Assessment*, *196*(12), 1182. DOI: 10.1007/s10661-024-13292-9 PMID: 39514026

Santos, F. A., Andre, R. S., Alvarenga, A. D., Alves, A. L. M. M., & Correa, D. S. (2025). Micro- and nanoplastics in the environment: A comprehensive review on detection techniques. *Environmental Science. Nano*, *12*(7), 3442–3467. DOI: 10.1039/D4EN00319E

Scheffer, M., Bascompte, J., Brock, W. A., Brovkin, V., Carpenter, S. R., Dakos, V., Held, H., van Nes, E. H., Rietkerk, M., & Sugihara, G. (2009). Early-warning signals for critical transitions. *Nature*, *461*(7260), 53–59. DOI: 10.1038/nature08227 PMID: 19727193

Schirinzi, G. F., Pérez-Pomeda, I., Sanchís, J., Rossini, C., Farré, M., & Barceló, D. (2017). Cytotoxic effects of commonly used nanomaterials and microplastics on cerebral and epithelial human cells. *Environmental Research*, *159*, 579–587. DOI: 10.1016/j.envres.2017.08.043 PMID: 28898803

Schneider, M., Stracke, F., Hansen, S., & Schaefer, U. F. (2009). Nanoparticles and their interactions with the dermal barrier. *Dermato-Endocrinology*, *1*(4), 197–206. DOI: 10.4161/derm.1.4.9501 PMID: 20592791

Schwabl, P., Köppel, S., Königshofer, P., Bucsics, T., Trauner, M., Reiberger, T., & Liebmann, B. (2019). Detection of various microplastics in human stool: A prospective case series. *Annals of Internal Medicine*, *171*(7), 453–457. DOI: 10.7326/M19-0618 PMID: 31476765

Schwarzfischer, M., Ruoss, T. S., Niechcial, A., Lee, S. S., Wawrzyniak, M., Laimbacher, A., Atrott, K., Manzini, R., Wilmink, M., Linzmeier, L., Morsy, Y., Lang, S., Rogler, G., Kaegi, R., Scharl, M., & Spalinger, M. R. (2024). Impact of nanoplastic particles on macrophage inflammation and intestinal health in a mouse model of inflammatory bowel disease. *Nanomaterials (Basel, Switzerland)*, *14*(16), 1350. DOI: 10.3390/nano14161350 PMID: 39195388

Schymanski, D., Goldbeck, C., Humpf, H.-U., & Fürst, P. (2018). Analysis of microplastics in water by micro-Raman spectroscopy: Release of plastic particles from different packaging into mineral water. Water Research, 129, 154 162. DOI: 10.1016/j.watres.2017.11.011

Scoditti, E., Tumolo, M. R., & Garbarino, S. (2022). Mediterranean Diet on Sleep: A Health Alliance. *Nutrients*, *14*(14), 2998. DOI: 10.3390/nu14142998 PMID: 35889954

Seggio, M., Arcadio, F., Radicchi, E., Cennamo, N., Zeni, L., & Bossi, A. M. (2024). Toward Nano- and Microplastic Sensors: Identification of Nano- and Microplastic Particles via Artificial Intelligence Combined with a Plasmonic Probe Functionalized with an Estrogen Receptor. *ACS Omega*, *9*(17), 18984–18994. DOI: 10.1021/acsomega.3c09485 PMID: 38708270

Segovia, M., & Garcia-Alfaro, J. (2022). Design, Modeling and Implementation of Digital Twins. *Sensors (Basel)*, *22*(14), 5396. DOI: 10.3390/s22145396 PMID: 35891076

Seiyaboh, E. I., Angaye, T. C. N., & Ogidi, O. I. (2019). Environmental and Health Risk Assessment of Particulate Matter Associated with Dusty Football Field. *Journal of Experimental and Clinical Toxicology*, *1*(2), 1–6. DOI: 10.14302/issn.2641-7669.ject-19-2595

Sekine, R., Moore, K. L., Matzke, M., Vallotton, P., Jiang, H., Hughes, G. M., Kirby, J. K., Donner, E., Grovenor, C. R. M., Svendsen, C., & Lombi, E. (2017). Complementary imaging of silver nanoparticle interactions with green algae: Dark-field microscopy, electron microscopy, and nanoscale secondary ion mass spectrometry. *ACS Nano*, *11*(11), 10894–10902. DOI: 10.1021/acsnano.7b04556 PMID: 29061049

Sendra, M., Saco, A., Yeste, M. P., Romero, A., Novoa, B., & Figueras, A. (2020). Nanoplastics: From tissue accumulation to cell translocation into Mytilus galloprovincialis hemocytes. *Journal of Hazardous Materials, 388*, 121788. DOI: 10.1016/j.jhazmat.2019.121788 PMID: 31813690

Sendra, M., Sparaventi, E., Novoa, B., & Figueras, A. (2021). An overview of the internalization and effects of microplastics and nanoplastics on marine organisms. *The Science of the Total Environment, 760*, 143953. DOI: 10.1016/j.scitotenv.2020.142024 PMID: 33207452

Sepe, R., Bollino, F., Boccarusso, L., & Caputo, F. (2017). Influence of chemical treatments on mechanical properties of hemp fiber reinforced composites. *Composites. Part B, Engineering, 133*, 210–217. DOI: 10.1016/j.compositesb.2017.09.030

Serra, A., Torres Maia, M., Tsiros, P., Minadakis, V., Riudavets-Puig, R., Perello-y-bestard, A., Nikiforou, F., Karakoltzidis, A., Di Lieto, E., Schaffert, A., Al-Abdulraheem, Z., Virmani, I., Dziubaniuk, O., Karhukorpi, S., Dokler, J., Zouraris, D., Mintis, D. G., Varsou, D.-D., Tsoumanis, A., & Greco, D. (2026). Review on Predictive Models and Integration Strategies for Holistic Impact Assessment of Chemicals and Materials. *Environmental Science & Technology, 60*(5), 3739–3766. DOI: 10.1021/acs.est.5c04489 PMID: 41603589

Sessitsch, A., Wakelin, S., Schloter, M., Maguin, E., Cernava, T., Champomier-Verges, M. C., Charles, T. C., Cotter, P. D., Ferrocino, I., Kriaa, A., Lebre, P., Cowan, D., Lange, L., Kiran, S., Markiewicz, L., Meisner, A., Olivares, M., Sarand, I., Schelkle, B., & Kostic, T. (2023). Microbiome Interconnectedness throughout Environments with Major Consequences for Healthy People and a Healthy Planet. *Microbiology and Molecular Biology Reviews : MMBR, 87*(3), e0021222. DOI: 10.1128/mmbr.00212-22 PMID: 37367231

Severino, A., Russo, B., Lavorato, C., Argurio, P., Figoli, A., Molinari, R., & Poerio, T. (2025). Integrated nanofiltration and photocatalytic processes for the removal of polystyrene nanoplastics waste in water. *Separation and Purification Technology, 286*, 131232. DOI: 10.1016/j.seppur.2024.131232

Shah, S. S., Noman, O., & Jaiswal, N. (2023). Unveiling the Gut Microbiome: How Junk Food Impacts the Gut. *Cureus, 15*(11), e49179. DOI: 10.7759/cureus.49179 PMID: 38130525

Shan, Q., Li, H., Chen, N., Qu, F., & Guo, J. (2020). Understanding the Multiple Effects of PCBs on Lipid Metabolism. *Diabetes, Metabolic Syndrome and Obesity, 13*, 3691–3702. DOI: 10.2147/DMSO.S264851 PMID: 33116719

Shao, K., Zou, R., Zhang, Z., Mandemaker, L. D. B., Timbie, S., Smith, R. D., Durkin, A. M., Dusza, H. M., Meirer, F., Weckhuysen, B. M., Alderete, T. L., Vermuelen, R., & Walker, D. I. (2025). Advancements in Assays for Micro- and Nanoplastic Detection: Paving the Way for Biomonitoring and Exposomics Studies. *Annual Review of Pharmacology and Toxicology*, *65*(1), 567–585. DOI: 10.1146/annurev-pharmtox-030424-112828 PMID: 39270670

Sharifa, M., Ghosh, T., Daher, O. A., Bhusal, P., Alaameri, Y. A., Naz, J., Ekhator, C., Bellegarde, S. B., Bisharat, P., Vaghani, V., & Hussain, A. (2023). Unraveling the Gut-Brain Axis in Multiple Sclerosis: Exploring Dysbiosis, Oxidative Stress, and Therapeutic Insights. *Cureus*, *15*(10), e47058. DOI: 10.7759/cureus.47058 PMID: 38022314

Sharma, P., & Sharma, P. (2024). Micro(nano)plastics: Invisible compounds with a visible impact. *F1000 Research*, *13*, 69. DOI: 10.12688/f1000research.142212.1 PMID: 38659492

Sharma, R., Patel, M., & Singh, A. (2023). Nanobioremediation: Synergistic effects of nanoparticles and microbes in pollutant removal. *Frontiers in Environmental Science*, *11*, 10204040. DOI: 10.3389/fenvs.2023.10204040

Shen, R., Yang, K., Cheng, X., Guo, C., Xing, X., Sun, H., Liu, D., Liu, X., & Wang, D. (2022). Accumulation of polystyrene microplastics induces liver fibrosis by activating cGAS/STING pathway. *Environmental Pollution*, *300*, 118986. DOI: 10.1016/j.envpol.2022.118986 PMID: 35167931

Shim, W. J., Hong, S. H., & Eo, S. E. (2017). Identification Methods in Microplastic Analysis: A Review. *Analytical Methods : Advancing Methods and Applications*, *9*(9), 1384–1391. DOI: 10.1039/C6AY02558G

Shi, Y., Chen, W., Yang, S., Fan, Y., & Lu, L. (2024). Freshwater microplastics governance and sustainable development: Pollution status, interactions, policies, and prospective studies. *Desalination and Water Treatment*, *320*, 100704. DOI: 10.1016/j.dwt.2024.100704

Shruti, V. C., & Kutralam-Muniasamy, G. (2019). Bioplastics: Missing link in the era of microplastics. *The Science of the Total Environment*, *697*, 134139. DOI: 10.1016/j.scitotenv.2019.134139 PMID: 32380615

Shruti, V. C., Kutralam-Muniasamy, G., Pérez-Guevara, F., Roy, P. D., & Elizalde Martínez, I. (2022). Occurrence and Characteristics of Atmospheric Microplastics in Mexico City. *The Science of the Total Environment*, *847*, 157601. Advance online publication. DOI: 10.1016/j.scitotenv.2022.157601 PMID: 35882345

Sidhu, A. K., Agrawal, S. B., Verma, N., Kaushal, P., & Sharma, M. (2025). Fungal-mediated synthesis of multimetallic nanoparticles: Mechanisms, unique properties, and potential applications. *Frontiers in Nanotechnology*, *7*, 1549713. Advance online publication. DOI: 10.3389/fnano.2025.1549713

Silva Trentin, A. W. de S., Visentin, C., Braun, A. B., & Thomé, A. (2021). Sustainability assessment of nanoscale zero-valent iron production methods. *Environmental Engineering Science*, *38*(2), 123–137.

Singh, A. K., Lou, H. H., Pike, R. W., Agboola, A., Li, X., Hopper, J. R., & Yaws, C. L. (2008). Environmental impact assessment for potential continuous process for the production of carbon nanotubes. *American Journal of Environmental Sciences*, *4*(5), 522–534. DOI: 10.3844/ajessp.2008.522.534

Skaba, D., Fiegler-Rudol, J., Dembicka-Mączka, D., & Wiench, R. (2025). Nanoplastics and Immune Disruption: A Systematic Review of Exposure Routes, Mechanisms, and Health Implications. *International Journal of Molecular Sciences*, *26*(11), 5228. DOI: 10.3390/ijms26115228 PMID: 40508037

Smith, M., Love, D. C., Rochman, C. M., & Neff, R. A. (2018). Microplastics in Seafood and the Implications for Human Health. *Current Environmental Health Reports*, *5*(3), 375–386. DOI: 10.1007/s40572-018-0206-z PMID: 30116998

Smith, N. R., & Levy, D. E. (2024). Budget impact analysis for implementation decision making, planning, and financing. *Translational Behavioral Medicine*, *14*(1), 54–59. DOI: 10.1093/tbm/ibad059 PMID: 37776567

Snehamayee, N., Somya, S., Kumar, S. C., Niranjan, M., Ranjan, S. B., & Kumar, M. N. (2026). Microplastics and Human Health: A Comprehensive Review on Exposure Pathways, Toxicity, and Emerging Risks. *Microplastics*, *5*(1), 8. DOI: 10.3390/microplastics5010008

Sol, D., Laca, A., Laca, A., & Díaz, M. (2020). Approaching the environmental problem of microplastics: Importance of WWTP treatments. *The Science of the Total Environment*, *740*, 140016. DOI: 10.1016/j.scitotenv.2020.140016 PMID: 32569912

Song, T.-T., Pan, E.-Y., & Tseng, H.-C. (2012). Nanorisk governance in Taiwan: Studying potential impacts on the environment, health, and safety. *IEEE Nanotechnology Magazine*, *6*(4), 6–11. DOI: 10.1109/MNANO.2012.2192653

Song, W., Zhang, J., Ling, C., Zhao, Y., Ji, R., Su, Y., & Xing, B. (2026). Formation and transformation of micro(nano)plastics: Mechanisms and environmental health implications. *Environment & Health*, *4*(5), 826–842. DOI: 10.1021/envhealth.5c00592 PMID: 42164879

Song, Y. K., Hong, S. H., Jang, M., Han, G. M., Jung, S. W., & Shim, W. J. (2015). Microplastics in the ocean: A review of the methods used for identification and quantification. *Environmental Science & Technology*, *49*(6), 3352–3360. DOI: 10.1021/es505342w

Song, Y., Hou, D., Zhang, J., O'Connor, D., Li, G., Gu, Q., Li, S., & Liu, P. (2018). Environmental and socio-economic sustainability appraisal of contaminated land remediation strategies: A case study at a mega-site in China. *The Science of the Total Environment*, *610–611*, 391–401. DOI: 10.1016/j.scitotenv.2017.08.016 PMID: 28806555

Song, Z., Yang, X., Chen, F., Zhao, F., Zhao, Y., Ruan, L., Wang, Y., & Yang, Y. (2019). Fate and transport of nanoplastics in complex natural aquifer media: Effect of particle size and surface functionalization. *The Science of the Total Environment*, *669*, 120–128. DOI: 10.1016/j.scitotenv.2019.03.102 PMID: 30878920

Sorensen, R. M., Kanwar, R. S., & Jovanović, B. (2023). Past, present, and possible future policies on plastic use in the United States, particularly microplastics and nanoplastics: A review. *Integrated Environmental Assessment and Management*, *19*(5), 1240–1254. DOI: 10.1002/ieam.4678 PMID: 36036190

Srivastava, A., Ravula, S., Bara, J. E., & Esfahani, M. R. (2025). Efficient extraction of polystyrene nanoplastics from water using an ionic liquid. *Environmental Science. Water Research & Technology*, *11*(5), 1151–1162. DOI: 10.1039/D4EW00909F

Stefaniuk, M., Oleszczuk, P., & Ok, Y. S. (2016). Review on nano zerovalent iron (nZVI): From synthesis to environmental applications. *Chemical Engineering Journal*, *287*, 618–632. DOI: 10.1016/j.cej.2015.11.046

Stine, J. S., Aziere, N., Harper, B. J., & Harper, S. L. (2023). A novel approach for identifying nanoplastics by assessing deformation behavior with scanning electron microscopy. *Micromachines*, *14*(10), 1903. DOI: 10.3390/mi14101903 PMID: 37893340

Stott, P. A., Lo, Y. T. E., Marsham, J. H., Obura, D., Oliver, T. H., Palmer, M. D., Ranger, N., Sharpe, S., & Sutton, R. (2026). We need a global assessment of avoidable climate-change risks. *Nature*, *650*(8103), 826–828. DOI: 10.1038/d41586-026-00544-6 PMID: 41741746

Su, C., Puls, R. W., & Krug, T. (2012). A two-and-a-half-year performance evaluation of field application of emulsified zero-valent iron for remediation of chlorinated volatile organic compounds in groundwater. *Environmental Science & Technology*, *46*(24), 12962–12970. DOI: 10.1021/es302725t

Suciu, N. A., Trevisan, M., Fragkoulis, G., Lamastra, L., Triachini, S., Abrantes, N., Norgaard, T., Harkes, P., Nunes, J. P., Silva, V., Navarro, I., de la Torre, A., Martínez, M. Á., Scheepers, P., & Martins Figueiredo, D. (2026). One Health perspective: An integrated in silico approach to assess the environmental fate of pesticides, exposure of aquatic and soil organisms, and risks for human health. *Environmental Research, 294*, 123805. DOI: 10.1016/j.envres.2026.123805 PMID: 41577111

Su, J., Wang, H., Man, S., & Yan, Q. (2026). Biochemical insights into the alleviated inhibition on nitrogen metabolism by microand nanoplastics at the biocathode of bioelectrochemical systems. *Bioresource Technology, 444*, 133938. DOI: 10.1016/j.biortech.2026.133938 PMID: 41490677

Sundas, A., Contreras, I., Mujahid, O., Beneyto, A., & Vehi, J. (2024). The Effects of Environmental Factors on General Human Health: A Scoping Review. *Healthcare (Basel), 12*(21), 2123. DOI: 10.3390/healthcare12212123 PMID: 39517336

Sun, X. D., Yuan, X. Z., Jia, Y., Feng, L. J., Zhu, F. P., Dong, S. S., Liu, J., Kong, X., Tian, H., Duan, J. L., Ding, Z., Wang, S. G., & Xing, B. (2020). Differentially charged nanoplastics demonstrate distinct accumulation in *Arabidopsis thaliana. Nature Nanotechnology, 15*(9), 755–760. DOI: 10.1038/s41565-020-0707-4 PMID: 32572228

Sun, X., Ren, Y., & Gao, G. (2021). Mechanisms and influencing factors of micro(nano)plastics removal by adsorption technology: A review. *Environmental Science and Pollution Research International, 28*(5), 5314–5327. DOI: 10.1007/s11356-020-10793-0 PMID: 32989695

Sutrisna, P. D., Riadi, L., Buana, P. C. W., Khoiruddin, K., Boopathy, R., Wenten, I. G., & Siagian, U. W. R. (2024). Membrane and membrane-integrated processes for nanoplastics removal and remediation. *Polymer Degradation & Stability, 220*, 110635. DOI: 10.1016/j.polymdegradstab.2023.110635

Tallec, K., Huvet, A., Di Poi, C., Gonzalez-Fernandez, C., Lambert, C., Petton, B., Le Goic, N., Berchel, M., Soudant, P., & Paul-Pont, I. (2018). Nanoplastics impaired oyster free-living stages, gametes and embryos. *Environmental Pollution, 242*, 1226–1235. DOI: 10.1016/j.envpol.2018.08.020 PMID: 30118910

Tan, D. S. Y., Akelew, Y., Snelson, M., Nguyen, J., & O'Sullivan, K. M. (2024). Unravelling the link between the gut microbiome and autoimmune kidney diseases: A potential new therapeutic approach. *International Journal of Molecular Sciences, 25*(9), 4817. DOI: 10.3390/ijms25094817 PMID: 38732038

Tang, F., Deng, M., Xu, C., Yang, R., Ji, X., Hao, M., Wang, Y., Tian, M., Geng, Y., & Miao, J. (2024). Unraveling the microbial puzzle: Exploring the intricate role of gut microbiota in endometriosis pathogenesis. *Frontiers in Cellular and Infection Microbiology, 14*, 1328419. DOI: 10.3389/fcimb.2024.1328419 PMID: 38435309

Tang, K. H. D., & Li, R. (2025). The effects of plastisphere on the physicochemical properties of microplastics. *Bioprocess and Biosystems Engineering, 48*(1), 1–15. DOI: 10.1007/s00449-024-03059-4 PMID: 38960926

Taniguchi, I., Yoshida, S., Hiraga, K., Miyamoto, K., Kimura, Y., & Oda, K. (2019). Biodegradation of PET: Current status and application aspects. *ACS Catalysis, 9*(5), 4089–4105. DOI: 10.1021/acscatal.8b05171

Tao, E., & Lang, D. (2024). Unraveling the gut: The pivotal role of intestinal mechanisms in Kawasaki disease pathogenesis. *Frontiers in Immunology, 15*, 1496293. DOI: 10.3389/fimmu.2024.1496293 PMID: 39664384

Tao, Y., Cui, W., Song, W., Xu, H., Zeng, J., & Li, H. (2025). Major adverse cardiovascular events in angiogenesis inhibitor-based cancer therapy: A systematic review and meta-analysis. *Journal of the American Heart Association, 14*(1), e042896. Advance online publication. DOI: 10.1161/JAHA.125.042896 PMID: 41181998

Tchounwou, P. B., & Zhou, J. (2014). *International Journal of Environmental Research and Public Health* Best Paper Award 2014. *International Journal of Environmental Research and Public Health, 11*(1), 1192–1194. DOI: 10.3390/ijerph110101192 PMID: 24434595

Thakur, R., Joshi, V., Sahoo, G. C., Jindal, N., Tiwari, R. R., & Rana, S. (2025). Review of mechanisms and impacts of nanoplastic toxicity in aquatic organisms and potential impacts on human health. *Toxicology Reports, 14*, 102013. DOI: 10.1016/j.toxrep.2025.102013 PMID: 40230517

Thome, A., Reddy, K. R., Reginatto, C., & Cecchin, I. (2015). Review of nanotechnology for soil and groundwater remediation: Brazilian perspectives. *Water, Air, and Soil Pollution, 226*(4), 1–20. DOI: 10.1007/s11270-014-2243-z

Thompson, R. C., Olsen, Y., Mitchell, R. P., Davis, A., Rowland, S. J., John, A. W., McGonigle, D., & Russell, A. E. (2004). Lost at sea: Where is all the plastic? *Science, 304*(5672), 838. DOI: 10.1126/science.1094559 PMID: 15131299

Thompson, R. C., Wright, S. L., & Galloway, T. S. (2013). The physical impacts of microplastics on marine organisms: A review. *Environmental Pollution, 178*, 483–492. DOI: 10.1016/j.envpol.2013.02.031 PMID: 23545014

Tian, S., Li, R., Li, J., & Zou, J. (2025). Polystyrene nanoplastics promote colitis-associated cancer by disrupting lipid metabolism and inducing DNA damage. *Environment International, 195*, 109258. DOI: 10.1016/j.envint.2025.109258 PMID: 39805171

Tomonaga, T., Higashi, H., Izumi, H., Nishida, C., Sato, K., Nakamura, Y., Morimoto, T., Higashi, Y., Kojima, T., Sakurai, K., Yatera, K., & Morimoto, Y. (2025). Comparison of lung disorders following intratracheal instillation of polystyrene microplastics with different surface functional groups. *Journal of Occupational Health, 67*(1), uiaf006. Advance online publication. DOI: 10.1093/joccuh/uiaf006 PMID: 39898983

Tran-Dinh, A., Diallo, D., Delbosc, S., Varela-Perez, L. M., Dang, Q. B., Lapergue, B., Burillo, E., Michel, J. B., Levoye, A., Martin-Ventura, J. L., & Meilhac, O. (2013). HDL and endothelial protection. *British Journal of Pharmacology, 169*(3), 493–511. DOI: 10.1111/bph.12174 PMID: 23488589

Tratnyek, P. G., & Johnson, R. L. (2006). Nanotechnologies for environmental cleanup. *Nano Today, 1*(2), 44–48. DOI: 10.1016/S1748-0132(06)70048-2

Trofin, F., Badescu, A. C., Iancu, L. S., Buzila, E. R., Anton-Păduraru, D.-T., Sima, C. M., Temneanu, O.-R., Matei, A., Bilha, S. C., Benea, I. A., & Dorneanu, O. S. (2026). Seeding the Future: How Feeding Mode Shapes the Infant Gut Microbiota. *Microorganisms, 14*(3), 719. DOI: 10.3390/microorganisms14030719 PMID: 41900478

Tse, Y. T., Chan, S. M., & Sze, E. T. (2022). Quantitative Assessment of Full Size Microplastics in Bottled and Tap Water Samples in Hong Kong. *International Journal of Environmental Research and Public Health, 19*(20), 13432. DOI: 10.3390/ijerph192013432 PMID: 36294013

Tsigalou, C., Tsolou, A., Stavropoulou, E., Konstantinidis, T., Zafiriou, E., Dardiotis, E., Tsirogianni, A., & Bogdanos, D. (2024). Unraveling the intricate dance of the Mediterranean diet and gut microbiota in autoimmune resilience. *Frontiers in Nutrition, 11*, 1383040. DOI: 10.3389/fnut.2024.1383040 PMID: 38818135

Tumrani, S. H., Sharma, B. P., Otho, A. R., Ayob, E. A., Amin, M. A., Soliman, M. M., Soomro, R. A., & Karakuş, S. (2025). Biodegradation of PLA in surface water and its behavior towards PFOA: Interactional characteristics and computational insights. *Colloid & Polymer Science, 303*(7), 1397–1408. DOI: 10.1007/s00396-025-05423-6

Tu, P., Xue, J., Niu, H., Tang, Q., Mo, Z., Zheng, X., Wu, L., Chen, Z., Cai, Y., & Wang, X. (2023). Deciphering Gut Microbiome Responses upon Microplastic Exposure via Integrating Metagenomics and Activity-Based Metabolomics. *Metabolites*, *13*(4), 530. DOI: 10.3390/metabo13040530 PMID: 37110188

Turner, H. C., Hori, Y., Revill, P., Rattanavipapong, W., Arai, K., Nonvignon, J., Jit, M., & Teerawattananon, Y. (2023). Analyses of the return on investment of public health interventions: A scoping review and recommendations for future studies. *BMJ Global Health*, *8*(8), e012798. Advance online publication. DOI: 10.1136/bmjgh-2023-012798 PMID: 37648275

Tyc, H. J., Kłodnicka, K., Teresińska, B., Karpiński, R., Flieger, J., & Baj, J. (2025). Micro- and Nanoplastics as Disruptors of the Endocrine System-A Review of the Threats and Consequences Associated with Plastic Exposure. *International Journal of Molecular Sciences*, *26*(13), 6156. DOI: 10.3390/ijms26136156 PMID: 40649932

Uchmanowicz, D., Styszko, K., Chen, X., Terribile, G., Jakhar, R., Sancini, G., & Pyssa, J. (2025). Microplastics in Airborne Particulate Matter: A Comprehensive Review of Separation Techniques, In Vitro Toxicity and Health Impacts. *International Journal of Molecular Sciences*, *26*(21), 10332. DOI: 10.3390/ijms262110332 PMID: 41226370

UN Environment Programme. (2021, October 21). *Drowning in plastics—Marine litter and plastic waste vital graphics*. UNEP. https://www.unep.org/resources/report/drowning-plastics-marine-litter-and-plastic-waste-vital-graphics

United Nations Environment Programme. (2023). Turning off the tap: How the world can end plastic pollution and create a circular economy. UNEP. https://www.unep.org/resources/turning-off-tap-end-plastic-pollution-create-circular-economy

United Nations Environment Programme. (2023, May). *Chemicals in plastics: A technical report*. United Nations Environment Programme; Secretariat of the Basel, Rotterdam and Stockholm Conventions. https://wedocs.unep.org/handle/20.500.11822/42366

Valachovičová, M., & Mišľanová, C. (2026). Microplastics—The Microbiota Interactions: Mechanisms, Multi-Omics Insights and Health Implications. *Applied Sciences (Basel, Switzerland)*, *16*(9), 4110. DOI: 10.3390/app16094110

Veclin, C., Desmet, C., Pradel, A., Valsesia, A., Reynaud, S., & Gigault, J. (2022). Effect of the surface hydrophobicity–morphology–functionality of nanoplastics on their homoaggregation in seawater. *ACS ES&T Water*, *2*(1), 88–96. DOI: 10.1021/acsestwater.1c00263

Velumani, K., Thomas, A. G., Shaik, M. R., Hussain, S. A., Shaik, B., Guru, A., & Issac, P. K. (2025). Functionalized Zinc Oxide Nanoparticles Conjugated With Artemisinin Against Luperox-Induced Oxidative Stress and Their Impact on Antioxidant Defense Mechanism. *Journal of Biochemical and Molecular Toxicology*, *39*(7), e70380. DOI: 10.1002/jbt.70380 PMID: 40586296

Velzeboer, I., Kwadijk, C. J. A. F., & Koelmans, A. A. (2014).Strong sorption of PCBs to nanoplastics, microplastics, carbon nanotubes, and fullerenes. Environmental Science & Technology, 48(9), 4869 4876.DOI: 10.1021/es405721v

Venâncio, C., & Oliveira, M. (2025). A Multilevel Risk Assessment Framework for Nanoplastics in Aquatic Ecosystems. *Water (Basel)*, *17*(2), 180. DOI: 10.3390/w17020180

Vercauteren, M., Lambert, S., Hoogerwerf, E., Janssen, C. R., & Asselman, J. (2024). Microplastic-specific biofilm growth determines the vertical transport of plastics in freshwater. *The Science of the Total Environment*, *910*, 168399. DOI: 10.1016/j.scitotenv.2023.168399 PMID: 37981154

Vergani-Junior, C. A., Ventura, M. A. V. C., & De-Souza, E. A. (2026). Adaptive inter-tissue proteostasis networks in aging and neurodegeneration. *Bioscience Reports*, *46*(1), BSR20254097. Advance online publication. DOI: 10.1042/BSR20254097 PMID: 41510837

Verma, A., Bhagchandani, T., Rai, A., Nikita, , Sardarni, U. K., Bhavesh, N. S., Gulati, S., Malik, R., & Tandon, R. (2024). Short-chain fatty acid (SCFA) as a connecting link between microbiota and gut-lung axis—A potential therapeutic intervention to improve lung health. *ACS Omega*, *9*(13), 14648–14671. DOI: 10.1021/acsomega.3c05846 PMID: 38585101

Verma, D., Okhawilai, M., Dalapati, G. K., Ramakrishna, S., Sharma, A., Sonar, P., Krishnamurthy, S., Biring, S., & Sharma, M. (2022). Blockchain technology and AI-facilitated polymers recycling: Utilization, realities, and sustainability. *Polymer Composites*, *43*(12), 8587–8601. DOI: 10.1002/pc.27054

Vethaak, A. D., & Leslie, H. A. (2016). Plastic debris is a human health issue. Environmental Science & Technology, 50(13), 6825 6826. https://doi.org/DOI: 10.1021/acs.est.6b02569

Vethaak, A. D., & Legler, J. (2021). The exposome in the context of emerging contaminants. *Environmental Research*, *197*, 111125. DOI: 10.1016/j.envres.2021.111125

Vianello, A., Jensen, R. L., Liu, L., & Vollertsen, J. (2019). Simulating human exposure to indoor airborne microplastics using a breathing thermal manikin. *Scientific Reports*, *9*(1), 8670. DOI: 10.1038/s41598-019-45054-w PMID: 31209244

Vignesh, K. S., Prapanchan, V. N., Selvan, V. I., Karmegam, N., Kim, W., Barcelo, D., & Govarthanan, M. (2024). Microplastics, their abundance, and distribution in water and sediments in North Chennai, India: An assessment of pollution risk and human health impacts. *Journal of Contaminant Hydrology*, *263*, 104339. DOI: 10.1016/j.jconhyd.2024.104339 PMID: 38564944

Vince, F., Aoustin, E., Bréant, P., & Marechal, F. (2008). LCA tool for the environmental evaluation of potable water production. *Desalination*, *220*(1–3), 37–56. DOI: 10.1016/j.desal.2007.01.021

Visalli, G., Facciolà, A., Pruiti Ciarello, M., De Marco, G., Maisano, M., & Di Pietro, A. (2021). Acute and Sub-Chronic Effects of Microplastics (3 and 10 µm) on the Human Intestinal Cells HT-29. *International Journal of Environmental Research and Public Health*, *18*(11), 5833. DOI: 10.3390/ijerph18115833 PMID: 34071645

Visca, A., Di Gregorio, L., Clagnan, E., & Bevivino, A. (2024). Sustainable strategies: Nature-based solutions to tackle antibiotic resistance gene proliferation and improve agricultural productivity and soil quality. *Environmental Research*, *248*, 118395. Advance online publication. DOI: 10.1016/j.envres.2024.118395 PMID: 38307185

Vogt, A., Combadiere, B., Hadam, S., Stieler, K. M., Lademann, J., Schaefer, H., Autran, B., Sterry, W., & Blume-Peytavi, U. (2006). 40 nm, but not 750 or 1,500 nm, nanoparticles enter epidermal CD1a+ cells after transcutaneous application on human skin. *The Journal of Investigative Dermatology*, *126*(6), 1316–1322. DOI: 10.1038/sj.jid.5700226 PMID: 16614727

Völkl, M. (2023). *From single cells to microtissues: Using advanced culture techniques to create a better understanding of microplastic toxicity* (Publication No. 7182) [Doctoral dissertation, University of Bayreuth]. EPub Bayreuth. https://doi.org/DOI: 10.15495/EPub_UBT_00007182

Wagner, S., Hüffer, T., Klöckner, P., Wehrhahn, M., Hofmann, T., & Reemtsma, T. (2018).

Wagner, S., Hüffer, T., Klöckner, P., Wehrhahn, M., Hofmann, T., & Reemtsma, T. (2018). Tire wear particles in the aquatic environment - A review on generation, analysis, occurrence, fate and effects. *Water Research*, *139*, 83–100. DOI: 10.1016/j.watres.2018.03.051 PMID: 29631188

Walker-Franklin, I., Villalobos Santeli, A., & Neal-Walthall, N. (2026). Tracking Microplastics From Source to Impact: A Review of Environmental Presence, Exposure, Remediation, and Health Risks. *Current Environmental Health Reports, 13*(1), 12. DOI: 10.1007/s40572-026-00531-z PMID: 41844949

Walker, T. R., & Fequet, L. (2023). Current trends of unsustainable plastic production and micro (nano) plastic pollution. *Trends in Analytical Chemistry, 160*, 116984. DOI: 10.1016/j.trac.2023.116984

Walser, T., Demou, E., Lang, D. J., & Hellweg, S. (2011). Prospective environmental life cycle assessment of nanosilver T-shirts. *Environmental Science & Technology, 45*(10), 4570–4578. DOI: 10.1021/es2001248 PMID: 21506582

Walsh, A. N., Dunlea, A. G., Reddy, C. M., & Ward, C. P. (2025). Characterization of Inorganic Additives in and Photochemically Liberated from Consumer Plastics: Implications for Global and Local Biogeochemical Cycles. *ACS Environmental Au, 5*(3), 298–307. DOI: 10.1021/acsenvironau.4c00130 PMID: 40416843

Wang, J., & Chen, C. (2016). Nanomaterials for environmental remediation. *Environmental Science. Nano, 3*(2), 240–266. DOI: 10.1039/C5EN00230C

Wang, J., Li, D., Guan, R., Wang, Y., Li, H., & Zhao, C. (2025). A computational framework for multiscale data fusion in assessing the associations between micro- and nanoplastics and human hepatotoxicity. *Environment International, 206*, 109937. DOI: 10.1016/j.envint.2025.109937 PMID: 41270658

Wang, L., & Hu, J. (2024). Unraveling the gut microbiota's role in salt-sensitive hypertension: Current evidences and future directions. *Frontiers in Cardiovascular Medicine, 11*, 1410623. DOI: 10.3389/fcvm.2024.1410623 PMID: 39091359

Wang, Y., & Qian, H. (2021). Phthalates and their impacts on human health. *Health Care, 9*(603), 603. Advance online publication. DOI: 10.3390/healthcare9050603 PMID: 34069956

Wang, Z., Wang, Y., Zhang, J., Feng, G., Miao, S., Lu, R., Tian, X., & Ye, Y. (2025). Antioxidant Intervention Against Microplastic Hazards. *Antioxidants, 14*(7), 797. DOI: 10.3390/antiox14070797 PMID: 40722901

Waring, R. H., Harris, R. M., & Mitchell, S. C. (2018). Plastic contamination of the food chain: A threat to human health? *Maturitas, 115*, 64–68. DOI: 10.1016/j.maturitas.2018.06.010 PMID: 30049349

Wiesinger, H., Wang, Z., & Hellweg, S. (2021). Deep dive into plastic monomers, additives, and processing aids. *Environmental Science & Technology, 55*(13), 9339–9351. DOI: 10.1021/acs.est.1c00976 PMID: 34154322

Winburn, M. R., Alvarado, M. F., & Cheung, C. L. (2025). Plasma-assisted destruction of polystyrene nanoplastics. *Nanoscale*, *17*(4), 2138–2146. DOI: 10.1039/D4NR02498B PMID: 39655503

Winkler, A., Santo, N., Ortenzi, M. A., Bolzoni, E., Bacchetta, R., & Tremolada, P. (2019). Does mechanical stress cause microplastic release from plastic water bottles? *Water Research*, *166*, 115082. DOI: 10.1016/j.watres.2019.115082 PMID: 31542550

Woodruff, K. P. (2017). Microfluidic platforms for high-throughput mammalian cell printing, transfection, and dosage-dependent studies (Master's thesis, École Polytechnique Fédérale de Lausanne). EPFL Infoscience. EPFL Infoscience Repository

World Health Organization. (2019). Microplastics in drinking-water: Evidence and recommendations. World Health Organization. https://www.who.int/publications/i/item/9789241516198

World Health Organization. (2021). *Guidelines for nanoplastic safety in drinking water.* World Health Organization. https://www.who.int/publications/i/item/9789240039944

Worms, I. A. M., Tharaud, M., Gasco, R., Montaño, M. D., Goodman, A., Slaveykova, V. I., Benedetti, M. F., Churchill, C. M., Fernando, S., Alasonati, E., Moens, C., & Cuss, C. W. (2025). Exploring environmental nanobiogeochemistry using field-flow fractionation and ICP-MS-based tools: Progress and frontiers. *Environmental Science. Nano*, *12*(8), 3821–3846. DOI: 10.1039/D5EN00096C PMID: 40708569

Wright, S., Levermore, J., & Ishikawa, Y. (2023). Application of Infrared and Near-Infrared Microspectroscopy to Microplastic Human Exposure Measurements. *Applied Spectroscopy*, *77*(10), 1105–1128. DOI: 10.1177/00037028231199772 PMID: 37792505

Wu, B., Wu, X., Liu, S., Wang, Z., & Chen, L. (2019). Size-dependent effects of polystyrene microplastics on cytotoxicity. Chemosphere, 221, 333 341. https://doi.org/DOI: 10.1016/j.chemosphere.2019.01.056

Wu, J., Jiang, R., & Zhang, Y. (2019). Effect of salinity and humic acid on the aggregation and toxicity of polystyrene nanoplastics with different functional groups and charges. *Environmental Pollution*, *245*, 836–843. DOI: 10.1016/j.envpol.2018.11.055 PMID: 30502713

Wu, J., Ye, Q., Li, P., Sun, L., Huang, M., Liu, J., Ahmed, Z., & Wu, P. (2023). The heteroaggregation behavior of nanoplastics on goethite: Effects of surface functionalization and solution chemistry. *The Science of the Total Environment*, *870*, 161787. DOI: 10.1016/j.scitotenv.2023.161787 PMID: 36706999

Xiao, L., & Schultz, Z. D. (2017). Spectroscopic imaging at the nanoscale: Technologies and recent applications. *Analytical Chemistry*, *90*(1), 440–458. DOI: 10.1021/acs.analchem.7b04151 PMID: 29028297

Xie, L., Ma, M., Ge, Q., Liu, Y., & Zhang, L. (2025). Machine learning advancements and strategies in microplastic and nanoplastic detection. *Environmental Science & Technology*, *59*(18), 8885–8899. DOI: 10.1021/acs.est.4c11888 PMID: 40293506

Xi, X., Ding, D., Zhou, H., Baihetiyaer, B., Sun, H., Cai, Y., Wang, N., & Yin, X. (2022). Effects of physicochemical factors on the transport of aged polystyrene nanoparticles in saturated porous media. *Chemosphere*, *289*, 133239. DOI: 10.1016/j.chemosphere.2021.133239 PMID: 34896420

Xi, Z., Nicolas, R., & Wei, J. (2025). An Edge-Deployable Multi-Modal Nano-Sensor Array Coupled with Deep Learning for Real-Time, Multi-Pollutant Water Quality Monitoring. *Water (Basel)*, *17*(14), 2065. DOI: 10.3390/w17142065

Xu, J., Zhang, L., Cheng, J., Shi, Y., Hou, K., An, Q., Li, J., Cui, B., & Chen, F. (2024). Enhanced removal of polystyrene nanoplastics by air flotation modified with dodecyltrimethylammonium chloride. *Chinese Chemical Letters*, *36*(5), 110240. DOI: 10.1016/j.cclet.2024.110240

Xu, J., Zhang, W., Luo, X., Wang, Y., Huang, Y., Li, Y., & Chen, L. (2020). Microplastic pollution in China's inland waters: A review of findings, methods, characteristics, effects, and management. *The Science of the Total Environment*, *721*, 137561. DOI: 10.1016/j.scitotenv.2020.137561

Xu, Y., He, Q., Liu, C., & Huangfu, X. (2019). Are micro- or nanoplastics leached from drinking water distribution systems? *Environmental Science & Technology*, *53*(16), 9339–9340. DOI: 10.1021/acs.est.9b03673 PMID: 31393716

Xu, Z., Zhang, Y., Xu, Z., Zhong, Y., Wang, L., Liu, C., Chen, A., Jiang, S., Sun, W., Yu, G., Li, Y., Li, C., Wang, Y., Yu, R., & Zhang, L. (2023). Dysregulation of gut health in zebrafish by differentially charged nanoplastic exposure: An integrated analysis of histopathology, immunology, and microbial informatics. *Environmental Science. Nano*, *10*(3), 933–947. DOI: 10.1039/D2EN00542E

Yang, C. S., Chang, C. H., Tsai, P. J., Chen, W. Y., Tseng, F. G., & Lo, L. W. (2004). Nanoparticle-based in vivo investigation on blood-brain barrier permeability following ischemia and reperfusion. *Analytical Chemistry*, *76*(15), 4465–4471. DOI: 10.1021/ac035491v PMID: 15283589

Yang, H., Zhang, K., Wang, Y., Liu, S., Guo, Y., Liu, W., Sun, J., Zhang, Z., Zhang, S., Li, S., Zhao, Y., Liu, T., Liu, J., Pei, L., Xi, S., & Shi, P. (2026a). Weighted network analysis of adverse outcome pathways decodes the multiscale mechanisms of environmental toxicity. *Environmental Science and Ecotechnology*, *29*, 100663. DOI: 10.1016/j.ese.2026.100663 PMID: 41660038

Yang, J., Guo, X., Li, T., Xie, Y., Wang, D., & Long, Y. (2023). Sulforaphane Inhibits Exhaustive Exercise-Induced Liver Injury and Transcriptome-Based Mechanism Analysis. *Nutrients*, *15*(14), 3220. DOI: 10.3390/nu15143220 PMID: 37513640

Yang, K., Liu, L., Zhu, G., Militky, J., Venkataraman, M., & Yao, J. (2026d). Perspective: A sandwich encapsulation strategy to mitigate PCM-based microplastic generation from fibrous materials. *Advanced Sustainable Systems*, *10*(2), e01645. Advance online publication. DOI: 10.1002/adsu.202501645

Yang, L., Kang, S., Luo, X., & Wang, Z. (2024). Microplastics in drinking water: A review on methods, occurrence, sources, and potential risks assessment. *Environmental Pollution*, *348*, 123857. DOI: 10.1016/j.envpol.2024.123857 PMID: 38537794

Yang, L., Zhang, Y., Kang, S., Wang, Z., Wu, C., & Zhang, J. (2021). Microplastics in soil: A review of methods, occurrences, sources, and potential risks. *The Science of the Total Environment*, *780*, 146546. DOI: 10.1016/j.scitotenv.2021.146546 PMID: 33770602

Yang, X., Zhou, Y., Li, Y., & Wang, S. (2024). International Law and Regulation of Marine Microplastics: Current Situation, Problems, and Development. *Sustainability (Basel)*, *16*(21), 9337. DOI: 10.3390/su16219337

Yee, M. S., Hii, L. W., Looi, C. K., Lim, W. M., Wong, S. F., Kok, Y. Y., Tan, B. K., Wong, C. Y., & Leong, C. O. (2021). Impact of Microplastics and Nanoplastics on Human Health. *Nanomaterials (Basel, Switzerland)*, *11*(2), 496. DOI: 10.3390/nano11020496 PMID: 33669327

Yelleti, G., Maripini, N., & Bolar Suryakanth, V. (2026). Preeclampsia and environmental epigenomics: The emerging role of air pollution, gut microbiome, and maternal exposures in disease programming. *Environmental Epigenetics*, *12*(1), dvag001. Advance online publication. DOI: 10.1093/eep/dvag001 PMID: 41704669

Yeo, L. S., & Julaihi, M. R. M. (2021). Magnetic ferrous fluid for microplastics extraction application. *Materials Science Forum*, *1030*, 138–145. . DOI: 10.4028/www.scientific.net/MSF.1030.138

Yong, C. Q. Y., Valiyaveettil, S., & Tang, B. L. (2020). Toxicity of Microplastics and Nanoplastics in Mammalian Systems. *International Journal of Environmental Research and Public Health, 17*(5), 1509. DOI: 10.3390/ijerph17051509 PMID: 32111046

Yong, X., & Du, K. (2023). Effects of shape on interaction dynamics of tetrahedral nanoplastics and the cell membrane. *The Journal of Physical Chemistry. B, 127*(7), 1652–1663. DOI: 10.1021/acs.jpcb.2c07460 PMID: 36763902

Yoshida, S., Hiraga, K., Takehana, T., Taniguchi, I., Yamaji, H., Maeda, Y., Toyohara, K., Miyamoto, K., Kimura, Y., & Oda, K. (2016). A bacterium that degrades and assimilates poly(ethylene terephthalate). *Science, 351*(6278), 1196–1199. DOI: 10.1126/science.aad6359 PMID: 26965627

Yu, H.-R., Sheen, J.-M., & Tiao, M.-M. (2024). The Impact of Maternal Nanoplastic and Microplastic Particle Exposure on Mammal's Offspring. *Cells, 13*(16), 1380. DOI: 10.3390/cells13161380 PMID: 39195272

Yu, T., Huang, X., Zhang, X. F., Li, K., Liu, S. P., Dai, N., Zhang, K., Zhang, Y. X., & Li, H. (2024). A review of nanomaterials with excellent purification potential for the removal of micro- and nanoplastics from liquid. *DeCarbon, 5*, 100064. DOI: 10.1016/j.decarb.2024.100064

Zaki, M. R. M., & Aris, A. Z. (2022). An overview of the effects of nanoplastics on marine organisms. *The Science of the Total Environment, 831*, 154757. DOI: 10.1016/j.scitotenv.2022.154757 PMID: 35339559

Zarus, G. M., Muianga, C., Hunter, C., & Pappas, R. S. (2020). A review of data for quantifying human exposures to micro- and nanoplastics and potential health risks. *The Science of the Total Environment, 756*, 144010. DOI: 10.1016/j.scitotenv.2020.144010 PMID: 33310215

Zhang, H., Wang, J., Zhou, B., Zhou, Y., Dai, Z., Zhou, Q., Christie, P., & Luo, Y. (2018). Enhanced adsorption of oxytetracycline to weathered microplastic polystyrene. Environmental Pollution, 243, 1550 1557. https://doi.org/DOI: 10.1016/j.envpol.2018.09.122

Zhang, F., Guo, C. G., Yang, C., Wang, F., Wang, W., & Zhang, L. (2025). Exposure-response associations of ambient heavy metal and persistent organic pollutant with all-cause and cause-specific mortality: A prospective cohort study. *Environment & Health, 3*(5), 493–503. DOI: 10.1021/envhealth.4c00191 PMID: 40400556

Zhang, G. S., & Liu, Y. F. (2018). The distribution of microplastics in soil aggregate fractions in southwestern China. *The Science of the Total Environment*, *642*, 12–20. DOI: 10.1016/j.scitotenv.2018.06.004 PMID: 29894871

Zhang, J., Xia, X., Huang, W., Li, Y., Lin, X., Li, Y., & Yang, Z. (2024). Photoaging of biodegradable nanoplastics regulates their toxicity to aquatic insects (*Chironomus kiinensis*) by impairing gut and disrupting intestinal microbiota. *Environment International*, *185*, 108483. DOI: 10.1016/j.envint.2024.108483 PMID: 38382402

Zhang, K., Shi, H., Peng, J., Wang, Y., Xiong, X., Wu, C., & Lam, P. K. (2018). Microplastic pollution in China's inland water systems: A review of findings, methods, characteristics, effects, and management. *The Science of the Total Environment*, *630*, 1641–1653. DOI: 10.1016/j.scitotenv.2018.02.300 PMID: 29554780

Zhang, K., Su, J., Xiong, X., Wu, X., Wu, C., & Liu, J. (2016). Microplastic pollution of lakeshore sediments from remote lakes in Tibet plateau, China. *Environmental Pollution*, *219*, 450–455. DOI: 10.1016/j.envpol.2016.05.048 PMID: 27238763

Zhang, L., Xue, M., Xin, Q., Tang, J., Liu, Y., Bian, Z., & Zheng, X. (2025). Cellular uptake of polystyrene nanoplastics with surface functionalization: An AIE-based quantitative approach. *Analytica Chimica Acta*, *1362*, 344184. DOI: 10.1016/j.aca.2025.344184 PMID: 40447404

Zhang, M., Yang, J., Cai, Z., Feng, Y., Wang, Y., Zhang, D., & Pan, X. (2019). Detection of engineered nanoparticles in aquatic environments: Current status and challenges in enrichment, separation, and analysis. *Environmental Science. Nano*, *6*(3), 709–735. DOI: 10.1039/C8EN01086B

Zhang, S., Zhao, Y., & Zhang, W. (2020). Occurrence, sources, and ecological risks of micro(nano)plastics in terrestrial environments: A review. *The Science of the Total Environment*, *739*, 139913. DOI: 10.1016/j.scitotenv.2020.139913

Zhang, W. X. (2003). Nanoscale iron particles for environmental remediation: An overview. *Journal of Nanoparticle Research*, *5*(3–4), 323–332. DOI: 10.1023/A:1025520116015

Zhang, W., Li, H., Zheng, H., Lan, H., & Zhuang, Y. (2022). Arsenic speciation and concentration in the urine of infants in Xiamen, China: A cross-sectional survey [Preprint]. *Research Square*. https://doi.org/DOI: 10.21203/rs.3.rs-1522535/v1

Zhang, W., Wang, Q., & Chen, H. (2022). Challenges in characterization of nanoplastics in the environment. *Frontiers of Environmental Science & Engineering*, *16*(1), 11. Advance online publication. DOI: 10.1007/s11783-021-1445-z

Zhang, X., Shen, L., & Zhang, L. (2013). Life cycle assessment of the air emissions during building construction process: A case study in Hong Kong. *Renewable & Sustainable Energy Reviews*, *17*, 160–169. DOI: 10.1016/j.rser.2012.09.024

Zhang, Y.. (2022). Economic optimization of nano-enabled remediation technologies. *Journal of Cleaner Production*, *377*, 134512. DOI: 10.1016/j.jclepro.2022.134512

Zhang, Y., Chen, Q., Zhang, W., Huang, Y., & Wang, J. (2021). Advances in nanoplastic detection: Methods and applications. *Analytical Chemistry*, *93*(10), 4165–4173. DOI: 10.1021/acs.analchem.0c04678 PMID: 33650848

Zhang, Y., Diehl, A., Lewandowski, A., Gopalakrishnan, K., & Baker, T. R. (2020). Removal efficiency of micro- and nanoplastics (180 nm–125 μm) during drinking water treatment. *The Science of the Total Environment*, *720*, 137383. DOI: 10.1016/j.scitotenv.2020.137383 PMID: 32325555

Zhang, Y., Zhang, D., & Zhang, Z. (2023). A Critical Review on Artificial Intelligence-Based Microplastics Imaging Technology: Recent Advances, Hot-Spots and Challenges. *International Journal of Environmental Research and Public Health*, *20*(2), 1150. DOI: 10.3390/ijerph20021150 PMID: 36673905

Zhan, Z. S., Zheng, Z. S., Shi, J., Chen, J., Wu, S. Y., & Zhang, S. Y. (2024). Unraveling colorectal cancer prevention: The vitamin D - gut flora - immune system nexus. *World Journal of Gastrointestinal Oncology*, *16*(6), 2394–2403. DOI: 10.4251/wjgo.v16.i6.2394 PMID: 38994172

Zhao, B., Richardson, R. E., & You, F. (2024). Advancing microplastic analysis in the era of artificial intelligence: From current applications to the promise of generative AI. *Nexus*, *1*(4), 100043. Advance online publication. DOI: 10.1016/j.ynexs.2024.100043

Zhao, L., & Peters, D. (2024). Innovations in inorganic nanomaterial synthesis and environmental management. *Inorganic Chemistry Communications*, *162*, 112611. DOI: 10.1016/j.inoche.2024.112611

Zhao, X., Liu, W., Cai, Z., Han, B., Qian, T., & Zhao, D. (2016). An overview of preparation and applications of stabilized zero-valent iron nanoparticles for soil and groundwater remediation. *Water Research*, *100*, 245–266. DOI: 10.1016/j.watres.2016.05.019 PMID: 27206054

Zhao, Y. M., Zhang, J. Y., Liaw, P. K., & Yang, T. (2025). Machine learning-based computational design methods for high-entropy alloys. *High Entropy Alloys Materials*, *3*(1), 41–100. DOI: 10.1007/s44210-025-00055-5

Zheng, P. C., Li, R., Lai, K. P., & Zhang, X. X. (2024). Biological exposure to microplastics and nanoplastics and plastic additives: Impairment of glycolipid metabolism and adverse effects on metabolic diseases. *Environmental Science and Pollution Research International*, *31*(51), 60778–60791. DOI: 10.1007/s11356-024-35220-7 PMID: 39412714

Zhou, A., Zhao, Y., Liu, M., Suyamud, B., Yuan, W., & Yang, Y. (2023). Occurrence and risk assessment of microplastics in the Lhasa River-a remote plateau river on the Qinghai-Tibet Plateau, China. *Environmental Monitoring and Assessment*, *195*(3), 433. DOI: 10.1007/s10661-023-11040-z PMID: 36856933

Zhou, X., Wang, Q., Wang, J., Li, H., Ren, J., & Tang, S. (2023). Quantification of Microplastics in Plastic-Bottled Chinese Baijiu Using Micro-FTIR in Imaging Mode. *Applied Sciences (Basel, Switzerland)*, *13*(20), 11142. DOI: 10.3390/app132011142

Zhu, H., Fan, X., Zou, H., Guo, R.-B., & Fu, S.-F. (2023a). Effects of size and surface charge on the sedimentation of nanoplastics in freshwater. *Chemosphere*, *336*, 139194. DOI: 10.1016/j.chemosphere.2023.139194 PMID: 37315858

Zhu, M., Zhang, Z., Zhang, T., Hofmann, T., & Chen, W. (2023b). Eco-corona dictates mobility of nanoplastics in saturated porous media: The critical role of preferential binding of macromolecules. *Environmental Science & Technology*, *57*(1), 331–339. DOI: 10.1021/acs.est.2c07376 PMID: 36574476

Zhu, Z. M., Wu, X. M., Hu, Y., Bian, X. L., Wang, Y. Q., & Zhu, Q. N. (2025). Alternative Splicing: Molecular Mechanisms, Biological Functions, Diseases, and Potential Therapeutic Targets. *MedComm*, *6*(12), e70545. DOI: 10.1002/mco2.70545 PMID: 41399527

Zoeller, R. T., Bansal, R., & Parris, C. (2005). Bisphenol-A, an environmental contaminant that acts as a thyroid hormone receptor antagonist in vitro, increases serum thyroxine, and alters RC3/Neurogranin expression in the developing rat brain. *Endocrinology*, *146*(2), 607–612. DOI: 10.1210/en.2004-1018 PMID: 15498886

Zolotova, N., Kirillova, M., Dzhalilova, D., Tsvetkov, I., Fokichev, N., & Makarova, O. (2025). Microplastics' Impact on the Development of AOM/DSS-Induced Colitis-Associated Colorectal Cancer in Mice. *International Journal of Molecular Sciences*, *26*(23), 11511. DOI: 10.3390/ijms262311511 PMID: 41373666

Zoppas, F. M., Sacco, N., Soffietti, J., Devard, A., Akhter, F., & Marchesini, F. A. (2023). Catalytic approaches for the removal of microplastics from water: Recent advances and future opportunities. *Chemical Engineering Journal Advances*, *16*, 100529. DOI: 10.1016/j.ceja.2023.100529

About the Contributors

Reyed M. Reyed, PhD Systems Architect, Environmental Intelligence, Translational One Health. Dr. Reyed M. Reyed works at the intersection of microbial biotechnology, computational systems science, and environmental intelligence, with a research orientation centered on the development of integrative analytical frameworks capable of connecting environmental exposure biology, microbiome dynamics, computational analytics, and adaptive One Health interpretation within unified systems-oriented models. Rather than approaching environmental and biomedical complexity through isolated disciplinary pathways, his work emphasizes cross-domain convergence among exposure science, systems toxicology, microbiome-associated signaling networks, AI-assisted biosurveillance, precision nutrition, and translational environmental governance. This systems-level orientation positions biological and ecological instability not as fragmented events, but as interconnected analytical layers requiring integrated computational interpretation. His conceptual contributions include the Polybiome Systems Medicine (PSM) framework, the Digital Nano-Plastic Science (DNPS) paradigm, the Nanoplastic Proteostatic Panflemmosis Metabolic Continuum (NPPMC), and the Reyed Precision Convergence Framework for Antimicrobial Resistance Governance (RPCF-AMR). Collectively, these architectures explore the transition from fragmented observational methodologies toward predictive, computation-aware environmental and biomedical systems. Within this volume, these principles are operationalized through a progression-based editorial structure linking nanoplastic exposure dynamics, physicochemical transformation systems, biological signaling perturbation, microbiome instability, computational environmental analytics, biotechnology-assisted mitigation, and adaptive One Health governance within a unified interpretative continuum. As volume editor, Dr. Reyed has structured the present work not as a conventional thematic collection, but as an editorially harmonized systems-science architecture. Chapter 1 establishes the principal conceptual and analytical foundation, while subsequent chapters progressively expand this framework across forensic tracing systems, environmental fate dynamics, toxicological interpretation, microbiome interaction networks, food-system exposure analytics, translational mitigation strategies, and techno-economic environmental assessment. His current scientific initiatives focus on computational infrastructures integrating environmental sensing, multi-omics interpretation, ecological variability mapping, and AI-assisted decision-support systems across interconnected biological and environmental domains. His recent work contributes to the development of environmental intelligence, cyber-biotechnology, and systems-oriented precision health through integrative computational and systems-science approaches. Conceptual Frameworks Polybiome Systems Medicine (PSM) Integrative systems framework connecting host–microbiota interaction networks, environmental perturbation dynamics, and precision health interpretation. Digital Nano-Plastic Science (DNPS) Computational environmental intelligence paradigm for modeling nanoplastic particulates as multidimensional biological and ecological signaling systems. Nanoplastic Proteostatic Panflemmosis Metabolic Continuum (NPPMC) Systems-level continuum describing the progression from chronic nanoplastic exposure and proteostatic disruption toward inflammatory and metabolic instability. Reyed Precision Convergence Framework for Antimicrobial Resistance (RPCF-AMR) Cross-domain governance architecture integrating environmental surveillance, molecular epidemiology, and computational prediction for adaptive antimicrobial resistance management. His research domains include: Environmental Exposure Biology, Computational Toxicology, AI-Assisted Biosurveillance, Microbiome Systems Science, Translational Precision Health, Cyber-Biotechnology, One Health Governance, Environmental Intelligence Systems In addition to his interdisciplinary research activities, Dr. Reyed contributes to international scientific publishing, systems-oriented editorial development, and cross-disciplinary scientific integration initiatives spanning biotechnology, environmental systems science, computational biology, microbiome science, and translational precision health. Affiliation: Genetic Engineering and Biotechnology Research Institute (GEBRI) City of Scientific Research and Technological Applications (SRTA-City) Alexandria, Egypt

Vugar Abdullayev is a doctor of technical sciences, head of scientific-research lab., Azerbaijan University of Architecture and Construction, Baku, Azerbaijan and assos. Prof., Azerbaijan State Oil and Industry University, Baku, Azerbaijan. He has completed his Ph.D in Computer Science in the year 2005. He is author of 65 scientific papers. His researchers related to the study of the cyber physical systems, IoT, big data and information technologies.

Swapan Banerjee is among the top nutrition researchers and a faculty member in hospital nutrition management and public health. He is also a licensed dietitian with national and internationally recognized certifications from UNICEF, FAO, FSSAI Level III, and CPD. He is an assistant professor in the Faculty of Management & Commerce (Hospital and Healthcare Management) at Poornima University. Dr. Banerjee has over 25 years of experience in the pharma and nutrition industry, academia, and research. He was attached to Netaji Sebayan Hospital and Ariadaha Vivekananda Mission Seva Pratisthan in West Bengal. Dr. Banerjee has 2 PG degrees, a Ph.D (subject: nutrition), and 25 global certificates from FSSAI, FAO, CPD, and SGS, including Harvard, Cambridge, and others. He has published more than 60 articles in high-indexed, peer-reviewed international journals, 15 book chapters, one book on thyroid diet, two patents, and authored four books on health and nutrition. He is the editor-in-chief of the Journal of Preventive Medicine and Holistic Health and an Editorial Board Member of more than 13 best International Journals, including Oxford Database, MDPI, PLOS One, and Springer. He has also participated in over 50 national and international conferences as a paper presenter, including as an invited speaker, moderator, and organizing chair. Dr. Banerjee is a member of the WHO Civil Society, Life Member of the Indian Society for Paediatric & Adolescent Endocrinology, Life Member of the International Academy of Nutrition Educators (IANE), Life Member of the Indian Dietetics Association (IDA), and Nutrition Society of India (NSI). He also participated in many programs on All India Radio, ABP TV, and online media outlets such as The Telegraph News, Republic News India, and 24x7 Headlines. Apart from other achievements, he received the 'Director Award' from a pharmaceutical company and three research excellence awards.

Yash bhardwaj B.Sc.(H) Forensic Science student Interested in finding new techniques to help investigators and scientists discover new and alternative techniques to overcome the challenges of currently use techniques in solving crime scenes

Rajashree K. C is a Biomedical Engineer with a Master's in Business Administration from Saveetha University, specializing in Hospital Management and Marketing, and holds a Ph.D. in Healthcare Management. An energetic and enthusiastic professor, she excels in academics and as a multitasker, organizing seminars, workshops, cultural programs, and other extra-curricular activities. She serves as the Coordinator of NSS Unit of Saveetha School of Management, leading them to achieve multiple accolades including the Best Club award. A prominent researcher with publications in Web of Science, Scopus, and UGC-listed journals, she actively enhances her expertise through conferences, workshops, MOOCs, and certifications, including Quality Coordination in Hospitals. Awarded the "Young Academician Award" and "Young Teacher Award," she is passionate about innovation, student skill development, and inspiring young minds through exemplary teaching and mentorship.

Ejeomo Christopher has a B.Sc in Industry Chemistry, PGD and MEngr. in Chemical Engineering, M.Sc in Environmental Chemistry. I have a Ph.D in Environmental Chemistry. My area of research interest is in persistant organic pollutants and nanoparticles.

Emek Moroydor Derun was born in Istanbul in 1976. She received her B.Sc. (1998), M.Sc. (2000), and Ph.D. (2005) degrees in Chemical Engineering from Yildiz Technical University, Istanbul, Türkiye, where she is currently a Professor in the Department of Chemical Engineering. Her research interests include waste management, lightweight concrete, semiconductive materials, and boron technology. She has authored numerous articles in peer-reviewed journals, contributed to international and national conference proceedings, and supervised several graduate theses. Prof. Dr. Derun has also led and participated in various research projects at national and international levels, advancing both academic knowledge and practical applications in chemical and environmental engineering.

M. Thanjaivadivel is a highly accomplished and skilled researcher in the field of Computer Science and Engineering. He completed his Bachelor's in Computer Science and Engineering from DMI College of Engineering affiliated with Anna University, followed by a Master's in Information Technology from VIT University, and a PhD in CSE from Vel Tech University. Dr. Thanjaivadivel has published numerous research papers in reputed international journals, conferences, and workshops.

Prerna Mehta is a distinguished academic and accomplished author with a robust educational background. She completed her BSc in Biotechnology at Rungta College of Science and Technology in Bhilai, Chhattisgarh, where she was honored with a gold medal for her exceptional performance. She then pursued a Master of Science in Biotechnology from Government V.Y.P.T. College, Durg, and obtained a B.Ed. degree from Ghanshyam Singh Arya Kanya College in Durg. Prerna's expertise covers a range of fields, including genetics, nanoparticles, biochemistry, and genetic engineering.Additionally, she is a lifetime member of the International Organization of Academic and Scientific Development (IOASD) and serves as a reviewer for the East African Scholars Journal of Biotechnology and Genetics. Prerna is also a lifetime member of The Indian Science Congress Association in Kolkata. As an accomplished author, she has written three books and holds utility patents for innovative projects: "A chamber to detect radiation and a method to shower melanin to protect the user against radiation" and "AI-Powered Rural Enhancement System for Tackling Hunger & Ensuring Water Conservation through Smart Resource M

Odangowei Inetiminebi Ogidi is a Lecturer, Acting Head, Department of Biochemistry and Sub-Dean, Faculty of Basic Medical Sciences, Bayelsa Medical University, Yenagoa, Bayelsa State, Nigeria. He holds a Bachelor of Science degree in Biochemistry, Master of Science and Doctor of Philosophy degrees in Biotechnology. He has also served as a Lecturer, Head of Departments of Science Laboratory Technology and Biochemistry, and Deputy Director, Centre for Continuing Education in Federal Polytechnic Ekowe, Bayelsa State. He has authored over 80 peer-reviewed academic articles in reputable local and international journals, published over 50 book chapters and presented over 40 conference papers. He is a member of various Journal Editorial Boards. He is indexed in World Scientist and University Ranking (AD Scientific Index ID https://www.adscientificindex.com/scientist/ogidi-odangowei-inetiminebi/4306173) amongst other international bodies. He have supervised over 30 ND and HND students and currently supervising over 20 undergraduate students. His research interest has been Medical/Pharmaceutical and Environmental Biotechnology/Biochemistry, Phytomedicine, Phytochemistry.

Lilian Maureen Oguoma, is an early career researcher, who is keen in innovations and novel findings. My research interests covers Pharmaceutical Microbiology and Biotechnology, focusing on Antimicrobial Agents and Resistance. Looking forward for collaborations and mentorship in this area of specialty. I tutor at the Federal University of Technology Owerri, Imo State, Nigeria. Have supervised some students based on their research findings and have publications and articles on some renowned Journals. I'm currently a PhD student in one of the Federal Universities in my country and very optimistic to reach a greater height in my field of study.

P. Praveen Kumar holds an MCom, MBA, and PhD in Finance from Pondicherry University. He secured the University 1st Rank in his Under Graduation. His teaching interests include Financial Accounting, Financial Management, Mergers & Acquisitions, Security Analysis & Portfolio Management, and other finance-related courses. His research interests extend to interdisciplinary areas.

Mehmet Burçin Pişkin is a faculty member in the Department of Bioengineering at Yıldız Technical University, Istanbul, Türkiye. His research focuses on nanomaterials, advanced functional materials, and their applications in bioengineering and environmental technologies. He has authored numerous peer-reviewed articles, presented at international and national conferences, and contributed to collaborative research projects at both national and international levels. Prof. Dr. Pişkin also supervises graduate students and continues to contribute to the advancement of bioengineering and materials science.

Sameer Saharan is a distinguished Forensic Science academic, biometric researcher, and cybersecurity specialist currently serving as Head of the Department (HoD) and Assistant Professor of Forensic Science at Mody University of Science and Technology, Rajasthan, India. He earned his Ph.D. in Forensic Science from Amity University, Haryana, with pioneering research focused on the identification and differentiation of forged fingerprints generated using polyvinyl acetate (PVA). His doctoral work significantly contributes to biometric fraud detection, fingerprint authentication systems, and forensic evidence reliability. Dr. Saharan's core research expertise lies in forensic fingerprint analysis, latent fingerprint development, forged fingerprint detection, biometric authentication, cyber forensics, digital forensics, DNA fingerprinting, and emerging forensic technologies. His work bridges classical forensic methodologies with modern analytical chemistry, artificial intelligence applications, and digital evidence investigation techniques. He has authored multiple peer-reviewed research papers published in Scopus and Web of Science indexed journals, covering areas such as chemical staining techniques for fingerprint detection, forensic biometrics, cybercrime investigation, and AI-assisted forensic analysis. In addition to journal publications, Dr. Saharan has contributed several international book chapters with reputed publishers such as Springer and IGI Global, focusing on cybersecurity, artificial intelligence in forensic systems, metaverse cybercrime forecasting, and advanced digital forensic frameworks. His interdisciplinary research approach integrates forensic science, cybersecurity, artificial intelligence, and digital investigation methodologies. Dr. Saharan also holds published patents in forensic innovation, including a system and method for differentiating real and forged fingerprints and a forensic compact microscope for determining warp–weft thread density in textiles. These innovations reflect his commitment to advancing forensic instrumentation and applied research in crime scene investigation and biometric security technologies. As an academic leader, he actively promotes outcome-based forensic education, research supervision, forensic laboratory modernization, and interdisciplinary collaboration. He regularly organizes forensic workshops, expert lecture series, crime scene simulation programs, and skill-oriented training sessions to enhance practical competencies in forensic science students. With strong expertise in forensic laboratory instrumentation, digital forensic tools, MATLAB-based forensic analytics, and cybersecurity platforms, Dr. Saharan continues to contribute to the advancement of forensic science research, biometric authentication systems, and digital evidence investigation in India and beyond.

Fatma Tuğçe Şenberber Dumanlı is a faculty member in the Department of Bioengineering at Yıldız Technical University, Istanbul, Türkiye. She received her B.Sc., M.Sc., and Ph.D. degrees in Chemical Engineering from Yıldız Technical University. Her research interests focus on boron chemistry, advanced materials, adsorption processes, and environmental applications. She has published numerous peer-reviewed articles, contributed to international and national conference proceedings, and participated in various research projects. Dr. Şenberber Dumanlı also actively continues to contribute to the advancement of chemical and bioengineering sciences.

Khushbu Katiyar is a Lecturer in the Department of Forensic Science at Medicaps University, Indore. With 6 years of teaching experience, she holds a B.Sc. (BZC) and an M.Sc. in Forensic Science from Govt. Holkar Science. She is currently pursuing a Ph.D. in Forensic Science with a focus on Forensic Toxicology. Her areas of specialization include Forensic Toxicology, Fingerprint Analysis, Forensic Biology, and Serology. Additionally, she has contributed to the field with one published paper and one book chapter.

Index

M

N

O

P

www.ingramcontent.com/pod-product-compliance
Lightning Source LLC
LaVergne TN
LVHW061530290926
849162LV00004B/737